Engineering Materials and Processes

Series Editor

Professor Brian Derby, Professor of Materials Science
Manchester Materials Science Centre, Grosvenor Street, Manchester, M1 7HS, UK

Other titles published in this series:

Fusion Bonding of Polymer Composites
C. Ageorges and L. Ye

Composite Materials
D.D.L. Chung

Titanium
G. Lütjering and J.C. Williams

Corrosion of Metals
H. Kaesche

Corrosion and Protection
E. Bardal

Intelligent Macromolecules for Smart Devices
L. Dai

Microstructure of Steels and Cast Irons
M. Durand-Charre

Phase Diagrams and Heterogeneous Equilibria
B. Predel, M. Hoch and M. Pool

Computational Mechanics of Composite Materials
M. Kamiński

Gallium Nitride Processing for Electronics, Sensors and Spintronics
S.J. Pearton, C.R. Abernathy and F. Ren
Publication due September 2005

Thermoelectricity
J.P. Heremans, G. Chen and M.S. Dresselhaus
Publication due May 2006

Computer Modelling of Sintering at Different Length Scales
J. Pan
Publication due June 2006

Computational Quantum Mechanics for Materials Engineers
L. Vitos
Publication due June 2006

Fuel Cell Technology
N. Sammes
Publication due December 2005

Shape Memory Alloy Thin Films
A. Ishida and S. Miyazaki
Publication due March 2007

Ehrenfried Zschech, Caroline Whelan and Thomas Mikolajick (Eds.)

Materials for Information Technology

Devices, Interconnects and Packaging

With 338 Figures

Springer

Ehrenfried Zschech, Dr. rer. nat.
Materials Analysis Department, AMD Saxony Limited Liability Company & Co. KG,
Wilschdorfer Landstraße 101, D-01109 Dresden, Germany

Caroline Whelan, Ph.D.
IMEC, Kapeldreef 75, B-3001 Leuven, Belgium

Thomas Mikolajick, Dr.-Ing.
Flash Technology Predevelopment, Infineon Technologies SC300,
Königsbrücker Straße 180, D-01099 Dresden, Germany

British Library Cataloguing in Publication Data
Euromat 2003 (2003 : Lausanne, Switzerland)
  Materials for information technology. - (Engineering
  materials and processes)
  1. Microelectronics - Materials - Congresses 2. Information
  technology - Congresses
  I. Title II. Zschech, Ehrenfried III. Whelan, Caroline
  IV. Mikolajick, Thomas
  621.3'815
ISBN-10: 1852339411

Library of Congress Control Number: 2005924304

Engineering Materials and Processes ISSN 1619-0181
ISBN-10: 1-85233-941-1
ISBN-13: 978-1-85233-941-8
Springer Science+Business Media
springeronline.com

Typesetting: Electronic text files prepared by authors
Production and cover design: LE-TEX Jelonek, Schmidt & Vöckler GbR, Leipzig, Germany
Printed in Germany
69/3141-543210 Printed on acid-free paper SPIN 11305507

Foreword

Euromat is the biennial meeting of the Federation of European Materials Societies (FEMS) constituted by its 24 member societies in Europe. The 2003 meeting took place in Lausanne, Switzerland, and was organised by the French, German and Swiss member societies:

- Société Française de Métallurgie et de Matériaux (SF2M)
- Deutsche Gesellschaft für Materialkunde (DGM)
- Schweizerischer Verband für die Materialtechnik (SVMT)

The scientific programme of the EUROMAT 2003 congress was divided into 15 topics that in turn were substructured into 47 symposia. The present volume of the Euromat Publication series refers to selected papers of the Topic A:

Materials for Information Technology

Peter Paul Schepp
Conference Organiser of EUROMAT 2003

Preface

Information technology (IT) is driving the need for research, development and introduction of new materials and concepts for applications requiring significantly increased electrical and optical functionality. In particular, microelectronic products must be improved continuously to meet performance demands while maintaining acceptable levels of reliability. For the semiconductor industry, the challenges to process technology and advanced materials are outlined in the International Technology Roadmap for Semiconductors (ITRS). Microelectronic and, in the longer term, nanoelectronic products of future technology generations rely on advanced materials for the so-called front-end-of-line and back-end-of-line processes on the waferlevel (die level) as well as for assembly and packaging. More than ever before, the dramatic productivity enhancement of the IT era will only be possible using advanced materials.

Transistor and interconnect scaling has brought microelectronics a long way. However, the traditional down-scaling of device and interconnect structures is currently leading to performance limitations that have to be overcome by new device architectures and advanced materials. Microprocessor applications are pushing material innovations in the gate stack of MOS transistors and in on-chip interconnect structures, which result in new and exciting challenges to materials scientists. Memory applications, on the other hand, are driving research and development activities in the field of new materials for challenging capacitor dimensions and for several different concepts of nonvolatile memories.

The scope of this book is to provide an overview of recent developments and research activities in the field of materials used for IT, with a particular emphasis on future applications. Topics are materials for silicon-based semiconductor devices (including high-k gate dielectric materials), materials for nonvolatile memories, materials for on-chip interconnects and interlayer dielectrics (including silicides, barrier materials and low-k dielectric materials), materials for assembly and packaging, etc. The latest results in materials science and engineering as well as applications in semiconductor industry are covered including the synthesis of blanket and patterned thin-film materials, their properties, composition, and structure. Computer modeling and analytical techniques to characterize thin-film structures are included. In this book, material transitions that are necessary to improve the product performance and to maintain the reliability of products for IT applications are highlighted.

In Chapter 1, the synthesis and properties of ultrathin films for devices are described. Material approaches that are focused on an increase in transistor performance are addressed. In particular, the reduction of gate leakage currents by the implementation of high-k materials (at first silicon oxynitrides, followed by metal oxides) for gate dielectrics in MOS transistors and the increase of the charge mobility in the transistor channel by strained silicon are discussed. Novel concepts such as selective air-gaps and self-assembled monolayers are also included as potential approaches to extend the boundaries of current microelectronic technology.

Chapter 2 provides an overview of various material aspects in nonvolatile semiconductor memories. Following a short description and classification of all relevant nonvolatile memory concepts found in research and development today, material optimization in nanocrystal memories that extend the more classical floating gate memories to much smaller feature sizes in the few tens of nanometers range are explained in detail. Recent developments in the three most important nonvolatile RAM concepts, namely ferroelectric memories, magnetoresistive memories and phase change memories, demonstrate the importance of introducing switching mechanism in new materials to achieve improved nonvolatile memory performance. Finally a review of organic memories illustrates the potential new engineering options that arise when organic switching materials are combined with today's advanced silicon electronics.

Advanced materials for on-chip interconnects play an important role for performance and reliability of microelectronic products. The scope of Chapter 3 is to provide an overview of recent developments and research activities in the field of materials for the so-called backend-of-line technology, with a particular emphasis on future applications and challenges to materials science and development. The integration of high-conductivity interconnects and of low-k insulating materials is a particular point of focus. Material-related topics are on-chip interconnect materials and barriers, low-k and ultra low-k dielectric materials, nanotubes, etc. The role of microstructure of metal interconnects, particularly texture and stress, on product performance and reliability is presented.

Advanced materials for assembly and packaging play an important role for performance and reliability of microelectronic products, too. In Chapter 4, recent research and development activities, particularly for wafer level packaging and 3D integration, are discussed. Material-related topics are solders, adhesives, encapsulation materials, molding compounds and others for assembly and packaging.

Advanced analytical techniques for thin-film and interface characterization are highlighted in Chapter 5. The continuous improvement of standard techniques and the development of new methods is essential for process development, process control and failure analysis in semiconductor industry. Scientific topics that are addressed range from typical out-of-fab laboratory techniques to synchrotron radiation-based analytics.

This book combines updated selected contributions presented at the EUROMAT 2003 conference in Lausanne, Switzerland, sponsored by the Federation of European Materials Societies FEMS and additional papers from internationally recognized experts in the field from all over the world. The editors would like to thank Peter Paul Schepp of the Deutsche Gesellschaft für Materialkunde DGM, Germany, who actively supported the EUROMAT 2003 topic "Materials for Information Technology" and who encouraged us to prepare this publication.

This book involved the effort of many people. It is primarily based on the strong support and excellent contributions from the authors of the papers. The efficient coordination of the book project by Anthony Doyle, Oliver Jackson and Kate Brown of Springer is gratefully acknowledged. The editors would like to thank Constanze Korn and Tatjana Schmidtchen, AMD Saxony, Dresden, Germany, for their excellent organizational and logistic support and for the enormous amount of work they did to get the camera-ready manuscripts to the publisher on time.

Ehrenfried Zschech   December 2004
Caroline Whelan
Thomas Mikolajick

List of Contributors

V. Beyer
Forschungszentrum Rossendorf
Abteilung Strukturdiagnostik
P.O. Box 51 01 19
D-01314 Dresden, Germany
V.Beyer@fz-rossendorf.de

R. Bez
STMicroelectronics, Central R&D
Non Volatile Memory Technology
Development
Via C. Olivetti 2
20041 Agrate Brianza (MI), Italy
roberto.bez@st.com

H. Brückl
University of Bielefeld
Nano Device Group
Universitaetsstrasse 25
D-33615 Bielefeld, Germany
brueckl@physik.uni-bielefeld.de

L. Carbonell
IMEC
Advanced Deposition & Removal
Technologies
Kapeldreef 75
B-3001 Leuven, Belgium
carbonel@imec.be

D. Chrastina
INFM and L-NESS Dipartimento di
Fisica, Politecnico di Milano
Polo Regionale di Como
Via Anzani 52,
I-22100 Como, Italy
danny@chrastina.net

T. Conard
IMEC
Material and Component Analysis
Department
Kapeldreef 75
B-3001 Leuven, Belgium
tconard@imec.be

A. Diebold
SEMATECH
Manager Analytical Technology
Infrastructure
2706 Montopolis Drive
Austin, Texas 78741-6499, USA
alain.diebold@sematech.org

T. Dimoulas
National Center for Scientific Research DEMOKRITOS
Institute of Materials Science, MBE
Laboratory
153 10, Athens, Greece
dimoulas@ims.demokritos.gr

M. Engelhardt
Infineon Technologies
Corporate Research
Otto-Hahn-Ring 6
D-81730 Munich, Germany
Manfred.Engelhardt@infineon.com

J. P. Guenau de Mussy
IMEC
Kapeldreef 75
B-3001, Leuven, Belgium
JeanPaul.GueneaudeMussy@imec.be

M. Hecker
Leibniz-Institute for Solid State and
Materials Research
Helmholtzstrasse 20
D-01069 Dresden, Germany
M.Hecker@ifw-dresden.de

P. Ho
University of Texas at Austin
Laboratory for Interconnect and Packaging, UT-PRC
10100 Burnet Road, Bldg 160
Austin, TX 78758, USA
paulho@mail.utexas.edu

M. Joulaud
MERCK KgaA - CEA Grenoble/LETI
17, Rue des Martyrs
38054 Grenoble, France
joulaudmi@chartreuse.cea.fr

J. Kittl
IMEC
Kapeldreef 75
B-3001 Leuven, Belgium
kittlj@imec.be

G. Knight
Ottawa-Carleton Institute of Physics
c/o Electronic Engineering
1125 Colonel By Drive
Ottawa, Ontario K1S 5B6, Canada
gknight@doe.carleton.ca

J. J.-Q. Lu
Center for Integrated Electronics
Rensselaer Polytechnic Institute
Troy, New York 12180, USA
luj@rpi.edu

G. Luyckx
ICI Belgium N.V.
Emerson & Cuming
Nijverheidsstraat 7
B-2260 Westerlo, Belgium
geert.luyckx@nstarch.com

M. A. Meyer
AMD Saxony LLC & Co. KG
Materials Analysis Department
P. O. Box. 11 01 10
D-01330 Dresden, Germany
Moritz-Andreas.Meyer@amd.com

T. Mikolajick
Infineon Technologies Dresden
Flash Technology Predevelopment
IFD MDC TD FL2
Koenigsbruecker Strasse 180
D-01099 Dresden
Thomas.Mikolajick@infineon.com

J. P. Morniroli
Laboratoire de Métallurgie Physique
et Génie des Matériaux
UMR CNRS 8517, Bâtiment C6
Cité Scientifique
59655 Villeneuve d'Ascq Cédex,
France
Jean-Paul.Morniroli@univ-lille1.fr

M. Nihei
Nanotechnology Research Center, Fujitsu Laboratories Ltd.
10-1 Morinosato-Wakamiya
Atsugi 243-0197, Japan
nihei.mizuhisa@jp.fujitsu.com

H. Oppermann
Fraunhofer IZM
Gustav-Meyer-Allee 25
D-13355 Berlin, Germany
oppermann@izm.fraunhofer.de

V. G. Polovinkin
Institute of Semiconductor Physics
pr. Lavrentyev, 13
630090 Novosibirsk, Russia
PVG@isp.nsc.ru

S. Privetera
Istituto per la Microelettronica e Microsistemi (IMM), CNR
Stradale Primosole 50
95121, Catania, Italy
stefania.privitera@imm.cnr.it

Z. Rahman
University of Limerick
Department of Physics
Plassey Technological Park
Castletroy, Limerick, Ireland
Zakia.Rahman@ul.ie

F. Shamiryan
IMEC
Kapeldreef 75
B-3001 Leuven Belgium
Denis.Shamiryan@imec.be

N. Rapún
University of Surrey
Chemistry Division, SBMS
Guildford, Surrey, GU2 7XH, UK
N.Rapun@surrey.ac.uk

G. Scarel
MDM-INFM National Laboratory
Via C. Olivetti 2
20041 Agrate Brianza (MI), Italy
giovanna.scarel@mdm.infm.it

D. Schmeißer
BTU Cottbus
Angewandte Physik / Sensorik
Konrad-Wachsmann-Allee 1
D-03046 Cottbus, Germany
dsch@tu-cottbus.de

J. Schuhmacher
IMEC
SPDT-ADRT-FUDIC
Kapeldreef 75
B-3001 Leuven, Belgium
schuhm@imec.be

F. Sediri
Faculté des Sciences de Tunis
Département de Chimie
Campus Universitaire
2092 Elmanar I, Tunisia
faouzi.sediri@ipeit.rnu.t

C. Whelan
IMEC
SPDT-ITTO
Kapeldreef 75
B-3001 Leuven, Belgium
whelan@imec.be

V. Sukharev
LSI Logic
1501 McCarthy Blvd. MS AE-187
Milpitas, CA 95035
vsukhare@lsil.com

Z. Tőkei
IMEC
Kapeldreef 75
B-3001, Leuven, Belgium
zsolt@imec.be

M. Töpper
Fraunhofer IZM
Gustav-Meyer-Allee 25
D-13355 Berlin, Germany
toepper@izm.fhg.de

R. Waser
Forschungszentrum Jülich
Departemt IFF & CNI
D-52425 Jülich
r.waser@fz-juelich.de

O. Winkler
Aachen University
Institute of Semiconductor Electronics
Sommerfeldstrasse 24
D-52074 Aachen, Germany
winkler@iht.rwth-aachen.de

Y. Yang
University of California Los Angeles
Department of Materials Science and Engineering,
Los Angeles, CA 90095, USA
yangy@ucla.edu

P. Zaumseil
IHP
Im Technologiepark 25
D-15236 Frankfurt (Oder), Germany
Zaumseil@ihp-microelectronics.com

I. Zienert
AMD Saxony LLC & Co. KG
Materials Analysis Department
P. O. Box 11 01 10
D-01330 Dresden, Germany
Inka.Zienert@amd.com

List of Contents

Chapter 1: Recent Advances in Thin-Film Deposition

Chapter 2: Material Aspects of Non-Volatile Memories

Chapter 3: Materials for Interconnects

Chapter 4: Materials for Assembly/Packaging

Chapter 5: Advanced Materials Characterization

Part I

Recent Advances in Thin-film Deposition

Molecular-beam Deposition of High-k Gate Dielectrics for Advanced CMOS

A. Dimoulas

MBE Laboratory, Institute of Materials Science, Athens / Greece

Introduction

Aggressive scaling to the nanometer range has moved the semiconductor and ICs industry into a period of revolutionary change in materials, device architecture, device processing and tooling. The main driving force behind these changes is the need to increase the on-state (drive) current, and therefore the high frequency performance of logic devices, in order to obtain intrinsic MOSFET gate delay times in the range of picoseconds (or terahertz performance levels) well before the end of the decade [1]. To achieve high drive currents in excess of 1 mA/μm (see Table 1), it is necessary to combine improvements in gate-length scaling, equivalent oxide thickness (EOT), threshold voltage and channel mobility. Intel, the leading ICs manufacturer, has announced [2] that 45 nm CMOS, which will be in production by year 2007 will have advanced high-k dielectric/metal electrode stacks in the gate. Using high-κ gate dielectrics with large physical thickness (around 3 nm) it is possible to scale down EOT to less than 1 nm maintaining the gate leakage current at acceptable levels below 1 A/cm^2 [3]. However, transistors with advanced high-κ gate stacks suffer from mobility degradation for reasons which are not fully understood at the present time. To overcome this problem, typically a starting chemical oxide (SiO_2) layer of at least 4 Å is first deposited [3] before the high-κ. This improves channel mobility but limits further scaling for more advanced technologies of the 32 and 22 nm node where EOT as low as 5 Å is required (Table 1). For such low EOT values, ultimately clean interfaces are needed which could be obtained by epitaxial oxides on Si. An alternative is to use mobility enhancing channels such as strained Si or replace completely Si by high mobility semiconductors such as Ge or III–V compounds.

McKee *et al.*, [4] at ONRL have pioneered the growth of high-k crystalline oxides on silicon by molecular beam epitaxy (MBE), working with the cubic perovskite $SrTiO_3$ (STO) on Si (001). Researchers from EPFL [5] and IBM-Zurich [5], IBM/T.J. Watson Res. Center [6, 7] and Toshiba [8] have shown that the fluorite-like oxides such as $La_2Zr_2O_7$, $(La_xY_{1-x})_2O_3$ and CeO_2, respectively, can be grown epitaxially on Si (111). However, for gate dielectric applications in Si-based transistor devices, it is necessary to grow epitaxial oxides directly on (001)-oriented Si, which has been the favourite substrate orientation of the semiconductor industry for many years. It has been shown that Y_2O_3 [9–12] and rare earth oxides with cubic symmetry such as Pr_2O_3 [13] can be grown crystalline on Si (001). However, they adopt an unfavourable orientation $(110)_{Re2O3}//(001)_{Si}$, which gives rise to twinning and other defects [11]. Until now, $SrTiO_3$ was the only known oxide grown by MBE on Si (001) with acceptable epitaxial quality, although rotated in-plane by 45° in order to match the substrate lattice parameters [4, 14, 15]. Very recently [16], A. Dimoulas *et al.*, at NCSR have shown that the pyrochlore $La_2Hf_2O_7$ (LHO) can be grown crystalline on Si (001) in a cube-on-cube local epitaxial mode forming ultimately clean interfaces with Si.

Table 1. Technology requirements for mainstream logic devices

<u>**Front-end Technology for High Performance Logic**</u>

Tech. nodes	90 nm	65 nm	45 nm	32 nm	22 nm
Production	2003	2005	2007	2009	2011
Gate stack	SiO_2/poly	SiO_2/poly	High-κ/metal Chemical oxide starting layer	High-κ/metal Direct epitaxy ? κ~30 ⟶	
Channel	Strained Si planar	Strained Si planar	3D Multi-gate?	-Strained Si -high mobility ⟶ Semiconductors (Ge, III-V)	
Physical gate length (nm)	37	25	18	13	9
EOT (nm)	1.2	0.8	0.7	0.6	0.5
Drive current NMOS (mA/µm)	1.2	1.5	1.9	2.1	2.4

Germanium MOSFETs with high-k gates may provide an alternative solution for future high performance logic devices (see Table 1) since the (low-field) mobility of carriers in Ge is three times higher than that in Si. In addition, activation annealing of dopants can be performed at lower temperatures (around 500°C), while the smaller energy gap of Ge allows supply voltage V_{dd} scaling for reduced power consumption. Finally, monolithic integration with III–V optoelectronics is possible, adding functionality to the chip. Self-aligned n-type Ge MOSFETs with Ge oxynitride/LTO/W gates [17] and p-type Ge MOSFETs with ZrO_2 [18], Al_2O_3 [19] and HfO_2 [20] high-k gates have been demonstrated. Materials issues have been investigated in detail [21–26], including surface treatment prior to growth [21], band offsets [25] of HfO_2 with Ge and stability of HfO_2 on Ge after high-temperature treatment [26].

In this article, the MBE growth, and the structural and electrical properties of crystalline and amorphous LHO high-k material and its interface with Si(001) are reviewed. In addition, the growth and electrical behaviour of HfO_2 on Ge(001) are reported.

Oxide Molecular-beam Epitaxy/(Deposition) Methodology

It is well-known that MBE/(MBD) is an excellent research tool mainly for compound semiconductor (*i.e.* III–V) and SiGe devices for specific applications. In recent years, MBE has emerged as a good candidate tool for the front-end processing of mainstream Si-based devices. Oxide MBE can be used to deposit high-k gate dielectrics on semiconductors. A schematic is shown in Figure 1 below.

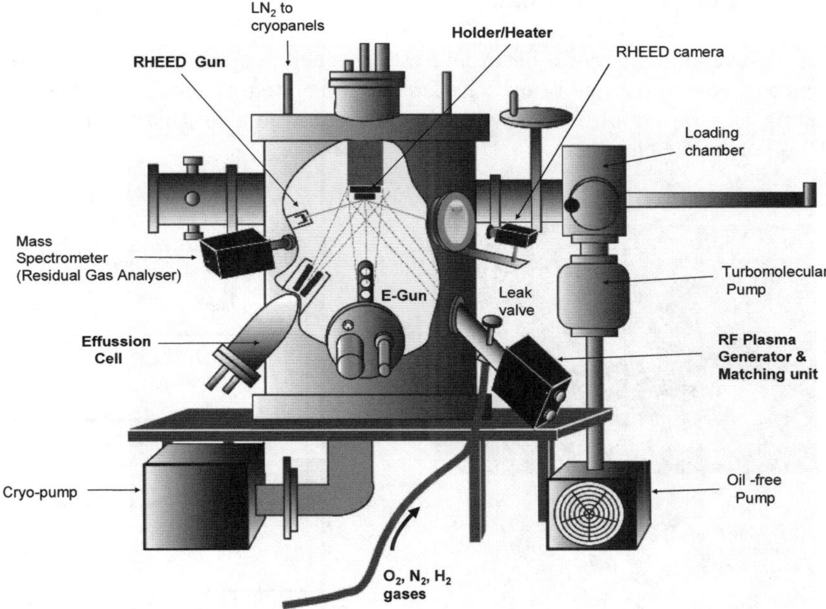

Figure 1. Molecular beam epitaxy UHV system and components. Base pressure ~5 x 10^{-10} Torr.

One of the main components is the RF plasma source used for the generation of atomic oxygen beams with thermal energies, which are reactive enough to oxidize the impinging metals on the substrate at relatively low O_2 partial pressure in the 10^{-6} Torr range. The source can produce also nitrogen and hydrogen atomic beams which can be used for surface treatment prior to growth in order to improve the electrical properties of the interfaces. Because MBE is a UHV technique it is possible to use reflection high energy electron diffraction (RHEED) for the *insitu* real-time monitoring of native oxide desorption, surface treatment and growth.

Lanthanum Hafnate on Silicon Substrates

The $La_2Hf_2O_7$ (LHO) material crystallizes in the cubic pyrochlore structure with space group symmetry Fd3m, which is very similar to that of Si (Figure 2 a). This similarity makes LHO a good candidate for epitaxial growth on Si (001), especially since there is good lattice matching between the two materials. The lattice constant of LHO is approximately twice that of Si, with the mismatch being only -0.74 % at room temperature and almost zero at around 800°C. The LHO films are deposited directly on clean, (2×1) reconstructed Si (001) surfaces at several temperatures T_g in the range between 60°C and 770°C as described in detail elsewhere [16]. Hf and La are co-evaporated from e-beam and effusion cell sources, respectively, in the presence of atomic oxygen beam. The latter is generated by the RF plasma source operating at a power of 350 W with O_2 flow equivalent to a partial pressure of 4×10^{-6} Torr in the chamber.

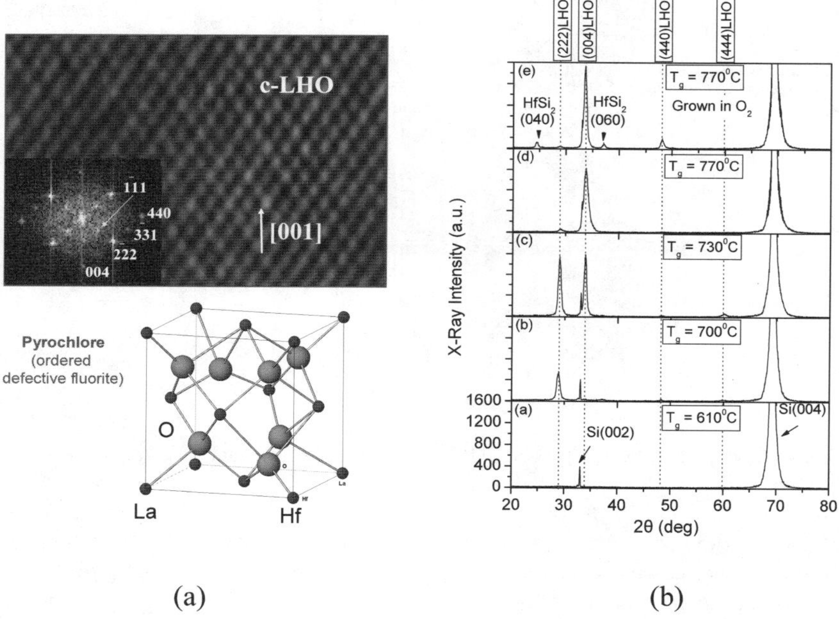

<table>
<tr><td>(a)</td><td>(b)</td></tr>
</table>

Figure 2. (a) HRTEM image of crystalline LHO grown on Si (100) at 770° C and the digital diffractogram in the inset. The super lattice spots (shown by arrows) indicate an ordered defective fluorite phase (pyrochlore). (b) XRD spectra of LHO grown on Si (100) at different temperatures.

The structural quality of thin LHO films and interfaces depends strongly on the growth temperature T_g. It can be seen from Figure 2 (b) that the material is amorphous for T_g lower than 600°C. At higher temperature of about 700°C, the material is polycrystalline with a predominant (111)LHO // (001)Si orientation. When T_g rises to about 770°C, the LHO film develops a strong preferential orientation

(001)LHO // (001)Si, indicating that this material tends to grow epitaxially when it is prepared at high temperature. When molecular oxygen is used, crystalline $HfSi_2$ phases are seen (see graph at the top of Figure 2 (b)). This means that reactive atomic oxygen is necessary in order to promote oxidation against competing silicide formation processes.

Crystalline LHO on Si(001) : Ultimately Clean Interfaces

The crystalline LHO grown at the highest temperature of 770°C is shown in Figure 3. The two materials can be distinguished between each other only from the difference in the contrast. Looking at the high magnification picture on the right, the fringes of the {111} lattice planes continue from the Si substrate all the way up to the LHO layer indicating clean interfaces with no interfacial oxide as required for ultimate scaling of devices. This also indicates cube-on-cube epitaxial growth with the $(001)_{LHO}$ // $(001)_{Si}$ and $[110]_{LHO}$ // $[110]_{Si}$ orientation relationships. However, the interface and surface morphology is rough, so that additional work is needed in order to optimize growth and obtain device quality layers.

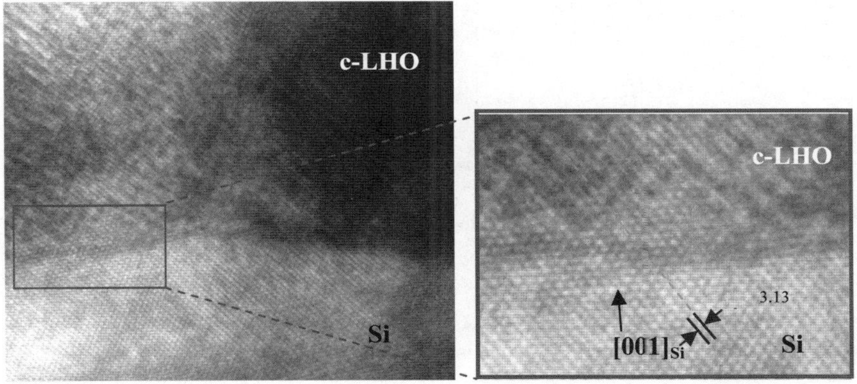

Figure 3. HRTEM micrograph showing ultimately clean interfaces between LHO and Si substrates. In the magnification the dotted line shows the lattice fringes of the {111} planes.

Electrical Properties of Amorphous LHO on Si(001)

Unlike the case of crystalline LHO, thin films of amorphous LHO deposited by MBD directly on Si are dense and continuous with a smooth surface morphology. They have a dielectric permittivity k_{LHO} of about 18 [27]. In order to investigate the electrical quality of the films at low EOT, Mo gates were sputtered and subsequently patterned by photolithography and etched using ammonia-based solution to obtain 50 μm square MIS capacitors. Standard data acquisition setup was used for admittance and gate current measurements, described elsewhere. The results for the thinnest (3.3 nm) samples are shown in Figure 4. The C-V and G-V data are fitted simultaneously using a multi-frequency model we have previously developed [18]. This is a self-consistent solution of Poisson–Schrodinger equation in the semiconductor which takes into account the charge due to interface states as boundary condition. The exchange of carriers between the semiconductor and the interface trap states is described by Shockley–Read–Hall statistics. From the fitting, the EOT, the flat band voltage shift ΔV_{FB}, the carrier capture cross section c_p, and the energy distribution of interface density of states D_{it} are estimated.

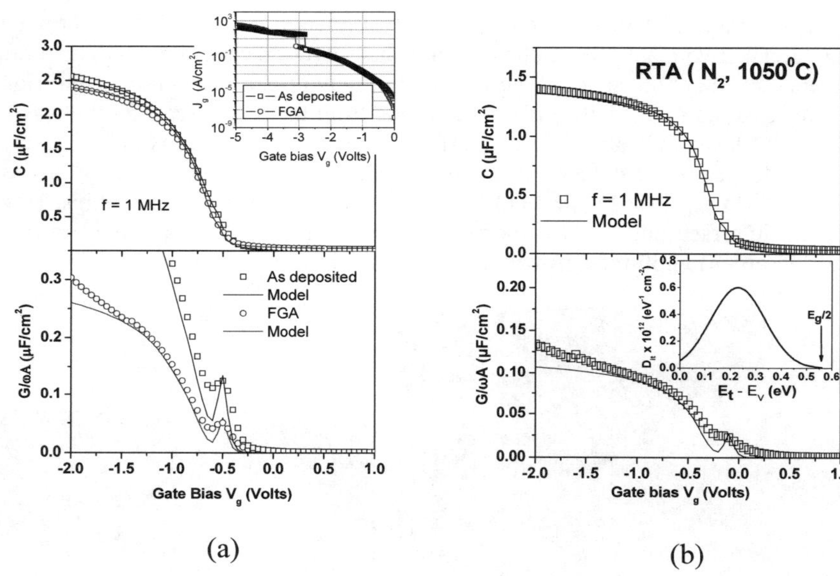

(a)   (b)

Figure 4. Electrical data obtained from a p-type MIS with 3.3 nm-thick LHO deposited at 450°C. (a) As-deposited and FGA-annealed C-V and G-V curves at 1 MHz. The gate current density J_g *vs* gate bias V_g in accumulation is shown in the inset. (b), the C-V and G-V curves after RTA treatment. The inset shows the energy distribution of D_{it} as determined from the simultaneous fitting of the C-V and G-V curves.

The EOT is estimated to be 0.94 nm for as-deposited samples and increases slightly to about 1 nm after forming gas anneal (FGA) at 450°C for 20 min. The maximum value of Dit is about 3×10^{12} cm^2 in as-deposited MIS, decreasing to about 1×10^{12} cm^2 after FGA as can be seen from the decrease of the conductance peak in Figure 4a. The flatband voltage shift ΔV_{FB} was found to be −0.5 V, the possible contributions in this shift coming from the workfunction difference $\Phi_{ms} \sim$ −0.26 V and positive fixed oxide charges in the LHO layer. As can be seen in the inset of Figure 4a, the a-LHO behaves as an excellent insulator with only 2×10^{-3} A/cm^2 at −1 V or 2×10^{-2} A /cm^2 at 1 V beyond flatband in accumulation. This is 4−5 orders of magnitude smaller that conventional SiO$_2$ devices with the same equivalent thickness of 0.94−1 nm. Rapid thermal annealing (RTA) in N$_2$ at

1050°C improves most of the electrical properties of a-LHO except from EOT. As seen in Figure 4 (b) the curves shift toward more positive voltage, corresponding to ΔV_{FB} of -0.23 V which can be explained in terms of workfuncion difference only. This value is very close to the ideal threshold voltage of -0.27 V in n-channel MOSFETs. The maximum density of states after RTA has dropped to 6×10^{11} cm^2 as can be seen from the inset of Figure 4 (b). In addition, no hysteresis is observed in the samples which have received an RTA annealing. However, the EOT is doubled compared to the as-deposited devices, probably because an interfacial low-k layer is formed. It should be noted however, that the high-temperature treatment in this work is not performed under controlled conditions which are typically employed during dopant activation in transistor processing. In addition, the trend in CMOS processing of high-κ gates is to perform dopant activation at lower temperature around 900°C, in which case EOT degradation could be avoided.

In summary, a-LHO with a very low leakage current of $\sim 2 \times 10^{-2}$ at 1V beyond flatband in accumulation and a low EOT around 0.94–1 nm could have an impact in future low standby devices of the 32 nm technology node which require EOT ~ 1.1 nm and gate leakage of 4×10^{-2} A/cm^2, provided that EOT degradation could be avoided by applying appropriate high-temperature treatment during CMOS transistor processing.

Hafnium Oxide on Germanium Substrates

Adequate preparation of Ge substrates prior to deposition is very important in order to obtain HfO$_2$ films and interfaces with good electrical quality [21]. As an UHV technique, MBD offers *insitu* desorption of the native oxide, which cannot be achieved very easily by other more standard methodologies such as CVD. The substrate is heated to 360°C in vacuum until a (2×1) reconstruction is obtained (see Figure 5) by RHEED, which is indicative of a clean Ge(100) surface. Subsequently, the substrate is heated to 225° C and is exposed to combined nitrogen and oxygen atomic beams for one minute under a weak Ge flux with the aim to produce an ultra-thin GeO$_x$N$_y$ starting layer.

The procedure for this processing step is as follows. O$_2$ and N$_2$ gases are mixed before they are introduced in the RF plasma source through a leak valve. By powering the source at ~ 400 W, a bright plasma mode consisting of nitrogen and oxygen atoms is generated in the discharge chamber and then escape into vacuum as atomic beams. Different compositions of the GeO$_x$N$_y$ layer can be obtained by varying the composition of the initial gas mixture which can be monitored by the ratio of O$_2$ and N$_2$ partial pressures in the chamber using a mass spectrometer. For the experiments reported here, a ratio $P_{O2}/P_{N2} \sim 2$ is used. It must be noted that the treatment of the clean Ge surface by O and N before HfO$_2$ deposition is necessary, otherwise the films are either leaky or do not show MOS behaviour.

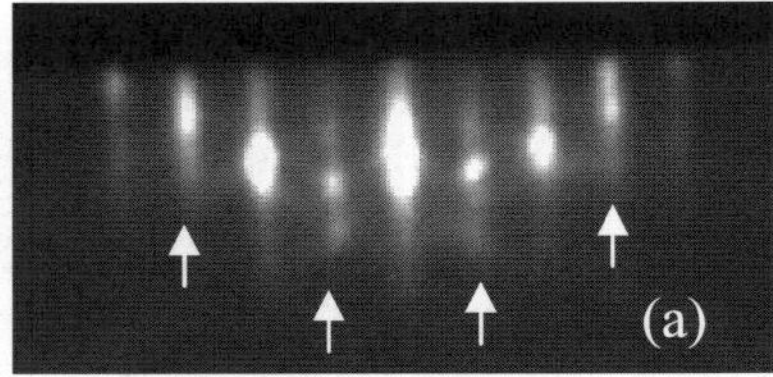
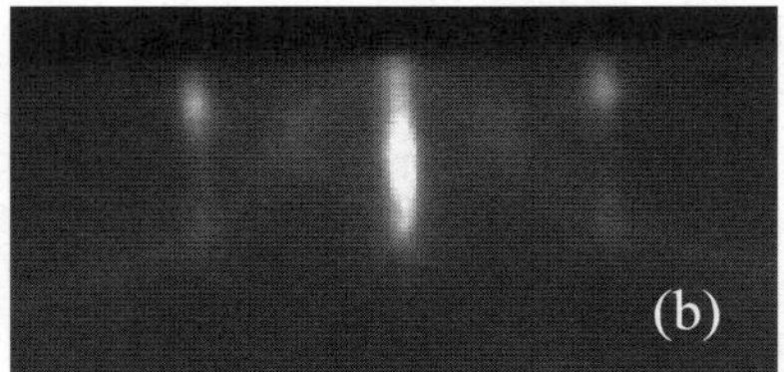

Figure 5. RHEED pattern of a clean Ge(100) surface. (a) along the [110] azimuth, (b) along the [100] azimuth, rotated by 45°. The weaker spots shown by arrows in (a) are diffraction spots due to (2×1) reconstruction of the surface.

After surface preparation, the substrate is heated to the desired temperature for HfO_2 deposition. Temperatures in the range between 60 and 360°C have been used, since above 360°C the GeO_xN_y layer is unstable. The HfO_2 is synthesized on the substrate by evaporating Hf from an e-beam source at a rate of 0.2 A/s in atomic oxygen beams generated by the RF plasma source. The partial pressure of oxygen P_{O2} during deposition is $2-4 \times 10^{-6}$ Torr.

The real images of the HfO_2 films in (110) Ge cross-section are shown in the micrographs of Figure 6 for two samples with thickness 3 and 8 nm which are deposited at 225°C. The thin sample is amorphous while the thick one is polycrystalline. This may be due to re-crystallization occurring during HfO_2 deposition since the thick sample is exposed to elevated temperature for a longer period of time. It is worth noticing from Figure 6 that the interfaces between HfO_2 and Ge are very sharp with the interfacial layer being very thin (1–2 monolayers). This could be expected given that GeO_2 is unstable, preventing reactions of Ge with oxygen at the interface.

Electrical Characterization of HfO₂/Ge M-I-S Capacitors

MIS capacitors are prepared by shadow mask and e-beam evaporation of 30 nm-thick Pt electrodes to define circular dots, with an area of $7 \times 1 \cdot 10^{-4}$ cm^2. Both Sb-doped n-type and Ga-doped p-type substrates were used with resistivity $\rho \sim 2-5$ $\Omega \cdot cm$. The back ohmic contacts are made using eutectic In–Ga alloy. Both p-type and n-type MIS show similar behaviour. Also, the main characteristics of MIS do not depend on the deposition temperature of HfO_2. Typical C-V and G-V curves for n-type MIS capacitors are shown in Figure 7a. The main characteristics are the rather large hysteresis of about 500 mV and the large ac conductance due to interface states which, however, decreases by a factor of three after forming gas anneal. Figure 7b shows the frequency dependence of the C-V curves for an n-type MIS capacitor. The most characteristic behaviour is the strong frequency dispersion in inversion (negative V_g). The high frequency C-V (*i.e.* 1 kHz) shows low-frequency behaviour with a high value of the inversion capacitance. This is different from the case of device quality silicon where the 1 kHz capacitance at inversion acquires a small value due to long response time of the minority carriers. The minority carrier response time τ_R at room temperature is controlled by generation-recombination processes in depletion, mediated by bulk semiconductor traps. In this case, $\tau_R \propto 1/n_i$ [28], where n_i is the intrinsic carrier concentration.

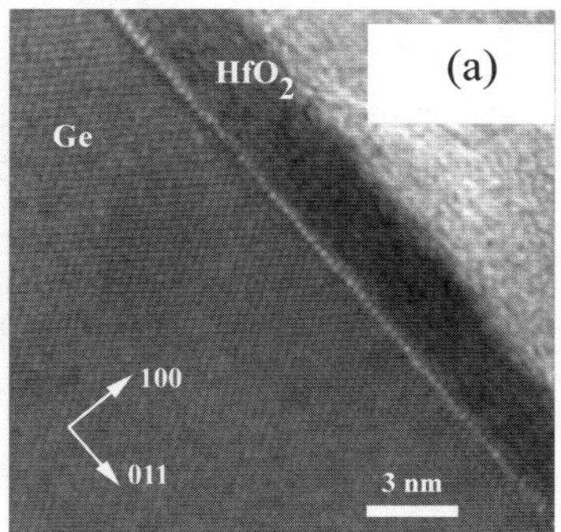

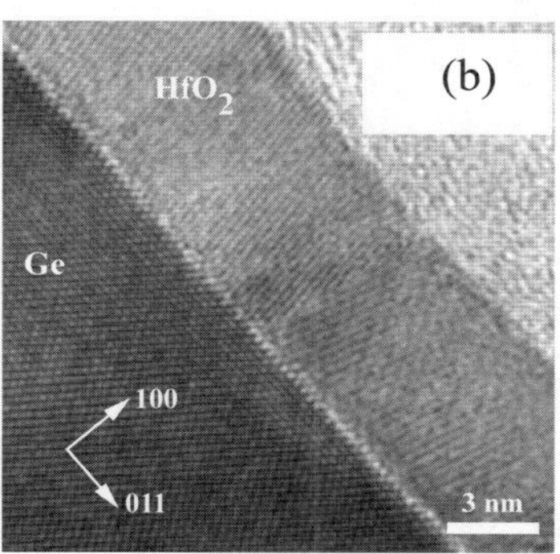

Figure 6. HRTEM micrograph in Ge (-110) cross section showing sharp HfO$_2$/Ge interfaces. (a) $t_{ox} \sim 3.1$ nm, (b) $t_{ox} \sim 8$ nm. The HfO$_2$ deposition temperature was $T_{dep} = 225°C$, for both samples.

In Si, $n_i \sim 1.45 \times 10^{10}$ cm^{-3} and τ_R ranges between the 0.01 and 1 s, depending on the quality of the material. However, in Ge, due to the smaller band gap, n_i is more than three orders of magnitude higher ($n_i \sim 2.4 \times 10^{13}$ cm^{-3}), therefore, the response time is expected to be at least three order of magnitude shorter, in the microsecond range, provided that the level of contamination is comparable to that found in device grade Si. This could explain the observed low frequency behaviour of the high frequency (1 kHz) CV curves in Figure 7b.

C-V data from p-type MIS capacitors with different HfO$_2$ thickness are given in Figure 8. As in the case of n-type MIS, there is strong frequency dispersion in inversion (positive V_g) and a weak dispersion in accumulation (negative V_g). The latter is attributed to a series resistance of ~ 50 Ω, which mainly affects the 1 MHz curve, lowering the accumulation capacitance.

By analyzing of the accumulation capacitance of the low frequency (20 Hz) curves, the EOT is estimated and plotted as a function of the physical oxide thickness in Figure 9a. From this plot, the interfacial thickness t_{int} is found to be about 0.28 ± 0.03 nm, in agreement with TEM observations, while the dielectric permittivity κ_{ox} of HfO$_2$ is estimated to be $\sim 2.47 \pm 0.7$. This is close to the expected bulk value and appreciably higher than that obtained from thin films of HfO$_2$ on Si, which typically ranges between 15 and 20. The gate current J_g as a function of gate bias V_g is shown in Figure 9b indicating that the HfO$_2$ on Ge behaves as an excellent insulator. The thinnest sample with EOT of 0.75 ± 0.1 nm exhibits a very low current of 4.5×10^{-4} A/cm^2 at -1V (accumulation), which are among the best values reported for either HfO$_2$/Ge or HfO$_2$/Si.

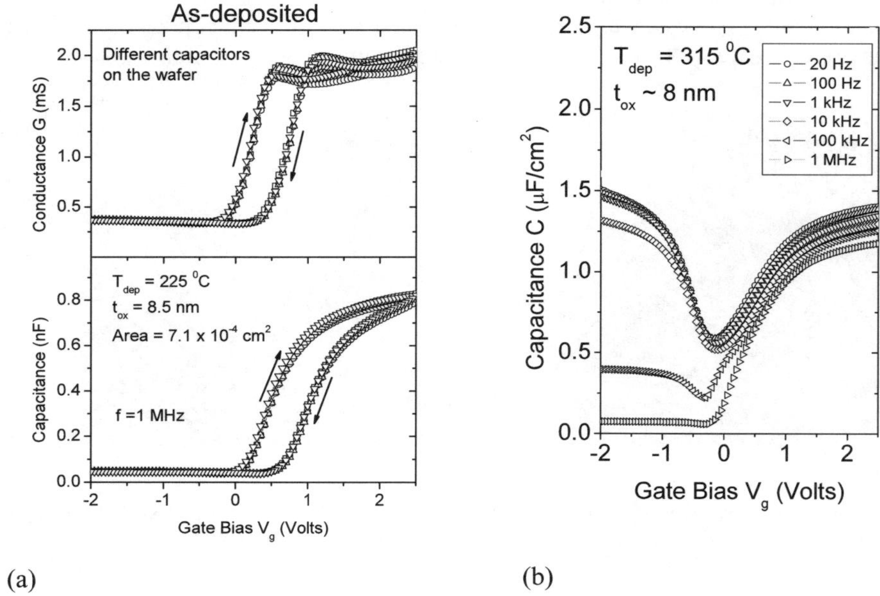

Figure 7. (a) Hysteresis of 1 MHz C-V and G-V curves of n-type MIS. (b) Frequency dependence of C-V curves for n-type HfO_2/Ge M-I-S capacitors.

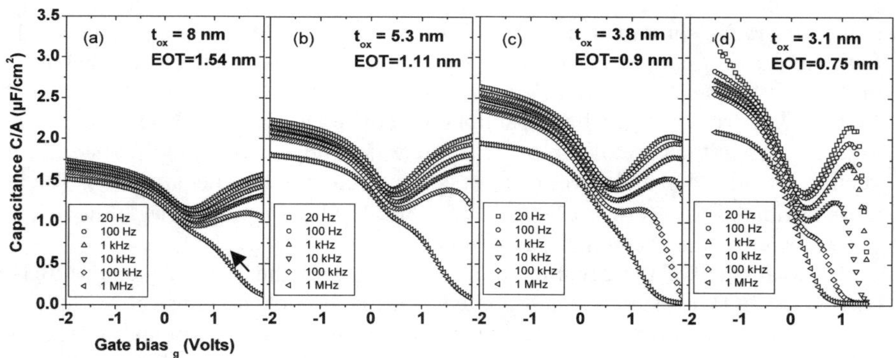

Figure 8. C-V curves for as-deposited p-type M-I-S with different thickness of HfO_2 dielectric. The scale and axes labels for graphs (b) to (d) are the same with the ones shown with the graph (a).

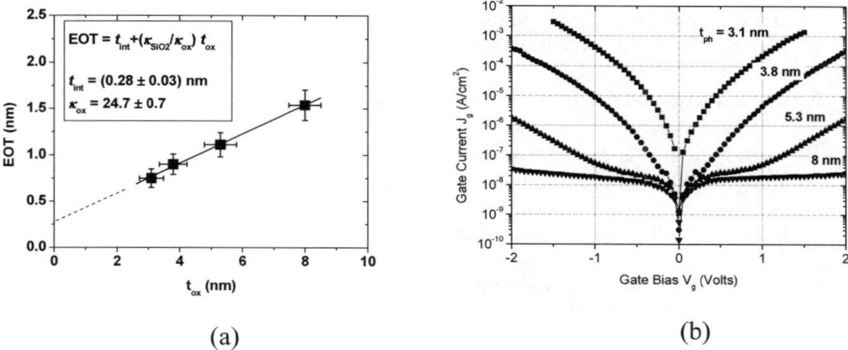

(a)   (b)

Figure 9. (a) Equivalent (EOT) *vs* physical oxide thickness t_{ox}. The dielectric permittivity κ_{ox} and the interfacial layer thickness t_{int} can be estimated from the slope and the intersection with the y-axis, respectively. (b) The gate current J_g *vs* the applied gate bias V_g for several MIS with different HfO_2 thickness.

Summary

Advanced high-κ gate dielectric stacks directly deposited on Si or high mobility semiconductors such as Ge by MBE may offer the solution for aggressive scaling of future nanoelectronic devices. A new high-k dielectric, the pyrochlore $La_2Hf_2O_7$, has been systematically investigated. This material can be prepared on Si(001) in a cube-on-cube epitaxial mode at high temperature around 770°C forming ultimately clean interfaces with the substrate. At lower temperature the material is amorphous having leakage current four to five orders of magnitude lower than conventional SiO_2 with the same equivalent oxide thickness (EOT) of about 1 nm. Further scaling is expected to be difficult since only moderate values of the dielectric permittivity k around 18 can be obtained. On the other hand, HfO_2 prepared by MBD on Ge(001) substrates forms sharp interfaces and has a relatively high $\kappa\sim25$ which is close to the expected bulk value. Low EOT values around 0.75 nm have been obtained at very low leakage current of 4.5×10^{-4} A/cm^2 at 1 V in accumulation. The observed low frequency behaviour of the high frequency (*i.e* 1 kHz) C-V curves in inversion was attributed to the high intrinsic carrier concentration in Ge due to the small energy band gap.

Acknowledgements

I would like to thank my students G. Vellianitis and G. Mavrou and my colleagues Drs. A. Travlos and N. Boukos at NCSR DEMOKRITOS, Prof. E. Evangelou of the University of Ioannina and Drs J.C. Hooker and Z.M. Rittersma of Philips Research Leuven for their contribution in this work.

References

1. 2003 Edition of the ITRS, International Technology Roadmap for Semiconductors (ITRS), Semiconductor International Association (2003), http://public.itrs.net/

2. R. Chau, International Workshop on Gate Insulators, Tokyo, Nov 6–8 (2003)

3. W. Tsai, L. Ragnarsson, P.J. Chen, B. Onsia, R.J. Carter, E. Cartier *et al.*, Symposium on VLSI Technology Digest of Technical Papers, Kyoto, 21–22 (2003).

4. McKee, F.J. Walker, M. Chisholm, Phys. Rev. Lett. 81, 3014 (1998)

5. J. W. Seo, J. Fompeyrine, A Guiller, G. Norga, C. Marchiori, H. Siegwart *et al.*, Appl. Phys. Lett. 83, 5211 (2003)

6. S. Guha, N.A. Bojarczuk, V. Narayanan, Appl. Phys. Lett. 80, 766 (2002)

7. N.A Bojarczuk, M. Copel, S. Guha, V. Narayanan, E.J. Preisler, F.M. Ross *et al.,* Appl. Phys. Lett. 83, 5443 (2003)

8. Y. Nishikawa, N. Fukushima, N. Yasuda, K. Nakayama, S. Ikegawa, Japan J. Appl. Phys. 41, 2480 (2002)

9. A. Dimoulas, A. Travlos, G. Vellianitis, N. Boukos, K. Argyropoulos, J. Appl. Phys. 90, 4224 (2001)

10. A. Dimoulas, G. Vellianitis, A. Travlos, V. Ioannou-Sougleridis, A.G. Nassiopoulou, J. Appl. Phys. 92, 426 (2002)

11. G. Apostolopoulos, G. Vellianitis, A. Dimoulas, M. Alexe, R. Scholz, M. Fanciulli *et al.,* Appl. Phys. Lett. 81, 3549 (2002)

12. V. Ioannou-Sougleridis, G. Vellianitis, A. Dimoulas, J. Appl. Phys. 93, 3982 (2003)

13. Fissel, H.J. Osten, E. Bugiel, J. Vac. Sci. Technol. B 21, 1765, and references therein (2003)

14. K. Eisenbeiser, J.M. Finder, Z. Yu, J. Ramdani, J.A. Curless, J.A. Hallmark, *et al.*, Appl. Phys. Lett. 76, 1324 (2000)

15. Norga, A. Guiller, C. Marchiori, J.-P. Locquet, H. Siegwart, D. Halley *et al.*, Mat. Res. Soc. Symp. Proc. 786, (2004)

16. Dimoulas, G. Vellianitis, G. Mavrou, G. Apostolopoulos, A. Travlos, C. Wiemer *et al.,* Appl. Phys. Lett. 85, October 11 issue (2004)

17. H. Shang, K.L. Lee, P. Kozlowski, C. D'Emic, I. Babich, E. Sikorski *et al.,* IEEE EDL 25, 135 (2004)

18. C. Chui, S. Ramanathan, B. Triplett, P. McIntyre, K. Saraswat, in IEDM Tech. Dig., 437–439 (2002)

19. D.S. Yun, C.H. Huang A. Chin, W.J. Chen, C.X. Zhu, B.J. Cho *et al.*, IEEE EDL 25, 138 (2004)

20. Ritenour, S.Yu, M.L. Lee, N. Lu, W. Bai, A. Pister *et al.*, IDEM Tec. Dig. 2003

21. Chui, H. Kim, P.C. McIntyre, K.C. Saraswat, IEEE EDL 25, 274 (2004)

22. H. Kim, C. Chui, K.C. Saraswat, P.C. McIntyre, Appl. Phys. Lett. 83, 2647 (2003)

23. D. Chi, C. Chui, K.C. Saraswat, B.B. Triplett, P.C McIntyre, J. Appl. Phys. 96, 813 (2004)

24. J.-H. Chien, N. A. Bojarczuk, Jr., H. Shang, M. Copel, J.B. Hannon *et al.*, IEEE Trans. Electron. Dev. 51, 1441 (2004)

25. V.V. Afanas'ev, A. Stesmans, Appl. Phys. Lett. 84, 2319 (2004)

26. E.P. Gusev, H. Shang, M. Copel, M. Gribelyuk, C. D'Emic, P. Kozlowski *et al.*, Appl. Phys. Lett. 85, 2334 (2004)

27. Apostolopoulos, G. Vellianitis, A. Dimoulas, J.C. Hooker, T. Conard, Appl. Phys. Lett. 84, 260 (2004)

28. E.H. Nicolian, J. Brews, in MOS Physics and Technology, Wiley, New York, 139 (1982)

LEPECVD – A Production Technique for SiGe MOSFETs and MODFETs

D. Chrastina[a], B. Rössner[a], G. Isella[a], H. von Känel[a], J. P. Hague[b], T. Hackbarth[c], H.-J. Herzog[c], K.-H. Hieber[c], U. König[c]

[a]INFM and L-NESS Dipartimento di Fisica, Politecnico di Milano, Como / Italy

[b]Department of Physics & Astronomy, University of Leicester, Leicester / UK

[c]DaimlerChrysler AG, Research and Technology, Ulm / Germany

Introduction

Crystalline silicon germanium alloys greatly extend the potential of silicon-based electronics. Both electron and hole conduction can be enhanced through control of strain, Ge content, and band-gap engineering [1, 2].

Compressive strain results if $Si_{1-x}Ge_x$ is deposited epitaxially on $Si_{1-y}Ge_y$ when $x>y$, due to the 4.17% mismatch between Ge and Si lattice constants. A compressively strained $Si_{1-x}Ge_x$ layer forms a quantum well for holes, as shown in the left-hand panel of Figure 1. $Si_{1-x}Ge_x$ layers with x up to about 0.5 can be grown pseudomorphically (without relaxing the strain imposed by the substrate) on Si, in thicknesses which are useful for electronic applications. However, pure Ge (x=1) cannot generally be deposited directly on a Si substrate (y=0) without a strain-induced transition to a three-dimensional growth (Stranski–Krastanow) mode after a few monolayers [3]. Furthermore, even in the case of two-dimensional (Frank-van der Merwe) growth there is a limit to how much strain can build up before the strained layer relaxes [4].

Therefore, the epitaxial two-dimensional growth of a strained Ge-rich layer ($x\sim1$) of useful thickness directly on a Si substrate is not possible. However, such a layer can be grown on a relaxed $Si_{1-y}Ge_y$ buffer layer, with $0.5{\leq}y<1.0$. The band profile in this case is shown in the right-hand panel of Figure 1. The Si substrate plus a relaxed buffer forms a virtual substrate (VS). A VS also allows the growth of a layer under tensile strain ($x<y$) which creates a quantum well for electrons, as shown in the left-hand panel of Figure 2 [2, 5]. Furthermore, tensile-strained Si surfaces are smoother than unstrained Si surfaces because tension increases the surface step energy [6].

Silicon germanium has become a commercially successful technology. Its most mature manifestation is the heterojunction bipolar transistor (HBT), which has the advantages that it is only a minor modification of the Si bipolar transistor and is easily integrated into standard complementary metal-oxide-semiconductor (CMOS) processes. This is known as BiCMOS. A thin layer of SiGe is deposited pseudomorphically to act as the base of the transistor, and as such this technology is easy to implement. Modern SiGe HBTs can now operate at frequencies above 300 GHz and are used in both radio frequency wireless and cable communications [2].

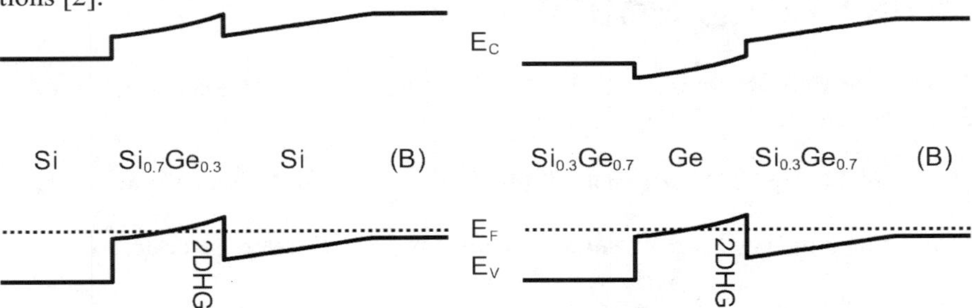

Figure 1. Band profiles of two compressively-strained systems. Left panel: a strained SiGe channel grown pseudomorphically on Si. Right panel: a strained Ge channel grown on a SiGe VS. The growth direction is from left to right, so in both cases, boron doping (B) is above the channel. Holes diffuse into the channel and an electric field is set up between the two-dimensional hole gas (2DHG) and ionized dopant atoms. The dashed line is the Fermi level E_F, and E_C and E_V are the conduction and valence bands, respectively. Band offsets and other parameters can be found in Ref. [5].

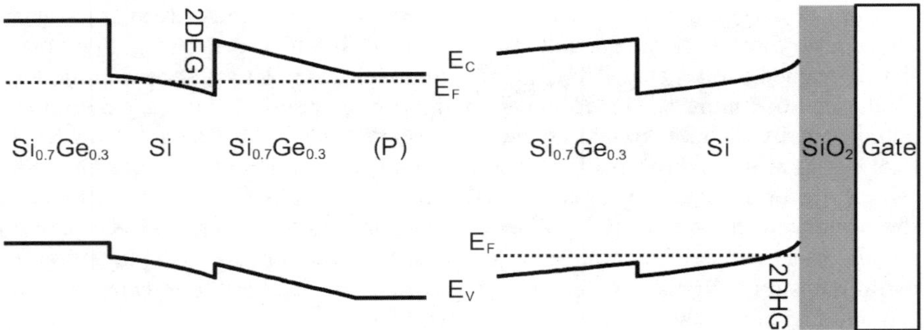

Figure 2. Band profiles of Si under tensile strain on a relaxed SiGe VS. Left panel: Si under tensile strain forms a quantum well for electrons, so electrons from a phosphorus-doped region (P) diffuse into the Si and a two-dimensional electron gas (2DEG) forms. Right panel: a strained Si surface channel can be biased with a gate, to induce a 2DHG [2, 5].

Strained Si on a VS on the other hand requires the VS to have low surface roughness, a high degree of relaxation, and low defect densities [1]. There exist two fundamentally different approaches for realizing VSs with acceptable properties. A well-established solution to this problem is the graded buffer. The basic concept of grading $Si_{1-x}Ge_x$ alloy layers to some final composition x, either linearly or step-wise, has proven to be highly successful [7].

Si or Ge quantum wells grown on VSs with thick graded buffers have been demonstrated to have excellent electrical properties [8–10], but there are disadvantages of thick buffers. Firstly, SiGe alloys are poor thermal conductors, leading to problems of heat dissipation for highly integrated circuits. Secondly, long range surface roughness (with ~1 µm correlation length) due to the so-called "cross-hatch" [11] may cause problems during device processing, and the large step height between regions covered by virtual substrates and bare Si regions hampers integration. Also, a large amount of precursor material is consumed, and in most MBE or CVD systems the growth time is at least a few hours.

For these reasons a second solution is being pursued, which involves thin constant-composition layers. Several concepts have been tried over the past few years to achieve highly relaxed SiGe layers as thin as a hundred nanometers [12, 13]. These methods fall into two categories.

The first involves growth of a Si buffer layer at unusually low substrate temperatures T_s (around 400°C) before the actual SiGe buffer is deposited at higher T_s [14-16]. This so-called low-temperature Si (LT-Si) layer contains a large number of point defects, the accumulation of which is believed to promote relaxation [17]. Alternatively, the SiGe buffer layer itself may be grown at very low T_s (below 200°C) with essentially a similar effect [18]. The disadvantage here is that long periods of thermal cycling may be needed to achieve a stable state of relaxation. Also, such low-temperature growth cannot be realized by chemical vapor deposition (CVD) since growth rates decrease exponentially as T_s is reduced. Most work in this field has therefore been carried out with molecular-beam epitaxy (MBE) which is not suitable for high volume Si or SiGe production.

The second approach is based on ion implantation of H or He into a strained SiGe layer, and subsequent thermal treatment [19, 20]. This leads to bubble formation, facilitating dislocation loop nucleation close to the Si/SiGe interface. Except for the extra processing steps, this method appears very attractive, but seems to be limited to buffers with Ge content below 30% [20-22]. In a variation of this, Ar ions can be implanted into the Si substrate before SiGe growth [23].

In this chapter we shall discuss an alternative method suitable for high-volume production of both thick graded buffers and thin buffers. This method is called low-energy plasma-enhanced chemical vapor deposition (LEPECVD) [24].

LEPECVD

The basic LEPECVD system is shown schematically in Figure 3. A Ta filament in a plasma source attached to the deposition chamber is current-heated to the point of thermionic emission. Ar gas is passed through the source into the chamber, and a high-intensity direct current arc is struck between the source and the chamber. The arc current is of the order of 20-80A but the arc voltage is less than 30 V due to the electron-rich conditions. The plasma is focused onto the substrate with magnetic fields, while the substrate is heated from behind with a graphite heater. Precursor gases are introduced through a ring just below the substrate: SiH_4 and GeH_4 are used for SiGe growth, and PH_3 and B_2H_6 (both diluted in Ar) are used for doping; H_2 can also be used to fine-tune the surface mobility of adatoms.

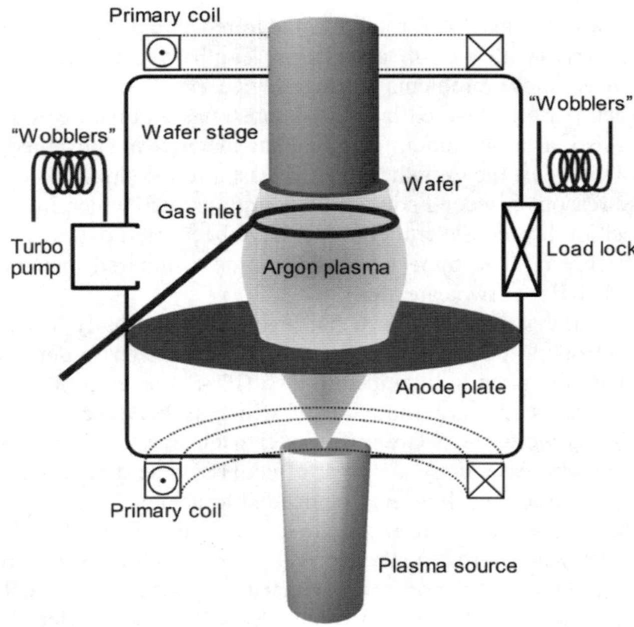

Figure 3. The low-energy plasma-enhanced chemical vapor deposition (LEPECVD) system. A low-voltage high-current dc discharge is sustained between the source and the anode plate. The substrate is exposed to the plasma, but the ion energies are too low to cause any damage.

Epitaxial growth rates for $Si_{1-x}Ge_x$ of any composition can be varied arbitrarily from 1 Ås^{-1} to almost 10 nms^{-1}, at substrate temperatures of 450-750°C. The growth rate is controlled both by varying the precursor gas flows and by changing the plasma arc current and the strength of the magnetic confinement field. The growth rate and film composition are effectively independent of substrate temperature. Around 20% of the precursor material is incorporated into the sample under plasma conditions optimized for high growth rates.

The development of LEPECVD was motivated by the need for relaxed SiGe alloy buffer layers epitaxially grown on Si wafers. In the case of chemical vapor deposition (CVD) without plasma, the gaseous precursors decompose on the growing surface. For Si growth at substrate temperatures below ~800°C, the growth rate is limited by H desorption from the growing surface [25]. In fact, there is an exponential dependence of growth rate on substrate temperature with an activation energy of around 200 kJmol^{-1}. The activation energy for desorption of H from a Ge surface is significantly lower, and this leads to a $Si_{1-x}Ge_x$ growth rate that is strongly dependent on x as well as temperature, in a non-trivial way. Typical growth rates are 10-100 nmmin^{-1} and less than 1% of the silicon and germanium precursor gases is deposited.

The low energy plasma improves matters in two ways. Firstly, the precursor gases are efficiently decomposed in the plasma column. Secondly, H is desorbed from the growing surface by ion bombardment. The energy of the ions is however too low (~10 eV) to cause any damage to the crystalline structure of the substrate.

MBE does not suffer from the problem of growth rate dependence on substrate temperature, but since it is a solid-source process it is unsuitable for the industrial growth of thick layers. Typically, the growth chamber vacuum needs to be broken to replenish the sources after ~5-10 VSs have been grown. SiGe MBE growth rates are also usually of the order of 10-20 $nmmin^{-1}$ unless special techniques are used [26].

Applications

Virtual Substrates

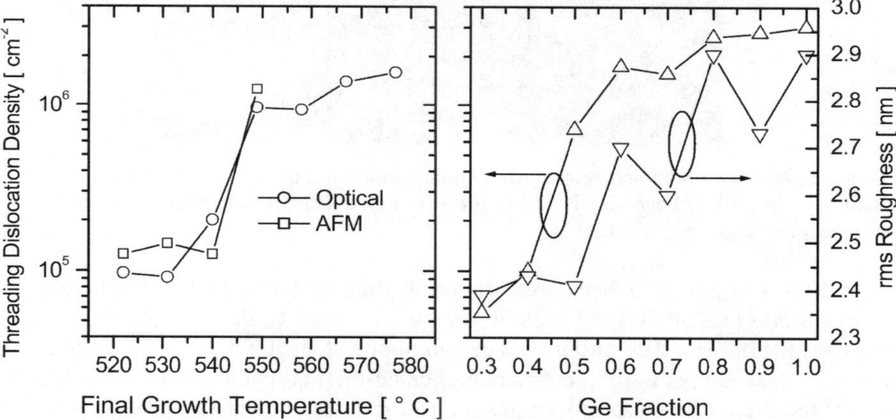

Figure 4. Left panel: Threading dislocation density on thick graded (*x*=0.4) VSs grown by LEPECVD as a function of final growth temperature. The grading rate was 7% per micron and the constant composition cap is 1 μm thick. Etch pits were counted either with an optical microscope or an AFM. Right panel: Surface roughness and threading dislocation density on thick graded VSs grown by LEPECVD, with optimized temperature profiles.

Growth rates in LEPECVD are not strongly dependent on substrate temperature, so the temperature profile during the growth of a thick VS can be varied to optimize the defect density, as shown in the left-hand panel of Figure 4 for thick VSs graded at 7% per micron to a final Ge content of 40%. Generally, the growth temperature should be reduced as the Ge content increases; it can be seen that with a suitably low final temperature, the threading dislocation density can be reduced to less than $10^5 cm^{-2}$. The right-hand panel of Figure 4 shows how the roughness and dislocation density of a VS tend to increase as a function of Ge content, up to 3×10^6 cm^{-2} for pure Ge. An atomic force microscopy (AFM) image of a thick graded VS (with a final Ge content of 30%) is shown in Figure 5. The rms roughness is 2.6 nm. For comparison, a 50% graded VS grown by UHVCVD has an rms roughness of 37 nm [29].

Figure 5. Atomic force microscopy (AFM) image of a 6 μm thick VS graded from 0 to 30% (sample 7208). The image size is 25×25 μm. Root mean square (rms) roughness is 2.6 nm and the total height range is 15.6 nm.

LEPECVD has also been used to grow thin buffers which are suitable for n-MODFETs [13, 30, 31]. At 500nm, the buffers are 5-10 times thinner than standard graded buffers. The buffers are grown in one step, at a high growth rate and a low substrate temperature. Neither complicated thermal cycling nor ion implantation is required. Chemical–mechanical polishing is also unnecessary.

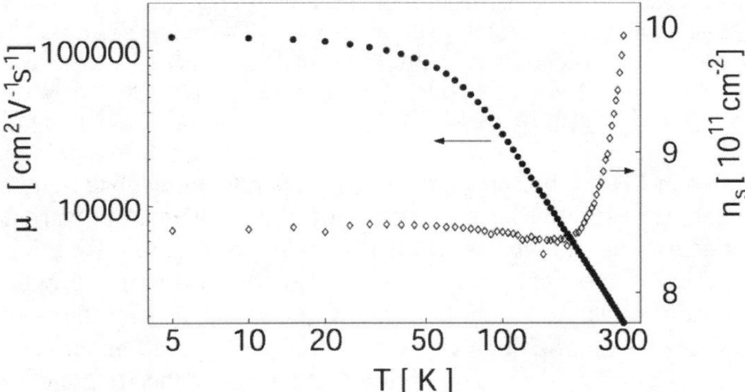

Figure 6. Hall mobility (filled symbols) and hole sheet density (open symbols) of a strained Ge p-MODQW structure on a relaxed linearly-graded SiGe buffer (sample 6745).

The combination of high growth rate and low substrate temperature allows a high degree of strain to accumulate in the $Si_{1-x}Ge_x$ layer during growth, and this may lead to a higher dislocation glide velocity which facilitates a high degree of relaxation. Molecular dynamics simulations suggest that the nature of the dislocation core is strain dependent [32], with more mobile cores forming under high strain.

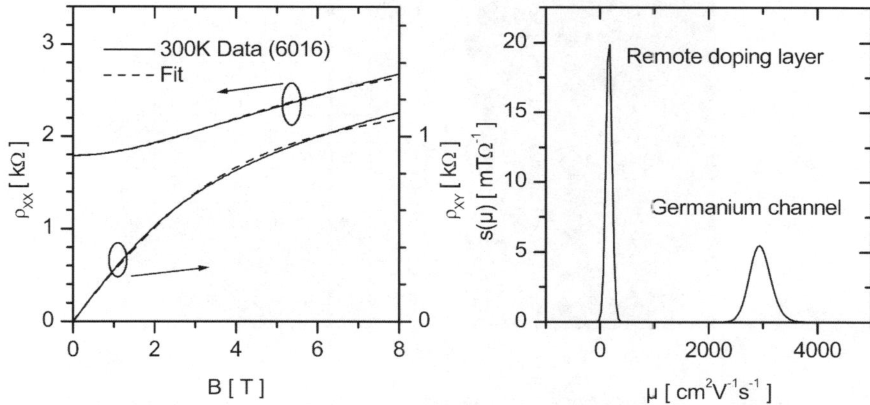

Figure 7. Left panel: Longitudinal and transverse magnetoresistance measured at 300K on a strained Ge p-MODQW structure, together with fit from a calculated mobility spectrum [27, 28] shown on the right. Right panel: Mobility spectrum obtained for sample 6016 at 300 K from the magnetoresistance shown on the left. Channel mobility is 2940 cm^{2}V^{-1}s^{-1} at a hole density of 5.7×10^{11} cm^{-2}.

Hole Mobility Enhancement

Traditionally, the speed of CMOS has been limited by the poor hole mobility of Si, relative to the electron mobility. Since Ge has one of the highest hole mobilities of any semiconductor, SiGe is an obvious choice for p-channel devices. A complete p-HMOSFET (heterostructure-MOSFET) layer structure with a Ge-rich channel has been grown by LEPECVD [33]. Transistors fabricated on this material demonstrate effective hole mobilities at room temperature which approach the Si electron mobility.

Alloy scattering means that the best hole mobilities should be seen in pure Ge channels [34, 35]. Modulation-doped p-channel quantum wells (p-MODQWs) grown by LEPECVD, with strained Ge channels on Si$_{0.3}$Ge$_{0.7}$ VSs, have a maximum low-temperature (2 K) mobility (Figure 6) of 120,000 cm^{2}V^{-1}s^{-1} at a sheet density of 8.5×10^{11} cm^{-2}. These structures were front-side doped (with the dopant above the channel). Optimization of the growth temperature and use of hydrogen means that the upper heterointerface was very smooth. The mobility is a factor of two greater than the best p-type Ge channel mobility result demonstrated on material grown by MBE [10]. There, the doping had to be introduced underneath the channel (probably leading to dopant segregation into the channel) since the upper interface was of lower quality due to strain-induced roughening. Our results also evince the quality of the VS, in terms of threading dislocation density and surface roughness.

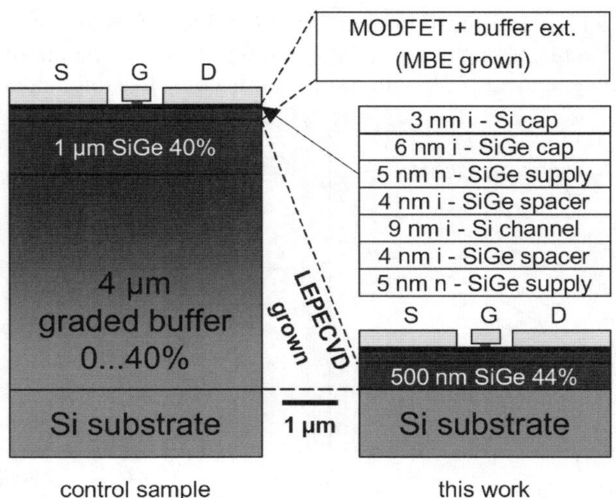

Figure 8. Schematic of the MODFET structures grown on a conventional thick graded VS (left) and on the novel thin VS (right), drawn to scale. The inset shows the stack of the active layers. A buffer extension of 150 nm was grown by MBE before the active layers.

Room temperature channel mobilities have been extracted from the magnetic field dependence of the conductivity and Hall coefficient (Figure 7). The right-hand panel of Figure 7 shows a mobility spectrum of a strained Ge p-MODQW structure on a relaxed linearly-graded SiGe buffer (sample 6016) at 300 K, found from the data in the left-hand panel. Two peaks are evident, corresponding to conduction in the strained Ge channel (2940 $cm^2V^{-1}s^{-1}$) and the boron-doped supply layer (180 $cm^2V^{-1}s^{-1}$). By integrating these peaks, the carrier sheet densities are found to be $5.7\times10^{11}cm^{-2}$ in the channel and 1.0×10^{13} cm^{-2} in the supply layer [27, 28]. Similar values for the channel mobility have been found by mobility spectrum analysis in strained Ge structures grown with other techniques [37, 38].

In addition, Si under tensile strain should feature enhanced hole mobility over unstrained Si due to a reduction of effective mass [34]. However, since tensile-strained Si on SiGe does not form a quantum well for holes then a surface channel must be induced using the field effect, as shown in the right-hand panel of Figure 2 [2].

Electron-channel Devices
Modulation-doped n-channel FETs (n-MODFETs) and n-MODQWs have been fabricated using a combined LEPECVD+MBE process: a buffer is grown by LEPECVD and then the electrically active structure is grown by solid-source MBE [39-41]. This combines the advantages of LEPECVD (high growth rates for high-quality VS production) with the advantages of MBE (good control of n-type doping and concentration profiles).

The active MODFET layer stack was grown by MBE at a constant temperature of 540°C and a rate of 0.25 nms^{-1} for SiGe or 0.15 nms^{-1} for Si. The double-sided modulation-doped structure is formed by a 9 nm strained Si channel, embedded in 4 nm doping setback SiGe layers and 5 nm Sb doped SiGe supply layers. Doping was achieved by secondary implantation (DSI) [42]. Device processing included isolation by dry mesa etching, deposition of SiO_2 as a field oxide, window opening by wet chemical etching, P implantation and thermal activation in the ohmic contact region, deposition and lift-off of the Ti/Pt/Au source and drain contacts and finally definition of the Pt/Au Schottky gate by means of electron-beam lithography. The smallest structures have a gate length l_G=70 nm and a source-drain distance d_{SD}=1 µm [30].

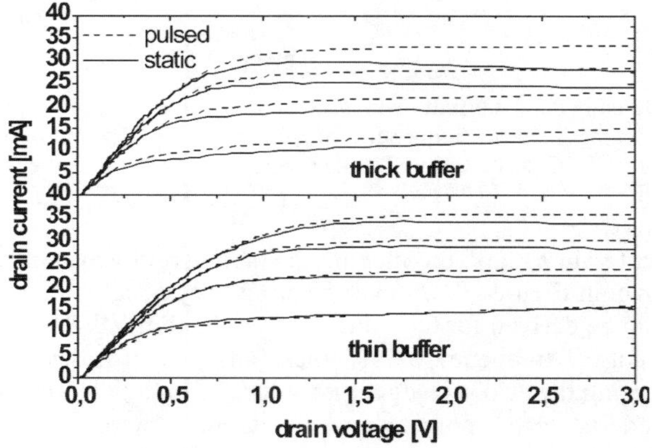

Figure 9. Comparison of the MODFET output characteristics on thin and thick relaxed LEPECVD-grown $Si_{0.6}Ge_{0.4}$ buffers (lower and upper curves, respectively) under static and pulsed conditions. The gate voltage V_G was varied between -0.6 and +0.6V in steps of 0.4V.

A comparison of thick and thin VS structures is shown in Figure 8, and device characteristics are shown in Figures 9 and 10. The lower and upper set of curves in Figure 9 show the I-V characteristics of 100 µm wide devices prepared on the thin and the thick LEPECVD-grown buffer, respectively. The measurements were performed both under static and pulsed conditions. The pulse width was 200 ns. Drain saturation currents I_{DSS} of more than 300 mA/mm and a maximum transconductance of g_m=230 mS/mm have been achieved in both cases. The divergence between the DC and pulsed currents is much larger for the thick buffer. At full enhancement (drain voltage V_D=3 V, gate voltage V_G=+0.6 V) a current difference of 17% can be derived from the uppermost curves, which has to be compared to only 6% in the case of the thin buffer. This indicates a significant reduction of self-heating from the tenfold decrease of buffer thickness. The difference in the self-heating effect by a factor of approximately three is in good agreement with the rough estimation that the thermal resistance of a device on a relaxed SiGe buffer is proportional to the square root of the buffer thickness [43]. Taking this into account a device temperature reduction of 70 K can be estimated by applying the thin VS.

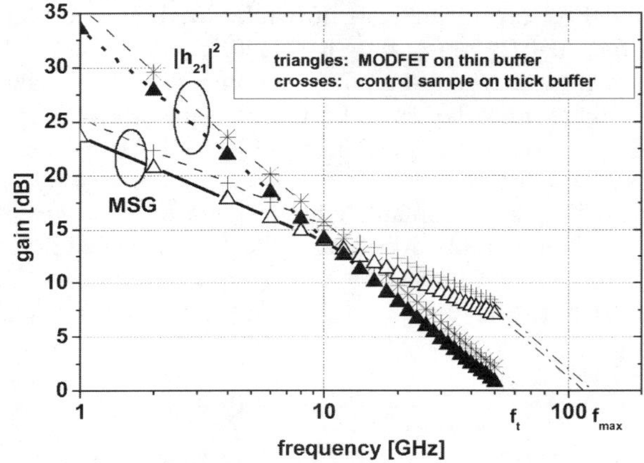

Figure 10. De-embedded current gain $|h_{21}|$ and maximum stable gain *MSG* spectra of 100 μm wide devices. Transit frequencies of f_T=55 GHz/63 GHz and maximum frequencies of oscillation $f_{max}(MSG)$=113 GHz/128 GHz can be derived for devices on thin and thick VS, respectively.

The current gain h_{21} and the maximum stable gain (MSG) of 100 μm wide devices are shown in Figure 10. By extrapolating the current gain, a transit frequency f_T=55 GHz can be derived for the transistors on the thin VS. The results are comparable to the cut-off frequency of the thick buffer control device having an f_T of 63 GHz. By using the acknowledged procedure of extrapolating MSG with a slope of 10dB/decade up to the point where the stability factor k reaches 1 and further extrapolation of the maximum available MAG with 20 dB/decade, maximum frequencies of oscillation $f_{max}(MAG)$=113 GHz and 128 GHz are achieved for the thin and thick buffer device. Assuming reflection-free input and output matching, cutoff frequencies of $f_{max}(U)$=138 GHz and 144 GHz can be derived from the 10 dB transits of Masons's gain U (for clarity not shown in Figure 10), respectively. High-frequency performance is not degraded significantly by a ten-fold reduction in buffer thickness; in fact, LEPECVD-grown thin VSs are competitive with thin VSs produced both low-temperature epitaxy and He implantation [13].

Conclusions

LEPECVD is a new process for epitaxy of SiGe at rates of several nanometres per second. Growth rates are independent of alloy composition and substrate temperature, giving the maximum freedom for optimization of growth parameters. Being a gas source process, it is suitable for industrial production of electronic device-grade material.

High quality n-MODFETs have been produced in mixed LEPECVD+MBE technology [41], and a novel thin VS design (which is made possible by the high growth rates available in LEPECVD even at low substrate temperature) allows access to the advantages of strained Si without the disadvantages of thick graded SiGe layers [30].

High effective mobilities have been demonstrated at room temperature in p-HMOSFETs [33], and exceptional mobilities have been demonstrated in p-MODQW structures at low temperature [36].

Acknowledgements

Financial support from GROWTH Program ECOPRO No. GRD2-2000-30064 and EC, frame of IST-1999-SIGMUND and German BMBF Project Ultra2 is gratefully acknowledged.

References

1. F. Schäffler. Semicond. Sci. Technol. 12, 1515 (1997)

2. D.J. Paul. Semicond. Sci. Technol. 19, R75 (2004)

3. Y.-W. Mo, D.E. Savage, B.S. Swartzentruber, M.G. Lagally. Phys. Rev. Lett. 65, 1020 (1990)

4. J.W. Matthews, A.E. Blakeslee. J. Cryst. Growth 32, 265 (1974)

5. M.M. Rieger, P. Vogl. Phys. Rev. B 48, 14276 (1993)

6. Y.H. Xie, G.H. Gilmer, C. Roland, P.J. Silverman, S.K. Buratto, J.Y. Cheng, *et al.* Phys. Rev. Lett. 73, 3006 (1994)

7. E.A. Fitzgerald, Y.-H. Xie, M.L. Green, D. Brasen, A.R. Kortan, J. Michel, *et al.* Appl. Phys. Lett. 59, 811 (1991)

8. K. Ismail, M. Arafa, K.L. Saenger, J.O. Chu, B.S. Meyerson. Appl. Phys. Lett. 66, 1077 (1995)

9. P. Weitz, R.J. Haug, K. von Klitzing, F. Schäffler. Surf. Sci. 361/362, 542 (1996)

10. Y.H. Xie, D. Monroe, E. A. Fitzgerald, P.J. Silverman, F.A. Thiel, G.P. Watson. Appl. Phys. Lett. 63, 2263 (1993)

11. R.M. Feenstra, M.A. Lutz, F. Stern, K. Ismail, P.M. Mooney, F.K. LeGoues, *et al.* J. Vac. Sci. Technol. B 13, 1608 (1995)

12. T. Hackbarth, H.-J. Herzog, M. Zeuner, G. Höck, E.A. Fitzgerald, M. Bulsara, *et al.* Thin Solid Films 369, 148 (2000)

13. M.E. Aguilar, M. Rodriguez, N. Zerounian, F. Aniel, T. Hackbarth, H.-J. Herzog, *et al.* Solid State Eletron. 48, 1443 (2004)

14. H. Chen, L.W. Guo, Q. Cui, Q. Hu, Q. Huang, J.M. Zhou. J. Appl. Phys. 79, 1167 (1996)

15. K.K. Linder, F.C. Zhang, J.-S. Rieh, P. Bhattacharya, D. Houghton. Appl. Phys. Lett. 70, 3224 (1997)

16. Y.H. Luo, J L. Liu, G. Jin, J. Wan, K. L. Wang, C. D. Moore, et al. Appl. Phys. Lett. 78, 1219 (2001)

17. Y. B. Bolkhovityanov, A. K. Gutakovskii, V. I. Mashanov, O. P. Pchelyakov, M. A. Revenko, L. V. Sokolov. Thin Solid Films 392, 98 (2001)

18. E. Kasper, K. Lyutovich, M. Bauer, M. Oehme. Thin Solid Films 336, 319 (1998)

19. S. Mantl, B. Holländer, R. Liedtke, S. Mesters, H.-J. Herzog, H. Kibbel, *et al.* Nucl. Instr. Meth. B 147, 29 (1999)

20. B. Holländer, S. Lenk, S. Mantl, H. Trinkaus, D. Kirch, M. Luysberg, *et al.* Nucl. Instr. Meth. B 175-177, 357 (2001)

21. J. Cai, P.M. Mooney, S.H. Christiansen, H. Chen, J.O. Chu, J.A. Ott. J. Appl. Phys. 95, 5347 (2004)

22. D. Buca, B. Holländer, H. Trinkaus, S. Mantl, R. Carius, R. Loo, *et al.* Appl. Phys. Lett. 85, 2499 (2004)

23. K. Sawano, S. Koh, Y. Shiraki, Y. Ozawa, T. Hattori, J. Yamanaka, *et al.* Appl. Phys. Lett. 85, 2514 (2004)

24. C. Rosenblad, H. von Känel, M. Kummer, A. Dommann, E. Müller. Appl. Phys. Lett. 76, 427 (2000)

25. J.M. Hartmann, V. Loup, G. Rolland, P. Holliger, F. Laugier, C. Vannuffel, *et al.* J. Cryst. Growth 236, 10 (2002)

26. R.B. Bergmann, L. Oberbeck, T.A. Wagner. J. Cryst. Growth 225, 335 (2001)

27. D. Chrastina, J. P. Hague, D.R. Leadley. J. Appl. Phys. 94, 6583 (2003)

28. H. von Känel, D. Chrastina, B. Rössner, G. Isella, J.P. Hague, M. Bollani. Microelectron. Eng. 76, 278 (2004)

29. M.T. Currie, S. B. Samavedam, T.A. Langdo, C.W. Leitz, E.A. Fitzgerald. Appl. Phys. Lett. 72, 1718 (1998)

30. T. Hackbarth, H.-J. Herzog, K.-H. Hieber, U. König, M. Bollani, D. Chrastina, *et al.* Appl. Phys. Lett. 83, 5464 (2003)

31. G. Isella, D. Chrastina, B. Rössner, T. Hackbarth, H.-J. Herzog, U. König, *et al.* Solid State Electron. 48, 1317 (2004)

32. A. Marzegalli, F. Montalenti, M. Bollani, L. Miglio, G. Isella, D. Chrastina, *et al.* Microelectron. Eng. 76, 289 (2004)

33. G. Höck, E. Kohn, C. Rosenblad, H. von Känel, H.-J. Herzog, U. König. Appl. Phys. Lett. 76, 3920 (2000)

34. M.V. Fischetti, S.E. Laux. J. Appl. Phys. 80, 2234 (1996)

35. C. W. Leitz, M. T. Currie, M. L. Lee, Z.-Y. Cheng, D. A. Antoniadis, E. A. Fitzgerald. J. Appl. Phys. 92, 3745 (2002)

36. B. Rössner, D. Chrastina, G. Isella, H. von Känel. Appl. Phys. Lett. 84, 3058 (2004)

37. M. Myronov, T. Irisawa, O. A. Mironov, S. Koh, Y. Shiraki, T. E. Whall, *et al.* Appl. Phys. Lett. 80, 3117 (2002)

38. R. J. H. Morris, T. J. Grasby, R. Hammond, M. Myronov, O. A. Mironov, D. R. Leadley, *et al.* Semicond. Sci. Technol. 19, L106 (2004)

39. M. Kummer, C. Rosenblad, A. Dommann, T. Hackbarth, G. Höck, M. Zeuner, *et al.* Mat. Sci. Eng. B 89, 288 (2002)

40. T. Mack, T. Hackbarth, U. Seiler, H.-J. Herzog, H. von Känel, M. Kummer, *et al.* Mat. Sci. Eng. B 89, 368 (2002)

41. M. Enciso-Aguilar, F. Aniel, P. Crozat, R. Adde, H.-J. Herzog, T. Hackbarth, *et al.* Electron. Lett. 39, 149 (2003)

42. H. Jorke, H.-J. Herzog, H. Kibbel. Appl. Phys. Lett. 47, 511 (1985)

43. K.A. Jenkins, K. Rim. IEEE Electron Device Lett. 23, 360 (2002)

Thin-film Engineering by Atomic-layer Deposition for Ultra-scaled and Novel Devices

G. Scarel, M. Fanciulli

MDM-INFM National Laboratory, Agrate Brianza / Italy

Introduction

High dielectric constant (κ) oxide thin films (*e.g.* Al_2O_3, HfO_2, ZrO_2, TiO_2, rare earth oxides) as well as ultra-thin SiO_2 or GeO_2, and low-κ dielectrics of very high quality are of paramount importance for the development of ultra-scaled microelectronic devices based on complementary metal-oxide-semiconductor structures (CMOS) [1] and for spintronics [2], molecular electronics [3], and high capacitance neuroelectronic [4] devices with thin but optimal interfacial layers in contact with the substrate [5]. A successful film deposition method for such applications is required to produce thin, smooth, conformal, and pure films with good stoichiometry and well-controllable thickness at low growth temperatures (100–400°C). Atomic layer deposition (ALD) is acknowledged to have the capability of satisfying these requirements [6] and good quality films of a wide range of oxides have been produced using this technique [1, 6, 7]. Here we illustrate how ALD allows further engineering of film properties (roughness, crystallinity, density, impurity content, interfacial layer, electrical properties) and structures (films on different substrates, multilayers, and nano-scale defined structures).

Material Properties

Material selection is needed before ALD growth in order to better engineer film properties. Among high-κ oxides there is great competition between suitable transition metal oxides (mainly HfO_2 and ZrO_2) and the whole series of rare earth oxides (from La to Lu). The latter exhibit the potential of fine tuning some structural, thermal, mechanical, vibrational, electric, and electronic properties due to the progressive filling of the f shell [8], and the variation of ionic radius and electronegativity [9] as the atomic number of the rare earth element increases (see Table 1). Among the research groups involved in the growth by ALD of oxides, many are investigating high-κ transition metal oxide films, but only a few are exploring the potentials of rare earth oxide films [7, 10]. The main difficulty in the ALD of rare earth oxides is the identification of suitable precursors.

Atomic Layer Deposition

ALD is viewed as a deposition method in which successive "showers" [13] of separated random fluxes of the anion and the cation precursors react with the surface. Usually a purging N_2 flux follows each "shower" to effectively remove the reaction by-products [14]. Many variables and both chemical and physical parameters play a significant role in the ALD process, therefore a simple and exhaustive model is hard to find. Among the proposed ones [15, 16], only few turn out to correctly describe some of the macroscopic parameters (*e.g.* growth per cycle versus growth temperature) for certain precursor combinations [16]. Selected ALD parameters and their exploitation to engineer film properties are briefly discussed below.

Table 1. Electronic properties of selected rare earth oxides. The band-gap (E_g) values are taken from Ref. [11]. The κ values are measured for ALD films [7]. The conduction band offset (CBO) values on silicon are from Ref. [12]. All the data for Lu_2O_3 were measured on ALD grown films [10]

oxide	La_2O_3	Ce_2O_3	Pr_2O_3	Nd_2O_3	Sm_2O_3	Eu_2O_3	Gd_2O_3	Tb_2O_3	Dy_2O_3	Er_2O_3	Yb_2O_3	Lu_2O_3
E_g (eV)	5.7	2.4	3.8	4.6	4.9	4.3	5.5	3.7	4.8	5.3	4.8	5.8
κ	-	-	-	10.4	10.0	11.1	8.9	-	8.4	10.0	-	12
CBO (eV)	2.3	-	-	-	-	-	3.1	-	-	-	-	2.1

Surface Functional Groups and Growth Temperature

In ALD the functional groups on the surface play a major role in determining the growth mechanisms and the film properties. Hydroxyl groups on silicon [17] and silica [18] are a good and well-known example. Their number evolves with temperature (T): single, geminal or H-bonded silanol (Si–OH) groups transform into siloxane (Si–O–Si) bridges with increasing T [17, 18], especially at and above 300° C. For the metal-chloride or metal-alkyl and water precursor combinations, silanol groups at T < 300° C are acknowledged to promote ligand exchange reactions [14, 16, 19, 20, 21], in which each silanol group is exchanged with the ligand of the incoming precursor. A large initial number of silanol groups favours a linear growth behaviour of thickness versus number of cycles and a Frank-van der Merwe layer-by-layer growth mechanism, whereas a small initial number of silanol groups favours a parabolic growth behaviour, correlated with a Volmer–Weber islands growth mechanism [22]. Siloxane bridges on the other hand, promote agglomeration reactions at T > 300° C [20], in which the oxide formation occurs through direct chlorination of the surface functional groups without need of the oxygen precursor. For the $HfCl_4$ or $ZrCl_4$ and H_2O precursor combinations, the as grown HfO_2 and ZrO_2 films have a larger crystallized fraction, are denser, rougher (see Figure 1), purer, and with better electrical properties [23, 24, 25] when the agglomeration reaction rather than the ligand exchange reaction occurs.

The density functional theory (DFT) investigation of the elementary reactions in ALD explains the improved properties in films deposited at high T with the shift of the overall equilibrium from the stable absorbed precursor state to the final state [19]. More in general, ALD is powerful because it proceeds on a large variety of surface functional groups, and not just on silanol groups or siloxane bridges. Indeed, successful studies were accomplished on Al_2O_3 films deposited by ALD in a temperature range between 80–180° C on polyethylene particles, and on various polymer films [26]. Successful growth of ZrO_2 at 300° C on carbon nanotubes was also reported [27]. These results are relevant for the fabrication of novel nano-scale defined structures, and support the concept of "area-selective" growth of high-κ dielectrics using organic monolayers that can either block or promote precursor reactions [28, 29].

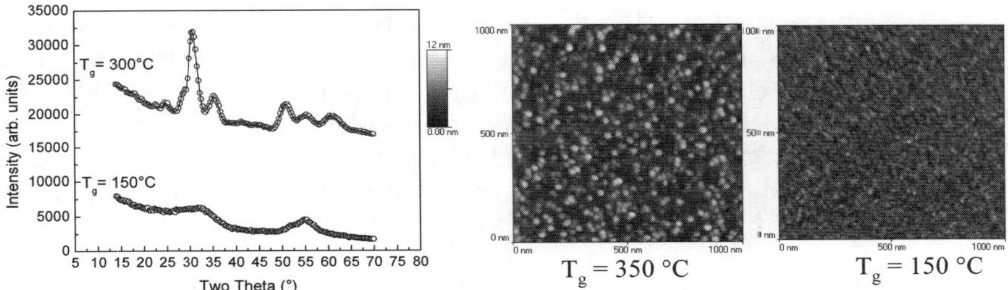

Figure 1. AFM images and corresponding XRD spectra (open circles) and fittings (continuous lines) of HfO_2 films, about 10 nm thick, deposited from $HfCl_4$ and H_2O at 150° C and at 350° C [24]

Precursor Combinations

The precursor combination in ALD significantly affect the film properties. Two examples are illustrated to support this statement:

1. The understanding of the surface morphology of films grown by ALD is still not totally clear. According to the random deposition (RD) model [13] the growth per cycle (GPC) — related to precursor ligand packing and coverage, and to available and consumed reactive sites — seems to have a significant influence. In the RD model, constant and high GPC values usually correlate with smooth films [13], whereas non-constant GPC values correlate with smooth films only when the GPC on the substrate is high. To validate this model, we examined HfO_2 films deposited at 375° C from (a) $HfCl_4$ and $Hf(O^tBu)_2(mmp)_2$, (b) $HfCl_4$ and O_3, (c) $HfCl_4$ and H_2O, and (d) $Hf(mmp)_4$ and $Hf(O^tBu)_2(mmp)_2$. For cases (a), (b), and (c), where the GPC is constant and measured to be respectively 0.17 nm/cycle, 0.10 nm/cycle, and 0.07 nm/cycle, the surface roughness, at a given thickness, is indeed lower for the film with higher GPC. For case (d), the GPC is not constant as a function of number of cycles. In this case, the GPC is rather low at the

beginning of the growth (0.07 nm/cycle was measured for a 3-nm thin film) and the corresponding films are rougher than those with the same thickness generated for cases (a) and (b). For a thick film (~ 10 nm), the large GPC (0.14 nm/cycle), correlates with a high roughness (rms = 1.4 nm), as expected from the RD model. In case (a), the GPC is high because both precursors supply Hf atoms. In case (b), O_3 molecules dissociate very efficiently and produce effectively many suitable reactive sites leading also to a high GPC [30, 31]. In case (c), the GPC is low probably because of increased precursor desorption and sub-monolayer growth occurring at high temperature [19]. Finally, for case (d), the mechanisms leading to the film formation and to the described GPC and roughness behaviour, are not yet understood.

2. The interfacial layer (IL) thicknesses of HfO_2 films deposited using the combinations (a) through (d) mentioned above were measured using transmission electron microscopy (TEM) and X-ray reflectivity (XRR). The results show that each precursor combination modifies differently the chemical oxide prepared on the Si(100) substrate prior to deposition leading, for combinations (c) and (d), to thinner interfacial layers than for combination (b) (Figure 2) [30].

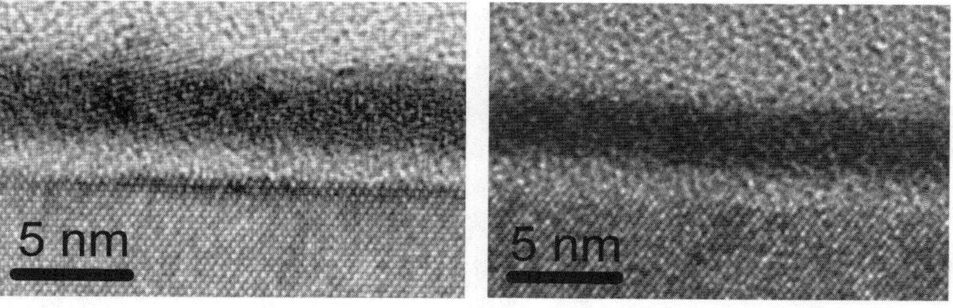

Figure 2. Comparison of TEM micrographs of thin HfO_2 films deposited from $Hf(mmp)_4$ and $Hf(O^tBu)_2(mmp)_2$ (left) and from $HfCl_4$ and O_3 (right)

Microelectronics

ALD is considered a mature technique for microelectronic device manufacture. Although high-κ dielectrics can find different applications in microelectronics, most of the effort is today concentrated in lowering the equivalent oxide thickness (EOT) below 1 nm without degrading the channel mobility as required for ultra-scaled CMOS devices [1]. Substrate preparation and ALD precursor combinations must be properly engineered to achieve these objectives. The growth temperature, for instance, might influence the κ value and thus the EOT (Figure 3). Despite the intrinsic problem related to the small band-gap, Ge has been also proposed instead of Si due its higher mobility. Moreover, the IL between the ALD deposited oxides and Ge is often thinner than on Si [32].

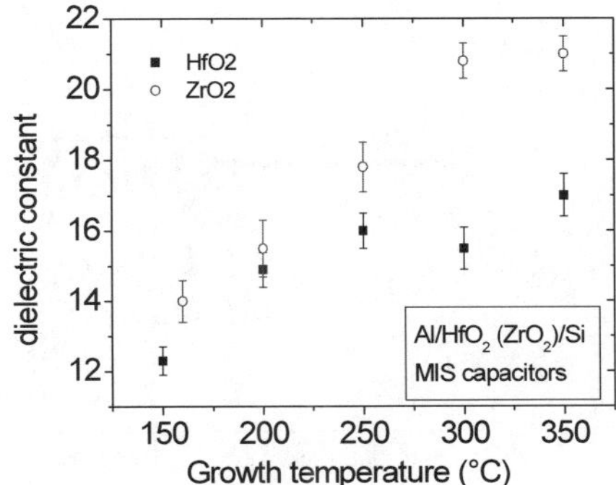

Figure 3. Dielectric constant as function of the growth temperature for HfO$_2$ and ZrO$_2$ films deposited from HfCl$_4$ and ZrCl$_4$, respectively, and H$_2$O [33]

Spintronics and Neuroelectronics

The promising properties of ALD grown layers suggest that this technique might be employed to fabricate devices for spintronics, such as ferromagnetic tunnel junctions (FTJ). A FTJ consists of two ferromagnetic contacts with the exchange coupling strength tuned by an insulating layer embedded between them. ALD offers the possibility to grow not only an efficient insulating tunnel oxide, but also the full FTJ structure. In this case, the ferromagnetic metal [34] or ferromagnetic metal-oxide, and the insulating oxide [35] layers could be deposited in sequence without breaking the vacuum, thus reducing or eliminating any detrimental interfacial layer (Figure 4) [36]. Very high-κ oxides deposited by ALD are promising also for neuroelectronic interfacing devices [4], in which silicon-based structures are used for the capacitive stimulation of neurons. TiO$_2$ layers, deposited by ALD, exhibit capacitances much larger than SiO$_2$ and low leakage currents. These properties must be optimized in electrolyte-insulator-semiconductor (EIS) devices. Since an interface layer cannot be avoided, at least a controlled one (Al$_2$O$_3$ deposited *in situ* or Si$_3$N$_4$ deposited *ex situ*) is required to maximize the overall capacitance of the structure [37].

Conclusion

ALD is shown to be a suitable technique for fabricating most advanced devices in microelectronics, spintronics, molecular electronics, and neuroelectronics. Growth temperature, surface preparation and functionalization, and precursor combinations, when properly selected, allow achieving the desired film properties. Improvements in ALD film quality require a deeper knowledge of the growth mechanisms involved, while substrates, precursors, and deposition parameters change.

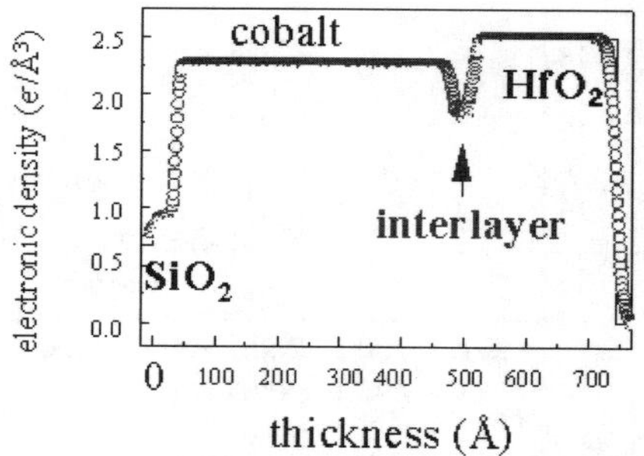

Figure 4. Electronic density profile versus thickness of a HfO$_2$/Co/SiO$_2$ stack. The dip between the Co and the HfO$_2$ layers corresponds to a Co oxide layer. This layer degrades the device performances and might be significantly reduced upon sequential deposition of both oxide and ferromagnetic layers [36]

Acknowledgements

We thank C. Wiemer, G. Tallarida, S. Spiga, S. Ferrari, E. Bonera, G. Seguini, and S. Baldovino (MDM-INFM National Laboratory) for their work on film characterization, and P. Fromherz and F. Wallrapp (Max-Planck Institut für Biochemie) for fruitful discussions on neuroelectronic interface devices. INFM, MAE, and MIUR are acknowledged for financial support.

References

1. G. D. Wilk, R. M. Wallace, and J. M. Anthony, J. Appl. Phys. 89, 5243 (2001)

2. I. Žutić, J. Fabian, and S. Das Sarma, Rev. Mod. Phys. 76, 323 (2004)

3. A. W. Gosh, P. S. Damle, S. Datta, and A. Nitzan, MRS Bull. 29, 391 (2004)

4. P. Fromherz, in: Nanoelectronics and Information Technology (Ed.: R. Waser), Wiley-VCH, Berlin 2003, p. 781

5. G. Lucovsky, Y. Wu, H. Niimi, V. Misra and J. C. Phillips, Appl. Phys. Lett. 74, 2005 (1999). A good and thin interface layer limits both the decrease of the gate oxide stack κ, and of the device mobility due to remote phonon and Coulomb scattering

6. M. Ritala and M. Leskelä, in: Handbook of Thin Film Materials (Ed.: H. S. Nalwa) vol. 1, Academic, San Diego CA 2002, p. 103

7. L. Niinistö, J. Päiväsaari, J. Niinistö, M. Putkonen and M. Nieminen, Phys. Stat. Sol. (a) 201, 1433 (2004), and J. J. Päiväsaari, M. Putkonen, and L. Niinistö, Thin Solid Films (in press)

8.   G. V. Samsonov and I. Ya. Gil'man, Poroshkovaya Metallurgiya 13, 73 (1974)

9.   S. Jeon and H. Hwang, J. Appl. Phys. 93, 6393 (2003)

10. G. Scarel, E. Bonera, C. Wiemer, G. Tallarida, S. Spiga, M. Fanciulli *et al.*, Appl. Phys. Lett. 85, 630 (2004), and G. Seguini, E. Bonera, S. Spiga, G. Scarel, and M. Fanciulli, Appl. Phys. Lett. (in press)

11. A. V. Prokofiev, A. I. Shelykh, and B. T. Melekh, J. All. Comp. 242, 41 (1996)

12. T. Hattori, T. Yoshida, T. Shiraishi, K. Takahashi, H. Nohira, S. Joumori *et al.*, Microelectron. Eng. 72, 283 (2004)

13. R. L. Puurunen, Chem. Vap. Dep. 10, 159 (2004)

14. L. Jeloaica, A. Estève, M. Djafari Rouhani, and D. Estève, Appl. Phys. Lett. 83, 542 (2003)

15. M. K. Gobbert, V. Prasad, and T. S. Cale, J. Vac. Sci. Technol. B 20, 1031 (2002), and M. Ylilammi, Thin Solid Films 279, 124 (1996)

16. R. L. Puurunen, Chem. Vap. Dep. 9, 249 (2003), and R. L. Puurunen, Chem. Vap. Dep. 9, 327 (2003)

17. Y. B. Kim, M. Tuominen, I. Raaijmakers, R. de Blank, R. Wilhelm, and S. Haukka, Electrochem. Solid-State Lett. 3, 346 (2000)

18. L. T. Zhuravlev, Colloids Surf. A Physicochem. Eng. Aspects 173, 1 (2000)

19. J. H. Han, G. Gao, Y. Widjaja, E. Garfunkel, and C. B. Musgrave, Surf. Sci. 550, 199 (2004)

20. A. Kytökivi and S. Haukka, J. Phys. Chem. B 101, 10365 (1997)

21. M. A. Alam and M. L. Green, J. Appl. Phys. 94, 3403 (2003), and M. L. Green, M.-Y. Ho, B. Busch, G. D. Wilk, T. Sorsch *et al.*, J. Appl. Phys. 92, 7168 (2002)

22. A. Satta, A. Vantomme, J. Schuhmacher, C. M. Whelan, V. Sutcliffe, and K. Maex, Appl. Phys. Lett. 84, 4571 (2004)

23. G. Scarel, C. Wiemer, S. Ferrari, G. Tallarida, and M. Fanciulli, Proc. Estonian Acad. Sci. Phys. Math. 52, 308 (2003), and G. Scarel, S. Ferrari, S. Spiga, G. Tallarida, C. Wiemer, and M. Fanciulli, J. Vac. Sci. Technol. A 21, 1359 (2003)

24. G. Scarel, S. Spiga, C. Wiemer, G. Tallarida, S. Ferrari, and M. Fanciulli, Mat. Sci. Eng. B 109, 11 (2004)

25. K. Kukli, M. Ritala, J. Aarik, T. Uustare, and M. Leskelä, J. Appl. Phys. 92, 1833 (2002)

26. S. M. George, ALD2004 Congress, Helsinki (Finland) August 16–18, 2004

27. A. Javey, H. Kim, M. Brink, Q. Wang, A. Ural, J. Guo *et al.*, Nat. Mater. 1, 241 (2002)

28. S. F. Bent, R. Chen, J. Hong, D. Porter, P. McIntyre, H. Kim *et al.*, ALD2004 Congress, Helsinki (Finland) August 16–18, 2004

29. Y. Xu and C. B. Musgrave, Chem. Mater. 16, 646 (2004)

30. C, Wiemer, C. Marchiori, G. Scarel, S. Spiga, S. Baldovino, S. Ferrari *et al.*, WoDim 2004, Kinsale (Ireland) June 28–30, 2004

31. G. Scarel, S. Spiga, C. Wiemer, G. Tallarida, S. Baldovino, E. Bonera *et al.*, E-MRS 2004 Spring Meeting, Strasbourg (France) May 24–28, 2004

32. M. Fanciulli, S. Spiga, G. Scarel, G. Tallarida, C. Wiemer, G. Seguini, Mat. Res. Soc. 786, 341 (2004), and E. P. Gusev, H. Shang, M. Copel, M. Gribelyuk, C. D'Emic, P. Kozlowski *et al.*, Appl. Phys. Lett. 85, 2334 (2004)

33. G. Scarel, S. Ferrari, S. Spiga, G. Tallarida, C. Wiemer, E. Bonera *et al.*, IV Silicon Workshop 2003, Genova (Italy) February 12–14 2003

34. B. S. Lim, A. Rahtu, and R. G. Gordon, Nat. Mater. 2, 749 (2003)

35. O. Nilsen, H. Fjellvåg, and A. Kjekshus, ALD2004 Congress, Helsinki (Finland) August 16–18, 2004

36. G. Scarel, C. Wiemer, G. Tallarida, and M. Fanciulli, MDM-INFM Biennial Report 2002–2003

37. F. Wallrapp and P. Fromherz, DPG-Spring Meeting, Regensburg (Germany) March 8– 12, 2004

Atomic-layer Deposited Barrier and Seed Layers for Interconnects

J. Schuhmacher[a], A. Martin[a,b], A. Satta[a], K. Maex[a,b]

[a]IMEC, Leuven / Belgium

[b]ESAT Department K.U. Leuven / Belgium

Introduction

Barrier and Seed Layers in Interconnects

In current semiconductor technologies, interconnects are fabricated according to a damascene metallization process flow. Intra metal dielectric (IMD) material is deposited. Photolithography and dry-etch processing of IMD material provides substrate structures with contact holes (to the underlying metal level) and trenches, to be filled with copper. In between the IMD and copper, a barrier layer is required to prevent copper from penetrating into the IMD in which it is highly mobile. A copper seed layer is deposited on top of the barrier layer to facilitate the electrochemical deposition (ECD) of copper. The barrier/seed layer also guarantees adhesion of the copper to the IMD. Barrier layers are less conductive than copper and should be thin and conformal to ensure a large cross section of the copper wire and a low via resistance at the via bottom. The typical situation is illustrated in Figure 1 by a transmission electron microscope image.

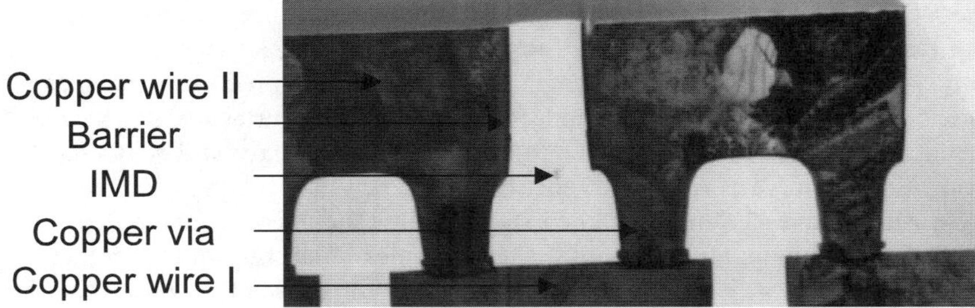

Figure 1. Cross section of a metallized damascene structure with metal levels I and II with a conformal atomic layer deposited barrier separating the copper wires and vias from the inter metal dielectric.

Physical vapour deposition (PVD) is currently used to deposit the barrier and seed layers. As it is an inherently directional deposition process, PVD layers are expected to be incompatible with very narrow features. For future technology nodes, scaling to smaller dimensions with atomic layer deposited barriers in the range of 5 nm and below are foreseen [1].

Characteristics of Atomic Layer Deposition (ALD)

The main advantage of ALD is its perfect step coverage, which is inherent to the technique and achieved by controlling a chemical reaction of the starting materials (precursors) at the substrate surface. An almost perfect step coverage of ALD TiN in a damascene trench is shown in Figure 2.

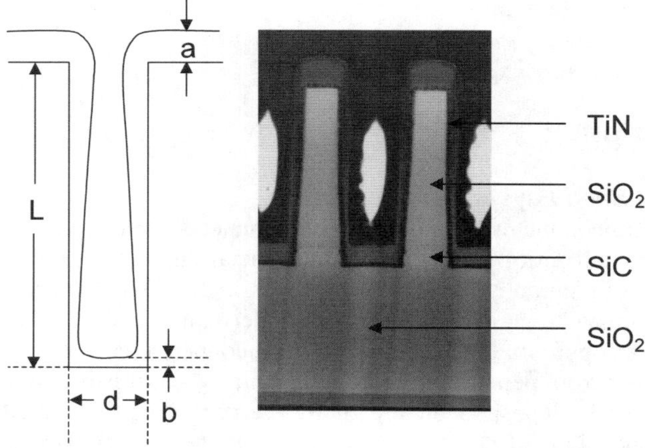

Figure 2. Left: definition of step coverage as the ratio of (a) maximum to (b) minimum thickness in a trench or hole feature with aspect ratio L/d; right: ALD TiN film with a step coverage close to 100% in damascene-type trenches with an aspect ratio of 2.5.

Atomic layer deposition is a surface-controlled technique to deposit material by a repetitive sequence of irreversible, self-terminating surface reactions. Each reaction results in a new surface condition for the following surface reaction. Practically, ALD is realized from the gas phase by reacting the precursors (such as metal compound- and non-metal compound vapours) with the surface one at a time. The chemical vapours are introduced into the reactor alternately and separated by an evacuation or purge period. The repetitive sequence of reactions is called an ALD cycle. During the ALD process, the reaction of the metal compound precursor with the surface results in an immobile surface complex consisting of the metal and the ligands left intact which serve as reactive sites for the subsequent source vapour.

ALD material is grown onto a substrate surface. During a transient growth period, this starting surface is transformed into a continually renewed, but constant ALD material surface. Frequently, the substrate surface is less reactive than the ALD film surface and the growth per cycle is reduced. In this case, the ALD material forms "islands" initiated at reactive groups or defects on the starting surface. If the entire substrate surface is non-reactive in the ALD environment, no deposition will occur. A heterogeneous substrate with reactive and non-reactive areas can be utilized to achieve area-selective depositions.

In this paper, ALD copper diffusion barrier layers will be addressed with respect to requirements set by the barrier application, the ALD processing and the substrate sensitivity of the technique. Specific illustrative examples are given with TiN- and WNxCy ALD.

Recent developments concerning ALD metal glue or seed layers are also briefly reported.

Requirements for ALD Copper Diffusion Barriers

The application of the ALD technique for interconnects requires a deposition process temperature of 400° C or below, sufficient adhesion of the barrier layer to the dielectric and copper, low solubility in copper and good electromigration resistance.

When depositing ALD barrier films for interconnects, the ALD process interacts with the surfaces of IMD, copper, and those of etch stop or hard mask layers. The ideal ALD process for a given damascene-type copper interconnect would have zero growth on copper, thus minimizing via resistance, but would grow without inhibition on the IMD-, etch stop- and hard mask-surfaces to allow for a homogeneous, continuous, thin film on the via and trench sidewall.

In practice, however, ALD selectivity towards the various substrates is far from ideal. The condition of the initial substrate surface controls the evolution of film growth for a given ALD processing condition. That is, the interrelation of substrate surface condition and ALD process chemistry determines the transient growth period. From the point of view of damascene metallization process flows, the IMD/ALD barrier choice has to be considered as a whole because the barrier property of a thin film of a given ALD process will depend on the surface condition of the dielectric concerned. Therefore, any surface modification introduced to the IMD during etching and resist removal can be predicted to have a significant impact on a subsequent ALD process.

To guarantee a low dielectric permittivity, IMD materials have, in general, a porous structure with pore sizes larger than ALD precursors. To avoid penetration of precursors inside the IMD, sealing of the IMD surface before ALD is a prerequisite. For many IMD materials, standard etching and resist removal processes modify the IMD sidewall and result in a sealed, ALD compatible surface.

ALD Processing of Copper Diffusion Barriers

ALD of "metal and nitride thin films" has been reviewed recently [2]. ALD barriers are classified according to barrier material and processing technique (Table 1). Among these materials, the focus resides on binary or ternary transition metal nitrides. From a processing perspective, thermal or plasma-enhanced (PE) processing are distinguished. Thermal processing allows for a simple reactor design and only simple reaction chemistry affects the substrate, while plasma-enhanced processing allows for lower growth temperatures as compared to thermal processes. The plasmas also have a densifying and purifying effect.

The materials considered for barriers are mainly nitrides of titanium, tantalum and tungsten. For binary systems processing is generally less complex than for the ternary systems. On the other hand, the composition in ternary systems may be correlated with the ratio of precursor exposures due to competitive surface reactions. This gives ternary ALD systems more processing flexibility. The binary compounds tend to form thin films with columnar grain structures. Since grain boundaries play an important role as diffusion pathways for copper, the more amorphous, ternary films are expected to behave as good diffusion barriers [2].

Table 1. Transition metal nitride ALD materials after [2]; PEALD: plasma-enhanced ALD

Technique	Binary system	Ternary system
PEALD	TiN, TaN, W_2N	$TiSi_xN_y$, $TaSi_xN_y$,
Thermal ALD	TiN, TaN, Ta_3N_5, W_2N, WN	$TiSi_xN_y$, $TiAl_xN_y$, N_xC_y,

Transition metals and transition metal nitrides are prepared from an inorganic or organic metal source and a further reactant.

Among the inorganic precursors for ALD processes, metal halides are popular since they are readily available, reactive and stable against self-decomposition. A drawback of metal halide precursors is their potential to corrode the substrate. An example of a transition metal nitride process is ALD of TiN prepared by thermal processing from titanium(IV) chloride ($TiCl_4$) and ammonia (NH_3). This process, carried out between 350° C and 400° C, gives a low resistivity film of about 200 $\mu\Omega\cdot$cm, but it is not useful for copper interconnects due to intolerable pitting of the copper substrates [3]. The pitting is driven by the formation of volatile copper(I) chloride. Depositions on exposed copper in damascene structures led to redeposition of CuCl onto the reactor walls and inside the deposited TiN film [4].

Corrosion is not expected for metal organic precursors. However, precursors of this type tend to decompose at comparatively low temperature leading to non-saturative reactions. For example, ALD TiN prepared from tetrakis(dimethylamido)titanium [$Ti(NMe_2)_4$] and NH_3 at 152° C showed non-ideal behaviour. For the $(Ti(N(CH_3)_2)_4$ surface reaction, a continuously increasing growth-per-cycle with longer exposures was observed [5]. Tantalum nitride can be prepared from pentakis(dimethylamido)tantalum and NH_3 in a thermal process [6]. For both half-reactions the growth-per-cycle saturates at $\approx$0.8 Å and from 200–300° C the growth-per-cycle and uniformity is only weakly temperature dependent. However, this material shows a high resistivity due to the low efficiency of ammonia to reduce Ta(V) to Ta(III).

The ligand exchanging transaminations yield highly resistive material. The formal surface reaction sequence by ligand exchange suggests Ta_3N_5 corresponding to:

(i) $x \parallel$-NH $+$ $Ta(N(CH_3)_2)_5$ $\rightarrow$ $(\parallel$-N$)_x Ta(N(CH_3)_2)_{5-x}$ $+$ x $HN(CH_3)_2$

(ii) $(\parallel$-N$)_x Ta(N(CH_3)_2)_{5-x}$ $+$ $(5-x)$ NH_3 $\rightarrow$ $(\parallel$-N$)_x Ta(NH_2)_{5-x}$ $+$ $(5-x)$ $HN(CH_3)_2$

with $x = 1,2,$ or 3; and $\parallel$- indicating a bond to the surface.

As described for the examples above, NH_3 is commonly used to deposit transition metal nitrides. In the $TiCl_4/NH_3$ TiN process, it functions both as reducing agent and as nitrogen source. In a traveling wave-type Pulsar 2000 reactor, the ALD-typical feature of an almost perfect linear increase in layer thickness with progressing ALD cycle number was observed (Figure 3a). For a given flow of 1200 sccm, the NH_3 reaction requires comparatively long pulse times between 1.5 and 3 s to saturate the surface (Figure 3b). Such long times limit the throughput of the process. Therefore, replacing NH_3 with more reactive reducing agents has been considered.

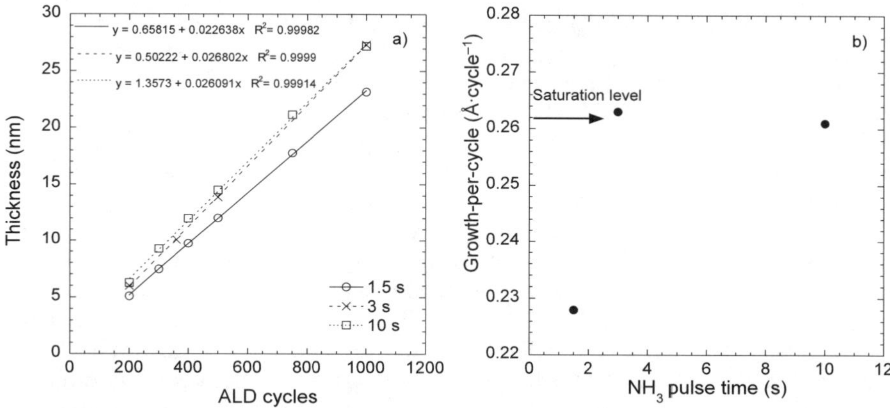

Figure 3. ALD TiN growth from NH_3 and $TiCl_4$ for different NH_3 pulse times with a flow of 1200 sccm NH_3.

ALD of TiN has been done with 1,1-dimethylhyrazine (DMHy) [7], as well as allylamines and *tert*-butylamine [8]. The use of DMHy or allylamine at 400° C was reported to give improved films over NH_3. However, at lower temperatures significant carbon and hydrogen impurities were detected. With *tert*-butylamine, low carbon content and low resistivity has been achieved, but required the addition of NH_3 for sufficient growth-per-cycle at low temperatures.

The ternary material $TiAl_xN_y$ was processed from $TiCl_4$, trimethylaluminum and NH_3. The growth-per-cycle was strongly dependent on temperature and reaction sequence [9]. A decreased resistivity and low chlorine content as compared to the binary $TiCl_4/NH_3$ system has been found. Another example for ternary barrier films is $TiSi_xN_y$, which consists of titanium silicon nitride material. When silane (SiH_4) and NH_3 were dosed together the growth-per-cycle and composition was strongly dependent on the SiH_4/NH_3 ratio. The growth-per-cycle decreased with increasing silicon content. However, if deposited in a sequence of surface reactions with first $Ti(NMe_2)_4$, second SiH_4 and third NH_3, a more stable composition of approximately $Ti_{0.32}Si_{0.18}N_{0.5}$ with a growth-per-cycle of 2.2 Å was obtained.

Compared to thermal processing techniques, the activation energy of the chemisorption, ligand exchange, and redox reactions are more readily overcome in plasma-enhanced ALD (PEALD) in which the non-metal precursor is activated by a plasma source. TaN with a resistivity of 400 $\mu\Omega\cdot$cm can be prepared from *tert-*

butylimidotris(dimethylamido)tantalum (NEt$_2$)$_3$Ta=N^{t}Bu) and hydrogen atoms produced by plasma [3]. A potential problem of PEALD is the impact of ions and radicals on the substrate material. In particular, the plasma could damage the IMD.

Considering the processing and material issues related to nitrides of titanium and tantalum, ALD tungsten nitrides (WN$_x$) and ALD tungsten nitride carbide (WN$_x$C$_y$) have been the focus of recent research effort. W$_2$N has been prepared from tungsten(VI) fluoride (WF$_6$) and NH$_3$ and showed a high resistivity of 4500 $\mu\Omega$·cm [11]. ALD WN has been deposited from bis(*tert*-butylimido)bis-(dimethylamido)tungsten [(^{t}BuN)$_2$W(NMe$_2$)$_2$] and NH$_3$ [12]. The density was 12 g·cm^{-3}, but the resistivity was still high (1500–4500 $\mu\Omega$·cm). During WN growth according to this process, the *tert*-butylimido ligands undergo decomposition and the ALD growth-per-cycle ranges from 0.2–0.8 Å for temperatures of 280–330° C, respectively. The (^{t}BuN)$_2$W(NMe$_2$)$_2$ precursor was also used to prepare tungsten carbide films (WC$_x$) in a PEALD process using H$_2$ plasma [13]. Depending on the plasma conditions, the composition of the deposited WC$_x$ film varied and its resistivity ranged from 295–22000 $\mu\Omega$·cm .

For WN$_x$C$_y$, the advantage of the low resistivity of tungsten carbide material can be achieved with a thermal ALD process using the precursors WF$_6$, NH$_3$, and triethylborane (Et$_3$B) [14]. For this process, compatibility and good adhesion to copper was reported [3]. A high density of 15.4 g·cm^{-3} was found. In barrier tests, a performance superior to ALD TiN or sputter-deposited Ta of ALD WN$_x$C$_y$ was attributed to the formation of a nanocrystalline, equiaxial non-columnar structure [15]. The resistivity of 8.5 nm WN$_x$C$_y$ on top of copper in the bottom of damascene structures was 375 $\mu\Omega$·cm [16].

ALD of Barrier Layers on Intrametal Dielectrics

During the growth transient, the substrate surface, characterized by specific functional groups, converts to the ALD process specific surface with newly created reactive species. Most IMD surfaces have a low reactivity (reactive site density) and island-type growth is observed. To guarantee the barrier property of the ALD film, at least a continuous layer on top of the substrate is required. The amount of ALD TiN (TiCl$_4$/NH$_3$) required to cover the starting surface is larger than a monolayer of the expected material (Figure 4) as was found in a study using Rutherford back-scattering spectroscopy (RBS) to determine the amount of deposited layers, and time-of-flight secondary ion mass spectrometry (ToF-SIMS) – for which the depth information is limited to 1–3 monolayers (<10 Å) – to determine the coverage of the substrate [17]. The early stage of film formation is dominated by a Volmer–Weber-type growth mode. The density of functional groups on the initial surface affects the density and the vertical dimension of TiN islands. During the deposition, the substrate is covered by three-dimensional island growth and coalescence of islands occurs. Only after complete coverage of the substrate a two-dimensional Frank–van der Merwe (layer by layer) growth may start. The coverage evolution corresponds to the reactive site density, which decreases from chemical oxide (grown at 20° C), to thermal oxide (grown at 750° C) to silicon carbide (SiC). The dominant reactive sites on these substrates are silanol groups ($\equiv$Si–OH), which react with $TiCl_4$ in ligand exchange reactions. At 350° C, one, two or three silanol groups can undergo ligand exchange with $TiCl_4$ molecules to form a ($\equiv$Si–O)$_{4-x}$–$TiCl_x$ surface complex (x=1,2,3) and HCl. At temperatures above 300° C, $\equiv$Si–OH groups are also chlorinated by $TiCl_4$ to give $\equiv$Si–Cl surface groups, and titanium(IV) hydroxy chlorides ("Ti(OH)$_x$Cl$_y$") intermediates, forming TiO_2 particles in a secondary process [18].

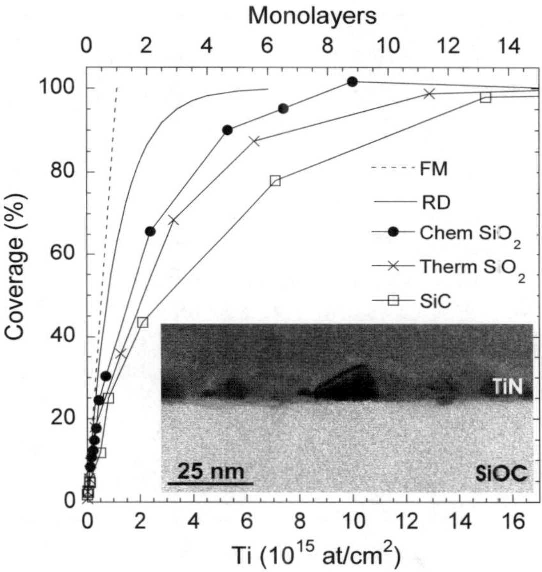

Figure 4. Surface coverage of ALD TiN films on different surfaces; closure on chemical oxide (circles), thermally grown oxide (crosses), SiC (squares), according to an ideal Frank–van der Merwe growth (dashed line) and according to a random deposition of TiN units onto the surface with equal probabilities for the substrate and TiN covered surface (solid line); TiN islands on SiOC:H-type IMD (inset picture).

Assuming a constant growth-per-cycle of the islands, the surface area of a growing island increases quadratically with the number of deposition cycles, and the growth curves are expected to fit to a second-order polynomial. A model for this type of substrate inhibited growth has proven useful to fit experimental data and allows the prediction of the point of formation of a continuous ALD layer [19]. An example of the importance of preparing the substrate surface for ALD growth is found in the case of WN_xC_y ALD barrier growth onto an organic, polymeric IMD [20]. The polymer surface was modified by exposing it to an inductively coupled plasma (ICP) of oxygen or a reactive ion etch (RIE) plasma of nitrogen. Compositional changes in the surfaces after plasma treatment were characterized by X-ray photoelectron spectroscopy (XPS) (Figure 5). Since all plasma-activated surfaces were exposed to air before the ALD was performed, the relatively high oxygen concentration of 15.7% found for the N_2 RIE treatment has been attributed to a reaction with air oxygen and moisture. The plasma treatments give rise to surfaces with hydrophilic functional groups containing nitrogen and oxygen, such as amino, carbonyl, or hydroxyl groups. Comparison of the C1s XPS spectra after treatment with the pristine case reveals that the plasma-activation process affects the aromaticity of the surface as evidenced by a decrease in the relative intensity of the C1s "shake up" feature at 292 eV. A suitable surface preparation requires minimal impact on the IMD.

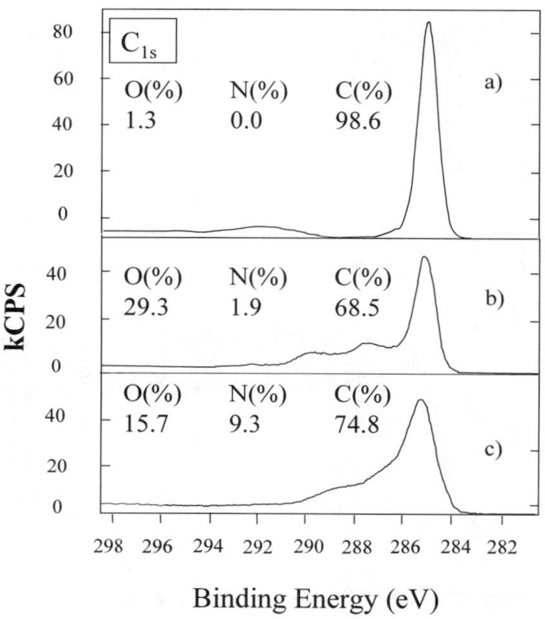

Figure 5. XPS composition and C1s spectra of a) pristine organic IMD surface, b) O_2 ICP exposed, and c) N_2 RIE exposed organic IMD surface.

In order to monitor thin film formation and substrate coverage in a comparative manner, a simple contact angle (CA) measurement can be used (Figure 6). Rehydrophobation of the surface occurs when the substrates are heated to the ALD temperature of 300–350 °C. As the substrate is covered with the ALD material, the contact angle drops sharply and converges towards a value characteristic of the thin film material. W growth on the pristine polymer surface is characterized by an extended transient period, and the growth curve from ALD cycles 20 to 80 can be accurately fitted with a second-order polynomial (Figure 6), indicating island-type growth. Plasma treating the dielectric enhances the ALD growth significantly. For the N-rich surface, the observed growth delay is reduced to only a few cycles. It can be speculated that N-containing groups in the substrate surface facilitate the initial stages of growth. While for the pristine polymer the point of formation of a closed ALD film was about 10 nm, the minimum thickness for closure was only 1.5 to 2.5 nm in the case of the N_2 RIE treatment. In the example given, the surface preparation gives considerable growth enhancement and allows for continuous ALD films with a thickness of only a few nm. However, the treatment does have compositional and structural effects on the IMD. Densification and introduction of polarity should be limited to the very top surface. To enable ALD, mild activation of the unreactive pristine IMD surface without increasing the permittivity or lowering the mechanical strength of the dielectric or other properties is required.

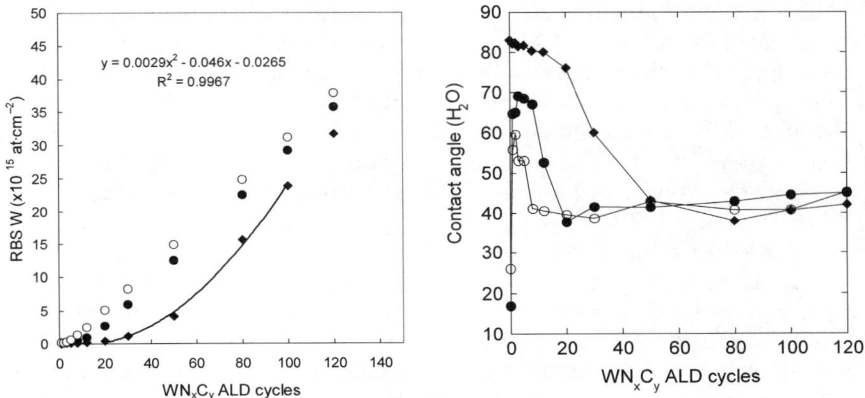

Figure 6. ALD WN_xC_y growth on pristine (diamonds), oxygen plasma treated (filled circles), and nitrogen plasma treated (open circles) organic IMD surface; W atoms (left) and water contact angle (right); The W growth from cycle 20 through 100 follows an island-type growth

ALD of Metal Layers

In damascene metallization, ALD metal layers are discussed as glue layers in their function to guarantee good adhesion to copper and as copper seed layers to guarantee sufficient conductance on the wafer for the ECD of copper. ALD copper has been prepared along two different routes; firstly, by ALD of copper with copper precursor and a reducing agent, and secondly, by ALD of copper oxide and subsequent reduction of the copper oxide layer. An example of the first route involves copper(I) chloride with hydrogen as reducing agent [21]. A theoretical investigation of the reaction mechanism confirms the finding that the reduction is rate limiting but it could not be decided, whether the reduction occurs by surface reaction of chemisorbed copper(I) chloride with chemisorbed hydrogen or through reduction of chemisorbed copper(I) chloride by hydrogen from the gas phase [22]. For the deposition of ALD copper oxide, copper(II) β-diketonate precursors have been reacted with oxygen/ozone at 110–210° C. The copper oxide layer can then be reduced to elemental copper in the environment of alcohols, aldehydes or carboxylic acids at 250–375° C and give a thin film with a specific resistivity of less than 10 μΩ·cm [23]. A novel class of organometallic compounds is used for deposition of transition metals [24]. Copper, cobalt, and nickel thin films can be deposited by ALD from homoleptic *N,N'*-dialkylacetamidinato metal compounds and hydrogen. On the potential ALD WN barrier material [12], copper grows only if the WN is precoated with cobalt demonstrating ALD cobalt as a useful glue layer in damascene metallization.

Ruthenium has been recently suggested as an alternative substrate for ECD of copper and attributed potential barrier properties [25]. This would allow the replacement or elimination of the Cu seed deposition in current damascene metallization schemes.

ALD of ruthenium films has been recently achieved from ruthenocene-type starting materials and oxygen. bis(cyclopentadienyl)ruthenium ($RuCp_2$) as well as the derivative Bis(ethylcyclopentadienyl)ruthenium [$Ru(EtCp_2)$] were used as ruthenium precursors. The films deposited with $RuCp_2$ in a temperature range of 275–400° C showed a growth-per-cycle of 0.1–0.5 Å and a decrease in specific resistivities from 18–10 $\mu\Omega\cdot$cm, respectively [26]. $Ru(EtCp_2)$ instead of ($RuCp_2$) gave a growth-per-cycle of 1.5 Å for films deposited at 270° C and a specific resistivity of 15 $\mu\Omega\cdot$cm [27]. The oxygen partial pressure during oxygen exposure determined whether ruthenium or ruthenium(iv) oxide film was obtained [27]. The ruthenium films could be used as an adhesion promoting glue layer between titanium nitride and copper deposited by metalorganic chemical vapour deposition (MOCVD) [27]. It is a concern that the oxygen originating from the ruthenium deposition may cause the formation of an insulating metal oxide interfacial layer with the underlying barrier metal. This can be avoided in a PEALD process using NH_3 plasma and $Ru(EtCp_2)$ [28]. In a film deposited at 270 ºC, no carbon or nitrogen impurities could be detected and the resistivity was 12 $\mu\Omega\cdot$cm. However, the saturated growth-per-cycle was only 0.38 Å and required $Ru(EtCp_2)$ pulse times of at least 5 s, which limits the throughput of such a process.

Summary

A thin film with barrier property contained in damascene process flows prevents the conducting copper from mixing with the dielectric. Due to control of chemical reactions at the surface and its excellent step coverage, ALD has drawn considerable attention for this application. A significant amount of processes for transition metal nitrides and metal films have been developed, which are mostly based on thermal or plasma-enhanced ALD. Both of these techniques have benefits and drawbacks to guarantee a substrate compatible process, as well as a dense and sufficiently conductive material. The substrate sensitivity is inherent to the ALD processing technique. Preparing a substrate surface that shows high reactivity towards a given ALD process is crucial to guaranteeing an intact film with a low thickness of 5 nm or less. Transition metal nitrides are mainly considered as ALD layers with barrier property. ALD metal layers of cobalt and ruthenium are useful as barrier adhesion promoters. Ruthenium films are also potential Cu seed and direct plating materials.

Acknowledgements

The authors would like to thank their colleagues at IMEC, in particular C.M. Whelan, Y. Travaly, D. Shamiryan, G. Beyer, V. Sutcliffe (TI affiliate), T. Abell (Intel affiliate). The authors also gratefully acknowledge R. Schuhmacher (IFA Tulln) and S. Haukka (ASM) for suggestions to improve the manuscript.

References

1. International Technology Roadmap for Semiconductors (ITRS), Semiconductor International Association (2003), http://public.itrs.net/

2. H. Kim, J. Vac. Sci. Technol. B 21(6), 2231 (2003)

3. K.-E. Elers, V. Saanila, P.J. Soininen, W.-M.Li, J.T. Kostamo, S. Haukka et al., Chem. Vap. Dep. 8(4), 149 (2002)

4. W. Besling, A. Satta, J. Schuhmacher, T. Abell, V. Sutcliffe, A. Martn-Hoyas et al., International Interconnect Technology Conference, IITC 2002, Proceedings, 288 (2002)

5. J.W. Elam, M. Schuisky, J.D. Ferguson, S.M. George, Thin Solid Films 436(2), 145 (2003)

6. C. Basceri, M.W. Miller, P. Castrovillo, V.D. Hou, K. Holtzclaw, J. Smythe et al., Advanced Metallization Conference, Proc. Vol., AMC 2003, 713 (2003)

7. M. Juppo, M. Ritala, M. Leskelä, J. Electrochem. Soc. 147, 3377 (2000)

8. M. Juppo, P. Alén, M. Ritala, T. Sajvaara, J. Keinonen, M. Leskelä, J. Electrochem. Solid State Lett. 5, C4 (2002)

9. M. Juppo, P. Alén, M. Ritala, M. Leskelä, Chem. Vap. Deposition 7, 211 (2001)

10. J.-S. Min, H.-S. Park,. S.-W. Kang. Appl. Phys. Lett. 75, 1529 (1999)

11. J.W. Klaus, S.J. Ferro, S.M. George, J. Electrochem. Soc. 147, 1175 (2000)

12. J.S. Becker, S. Suh, S. Wang, R.G. Gordon, Chem. Mater. 15, 2969 (2003)

13. D.-H. Kim, Y.J. Kim, Y.S. Song, b.-T. Lee, J.H. Kim, S. Suh et al., J. Electrochem. Soc. 150, C740 (2003)

14. W.M.Li, K. Elers, J. Kostamo, S. Kaipio, H. Huotari, M. Soininen et al., International Interconnect Technology Conference, IITC 2002, Proceedings, 191 (2002)

15. S.-H. Kim, S.S. Oh, H-M. Kim, D.-H. Kang, K.-B. Kim, W.-M. Li, S. Haukka et al., J. Electrochem. Soc. 151(4), C272 (2004)

16. J. Schuhmacher, G. Beyer, I. Vos, V. Sutcliffe, Zs. Tőkei, W. Besling, K. Maex, Advanced Metallization Conference, Proc. , AMC 2002, 759 (2002)

17. A. Satta, A. Vantomme, J. Schuhmacher, C.M. Whelan, V. Sutcliffe, K. Maex, Appl. Phys. Lett. 84(22), 4571 (2004)

18. S. Haukka, E.-L. Lakomaa, O. Jylhä, J. Vilhunen, S. Hornytzkyi, Langmuir 9, 3497 (1993)

19. R. L. Puurunnen, W. Vandervorst, J. Appl. Phys., (in press)

20. A. Martin Hoyas, J. Schuhmacher, D. Shamiryan, J. Waeterloos, W. Besling, J.P. Celis, K. Maex, J. Appl. Phys. 95, 381 (2004)

21. P. Mårtensson, J.-O. Carlsson, Chem. Vap. Deposition, 3, 45 (1997)

22. P. Mårtensson, K. Larsson, J.-O. Carlsson, Appl. Surf. Sci., 148, 9 (1999)

23. J. Kostamo, V. SAAnila, M. Tuominen, S. Haukka, K.-E. Elers, M. Stokhof *et al.*, presented at AVS Topical Conference ALD2002 (2002)

24. B.S. Lim, A. Rahtu, R.G. Gordon, Nat. Mater., 2, 749 (2003)

25. O. Chyan, T.N. Arunagiri, T. Ponnuswamy, J. Electrochem. Soc., 150(5), C347, (2003)

26. T. Aaltonen, P. Alén, M. Ritala, M. Leskelä, Chem. Vap. Deposition, 9, 45 (2003)

27. O.-K. Kwon, J.-H. Kim, H.-S. Park, and S.-W. Kang, J. Electrochem. Soc., 151(2), G109, (2004)

28. O.-K. Kwon, S.-H. Kwon, H.-S. Park, S.-W. Kang, J. Electrochem. Solid State Lett. 7, C46 (2004)

Copper CVD for Conformal Ultrathin-film Deposition

M. Joulaud[a, b], P. Doppelt[b]

[a] MERCK KgaA, CEA-Leti/D2NT/LBE, Grenoble / France

[b] CECM-CNRS, Vitry-sur-Seine / France

Introduction

Since the beginning of the 1990s, copper replaced aluminum as interconnects metal because of its lower resistivity, higher thermal conductivity and higher resistance to electromigration [1, 2].

The electrochemical deposition (ECD) is the best deposition technique to metallize interconnects with pure copper at a low temperature and high deposition rate with low-cost equipment. However, the barrier deposited prior to copper in order to avoid copper diffusion in the dielectric is too resistive (R = 250 $\mu\Omega$.cm) to allow a copper ECD. Thus, a thin conductive film of copper (R = 2 to 2.5 $\mu\Omega$.cm) called seed layer is deposited on the diffusion barrier in order to allow the copper ECD. The scheme of the different layers composing a copper interconnect is represented in Figure 1.

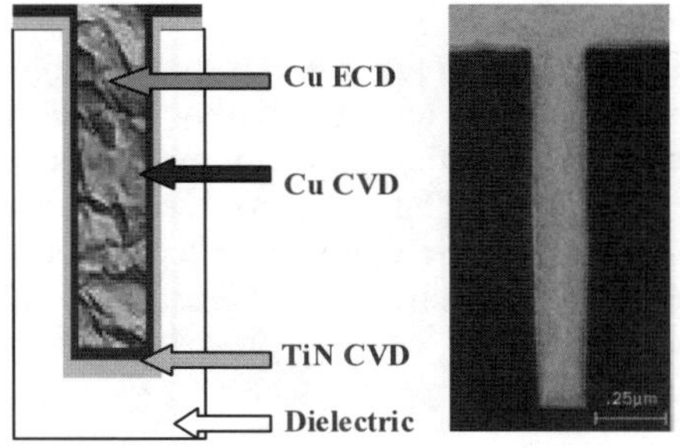

Figure 1. Stack of a line in damascene technology

The seed layer has to be conformal and thin in order to leave enough free space to allow a defect-free copper filling of interconnects. Seed layers are commonly deposited by ionized PVD. This deposition technique already used for aluminum technology allows a pure copper deposit with a relatively high deposition rate. However, this deposition technique is not a conformal one because of the directional aspect of the deposition. It is expected that this lack of uniformity of the deposit will be critical for a defect-free filling by ECD of high aspect ratio and narrow features (Figure 2). Copper CVD appears as a good alternative to the ionized PVD techniques since the deposition mechanisms are based on gaseous chemical reaction of precursors and by the way Brownian motion of the precursor molecules at the substrate surface. Thus, copper CVD films are conformal whatever the geometry of the substrate is.

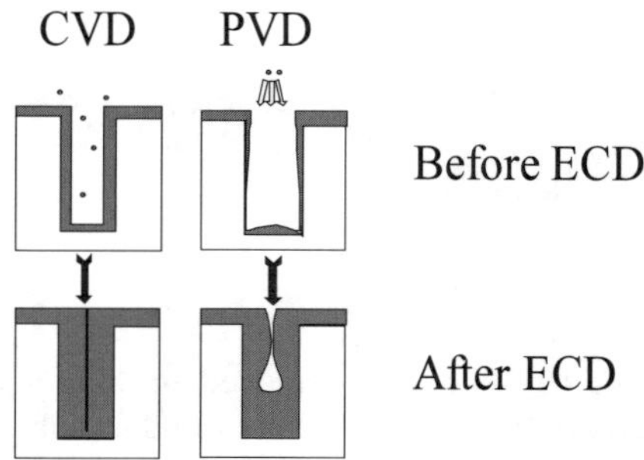

Figure 2. Influence of the seed layer quality on the filling efficiency of the ECD

Among all precursors synthesized for copper CVD, (hfac)CuL compounds (hfac = hexafluoroacetylacetonate) are the best candidates to deposit high-purity copper films at low temperature [3, 4]. We will focus more precisely on one of the most promising ones: (hfac)Cu(MHY), (MHY = 2-methyl-1-hexen-3-yne, trade name Gigacopper) [5-9]. The deposition reaction is a disproportionation [10, 11]. It allows the deposition of a pure copper atom without decomposition of the precursor organic parts and with no contamination of the film due to the by-products since they are volatile.

As said before, it is important to minimize as much as possible the thickness of the deposit to keep a free space big enough to allow a conformal filling of the features by ECD. The deposition reaction is driven by the nucleation/growth mechanism as for all chemical deposit. Thus, it is necessary to enhance the nucleation density and the lateral growth of the grains in order to get the coalescence of the film (complete film) at a minimum thickness as illustrated on Figure 3.

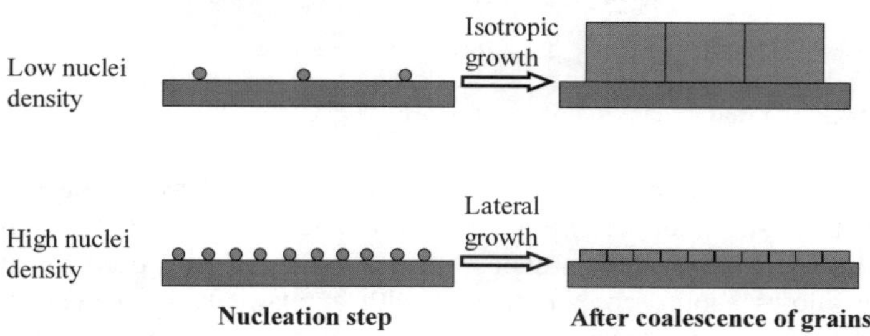

Figure 3. Influence of nuclei density and growth regime on the minimum film thickness

To improve the nucleation and the lateral growth means to enhance the chemical interaction between the substrate and the precursor. It can be done by using a specific ligand which is supposed to interact with the substrate as it is the case for (hfac)Cu(MHY). MHY ligand has a free double bond which may enhance precursor adsorption onto the substrate surface [6] (Figure 4).

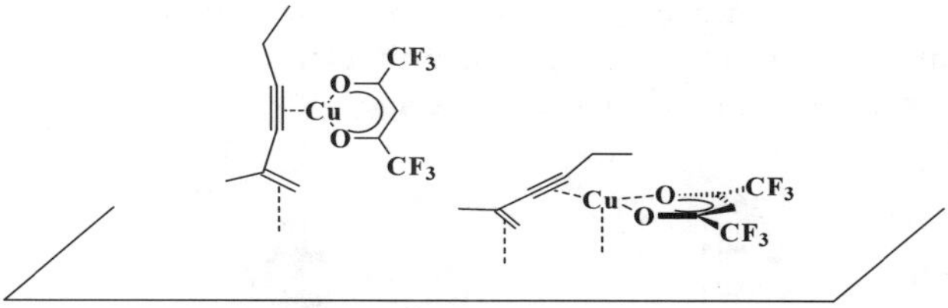

Figure 4. Adsorption scheme of (hfac)Cu(MHY)

Another way to improve substrate surface/precursor interaction is to make a chemical or physical treatment of the surface [12–14]. It was determined that the injection of water during the deposition increases the deposition rate, reduce the incubation time and the roughness of the deposit.

In the following, we focus on water and H_2 plasma treatment influences on nucleation and lateral grain growth of copper CVD deposited with (hfac)Cu(MHY). An understanding of the role of water is then presented. Finally, a specific recipe developed using the best settings for water addition is presented and compared to a "standard" deposition process.

Experimental Methods

Gigacopper, (hfac)Cu(MHY), is commercially available from Merck KgaA. Copper films were deposited on 200 mm wafers (8") priory coated with 10 nm CVD TiN as diffusion barrier. The equipment used was a P5000 Mark II (Applied Material) processing 2x6 wafers lots. The setting of the experiments were $P_{reactor} = 1$ Torr, precursor flow = 0.70 sccm, $T_{injector} = 55°C$, carrier gas flow = 200 sccm, water addition = 0.3 to   2 sccm in 150 sccm N_2 and $T_{susceptor} = 150°C$ to $190°C$.

The substrate temperature was chosen in the upper part of the kinetic range (190°C) in order to get the highest deposition rate with a good uniformity (190°C). Film thickness was calculated by weight and resistivity was calculated with a 49-point resistivity map measured with a CDE Rasemap.

Discussion

For the study of the influence of water on deposition rate and resistivity, a 160-s deposition was performed at 190°C with and without water added during the deposition. The results are presented in Figure 5 and compared to those obtained for a 160-s deposition at 250°C.

It appears that the deposition rate reached with water addition during the deposition is nearly as high as the deposition rate measured at 250°C, whereas the resistivity of the film is lower in the first case. Notice that the resistivity of the film deposited without water at 190°C was not presented since the film was not continuous.

However, the film uniformity was quite bad with a deviation of the thickness of around 20%, which is in accordance with the observation and Yand *et al.* [15]. They determined that water addition is mainly effective at the first second of the deposition corresponding to the nucleation stage.

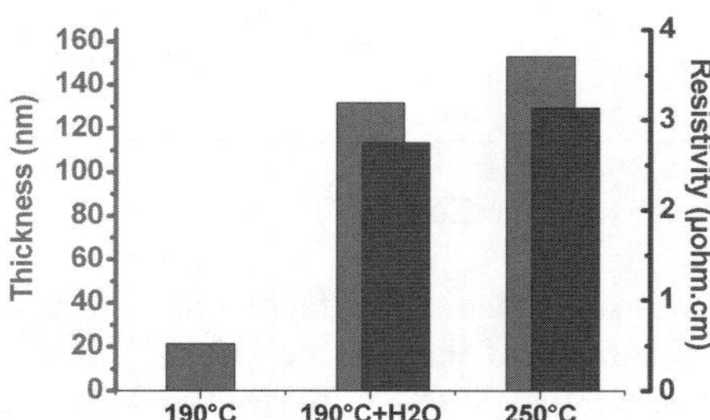

Figure 5. Influence of water on film thickness and resistivity for a 160-s deposition duration

Thus, water was added only during the first 30 s of deposition evaluated as the nucleation duration. A water flux from 0.4 to 2 sccm was evaluated. By comparing Figure 5 with Figure 6, it appears that a 1-sccm water addition at the first 30 s rather than a 0.4-sccm addition during the entire deposition increases the uniformity and does not deteriorate the deposition rate confirming the Yand *et al.* observations. The maximum deposition rate and the best uniformity in thickness are reached for a 2-sccm water injection.

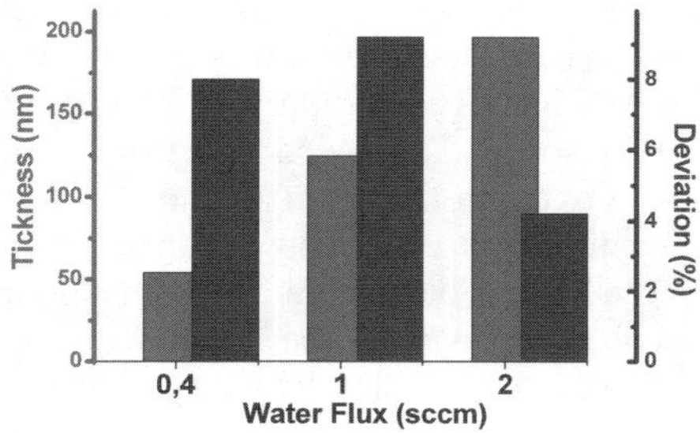

Figure 6. Influence of water flow on thickness and uniformity of the deposit for water addition during the first 30 s of a 160-s deposition at 190°C

In order to understand the role of water during the nucleation phase, only the first 30 s of deposition was investigated at 190°C. Water was added before or during deposition. SEM pictures of 30-s deposition with different water addition sequence are presented in Figure 7. Clearly, water added before or during the deposition increases the nuclei density. Moreover, water added before deposition not only increases the nuclei density but also the nuclei size. Various data about the nuclei are given in Table 1.

The nuclei density and the Feret's diameter of all nuclei were calculated by image analysis. The Feret's diameter given by the image analysis software is the diameter of a disc having the same area as the observed nuclei. It was observed that the nuclei Feret's diameter distribution follows a log-normal distribution law. Thus, the average Feret's diameter was calculated by the 50th percentile. The grain thickness was determined by AFM measurements.

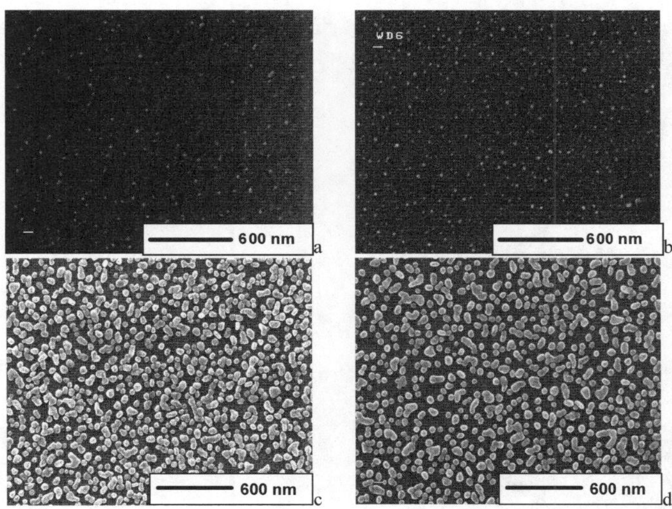

Figure 7. 30-s deposition at 190°C without water during deposition (a), with water during deposition, with water before deposition without (c) and with (d) a purge between water addition and deposition

Table 1. Influence of the water addition sequences on the copper nuclei characteristics

	Average Feret's diameter (nm)	Feret's diameter deviation (%)	Nuclei density (nuclei/cm^2)	Nuclei thickness (nm)
No H$_2$O	27.5	1.22	3.80 x 10^9	
H$_2$O during deposition	27.6	1.14	1.70 x 10^{10}	23.3
H$_2$O before deposition (no purge)	61.7	1.43	1 x 10^{10}	24.6
H$_2$O before deposition (with a purge)	61.7	1.38	1 x 10^{10}	21.3

For the same nuclei thickness, the average Feret's diameter is higher for water added before than during precursor injection. Thus, water addition before deposition enhances the lateral growth of nuclei (Figure 8).

Since water has an influence on copper deposition even if a purge is applied between water injection and precursor injection, it appears that water is chemically adsorbed at the TiN surface. One possible explanation is the water reaction with free Ti or Ti–H bonds possibly present at the TiN surface giving Ti–OH groups. Then these groups are potential sites for precursor adsorption and subsequent dissociation of the ligand MHY.

A 10-s 350 W H_2 plasma was applied at the TiN surface before water injection in order to increase Ti–H bonds or free surface Ti bonds. It appears in Figure 9 that the H_2 plasma treatment increases the nuclei density with no increase of the substrate roughness or nuclei thickness.

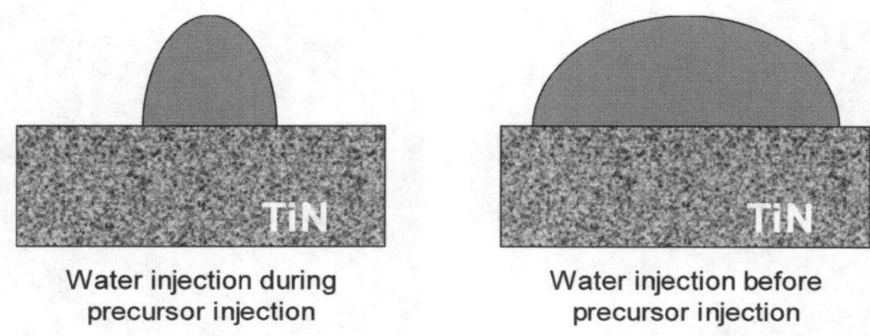

Figure 8. Illustration of the influence of water addition sequence on the nuclei shape

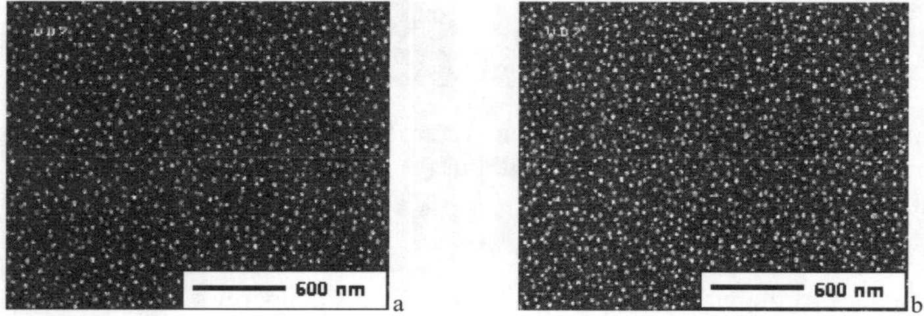

Figure 9. SEM pictures of a 30-s (hfac)Cu(MHY) injection at 150°C without (a) and with (b) a 10-s H_2 plasma pre-treatment

However, the positive effect of water seems to be limited for longer deposition duration. In order to solve this limitation, a pulse recipe was developed. The susceptor temperature was fixed at 150°C in order to have a better control of the film growth and a H_2 plasma treatment as developed before was applied prior to water deposition.

The sequence of the pulse recipe is then a 10-s water injection, 30-s precursor injection, both repeated four times. This sequence was followed by a 100-s deposition in order to be sure that the complete coalescence of the film was reached and then compared to a deposition of the same duration with just a single water exposition before deposition.

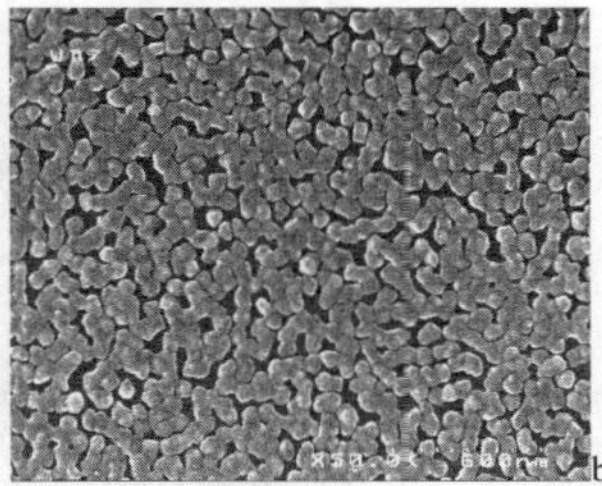

Figure 10. SEM pictures of a film deposited with a pulse recipe alternating water injection and precursor injection (a) and a single injection recipe (b) (total injection duration: 300 s)

As shown in Figure 10, the pulse recipe is continuous whereas the one obtained with the single injection recipe is not. In the case of the pulse recipe, the thickness is 53 nm whereas, in the case of a continuous precursor injection, it is 67 nm. Thus, the coalescence of grains was reached faster using a pulse recipe.

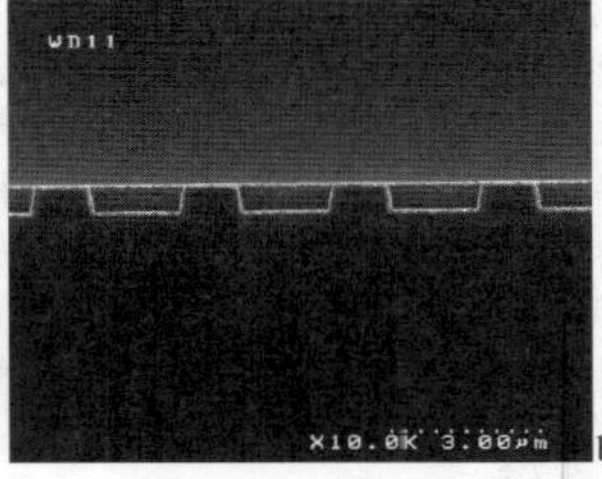

Figure 11. SEM pictures of a 47 nm thin copper seed layer deposited with (hfac)Cu(MHY) using four pulses of water then precursor injection at 170° C in a 1-μm wide line

The pulse recipe was experimented on patterned wafers. The deposition temperature was fixed to 170° C in order to reach a higher deposition rate. The precursor injection pulse was fixed to 15 s and the number of water and precursor pulses was fixed to four as before and no additional precursor injection was added. The thickness of the deposit is 47 nm. As shown in Figure 11, the copper seed layer deposited with the pulse recipe is conformal with a good step coverage. Even in narrow features, a conformal film is deposited at the bottom and on the sidewall of the interconnects (Figure 12).

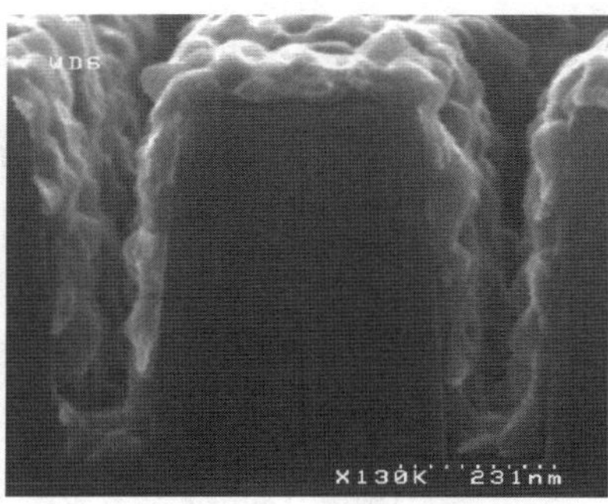

Figure 12. SEM pictures of a 47-nm thin copper seed layer deposited with (hfac)Cu(MHY) using four pulses of water then precursor injection at 170°C in a 0.16-μm wide feature

Conclusion

Chemical and physical substrate surface treatments for thin copper film deposited by CVD were evaluated. It was observed that lateral growth can be enhanced by using substrate pre-treatments as water exposition or H_2 plasma treatment in order to decrease the minimum thickness of copper film deposited by CVD. A specific recipe inspired of pulse CVD was developed with (hfac)Cu(MHY) introducing water between each precursor pulse. The improvement due to this recipe on the minimum film thickness was demonstrated. An ultrathin film of 47 nm was conformally deposited in narrow features with the pulse recipe developed in this study. Water pre-treatment of the substrate improves the lateral growth of copper grains.

Acknowledgements

Thanks are to the colleagues at LETI, in particular to T. Mourier for valuable contributions. Special thanks to A. Wirth and D. Mayer from Merck KgaA for their continuous support.

References

1.   D. C. Edelstein, G. A. Sai-Halasz, Y.-J. Mii, IBM J. Res. Develop. 39(4), 383–400 (1995)

2.   S. P. Murarka, S. W. Hymes, Crit. Rev. Solid State Mater. Sci., 20(2), 87–124 (1995)

3.   P. Doppelt, T. H. Baum, MRS Bulletin 19(8), 1994, 41–48

4.   M. J. Hampden-Smith, T. T. Kodas, Polyhedron 14(6), 1995, 699–732 (1995)

5.   P. Doppelt, French Patent 97 03 029, US Patent 6,130,345

6.   P. Doppelt, T.Y. Chen, R. Madar, J. Torres, Proceedings AMC98, MRS, 117–121 (1999)

7.   T. Y. Chen, J. Vaissermann, B. Ruiz, J. P Sénateur, P. Doppelt, Chem. Mater. 13, 3993–4004 (2001)

8.   M. Joulaud, C. Angekort, P. Doppelt, T. Mourier, D. Mayer, Microelectron. Eng. 64, 107–115 (2002)

9.   J. Zhang, D. Denning, G. Braeckelmann, G. Hamilton, J. J. Lee, R. Vankatraman, B. Fiordalice, E. Weitzman, Proceedings MRS Symposium 564, 243–249 (1999)

10. S. L. Cohen, M. Liehr, S. Kasi, Appl. Phys. Lett. 60(1), 50–52 (1992)

11. G. S. Girolami, P. M. Jeffries, J. Am. Chem. Soc. 115, 1015–1024 (1993)

12. V. Vezin, Y. Kojima, T. Hoshino, G. Chung, Japan J. Appl. Phys. 41 part. 2 n°8B, L903–L906 (2002)

13. C. L Chang, C. L. Lin, M. C. Chen, Japan J. Appl. Phys. 43 n°5, 2442–2446 (2004)

14. O. K. Kwon, J. H. Kim, H. S. Park, S. W. Kang, J. Electrochem. Soc. 149(2), G109–G113 (2002)

15. D. Yang, J. Hong, D. F. Richards, T. S. Cale, J. Vac. Sci. Technol. B 20(2), 495–506 (2002)

Pushing PVD to the Limits – Recent Advances

Zs. Tőkei

IMEC, Leuven / Belgium

Introduction

Physical vapour deposition (PVD) is dead. The dooming phrase for PVD has been outspoken on many occasions in the interconnect arena. The transition from aluminium to copper metallization at around the 0.18-µm technology node brought additional challenges. Ionised physical vapour deposition (I-PVD) was introduced in order to cope with the composite via and trench recesses of a typical dual damascene architecture [1,2]. First generation tools were successfully used for this technology node, but smaller dimensions appeared to be out of reach [3]. Significant resources were and are being deployed for identifying and optimising alternative deposition techniques, such as for example chemical vapour deposition or atomic layer deposition. At the same time, due to efforts in hardware design and process development, an important evolution of PVD is observed [4]. Dimensions of 65 nm are viable today by advanced deposition techniques and 45-nm node dimensions are being optimised actively. In the course of PVD evolution more and more "knobs" were introduced to control the deposition flux as close as possible to the substrate (cf. classical PVD, collimated PVD, I-PVD, high magnetron power PVD, *etc.* [2]). Where is the limit? There must be one (maybe even the 45-nm technology node), but it is difficult to derive this limit from first physical principles. In this short contribution the most important characteristics of PVD based copper metallization are described along with references to the state of the art.

Ta-based Barrier Deposition by PVD

Although copper drift diffusivity tends to be lowered in most of the low-k materials [5], it is not low enough to omit diffusion barriers that encapsulate copper wires. Possible material choices are varied, including, for example, nitrides, oxynitrides and silicides of titanium, tungsten, tantalum or cobalt. A barrier thin film has to satisfy numerous requirements before gaining acceptance in IC-fabrication. Today's mainstream metallization is tantalum based. Without being exhaustive with the list, the following criteria are of importance: resistivity, in-film stress, adhesion to various dielectrics, barrier defectivity, defectivity in combination with copper, impact on copper filling, influence on chemical mechanical polishing (CMP), barrier properties against copper migration, impact on via resistance, impact on RC-delay, influence onto stress and electro migration, effect on front-end of line (FEoL) devices, *etc.* Tantalum can be deposited in tetragonal beta and cubic alpha phase or a mixture of those [6]. Some properties of these two different phases are compared in Figure 1. From this non-exhaustive list it becomes clear that selection of a particular barrier for a given technology cannot be based on a single property and is often a challenging task.

Property	Ta(N)/Ta	Ta	Remark
Resistivity (µΩcm)	++	0	(30-50 vs. 180 µΩcm)
Stress (Gpa)	-	-	(-1.6 vs. -2.2 GPa)
Defectivity of Cu	-	+	self-anneal
Defectivity of barrier	++	++	
Filling performance	0	0	
Step coverage	-	-	I-PVD is 'line-of-sight'
Adhesion to SiO$_2$	+++	+++	
Adhesion to low-k	++	+++	(better adhesion is due to reactivity)
Barrier on SiO$_2$	+++	++	
Barrier on low-k	++	-	Reactivity with low-k
Via resistance	+	0	(above 90 nm)
RC	+	0	
EM	+++	+	
CMP performance	+	0	
Effect on FEOL	0	N/A	

--- : major drawback
--: bad
-: can be improved
0: neutral
+: good
++: better
+++: best

Figure 1. Comparison of alpha (Ta(N)/Ta) and beta Ta integrated into copper low-k interconnects. Properties of higher importance (arbitrary to some extent) are highlighted

The metal barrier deposition is typically aspect ratio (AR) dependent and non-conformal [2]. On one hand, AR dependence can be regarded as "an intrinsic" PVD property and is considered as one of the major limitations for barrier scaling. On the other hand, an evolution towards a more conformal deposition was observed [4]. In Figure 2 the reported sidewall thickness value corresponds to the minimum measured on the sidewall, while the bottom thickness represents the maximum value.

The performance of sub-100-nm on-chip interconnects is challenged by copper line resistance increase, which is to some extent related to the increasing relative volume occupied by the high resistivity metal diffusion barrier. This is the main drive to reduce the barrier thickness possibly to 2.5 nm by 2016 [1]. Can the PVD barrier step coverage become perfectly conformal, which would enable unlimited scaling? PVD deposition is "line-of-sight" in nature and a perfectly conformal coverage cannot be realized, but a quasi-conformal can be achieved at local interconnect level with an aspect ratio of about 1.7. It is based on a redistribution of the deposited non-conformal barrier material by adding an energetic Ar-bombardment to the deposition step [4,8,9] as indicated in Figure 3.

In sub-100-nm interconnects it is not enough to consider the minimum or maximum thickness value at a given surface, but the overall barrier thickness variation within the recess has to be minimized. Several different metal pitches with 50% pattern density were investigated and it was found that for AR~2 the variation within the recess is independent of feature size (Figure 4). In this example the variation was reduced from 73% 1 sigma to 25% 1 sigma (or to 1.3 nm).

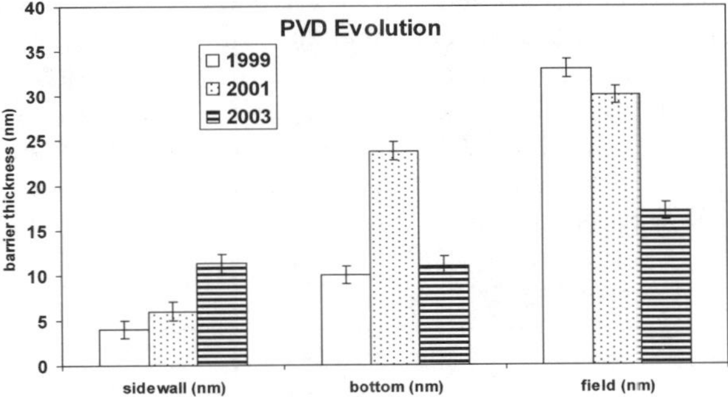

Figure 2. Step coverage evolution for different barrier technologies. 0.25-µm oxide trenches (AR=4) were used as test vehicle [4]

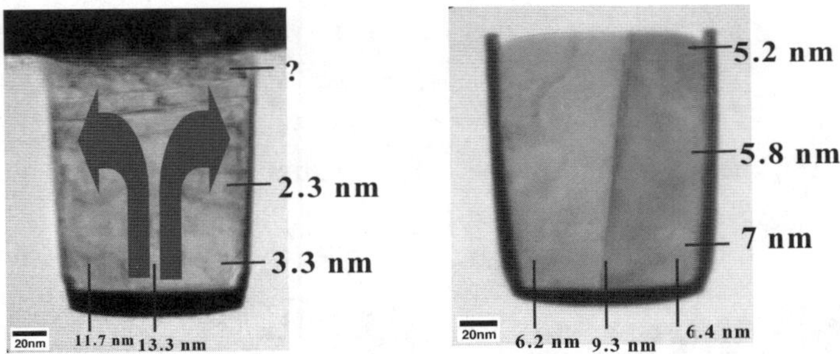

Figure 3. Quasi-conformal step coverage (right) of PVD Ta(N)/Ta using re-distribution of "thick" bottom deposits (left). The barrier was inlayed into Aurora ULK dielectric (Aurora is a trademark of ASM International)

It was shown that a quasi-conformal coverage can be maintained down to the narrowest investigated metal pitches (100 nm at the time of writing) [8]. If deposition rates were reduced and the thickness variation along the sidewall was decreased to 1 nm, for example, that could push PVD towards 2013 as the final destination for copper interconnects from the step-coverage perspective.

Apart from coverage requirements, the crystalline phase at recess bottom is of importance, because it can influence the copper grain structure, hence copper resistivity and electro migration [10]. Figure 5 shows that the most often desired alpha Ta phase can be maintained on a trench recess bottom down to 65-nm technology node dimensions and possibly beyond if needed.

PVD deposition is controlled close to the wafer surface (plasma sheath), but not on the wafer surface itself. Therefore, it is non-selective towards the substrate and deposition onto various dielectrics is feasible without extensive surface preparation. The structural integrity of the barrier, however, can strongly be impacted by the introduction of porous substrates [7,11–14]. In early years of copper porous low-k development the barrier deposition was adjusted to achieve compatibility with porous substrates. It became clear that an additive technique, such as PVD, is limited in sealing performance on mesoporous (pore size larger than 2 nm) surfaces. Though PVD deposition does provide some window [4,7], less than 5-nm barrier thickness on these substrates remains a challenge. Other sealing techniques prior to barrier deposition and alternative integration schemes are being investigated to make these porous substrates compatible with barrier technology.

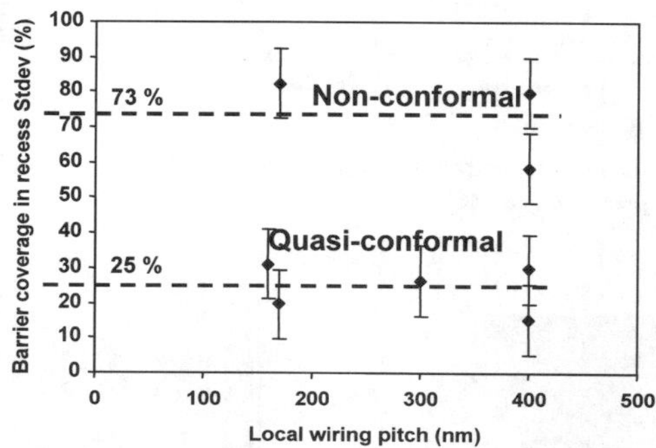

Figure 4. Thickness variation of the barrier within AR~2 recesses for a quasi-conformal and a non-conformal barrier process

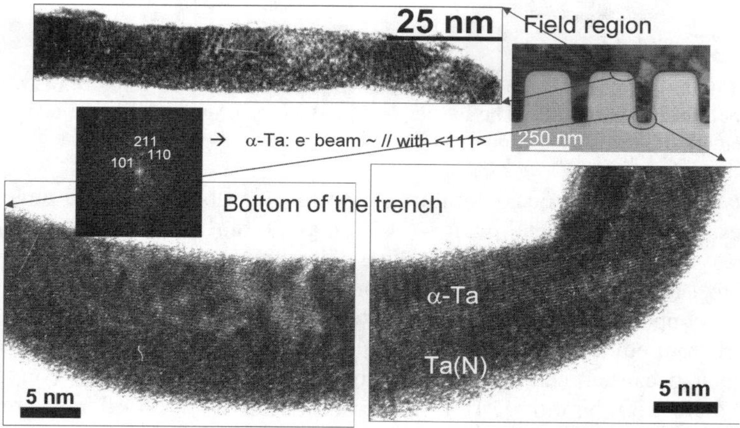

Figure 5. 100-nm-wide AR=3.3 trenches filled with Cu and α-Ta on bottom

For dual damascene integration thin barrier coverage on via bottom or eventually bottomless barriers might be required [6,9]. The optimization of these solutions is ongoing.

Copper Seed Layer Deposition by PVD

Copper seed deposition follows the barrier deposition without vacuum break in order to avoid surface oxidation of the barrier surface and ensure optimal barrier/copper interface. Copper adheres to both alpha or beta tantalum well. Adhesion to Ta(N) is somewhat reduced due to the nitrogen content in the barrier layer. A defect free barrier layer is essential in enabling high quality copper seed deposition, since copper does not adhere to most of the relevant dielectrics. The seed layer is typically deposited by PVD variants [2,15]. The necessity of this layer stems from the high resistivity of Ta-based barriers, which hampers direct plating onto the barrier. It has no functionality in the ultimate interconnect. The seed layer is a sacrificial layer that provides nucleation for subsequent electroplating and is intermixed with the plated copper. Accordingly, the requirements are different from that of the barrier layer. It does not necessarily have to be conformal, but should be suited for voidless copper fill of trenches and vias as shown in Figure 6.

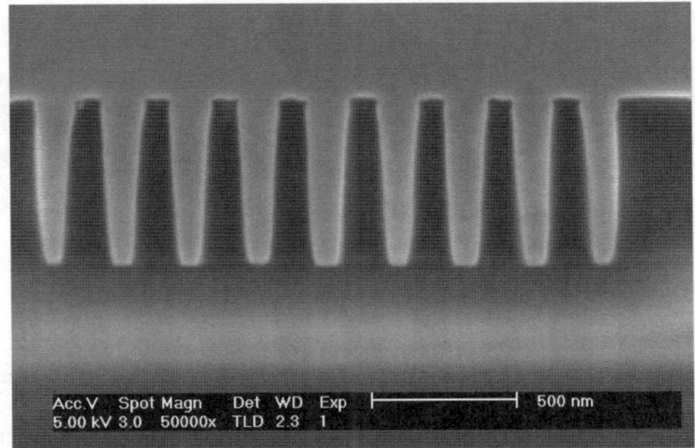

Figure 6. High aspect ratio 120-nm-wide and 700-nm-deep structures filled with Cu

The challenges in copper seed layer deposition do not come from the physics standpoint. Many different advanced PVD sources are already available. In principle, the PVD technique could be used till end of the roadmap dimensions, because the AR of features is expected to stay almost constant [1]. Or the least what can be stated is that the process window is larger than for Ta-based PVD barrier deposition. Nevertheless, the technological challenge is enormous. Owing to the finite geometry of magnetron sources the deposited film can show center-to-edge asymmetry over the wafer [10]. Continuous improvement in source design is expected to alleviate this shortcoming. Another important challenge is overhang formation at recess openings. First generation I-PVD techniques often resulted in significant overhang formation. Due to the presence of copper seed overhang premature closure during copper plating occurs, which results in massive voiding of vias and trenches. Next generation processes provide virtually no overhang, hence a larger process window (Figure 7). Extendibility into the sub-100-nm domain became a reality [16]. Of course, implementation into advanced nodes has to be accompanied by seed layer thickness scaling, which could result in marginal coverage on sidewalls. On one hand, tight process control is needed to ensure robust, good quality copper lines. On the other hand, it was shown that plating is possible on as thin as 3-nm-thick seed layers.

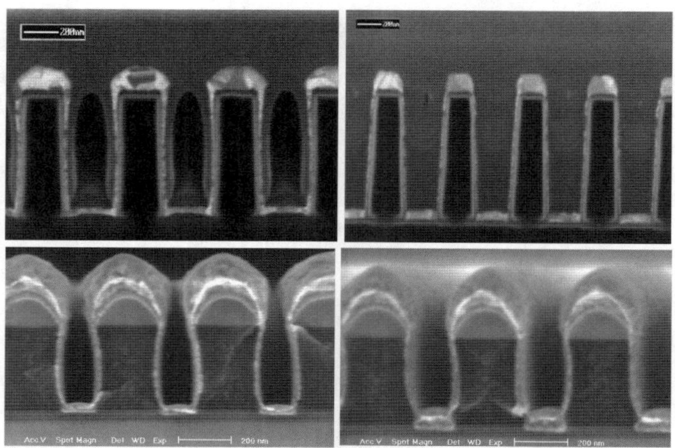

Figure 7. Example of excessive (left) and reduced copper overhang (right)

Performance and Reliability

Copper metallization has a strong impact onto the overall interconnect performance and reliability [10,17–20]. Copper itself is not causing the problem in interconnect structures as long as it is properly encapsulated into metallic and dielectric diffusion barriers. The compatibility of barrier metals with intermetal dielectrics is often overlooked. Full characterization of metallic and dielectric barriers on micro- and meso-porous low-k materials is mandatory. All of the involved barriers have to scale in thickness. The metallic barrier thickness directly impacts on copper resistivity, because the high resistivity metal can occupy important volume fraction of the copper wire [8]. Electromigration kinetics and thermodynamics are strongly influenced by the barrier phase selection [10]. Barrier integrity plays a most important role in line-to-line leakage [17] and time dependent dielectric breakdown (TDDB) reliability [18,19]. If a not fully dense metal diffusion barrier is integrated it will result in significant interconnect lifetime degradation [17–19]. TDDB has never been an issue for Cu–SiO_2 interconnects, but for sub-100-nm copper/barrier/low-k systems it can become a challenge. From PVD barrier deposition perspective, an optimised and a non-optimised barrier scheme are compared in Figure 8. The TDDB reliability was measured on passivated single damascene 80/80-nm width/spacing 1-cm long meander comb structures inlayed into a SiCO:H interlevel dielectric. It appears that the 10-year minimum requirement is not met as per se, but optimised schemes provide a safe reliability margin.

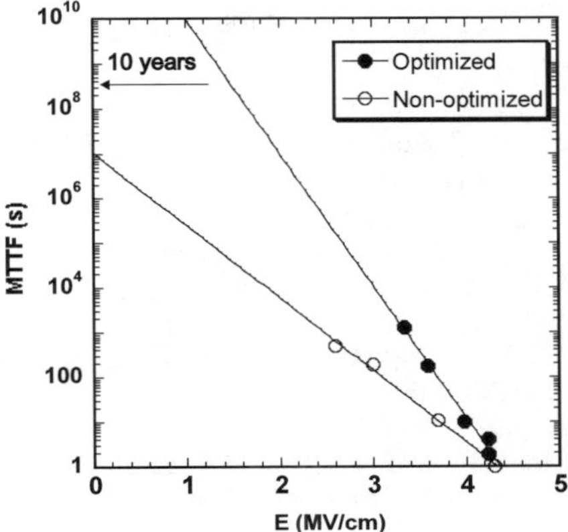

Figure 8. TDDB comparison of 80/80-nm width/spacing meander comb structures at 100° C using different PVD barrier treatments

Conclusion

PVD deposition served well and will continue to serve metallization of advanced copper low-k interconnects in sub 100-nm-dimensions. How long? As long as better, more robust and more economical solutions emerge that can be integrated, provide performance and are reliable.

Acknowledgements

O. Richard, H. Bender and Christel Drijbooms are acknowledged for careful materials analysis. Thanks go to the entire copper/low-k and reliability team as well.

References

1.　International Technology Roadmap for Semiconductors (ITRS), Semiconductor International Association (2003), http://public.itrs.net/

2.　S.M. Rossnagel, in: J.A. Hopwood (Ed.), Ionized Physical Vapor Deposition, Academic Press, New York, p. 37 (2002)

3.　Zs. Tőkei, D. McInerney, M. Baklanov, G.P. Beyer, K. Maex, Interconnect Technology Conference, IITC 2001, Proceedings, p. 213 (2001)

4.　Zs. Tőkei, F. Iacopi, O. Richard, J. Waeterloos, S. Rozeveld, E. Beach *et al.,* Microelectron. Eng. 70, 352 (2003)

5.   F. Lanckmans, K. Maex, Microel. Eng. 60, 125 (2002)

6.   S.M. Rossnagel, J. Vac. Sci. Technol. B 20(6), 2328 (2003)

7.   Zs. Tőkei, M. Patz, M. Schmidt, F. Iacopi, S. Demuynck, K. Maex, Microelectron. Eng. 76, 70 (2004)

8.   W. Wu, R. Jonckeere, Zs. Tőkei, S. H. Brongersma, M. Van Hove, K. Maex, J. Vac. Sci. Technol. B 22(4), L11 (2004)

9.   H. Yamaghishi, Zs. Tőkei, G.P. Beyer, R. Donaton, H. Bender, T. Nogami *et al.*, Advanced Metallization Conference, Proc. AMC 2000, p. 279

10. S. Demuynck, Zs. Tőkei, C. Bruynseraede, J. Michelon, K. Maex, Advanced Metallization Conference, Proc. AMC 2003, p. 355

11. Zs. Tőkei, J. Waeterloos, F. Iacopi, R. Caluwaerts, H. Struyf, J. Van Aelst *et al.*, Advanced Metallization Conference, Proc. AMC 2001, p. 307

12. D. Shamiryan, M.R. Baklanov, Zs. Tőkei, F. Iacopi, K. Maex, Advanced Metallization Conference, Proc. AMC 2001, p. 279

13. F. Iacopi, Zs. Tőkei, D. Shamiryan, Q.T. Le, S. Malhouitre, M. Van Hove *et al.*, Advanced Metallization Conference, Proc. AMC 2001, p. 61

14. F. Iacopi, Zs. Tőkei, Q. T. Le, D. Shamiryan, T. Conard, B. Brijs et al., J. Appl. Phys. 92(3), 1548 (2002)

15. P.J. Stout, D. Zhang, S. Rauf, P.L.G. Ventzek, J. Vac. Sci. Technol. B 20(6), 2421 (2002)

16. Zs. Tőkei, S. Demuynck, I. Vervoort, B. Mebarki, T. Mandrekar, S. Guggilla *et al.*, Advanced Metallization Conference, Proc. AMC 2002, p. 403

17. F. Iacopi, Zs. Tőkei, M. Stucchi, F. Lanckmans, K. Maex, IEEE Electron. Dev. Lett. 24(3), 147 (2003)

18. Zs. Tőkei, V. Sutcliffe, S. Demuynck, F. Iacopi, P. Roussel, G.P. Beyer *et al.*, International Reliability Phyiscs Symposium IRPS2004, Proc., p.326

19. Zs. Tőkei, D. Kelleher, B. Mebarki, T. Mandrekar, S. Guggilla, K. Maex, Microelectron. Eng. 70, 258 (2003)

20. Zs. Tőkei, F. Lanckmans, G. Van den bosch, M. Van Hove, K. Maex, H. Bender et al., International Symposium on the Physical and Failure Analysis of Integrated Circuits, IPFA 2002, Proc., p. 118

Surface Engineering Using Self-assembled Monolayers: Model Substrates for Atomic-layer Deposition

C. M. Whelan[a], A.-C. Demas[a], A. Romo Negreira[a], T. Fernandez Landaluce[a], J. Schuhmacher[a], L. Carbonell[a], K. Maex [a, b]

[a]IMEC, Leuven / Belgium

[b]ESAT Department K.U. Leuven / Belgium

Introduction

The trend towards increasing functional density in integrated circuits cannot be supported by traditional materials [1]. Device performance dictates a transition from SiO_2 to an insulator with lower dielectric constant (k) and Cu instead of Al and its alloys for lower resistance wiring [2, 3]. Such interconnect metallization requires the introduction of a barrier layer to prevent Cu diffusion under electrical bias. It is, however, difficult to obtain conformal barriers of the thicknesses (< 5 nm) foreseen in future device architectures without resorting to unconventional methods. To this end, the unique self-limiting and inherently conformal method of atomic layer deposition (ALD) of films from W, Ti. and Ta compounds is being investigated [4]. Based on sequential saturated gas phase-surface reactions, the growth rate in ALD is in theory layer-by-layer, controlled by the number of deposition cycles [5]. In practice, the early stages of film formation may be non-linear, involving three-dimensional growth depending on substrate reactivity [6]. The heterogeneous nature of the substrates of interest, low-k dielectric materials, further obscures the role of the initial surface on ALD-mediated growth mechanisms.

One approach to understanding the surface reactions relevant to metal deposition involves simplifying the substrate [7]. In previous work, the atomically controlled surface chemistry of self-assembled monolayers (SAMs) has been used as a means of studying Cu growth at organic surfaces relevant to interfacial adhesion enhancement and diffusion barrier development [8, 9]. Similarly, in this review, surface engineering using SAMs is demonstrated as a means of creating model substrates for ALD, a promising method of depositing diffusion barrier material for future technology nodes.

Formation and Characterization of Self-assembled Monolayers

SAMs are molecular assemblies formed by spontaneous adsorption of an active surfactant from the gas or liquid phase onto a solid substrate with appropriate bonding sites [10]. The first step in self-assembly is driven by bond formation of the functionalized headgroup onto specific substrate surface sites, *e.g.*, thiols on metals and semiconductors, acids on metal oxides, and silanes on hydroxylated surfaces. Lateral interactions between the hydrocarbon segments of the molecules, generally comprising all trans extended alkyl chains oriented close to the surface normal, causes them to pack densely resulting in the formation of a highly ordered layer. A single uniform terminal group, such as a methyl, phenyl, amine, carboxylic acid, or alcohol unit, is exposed at the monolayer–liquid or monolayer–gas interface. Thus SAMs provide a means of creating highly ordered surfaces of predefined structure and chemical composition [7–10].

Table 1. Name, chemical formula, abbreviation, and water contact angle (CA) measured for as-prepared SAMs of methyl-, cyano-, and bromo-terminated alkyltrichlorsilanes formed on silicon dioxide.

Name	Chemical formula	Abbreviation	CA°
Bromoundecyltrichlorosilane	$Br(CH_2)_{11}SiCl_3$	Br–SAM	86.6
11-Cyanoundecyltrichlorosilane	$CN(CH_2)_{11}SiCl_3$	CN–SAM	67.6
Octyltrichlorosilane	$CH_3(CH_2)_7SiCl_3$	CH_3–C_7–SAM	111.5
Decyltrichlorosilane	$CH_3(CH_2)_9SiCl_3$	CH_3C_9–SAM	113.4
Dodecyltrichlorosilane	$CH_3(CH_2)_{11}SiCl_3$	CH_3–C_{11}–SAM	112.8
Hexadecyltrichlorosilane	$CH_3(CH_2)_{15}SiCl_3$	CH_3–C_{15}–SAM	111.5
Octadecyltrichlorosilane	$CH_3 (CH_2)_{17}SiCl_3$	CH_3C_{17}–SAM	110.8

The present work focuses on the formation and properties of SAMs derived from methyl-, cyano-, or bromo-terminated alkyltrichlorsilanes (Table 1) adsorbed on oxidised silicon. Sample preparation was optimised with a view to developing a method suitable for 8-inch wafer level processing in a clean room environment. The SAMs were prepared by immersion of a SiO_2 substrate, < 1 nm SiO_2 on Si(100), in a dilute solution (10^{-3} M in toluene) of silane (used as received from Gelest Inc.) for 1 h at ambient temperature. Adsorption in the presence of a critical amount of physisorbed water facilitates hydrolysis of the trichlorosilane headgroup resulting in the replacement of chlorine atoms by hydroxyl groups (Figure 1). The hydroxyl groups of the self-assembling molecules interact with each other and with the surface hydroxyl groups on the substrate to form a silicon-oxygen-silicon network thus anchoring the organic layer with its hydrocarbon chains oriented close to the surface normal. The SAM modified wafers were then rinsed with copious amounts of solvent before being dried under nitrogen flow.

The as-prepared SAMs were of reproducibly high quality. They were macroscopically ordered with the expected termination as evidenced by water contact angle measurements. The values summarized in Table 1 are in excellent agreement with literature values [11, 12]. The well-defined chemical compositions of the SAMs were verified by atomic percentage determination using X-ray photoelectron spectroscopy (XPS). For example, in the case of the Br–SAM, C 1s photoemission at 285.1 eV can be attributed to the aliphatic methylene units with broadening towards higher binding energies due to a contribution at ~286.5 eV from the most electron deficient carbon atom bonded to bromine. The observation of Br $3d_{3/2}$ photoemission at 71 eV is in agreement with a previous study of a Br-terminated SAM [12, 13].

As shown schematically in Figure 1, Br–SAM adsorption on SiO_2 is predicted to take place through the hydrolysis of the silicon–chlorine bonds to form silicon–hydroxyl groups. In core level scans from 190 to 210 eV, we find no evidence of a Cl 2p peak. Its absence is an indication that all silicon–chlorine bonds have been hydrolyzed, thus favoring monolayer formation via silicon–oxygen–silicon bonding resulting from a condensation reaction between the Br–SAM hydroxyl groups and the oxidized silicon surface. The absence of chlorine also implies that precursor polymerization on the surface, a consequence of excess water, has not occurred.

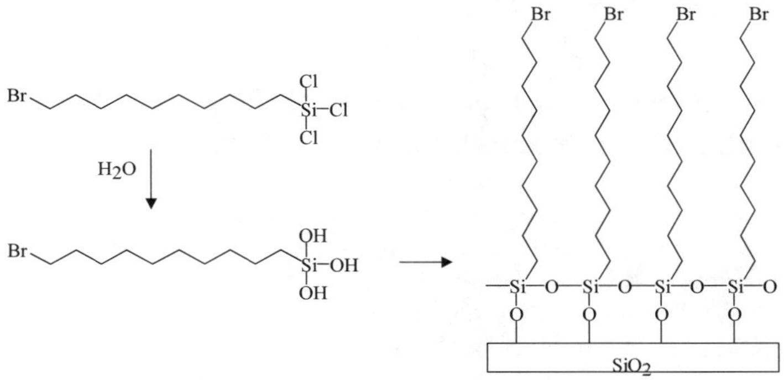

Figure 1. Schematic diagram of the formation of a bromo-terminated self-assembled monolayer.

Finally, in agreement with literature values for similar SAM surfaces, atomic force microscopy (AFM) images are generally featureless, with root-mean-square (RMS) roughnesses ranging from 0.093 nm for CH_3–C_{10}–SAM (which lies in between the RMS of 0.07 and 0.106 nm measured from hydrogen terminated-Si(100) and SiO_2 on Si(100), respectively) to 0.455 nm for CN–SAM [11].

Atomic Layer Deposition

An in-depth description of ALD can be found elsewhere [5]. The basic sequences of ALD for formation of a binary compound are shown schematically in Figure 2. While the principle remains the same, the growth of a WC_xN_y film involves a more complex, tertiary system requiring three precursors; triethylborane, tungsten hexaflouride (WF$_6$), and ammonia (NH$_3$). In our experiments, the deposition was performed using an ALCVDTM Pulsar 2000 reactor integrated with an automated wafer handling platform (ASM PolygonTM 8200). A precursor (mixed with a nitrogen carrier gas flow) pulse sequence of (C$_2$H$_5$)$_3$B, WF$_6$, and NH$_3$ represents one deposition cycle. Excess precursor gas was removed by flowing nitrogen after each precursor pulse. The deposition temperature was 300° C.

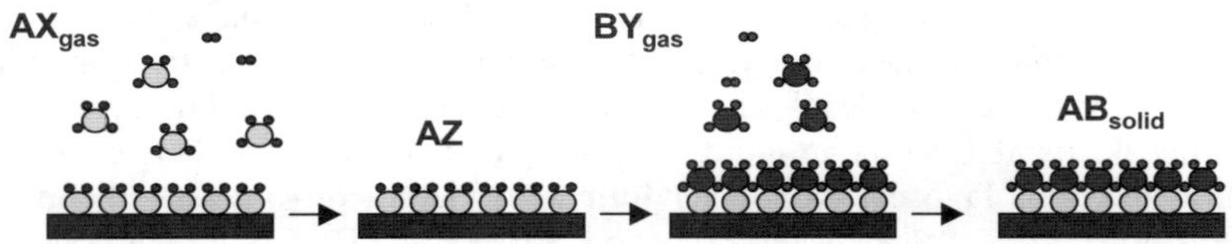

Figure 2. Basic sequences of ALD for an AB compound, from compound gas phase reactants AX$_{gas}$ and BY$_{gas}$ comprising the introduction of AX$_{gas}$ onto the substrate surface with formation of a chemisorbed layer AZ and, after a purging step to remove unreacted species and byproducts, the introduction of BY$_{gas}$ onto AZ surface with formation of AB$_{solid}$ surface layer, with release of byproducts.

Self-assembled Monolayer Compatibility with Atomic Layer Deposition Processing Temperature

Thermal desorption spectroscopy (TDS) was used to examine the compatibility of the SAMs with the processing temperature of 300°C required for ALD of WC$_x$N$_y$. Here we restrict our discussion to the thermal stability and decomposition pathway of Br-SAM. While only selected data (mass 68 desorption) is shown in Figure 3, our interpretation is based on the analysis of masses from 11 to 100 a.m.u. The decomposition process starts at 450°C as evidenced by desorption of high-molecular-weight hydrocarbon fragments created by alkyl chain carbon–carbon bond cleavage. A maximum in desorption intensity occurs at 550°C. After annealing to 650°C the monolayer has completely desorbed. These results are in excellent agreement with a previous study of the decomposition of alkyltrichlorosilane-derived monolayers on SiO$_2$ [14]. It has been suggested that above 540°C only methyl groups remain adsorbed implying that desorption of the SAMs is primarily the result of carbon–carbon rather than silicon–carbon bond cleavage [14]. While methyl groups can be considered to decompose via methyl radical or methane desorption, our analysis of masses 15 and 16 failed to reveal evidence of such reactions, suggesting decomposition to surface carbon and hydrogen.

TDS characterization of each of the SAMs indicates a thermal stability well within the limits of 300° C required for ALD processing. The decomposition process starts between 400 and 500° C with desorption maxima between 500 and 600° C depending on the SAM. Furthermore, all SAMs withstand multiple ALD cycles.

Atomic Layer Deposition on Self-assembled Monolayers

Having established the expected chemical composition, suitable thermal stability, and smooth surface morphology of the SAMs, they were then used as substrates for ALD. Multi-technique characterization including X-ray, fluorescence, mass, and electron spectroscopies and high-resolution microscopies were used to evaluate various ALD/SAM systems in terms of growth mechanism, chemical composition, interfacial structure and surface morphology.

XPS was used to monitor core level spectra as a function of increasing number of ALD cycles. Again, we focus on the results for Br–SAM as an example. A decrease in Br 3d intensity was observed to coincide with the appearance of W 4f and N 1s peaks assigned to formation of a WC_xN_y film. The fact that Br is still observed following 50 ALD cycles unambiguously confirms the compatibility of the SAM to the processing conditions. Obviously, the metal overlayer eventually attenuates photoemission from the SAM to the extent that the Br 3d signal can no longer be detected. Nevertheless, TDS analysis in Figure 1.3 shows that desorption characteristic of the as-prepared Br–SAM sample is maintained following 50, and even 200 (higher numbers were not analyzed), cycles of ALD.

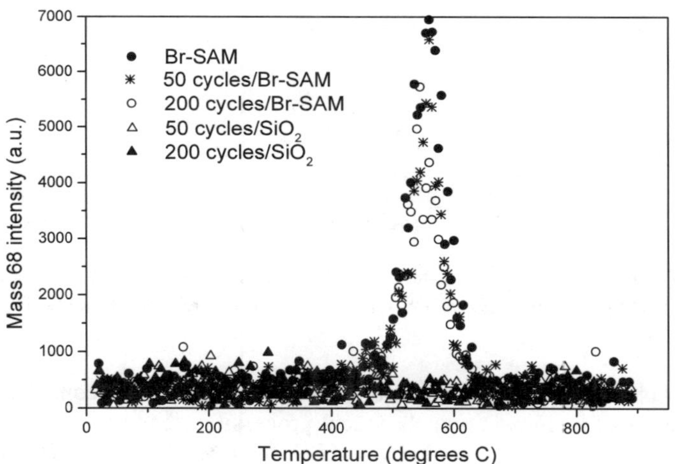

Figure 3. Thermal desorption spectra for mass 68 from Br–SAM before (●●●) and after deposition of 50 (***) and 200 (OOO) cycles of WC_xN_y ALD. The corresponding data for 50 (ΔΔΔ) and 200 (▲▲▲) cycles of ALD on SiO_2 are shown for comparison.

Figure 4 compares the X-ray fluorescence (XRF) determined tungsten content as a function of number of ALD cycles on Br–SAM, CN–SAM, and CH_3–C_{17}–SAM. Clearly, the offset from linearity varies depending on the SAM substrate. On CN–SAM and Br–SAM, WC_xN_y film closure (estimated from sheet resistance, contact angle measurements, and XRF derivative curves) occurs between ~100 and ~150 ALD cycles, respectively. For a 200 cycle deposition on Br-SAM, 14.1 CPS corresponding to 3.8×10^{16} atoms of tungsten are observed compared with 20.1 CPS or 5.4×10^{16} atoms on CN–SAM. The corresponding thicknesses estimated by ellipsometry of 10.3 and 14.2 nm, respectively, imply the growth of a lower density film on the Br-terminated compared with the cyano-terminated surface. These differences translate to material properties that influence the electrical performance of the film as evidenced by resistivities (ρ) of 504 and 444 $\mu\Omega$ cm, respectively. However, substrate-induced differences in the early stages of film growth eventually disappear as the coverage increases. Following 500 cycles of ALD, WC_xN_y films of similar thicknesses, 32.1 and 35 nm with similar ρ values of 360 and 347 $\mu\Omega$ cm are found for films grown on Br–SAM and CN–SAM, respectively.

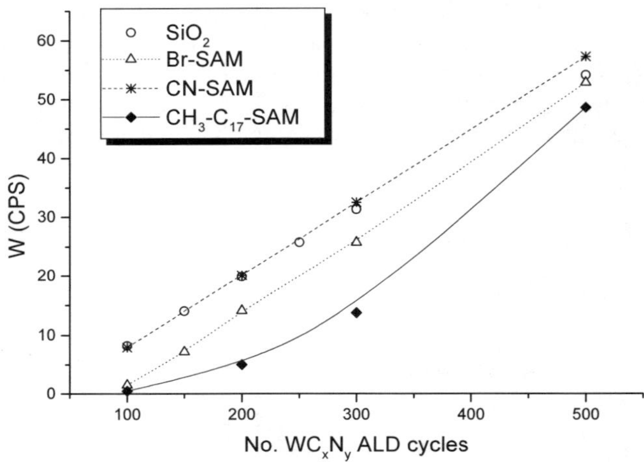

Figure 4. Tungsten content determined by XRF following ALD of WC_xN_y on Br–SAM ($\triangle\triangle\triangle$), CN–SAM (***) and CH_3–C_{17}–SAM ($\blacklozenge\blacklozenge\blacklozenge$). The corresponding data for SiO$_2$ ($\circ\circ\circ$) is shown for comparison.

The morphology of the diffusion barrier layer must also be considered, in particular in terms of its impact on subsequent processing steps such as Cu seed layer growth. Following 200 cycles of WC_xN_y ALD, the RMS for the Br–SAM increases from 0.272 to 0.879 nm. In contrast, on CN–SAM the RMS value increases only slightly from 0.455 to 0.497 nm.

The Influence of Alkyl Chain Length

CH_3–C_{17}–SAM results in markedly different growth with WC_xN_y film closure requiring ~400 deposition cycles. Upon examination of the WC_xN_y/CH_3–C_{17}–SAM interface we find that it is microscopically rough and discontinuous compared with an atomically sharp interface on CN–SAM, as revealed by transmission electron microscopy (TEM) and time-of-flight secondary ion mass spectrometry depth profiling analysis. Furthermore, while the as-prepared CH_3–C_{17}–SAM shows a relatively low RMS value of 0.173 nm, it gives rise to the roughest WC_xN_y film, *e.g.* RMS of 4.728 nm following 200 cycles of ALD. As this system varies considerably from CN–SAM and Br–SAM in terms of alkyl chain length, this raises the question of the influence of chain length on the observed growth behaviour. In Figure 5 we show a series of tungsten growth curves for methyl-terminated SAMs with different alkyl chains lengths. The offset from linear growth increases with increasing chain length. However, this influence is not exclusive as seen by comparing CN–SAM (comprising 11 CH_2 units) with CH_3–C_{11}–SAM. In fact, the growth is similar to that observed on CH_3–C_7–SAM rather than CH_3–C_{11}–SAM. Evidently, the growth is enhanced on the cyano-terminated SAM. This implies that while the chain length does influence growth, the terminal group structure and chemistry is also important.

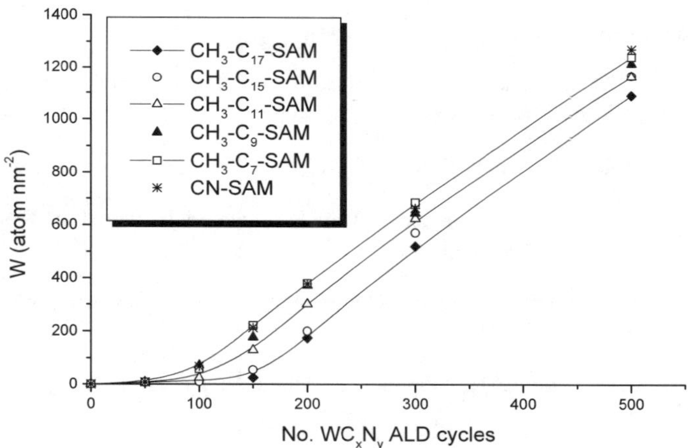

Figure 5. Tungsten content determined by XRF following ALD of WC_xN_y on CH_3–C_{17}–SAM (♦♦♦), CH_3–C_{15}–SAM (○○○), CH_3–C_{11}–SAM (ΔΔΔ), CH_3–C_9–SAM (▲▲▲), and CH_3–C_7–SAM (□□□). The corresponding data for CN-SAM (***) is shown for comparison.

Conclusion

The role of the initial surface on ALD of WC_xN_y can be investigated using SAMs as model substrates. SAMs provide films of known chemical composition with macroscopically well-ordered surfaces. With thermal stability extending from 500–600° C, such substrates withstand multiple cycles of ALD at 300°C. Comparison of WC_xN_y grown on SAMs with methyl-, bromo-, and cyano-terminal groups and alkyl chain lengths ranging from 7 to 17 methylene units shows differences in tungsten coverage, film thickness, density, resistivity, and interfacial and surface structure. The terminal group and the alkyl chain length both influence the observed growth behaviour. The cyano-terminated SAM is most promising for WC_xN_y growth. Overall, this study demonstrates that via tunable structure and surface chemistry, SAMs represent a novel approach to surface engineering and provide model substrates for atomic layer deposition.

Acknowledgements

This work was carried out within IMEC's Industrial Affiliation Cu/low-k BEOL Program.

References

1.   International Technology Roadmap for Semiconductors (ITRS), Semiconductor International Association (2003), http://public.itrs.net/.

2.   P. Murarka, Mater. Sci. Eng. R19, 87 (1997).

3.   K. Maex, M.R. Baklanov, D. Shamiryan, F. Lacopi, S.H. Brongersma, Z.S. Yanovitskaya, J. Appl. Phys. 93, 8793 (2003).

4.   D. Volger, P. Doe, Sol. State Technol. 46, 35 (2003).

5.   M. Ritala, M. Leskelä, Handbook of Thin Film Materials, edited by H.S. Nalwa, 1, 103-159 Academic Press, San Diego, (2002).

6.   A. Satta, A. Vantomme, J. Schuhmacher, C.M. Whelan, V. Sutcliffe, K. Maex, Appl. Phys. Lett. 84, 4571 (2004).

7.   G.L. Fisher, A.E. Hooper, R.L. Opila, D.L. Allara, N. Winograd, J. Phys. Chem. B 104, 3267 (2000).

8.   P.G. Ganesan, A.P. Singh, G. Ramanath, Appl. Phys. Lett. 85, 579 (2004).

9.   C.M. Whelan, J. Ghijsen, J.-J. Pireaux, K. Maex, Thin Solid Films 464–465, 388 (2004).

10. F. Schreiber, Prog. Surf. Sci. 65, 151 (2000) and references therein.

11. P.E. Laibinis, G.M. Whitesides, J. Am. Chem. Soc. 114, 1990 (1992).

12. M.-T. Lee, G.S. Ferguson, Langmuir 17, 762 (2001).

13. C.M. Whelan, A.-C. Demas, J. Schuhmacher, L. Carbonell, K. Maex, Mater. Res. Soc. Symp. Proc. 812, F2.2.1 (2004).

14. G.J. Kluth, M.M. Sung, R. Maboudian, Langmuir 13, 3775 (1997).

Selective Airgaps: Towards a Scalable Low-k Solution

J. P. Gueneau de Mussy [a, b], G. Beyer [a], K. Maex [a, b]

[a] IMEC, Leuven / Belgium

[b] ESAT Department K.U. Leuven / Belgium

Introduction

As integrated circuit (IC) technology shrinks, the interconnect is becoming a significant restricting factor in terms of dynamic power consumption [1], propagation delay, and signal errors [2]. At the local interconnects level, where metal lines have the smallest dimensions, the propagation delay is dominated by the interline capacitance [3]. As a consequence, materials with lower permittivity (κ) than that of SiO_2 ($\kappa{\sim}4$) [4] are introduced as intra-metal dielectrics (IMD). The integration of low-κ ($3<\kappa<4$) and ultra low-κ ($3<\kappa<2$) [5] materials into production processes creates challenges associated mainly with (i) diffusion of reactive chemical species into the porous matrix [7], leading to localized increase of the low-κ permittivity [6] and (ii) adhesion, making these materials difficult or even impossible to integrate in future processes. Consequently, a strong interest exists to integrating air ($\kappa{=}1$), having the lowest existing permittivity, between adjacent metal lines.

Nevertheless, in order for airgaps to be a real technological alternative to inter-metal dielectrics two main conditions should be satisfied:

1. Airgap integration should be scalable, *i.e.* compatible with future technology generations. This supposes a sustained reduction of the IMD κ value and a very limited degradation of the interconnects' reliability.

2. Air cavities should be placed only where really required, *i.e.* between narrowly spaced metal lines. This implies leaving the rest of the interconnects surrounded by dielectric material, which insures the stack's mechanical stability.

Precisely placing airgaps between narrowly spaced interconnects at the local level has been explored by various research teams.

One approach is to use an extra lithography step followed by reactive ion etch (RIE) and non-conformal dielectric deposition. This method has been explored with the aim of airgap integration between aluminum [8–10] and copper interconnects [11–14].

Another approach relies on removing the IMD layer by RIE after chemical mechanical polishing (CMP) followed by non-conformal dielectric deposition and planarization operations [15–16]. It is important to note that the aforementioned processes rely on various extra microprocessing steps making them complex and costly. In addition, these approaches are hardly scalable without compromising their electrical performance.

A third possibility to precisely embed airgaps is by inducing chemical selectivity. This approach relies on changing the structural properties of one fraction of the dielectric making it reactive to a certain chemical compound, while the rest of the material remains inert to it [17–19]. Delicate compositional changes can be induced in dielectrics containing carbon and silicon by means of oxygen plasma exposure. Subsequently, the whole stack is exposed to a suitable chemistry that removes the modified material, thus generating air cavities.

Scalable Approach

The scalability of an IMD material can be easily extracted by means of the simple analytical parallel plate capacitor model. This model determines the fraction of the dielectric that has been modified having a different κ value from the bulk material [17, 20] and is summarized by the following formula:

$$\frac{1}{C_{pp}} = \frac{1}{\varepsilon_0 k \, A}(S - S')$$

where C_{pp} is the parallel plate capacitance contribution, ε_0 is the permittivity constant, k is the native κ value of the IMD, A is the capacitor's area, S stands for the spacing between two adjacent metal lines and S' is the electrical equivalent of the actual modification.

If $1/C_{pp}$ shifts downwards the material suffers from an increase in capacitance. This is typically the case of low-κ materials undergoing diffusion of reactive chemical species into the porous matrix or sidewall damage as shown by the black arrow in Figure 1. For selective airgaps $1/C_{pp}$ shifts upwards, meaning a decrease in capacitance as compared to low-κ materials (indicated by the white arrow in Figure 1). When air is incorporated between the metal lines, the IMD permittivity will decrease as the inter-metal spacing shrinks eventually reaching 1, when full air cavities are formed between the copper lines (intersection of dashed $k_{\text{low-}\kappa}$ line and plain k_{air} line in Figure 1).

The aforementioned model clearly illustrates the advantage in performance of integrating airgaps as compared to available low-κ materials in future interconnect generations.

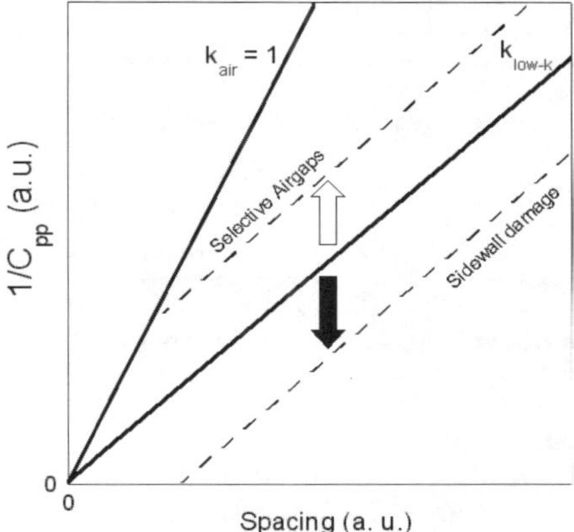

Figure 1. Plot representing the inverse of the parallel plate capacitance ($1/C_{pp}$) versus the distance between two adjacent interconnects (Spacing).

Process Considerations

As discussed previously, the selective airgap process requires (i) local change of the IMD's chemical composition and (ii) selective dissolution of the modified dielectric.

Modification of the dielectric material at the nanolevel can be achieved by carefully tuning the resist strip operation after the patterning process in a damascene stack. This modified dielectric must also react selectively with a chemical species in order to be easily etched away after chemical mechanical polishing. Films containing silicon and carbon compounds satisfy conditions (i) and (ii) when exposed to an HF solution. Examples of such films are standard microfabrication materials from the organo-silicon glass (OSG) family, such as SiOC:H and SiC.

Two approaches for selective airgaps formation at the interconnect level have been filed so far [22, 23]. The first filed approach considers employing a SiC hard mask where the oxidized SiC sites at the SiC/TaN interface act as preferential pathways for diffusion of HF and etching of the underlying SiO_2 dielectric (see Figure 2). The second one consists of employing SiOC:H as an IMD material. The carbon-depleted sidewalls of the SiOC:H are then selectively dissolved by HF leaving the bulk unmodified SiOC:H material in place (see Figure 3).

A common feature of both approaches is the generation of a defective, silicon oxide film (SiO_x). This low cross-linked film is dissolved when exposed to HF at a higher rate than SiOC:H and SiC [24], acting as a sacrificial layer.

In the first selective airgap approach mentioned, SiO_x results from the oxidation of the Si–C–Si bonds existing in the SiC film during oxygen-containing plasma exposure in the following way [21]:

$$\equiv\!Si\text{-}CH_n\text{-}Si\!\equiv\ +\ (2 + n/2)O \rightarrow\ \underbrace{\equiv\!Si\text{-}O\text{-}Si\!\equiv}_{SiO_x}\ +\ CO\ +\ (n/2)H_2O$$

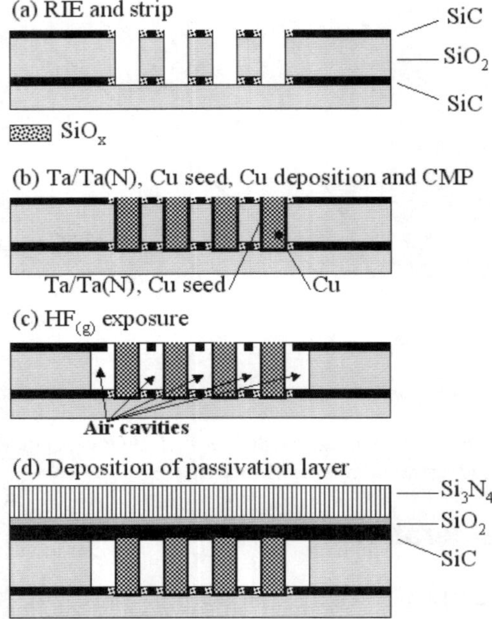

Figure 2. First approach for selective airgaps based on local SiC oxidation.

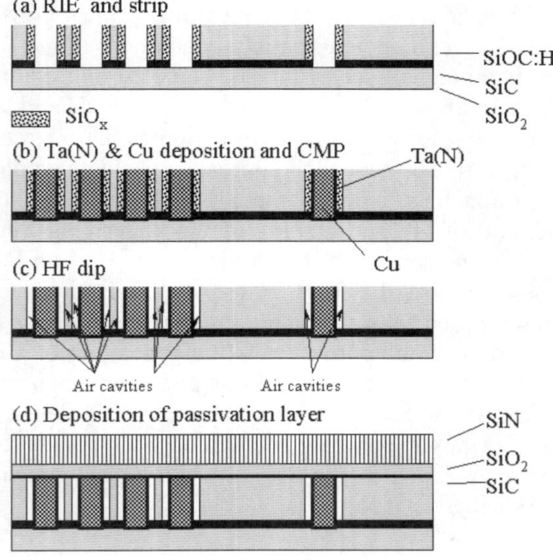

Figure 3. Second approach for selective airgaps based on local SiCO:H oxidation.

For the second selective airgap approach, the methyl groups from the SiOC:H layer are oxidized in the oxygen-containing plasma leading to the formation of SiO_x as follows [21]:

$$\equiv Si\text{-}CH_3 + H_3C\text{-}Si\equiv + 7O \rightarrow \equiv Si\text{-}OH + HO\text{-}Si\equiv + 2CO$$

$$+ 2H_2O$$

$$\equiv Si\text{-}OH + HO\text{-}Si\equiv \rightarrow \underbrace{\equiv Si\text{-}O\text{-}Si\equiv}_{SiO_x} + H_2O$$

The removal of the sacrificial material should be carried out with care in order to avoid attacking neighboring materials and moisture adsorption within the air cavities. Moisture leads to an increase in the IMD κ value, while a degradation of neighboring materials may lead to a lower reliability. Suitable systems to achieve SiO_x removal inside small dimensions are anhydrous HF/CH_3OH and anhydrous HF / supercritical CO_2 mixtures. Figure 4 gives an example of (a) full and (b) partial airgaps obtained from the aforementioned approaches.

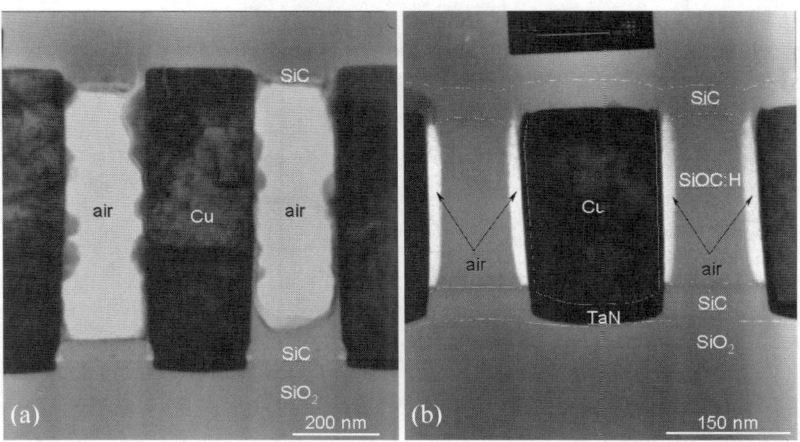

Figure 4. Selective airgaps obtained by the first (a) and second (b) approach respectively. See text for details.

Performance

The electrical performance of airgaps greatly depends on the fraction of cross-sectional air area. The higher the proportion of air encapsulated between the copper lines the smaller the interline capacitance. Unfortunately, incorporating full airgaps between copper lines may lead to an increase in leakage current and to degradation of the mechanical properties of the integrated stack. As a consequence, a compromise is required to get a relatively good electrical performance together with a high mechanical and electrical reliability.

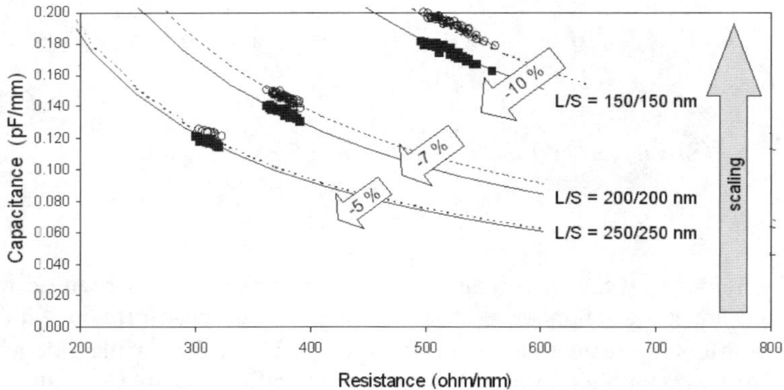

Figure 5. Drop in resistance × capacitance (RC) delay due to incorporation of airgaps.

From this perspective the selective airgap approach relying on SiOC:H damage by plasma exposure gives the best scenario. This alternative gives a capacitance drop, which increases as dimensions scale down (Figure 5), without compromising the system's reliability as shown in Figure 6.

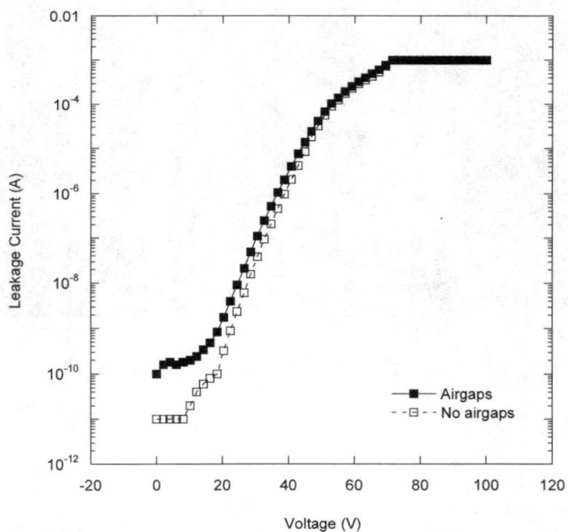

Figure 6. Leakage versus voltage curves at 100° C.

Conclusions

Unless a dramatic improvement in the properties of low-κ materials takes place, selective airgaps remain an attractive method for further decreasing the dielectric permittivity in future technology generations. Selective airgaps have the great advantage of being scalable, but their good electrical performance goes together with slightly higher leakage currents.

Acknowledgements

We thank Dr. Olivier Richard (Materials Components Analysis group), Rudy Caluwaerts and Matthieu Sales (Interconnect Modules and Integration) for the sample characterization. We are also grateful to IMEC's pilot line for the wafer processing.

References

1. J.M. Rabaey, in Digital Integrated Circuits: A Design Perspective, 1St. ed. (Ch. 3) p. 144, Prentice-Hall, Upper Saddle River, New Jersey, USA (1996).

2. M. Stucchi and K. Maex, IEEE Trans. Instrum. Meas. 51, 537 (2002).

3. J. M. Rabaey, A. Chandrakasan and B. Nikolic., in Digital Integrated Circuits: A design perspective, 2nd ed., Prentice-Hall, Example 5.10 (2003).

4. International Technology Roadmap for Semiconductors (ITRS), Semiconductor International Association, http://public.itrs.net/ (2003).

5. K. Maex , M.R. Baklanov, D. Shamiryan, F. Iacopi, S.H. Brongersma and Z.S. Yanovitskaya. J. Appl. Phys. 93, 8793, Table IV, p. 8837 (2003).

6. M. Lepage, D. Shamiryan, M. Baklanov, H. Struyf, G. Mannaert, S. Vanhaelemeersch et al., Proceedings of the International Interconnects Technology Conference, p. 174 (2001).

7. E.T. Ryan, J. Martin, K. Junker, J.J. Lee, T. Guenther, J. Wetzel et al., Proceedings of the International Interconnects Technology Conference, p. 27 (2002).

8. B. Shieh, K.C. Saraswat, J.P. Mc Vittie, S. List, S. Nag, M. Islamraja et al., IEEE Electron. Device Lett. 19, 16 (1998).

9. B. Shieh, L.C. Bassman, D.-K. Kim, K.C. Saraswat, M.D. Deal, J.P. Mc Vittie, R.S. List, S. Nag and L. Ting, Proceedings of the International Interconnects Technology Conference, p. 125 (1998).

10. B. Shieh, M.D. Deal, K.C. Saraswat, R. Choudhury, C.-W. Park, V. Sukharev, W. Loh, P. Wright in Proceedings of the International Interconnects Technology Conference, p. 203 (2002).

11. O. Demolliens, P. Berruyer, Y. Morand, C. Tabone, A. Roman, M. Cochet *et al.*, Proceedings of the International Interconnects Technology Conference, p. 198 (1999).

12. O. Demolliens, Y. Morand, M. Fayolle, M. Cochet, M. Assous, H. Felids *et al.*, Proceedings of the International Interconnects Technology Conference, p. 276 (2000).

13. V. Arnal, J. Torres, P. Gayet, R. Gonella, P. Spinelli, M. Guillermet *et al.*, Proceedings of the International Interconnects Technology Conference 2001, p. 298.

14. Z. Gabric, W. Palmer, G. Schindler, W. Steinhogl and M. Traving Proceedings of the International Interconnects Technology Conference 2004, p. 151.

15. J. Noguchi, T. Fujiwara, K. Sato, T. Nakamura, M. Kubo, S. Uno, K. *et al.*, Proceedings of the International Interconnects Technology Conference 2003, p. 68.

16. J. Noguchi, K. Sato, N. Konishi, S. Uno, T. Oshima, U. Tanaka *et al.*, Proceedings of the International Interconnects Technology Conference 2004, p. 81.

17. J.P. Gueneau de Mussy, O. Richard, G. Beyer and K. Maex, Electrochem. Solid State Lett., in press.

18. J.P. Gueneau de Mussy, G. Beyer and K. Maex, Proceedings of the MAM Conference, Brussels, March (2004).

19. J.P. Gueneau de Mussy, O. Richard, G. Beyer and K. Maex, Electrochem. Solid State Lett., in press

20. F. Iacopi, M. Stucchi, O. Richard and K. Maex, Electrochem. Solid-State Lett., 7, G79 (2004).

21. T. Furusawa, D. Ryuzaki, R. Yoneyama, Y. Homma, Electrochem. Solid State Lett., 4 (3) G31–G34 (2001).

22. J.P. Gueneau de Mussy, G. Beyer and K. Maex, patent US 0074960 A1 and EP 1 521 301 A1, April 2005.

23. G. Beyer, J. P. Gueneau de Mussy and K. Maex, patent US 0074960 A1 and EP 1 521 302 A1, April 2005.

24. M. Knotter, J. Am. Chem. Soc. 122, 4345 (2000).

Silicides – Recent Advances and Prospects

J. A. Kittl[c], A. Lauwers[a], O. Chamirian[a], M. A. Pawlak[a], M. Van Dal[d], A. Veloso[a], K. G. Anil[a], G. Pourtois[a], M. De Potter[a], K. Maex[a, b]

[a]IMEC, Leuven / Belgium

[b]ESAT Department K.U. Leuven / Belgium

[c]Texas Instruments Inc., Leuven / Belgium

[d]Philips Research Leuven / Belgium

Introduction

Silicides have been used in self-aligned (SALICIDE) processes for several genera-tions of complementary metal-oxide semiconductor (CMOS) devices, to reduce the sheet resistance and provide stable ohmic contacts with low contact resistivity on gate, source and drain areas [1–9]. In the SALICIDE process, a deposited metal layer is reacted thermally (typically using rapid thermal processing, RTP) with ex-posed Si on gate, source and drain areas to form silicide films. Unreacted metal and capping layers are removed in a wet etch which is selective to the grown silicide, leaving silicide films on the gate, source and drain areas and no metallic films on isolation and spacers. In one-step RTP processes, the thermal budget of the single RTP step used is high enough to obtain the target silicide phase. In contrast, in two-step RTP processes, a milder thermal budget is used in a first RTP step (RTP1) after which the silicide films are typically in a high resistivity phase. Selec-tive removal of unreacted metal and capping layers follow, after which a second RTP step is used to transform the silicide into the target low resistivity phase.

One of the key considerations for SALICIDE processes is the ability to silicide small structures achieving low sheet resistance. This is particularly important for narrow poly-Si lines, especially in designs where they serve also as local intercon-nects, since their sheet resistance may impact circuit speed. This consideration has been the main reason behind the migration from Ti to Co [1, 4-6] and now towards Ni silicides [8] as lateral circuit dimensions were scaled. For silicidation of source and drain (S/D) areas, low contact resistivity and low leakage on shallow junctions are key requirements, but low sheet resistance is also needed. As junction depths continue to be scaled down, the thickness of silicide films is also reduced. Silicides with low Si consumption, smooth interfaces and low processing temperature to avoid dopant deactivation are desired for S/D applications. Use of selective growth processes for raised S/D implementations may relax these constraints [5]. Silicide growth on spacers and isolation and associated bridging issues, ability to selec-tively etch the metal without removing the silicide, sensitivity to oxidation and to interface conditions, thermal stability and impact on reliability are also important considerations for integration of silicides into CMOS technologies.

The list of silicides considered traditionally for SALICIDE applications is not very extensive. $TiSi_2$ in the C54 phase, $CoSi_2$ and NiSi have been the most investigated [1–13], being the silicide phases with lowest resistivity (15–20 $\mu\Omega\times cm$). Co-based [8, 14] and Ni-based [8, 9, 15] alloy silicides have also received attention in the past few years. For future generations, metal gates are expected to replace partially silicided poly-Si gates in order to eliminate depletion issues [5], and fully silicided gates are one of the main metal gate candidates [9, 16, 17]. For this application, the work function of the silicide and its stability in contact with the gate dielectric become critical properties to be considered. For source and drain silicidation, compatibility with Si-Ge will need to be addressed for many of the strained channel device approaches.

Scalability and Nucleation Density

One of the most challenging issues confronted through several CMOS generations has been the scalability of silicide processes as device dimensions are shrunk. Reduction of device lateral dimensions has forced a migration of materials. $TiSi^2$ in the low resistivity C54 phase was the material of choice at the 0.35-μm technology node. C54 $TiSi^2$ grows by transformation from the high resistivity C49 $TiSi^2$ phase in a nucleation-controlled process. The low nucleation density of this transformation made it unpractical for applications in which device dimensions were smaller or similar to the typical distance between C54 nuclei [2, 3]. $TiSi^2$ on small devices (< 0.25 μm) remained predominantly in the high resistivity C49 phase resulting on a linewidth dependent sheet resistance ("linewidth effect"). Rapid thermal processing, pre-amorphization [3] and addition of metallic impurities such as Mo [10, 11] were used to extend the capability to smaller linewidths, but eventually the industry moved towards $CoSi_2$ [1, 4-6], which is currently widely used in production at the 130-nm node. In the last few years, however, it became apparent that Co silicide in turn has a linewidth effect, which is observed as linewidths are scaled below 40 nm [8] (Figure 1a).

Comparison of wide and narrow lines after a typical Co silicidation process showed evidence of agglomeration on narrow lines. However, the root cause of the sheet resistance increase at small linewidths is similar to that for the $TiSi_2$ case. Transformation from high resistivity CoSi to low resistivity $CoSi_2$ becomes more difficult as feature sizes are reduced [12] (Figure 1b). This in turn can be attributed to low nucleation density: $CoSi_2$ nucleation sites are scarce on features in the 40 nm length scale. At higher temperatures, CoSi films on narrow lines agglomerate rather than transform to $CoSi_2$ (Figure 2) Addition of Ni was proposed as a way to reduce the nucleation temperature of the di-silicide phase [14]. However, Co–Ni alloy silicides were also found to have similar linewidth effects to $CoSi_2$ (Figure 1a).

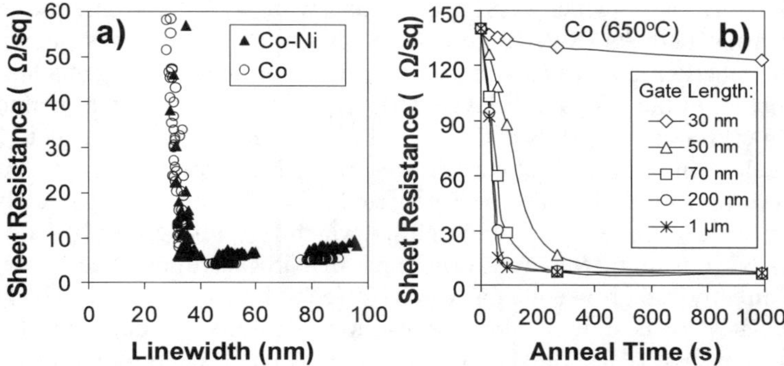

Figure 1. a) Sheet resistance as a function of linewidth for Co and Co–Ni silicided n+ gates. b) Sheet resistance vs. anneal time at 650°C of CoSi/poly-Si gates. Silicide films on 30-nm gates remain predominantly in the high resistivity CoSi phase, while for longer gate lengths they transform to low resistivity $CoSi_2$.

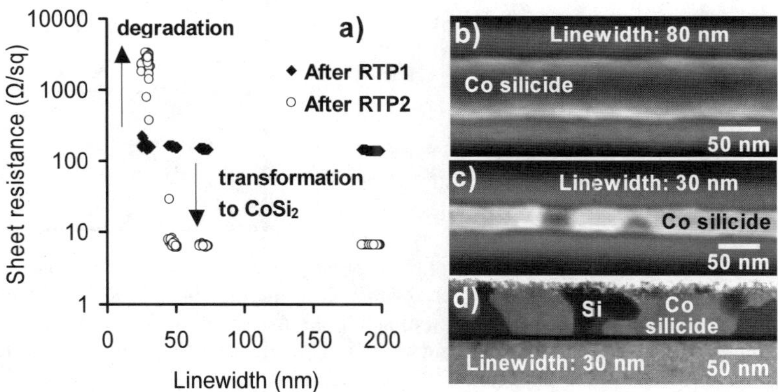

Figure 2. a) Sheet resistance vs. linewidth for Co silicided gates. Sheet resistance after the first RTP step corresponds to CoSi. A second RTP step results in transformation to $CoSi_2$ for wider lines and in agglomeration for narrower lines. b), c) top view and d) cross section along the line after two-step RTP Co silicidation. Agglomeration and inversion is observed on the 30-nm lines, while Co silicide films on the 80-nm lines remain smooth.

Ni-based Silicides

One of the main reasons for which NiSi is now considered as the main candidate for scaled CMOS applications is its ability to silicide small structures effectively (Figure 3). Nucleation issues have not been observed at linewidths down to 25 nm.

The target (low resistivity) phase for applications in the Ni–Si system is NiSi. In the Ni/Si reaction, metal rich phases are formed first. Ni_2Si grows with diffusion limited kinetics, after which NiSi grows, also with diffusion-limited kinetics, at higher thermal budgets [8-9] (Figure 4). Ni is the main moving species in both cases. The reaction kinetics can be affected by the presence of dopants and by the microstructure of the reacting Si layer. They were found to be faster on amorphous Si than on crystalline Si [8]. N and B were found to slow down the growth kinetics [8, 9]. Nucleation of $Ni_{31}Si_{12}$ at early stages of the Ni/Si reaction and of Ni_3Si_2 before the growth of NiSi has also been reported [13]. The thermodynamically stable Ni silicide phase on a Si substrate is $NiSi_2$, which typically grows by transformation from NiSi at high temperatures. Grains of $NiSi_2$ have also been observed to nucleate directly and grow epitaxially on (100) Si at low temperatures [7-9] (Figure 5), particularly for B and BF_2 doped substrates, resulting on rough interfaces that can induce leakage issues on shallow junctions. A shallow pre-amorphization implant can be used to avoid the epitaxial nucleation of $NiSi_2$ [9].

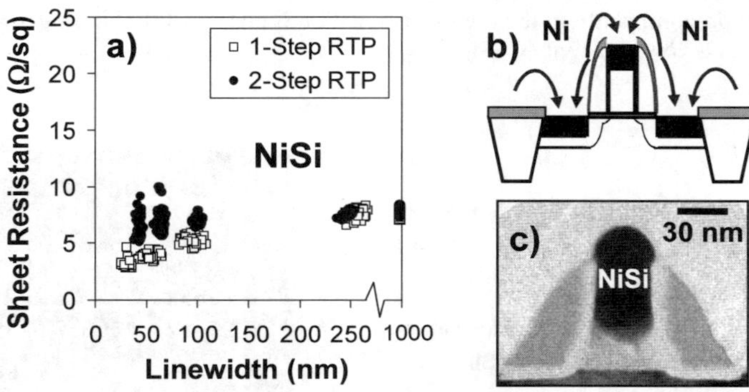

Figure 3. a) Sheet resistance vs. linewidth for Ni-silicided n+ gates, comparing a one-step RTP process to a two-step RTP process with a low thermal budget RTP1, b) transistor schematic representing excess silicidation of small features by Ni diffusion from Ni films on surrounding areas, and c) excessive silicidation: a thick silicide film grown on a narrow gate from a 10-nm-thick Ni film (expected NiSi thickness of 22 nm for planar reaction).

Although NiSi forms readily on small structures (at features sizes down to 25 nm), geometry effects need to be considered. A reverse linewidth effect (decreasing sheet resistance with decreasing linewidth) has been observed when a single RTP step is used for silicidation [8, 9] (Figure 3a). This is caused by excessive silicidation (growth of thicker silicides) on small structures due to Ni diffusion from films on surrounding isolation or spacer areas (Figure 3b, c). Excessive silicidation can be detrimental for devices, on small S/D areas it can result in higher junction leakage and on narrow gates it can result in local full silicidation.

A way of controlling Ni silicidation on small features is by using a two-step RTP silicidation process with a low thermal budget first RTP step [8, 9] (Figure 3a). In the first RTP step the reaction and silicide thickness are limited kinetically, obtaining a Ni-rich silicide film of controlled thickness. Excess Ni is then removed in a selective etch, followed by a higher temperature anneal to transform or complete the transformation into NiSi. The control of excessive silicidation by a low thermal budget RTP1 can also result in lower junction leakage, as shown in Figure 6a for small n+/p island diodes.

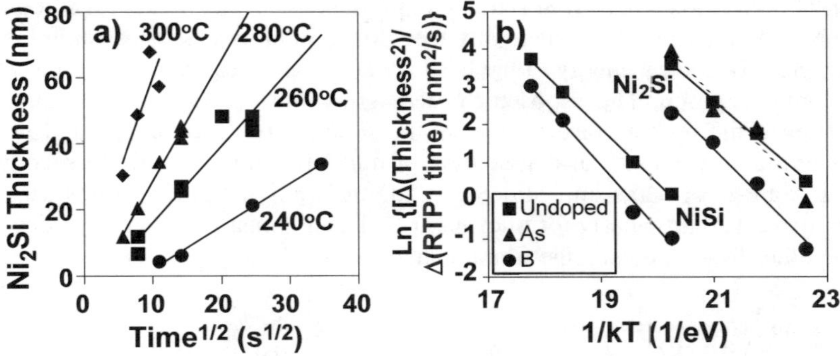

Figure 4. a) Ni_2Si film thickness vs. square root of RTP anneal time for the reaction of Ni with polycrystalline Si showing diffusion limited growth kinetics. Similarly, diffusion limited growth kinetics were obtained for the growth of NiSi on poly-Si and for the growth of both Ni_2Si and NiSi on (100) Si. b) Arrhenius plots of the Ni/poly-Si reaction kinetics showing the effect of dopants on the growth of Ni_2Si and of NiSi.

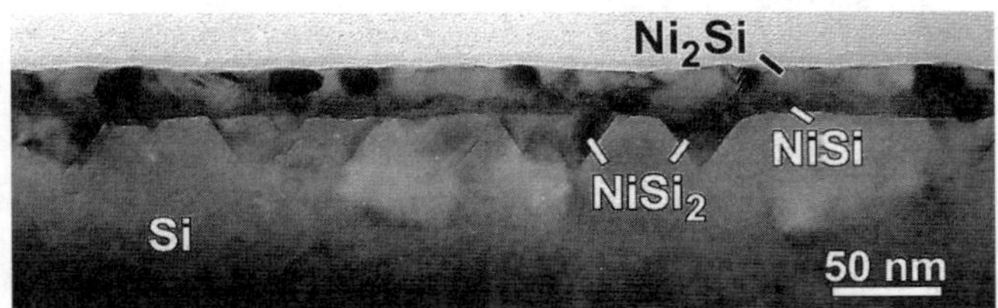

Figure 5. TEM cross section of the reaction of Ni with p+ (100) Si at 310°C showing the presence of epitaxial $NiSi_2$ grains.

Another advantage of NiSi over $CoSi_2$ is its lower contact resistance (Figure 6b), which can be attributed in part to its lower processing temperature that prevents dopant deactivation, an issue found for higher temperature $CoSi_2$ processes.

One of the main problems found for implementation of NiSi is its low thermal stability [8, 9, 13] (Figures 7, 8 and 9). NiSi films can degrade under thermal treatments by two mechanisms: morphologically (agglomeration) or by transformation to $NiSi_2$. The first is characterized by a continuous increase in sheet resistance with anneal time (Figure 7a). Scanning electron microscopy (SEM) inspection confirmed that films with this type of sheet resistance evolution degraded morphologically, an example is shown in Figure 9c. It was found that thin NiSi films (few tens of nm) degrade morphologically while still in the monosilicide phase. This was verified by transmission electron microscopy (TEM) and X-ray diffraction (XRD). After agglomeration, the surface of the samples remains quite flat, with NiSi grains alternating with exposed Si that has re-grown to the surface (Figure 9b). In contrast, the NiSi-Si interface becomes very rough. This suggests that the silicide-Si interface energy is significantly lower than the silicide surface energy. The degradation of NiSi films on (100) Si starts with grain boundary grooving predominantly at the NiSi/Si interface. This stage of the agglomeration process is driven by reduction of grain boundary energy in the NiSi films. As grooves reach the film surface, Si becomes exposed. The exposed Si areas grow subsequently driven mainly by reduction of the high surface energy of the silicide. The mechanism and driving forces at the initial stages of degradation of thin NiSi films on poly-Si seem to be similar to those of films on (100) Si. In addition, poly-Si grain growth seems to also play a role, particularly for thicker films. The dominant driving force depends, among other factors, on silicide film thickness and poly-Si grain size.

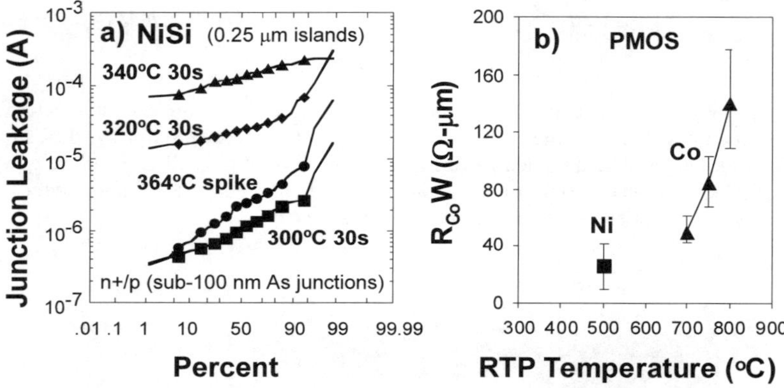

Figure 6. a) Junction leakage distributions for active area n+/p island diodes after two-step RTP Ni silicidation. Anneal conditions indicated correspond to the first RTP step. Low leakage is obtained with moderate RTP1 thermal budgets, which control the silicide thickness. b) Silicide to p+ Si contact resistance as a function of temperature of the second RTP step for Ni and Co silicides. High processing temperatures used in Co silicidation result in dopant de-activation in the underlying junction and contact resistance degradation.

Thicker NiSi films (~60 nm) annealed isothermally by RTP at high temperatures degrade by transformation to $NiSi_2$, while at lower temperatures they degrade first morphologically [8, 9]. Films that degraded by transformation to $NiSi_2$ showed an initial increase in sheet resistance with anneal time, after which the sheet resistance remained stable upon further annealing at a value of about twice the initial sheet resistance (Figure 7b).

Figure 8 shows Arrhenius plots of degradation times, τ (20% increase in sheet resistance), for NiSi films. The mechanism of degradation was agglomeration except for data points labeled as transformation to $NiSi_2$. Degradation times were longer for thicker films, indicating thermal stability improves with increasing film thickness. Activation energies for agglomeration are ~2.4 eV (100) Si and ~3 eV on poly-Si. The activation energy for the NiSi to $NiSi_2$ transformation is estimated to be higher than that of agglomeration. This would explain the observation that thick NiSi films degrade morphologically at low temperatures, and by transformation to $NiSi_2$ at higher temperatures. Transformation to $NiSi_2$ was observed at lower temperatures on poly-Si than on (100) Si.

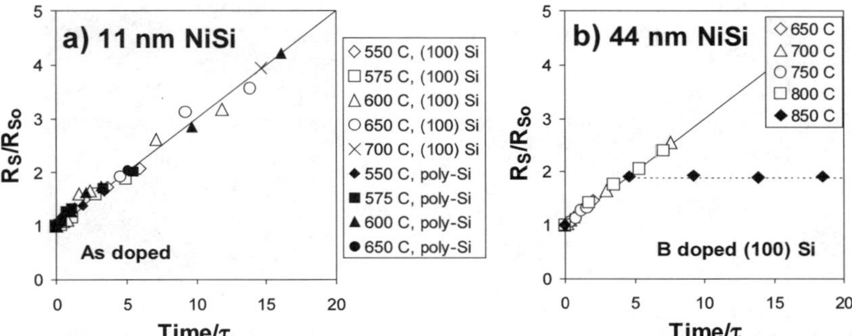

Figure 7. Sheet resistance (RS) normalized to initial value (RS0) vs. anneal time for a) 11 nm NiSi on n+ Si, and b) 44 nm NiSi on p +Si. Anneal times were normalized to degradation times, τ (20% increase in RS), shown in Figure 8.

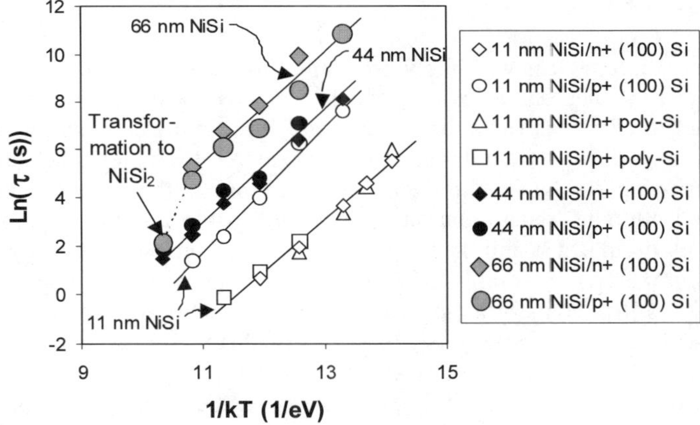

Figure 8. Arrhenius plots of degradation times, τ (20% increase in sheet resistance) of NiSi films. Data points correspond to the process of agglomeration, with activation energies of ~2.4 eV, except those labeled as "transformation to $NiSi_2$" (for which τ values are upper limits, since the transformation was too fast for the time scales used in the experiment).

Alloying with Pt was shown to improve the thermal stability of NiSi films both against transformation into $NiSi_2$ [15] and against agglomeration [8, 9] (the latter case shown in Figure 9). Pt at concentrations of ~7% was found to be predominantly in solution in the mono-silicide phase [8, 9].

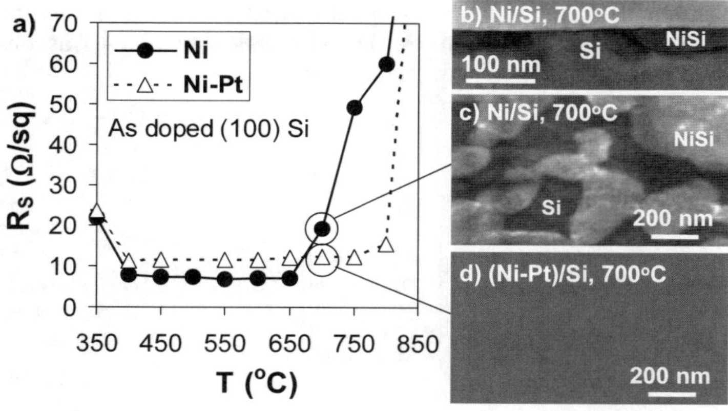

Figure 9. a) Sheet resistance vs. temperature, for the 30 s RTP reaction of 10 nm Ni and 10 nm Ni–Pt films with n+ (100) Si, showing an improvement in thermal stability with the addition of Pt. The NiSi films show significant morphologically degradation at 700°C (b and c), while (Ni–Pt)Si films remain smooth (d).

Germano-silicidation

Next generation devices will likely use some form of strained Si channel engineering for performance enhancement, which may involve the presence of Ge in the source and drain areas (used both in several local and global strain approaches). Compatibility with Si–Ge will need to be considered for material selection. For applications using Si–Ge, Co silicide is inadequate due to the lack of solubility of Ge in $CoSi_2$ [18]. The transformation temperature to the di-silicide phase increases with the addition of Ge and the transformation is accompanied by expulsion of Ge [18]. In contrast, Ge is soluble in NiSi, making Ni attractive for applications using Si–Ge alloys [9]. Transformation curves for the reaction of 10–nm Ni films with As and with B doped epitaxial Si–20%Ge on (100) Si are shown in Figure 10 and compared to those for Si. The low sheet resistance valley corresponds to Ni mono-germanosilicide (with a slightly higher resistivity than NiSi). The process window for low sheet resistance is narrowed from both sides by the addition of Ge, but still acceptable. The formation of the low resistivity mono-germanosilicide phase is shifted to higher temperatures and thermal stability is degraded by the presence of Ge, which induces an earlier onset (lower temperature) of agglomeration [9].

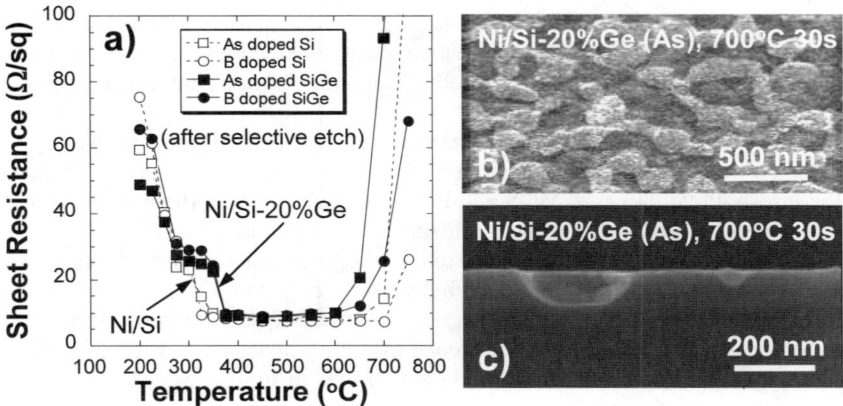

Figure 10. a) Sheet resistance vs. reaction temperature (30-s RTP) for 10-nm Ni on As-doped and on B-doped epitaxial Si–20%Ge/(100) Si compared to pure Si. b) Top-view and c) cross-sectional SEM images of Ni/Si–20%Ge/(100)Si after 700° C 30-s RTP reaction showing agglomeration.

The kinetics of agglomeration of Ni(SiGe) and NiSi films were studied, following the degradation of sheet resistance with anneal time at various temperatures. A lower activation energy (~1.6 eV) was found for the agglomeration of Ni(SiGe) films in comparison to NiSi (~2.4 eV) as shown in Figure 11a. XRD studies (Figure 11b) indicate that at lower reaction temperatures (*e.g.* 450°C) the Ge content in the Ni mono-germanosilicide films is similar to that in the substrate (~20%, assuming Vegard's law). However, at higher temperatures (700°C), a significant amount of Ge is expelled from the Ni(SiGe) grains (Figure 11b). It is likely that Ge segregation plays a role on the earlier onset and lower activation energy of agglomeration of Ni(SiGe) films.

Fully Silicided Gates

Recently, there has been significant interest on the application of silicides as metal gate electrodes, and in particular, on Ni FUSI (fully-silicided) gates [9, 16, 17]. From the processing point of view, it is a variation of the Ni self-aligned silicidation process in which the silicide fully consumes the Si in the gate. Ni FUSI appears as an attractive metal gate candidate that allows maintaining several aspects of the flows from previous generations (such as Si gate pattern and etch, and self-aligned silicide processes), appearing as a less disruptive candidate than other metal gate systems. In a typical Ni fully silicided (FUSI) gate process, an amorphous or polycrystalline Si gate is first implanted with dopants and may receive an optional activation anneal. Ni is then deposited and reacted to completely silicide the gate down to the dielectric interface. The ratio of Ni to Si is chosen to ensure that a single-phase silicide is in contact with the gate dielectric. Since Ni^{2}Si and NiSi grow sequentially by a diffusion limited growth mechanism, by choosing a slightly Ni-richer ratio than required for NiSi (*i.e.* Ni/Si > 0.55), it is possible to obtain full silicidation reproducibly with NiSi as the bottom layer — controlling the work function — with a thin Ni_2Si layer at the top of the gate. Dopants are segregated to the surface and snow-plowed to the gate dielectric interface during silicide growth (Figure 12a).

A key property that has attracted attention to NiSi FUSI gates is the effect of dopants on their effective work function (WF) [9, 16], which may allow for tuning of the threshold voltage of NMOS and PMOS devices without the need for two different metals. Figures 12b and 13 illustrate the dependence of the WF of NiSi FUSI/SiO_2 capacitors on dopants implanted into poly-Si before silicidation. A total shift in WF of ~600 mV was obtained between Sb doped and B doped Ni FUSI capacitors. It is speculated that the origin of the shift is the modification of the NiSi/SiO_2 interface dipole by the presence of a high concentration of dopants segregated to this interface during silicidation. Initial results of calculations using a density functional theory (DFT) molecular modeling technique support this view, showing a good correlation between calculated WF shifts resulting from the modification of the interface dipole by the addition of dopants and experimental data.

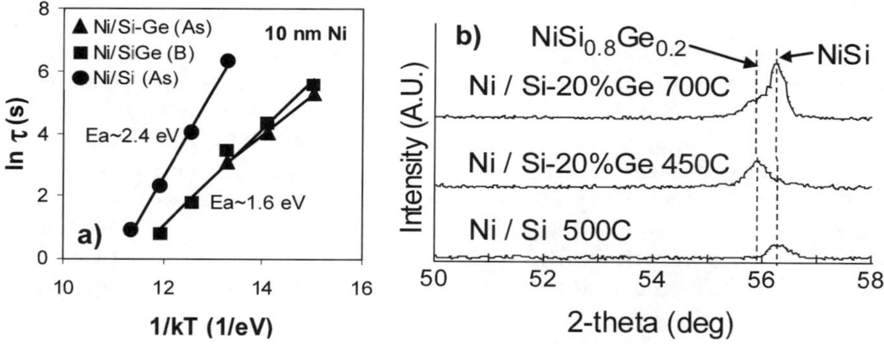

Figure 11. a) Arrhenius plot of degradation times (20% increase in sheet resistance) showing lower activation energy for agglomeration in the presence of Ge. b) XRD showing a shift from the NiSi peak position with the addition of Ge (~20% Ge assuming Vegard's law) for Ni/Si-20% Ge reacted at 450° C, and a bi-modal distribution indicating segregation of Ge out of the NiSi grains at 700° C.

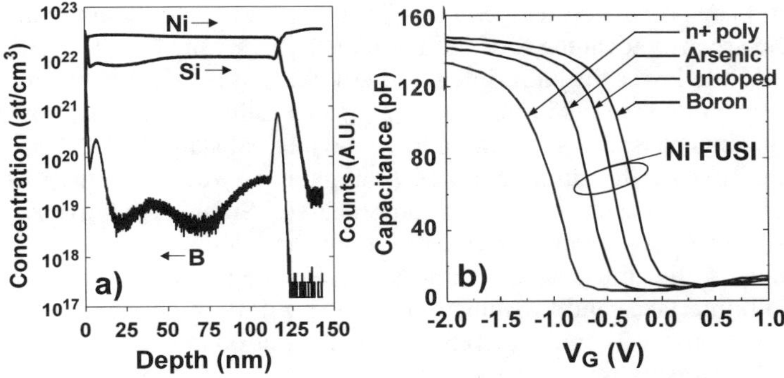

Figure 12. a) SIMS (secondary ion mass spectrometry) profile after full Ni silicidation of B-doped poly-Si gates on SiO$_2$ gate dielectric, showing dopant pile-up at the NiSi/SiO$_2$ interface. b) C-V characteristics of As-doped poly-Si/SiO$_2$ and of fully Ni-silicided/SiO$_2$ capacitors obtained from the reaction of Ni with undoped, As-doped and B-doped poly-Si.

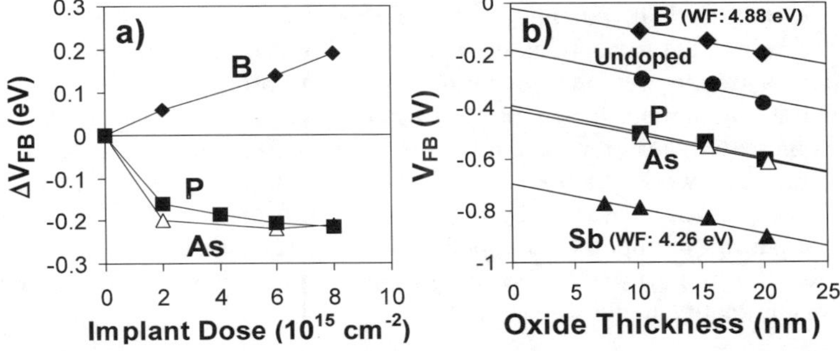

Figure 13. a) Flat-band voltage shift vs. implant dose for Ni FUSI/SiO$_2$ capacitors. b) Oxide thickness series for Ni fully silicided/SiO$_2$ capacitors, showing shifts of the effective work function with dopants with little effect on fixed charges at the SiO$_2$/Si interface.

Summary and Prospects

The evolution of silicidation processes and the change of silicidation materials have been driven for several CMOS generations by scaling of the transistor dimensions. The transitions from Ti to Co silicide and recently towards Ni-based silicides have been mainly caused by limitations on the growth of the target silicide phase on small structures, in turn caused by low silicide nucleation density. This was the case for C54 TiSi$_2$ at gate lengths <0.25 µm, further extended with techniques such as pre-amorphization implants that enhanced nucleation density, and for CoSi$_2$ at gate lengths below 40 nm. NiSi in contrast has no nucleation issues evident down to 25-nm gate lengths, and is viewed as the main candidate for SALICIDE processes for future generations. In fact, silicidation of small features can be excessive if not controlled properly due to diffusion of Ni from surrounding areas. Two-step RTP processes with a low thermal budget first RTP step are adequate to control this effect. Low processing temperatures, low contact resistivity and compatibility with Si–Ge are also advantages of NiSi over Ti and Co silicides. Thermal stability is however quite lower for NiSi. Integration of NiSi into CMOS flows requires the use of low thermal budgets in all processing steps following silicidation. Addition of alloying elements such as Pt can improve the thermal stability of NiSi.

Moving forward, the introduction of metal gates may decouple the S/D silicidation process from the gate stack formation processes. Silicides will in any case continue to be used as contacting materials in source and drain areas. With decreasing junction depths, the ability to form thin low resistance films with low Si consumption, smooth interfaces and low junction leakage will become more critical. These requirements, however, may be relaxed by the introduction of raised S/D approaches. As parasitic resistances become dominant in small devices, the need for low processing temperatures and low contact resistivity will also be critical. With the likely incorporation of Si–Ge in S/D areas for mobility enhancement through local or global strain approaches, compatibility with Si–Ge will also be required.

Silicides are also among the main metal gate candidates. The properties considered on material selection will vary for this application, with work function and gate stack stability and reliability becoming critical. The low resistivity requirement may be relaxed to some extent since the silicided gate will likely be thicker than silicides used in previous generation SALICIDE processes.

Although the number of silicides now being explored for S/D or gate applications has increased, NiSi remains, however, as one of the main candidates for both applications. The work function of NiSi was shown to change with the addition of dopants incorporated to the gate before silicidation. A change of ~600 mV was obtained comparing Sb doped to B doped Ni FUSI gates. The ability to tune the WF of NiSi and the similarity of the Ni FUSI process to previous SALICIDE processes makes it an attractive metal gate candidate.

References

1.   K. Maex, Mater. Sci. Eng. R11, 53 (1993).

2.   J.A. Kittl, D.A. Prinslow, P.P. Apte, M.F. Pas, Appl. Phys. Lett. 67, 2308 (1995).

3.   J.A. Kittl, Q.Z. Hong, M. Rodder, D.A. Prinslow, G. Misium, 1996 Symp. VLSI Tech., 14 (1996).

4.   J.A. Kittl, Q.Z. Hong, C.P. Chao, I.C. Chen, N. Yu, S. O'Brien *et al.*, 1997 Symp. VLSI Tech., 103 (1997).

5.   J.A. Kittl, W.T. Shiau, D. Miles, K.E. Violette, J. Hu, Q.Z. Hong, Solid State Technology 42(6), 81–92 and 42(8), 55–62 (1999).

6.   K. Maex, A. Lauwers, P. Besser, E. Kondoh, M. de Potter, A. Steegen, IEEE Trans. Electron. Devices 46, 1545 (1999).

7.   A. Lauwers, P. Besser, T. Gut, A. Satta, M. de Potter, R. Lindsay *et al.*, Microelectron. Eng. 50, 103 (2000).

8.   J.A. Kittl, A. Lauwers, O. Chamirian, M. Van Dal, A. Akheyar, M. De Potter *et al.*, Microelectron. Eng. 70, 158 (2003).

9.   J.A. Kittl, A. Lauwers, O. Chamirian, M.A. Pawlak, M. Van Dal, A. Akheyar *et al.*, Mater. Res. Soc. Symp. Proc. 810, 31 (2004).

10. J.A. Kittl, S.B. Samavedam, M.A. Gribelyuk, Appl. Phys. Lett. 73, 900 (1998).

11. J.A. Kittl, Q.Z. Hong, M. Rodder, T. Breedijk, 1997 IEDM Tech. Digest 111 (1997); and IEEE Electron. Device Lett. 19, 151 (1998).

12. M. Van Dal, M. Kaiser, J.G.M. Van Berkum, J.A. Kittl, C. Vrancken, M. de Potter *et al.*, in J. Appl. Phys., in press

13. C. Lavoie, F.M. d'Heurle, C. Detavernier, C. Cabral Jr., Microelectron. Eng. 70, 144 (2003).

14. C. Detavernier, R.L. Van Meirhaeghe, F. Cardon, K. Maex, Phys. Rev. B 62, 12045 (2000).

15. D. Mangelinck, J.Y. Dai, J.S. Pan, S.K. Lahiri, Appl. Phys. Lett. 75, 1736 (1999).

16. M. Qin, V.M.C. Poon, S.C.H. Ho, J. Electrochem. Soc. 148, 271 (2001).

17. W.P. Maszara, Z. Krivokapic, P.King, J.S.Goo, M.R.Lin, IEDM Tech. Dig. 367 (2002).

18. R.A. Donatan, K. Maex, A. Vantome, G. Langouche, Y. Morciaux, A. St. Amour *et al.*, Appl. Phys. Lett. 70, 1266 (1997).

TEM Characterization of Strained Silicon

J. P. Morniroli[a], P. H. Albarède[b] D. Jacob[c]

[a]LMPGM, Villeneuve d'Ascq / France

[b]Altis Semiconductor, TEM Lab, Corbeil-Essonnes / France

[c]LSPES, Villeneuve d'Ascq / France

Introduction

Strains present in silicon devices arise from various origins. In electronic components (MOS, CMOS), strains are non-intentionally produced during the technological steps of the growth process, such as oxidation (for example, insulator structures) or deposition of materials with different mechanical properties from those of silicon (for example, nitride lateral spacer). In the case of IV–IV heterojunctions, made of SiGe(C) alloys epitaxially grown on a Si or Ge substrate, strains are directly induced by the lattice parameter mismatch between the deposited layer and the substrate.

Strains lying in the active areas of the devices usually damage the electric properties of the components, but their effective influence is still badly known. Nevertheless, in the case of IV–IV heterojunctions, strains can be used to increase the device characteristics and particularly its transport properties [1]. It is thus an important challenge to precisely characterize strains in Si devices.

Strains present in devices are essentially evaluated from more or less complex simulations of the processes responsible for their appearance using, for instance, finite element modelling. However, due to the reduction of component dimensions, these simulations still need to be validated by comparison with direct characterization methods with high spatial resolution such as transmission electron microscopy (TEM). The aim of this paper is to review the various TEM methods available for strain determination in Si devices and more generally in strained epitaxial layers. Both images (conventional or high resolution) and diffraction patterns can be used. Preparation of thin foils for transmission electron microscopy constitutes a major step prior to characterizations. Some considerations on the specimen preparation methods will be given below.

Description of Strain in Epilayers

We briefly describe here the strains present in the simple case of an epitaxially grown layer (Figure 1a) and the possible stress relaxation mechanisms.

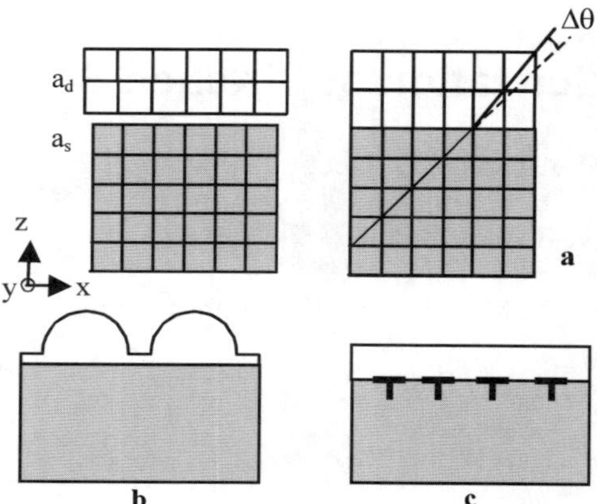

Figure 1. a) Deposition of a thin layer (with lattice parameter a_d) on a substrate (with lattice parameter a_s). The layer expands in the growth direction z. The consequent rotation $\Delta\theta$ of a diagonal lattice plane is also represented. b) Elastic relaxation by 3D growth. c) Plastic relaxation by formation of misfit dislocations at the interface.

Let us consider a thin layer (with initial lattice parameter a_d) deposited on a thick and flat oriented substrate (with lattice parameter a_s). To achieve the lattice match, the layer is biaxially strained in the growth plane and expanded in the perpendicular growth direction (it is generally assumed that the substrate remains unstrained since it is usually much thicker than the layer). Strains in the x and y directions of the interface plane are given by $e_{//} = (a_s-a_d)/a_d$ ($e_{//}$ can either be negative or positive for compressive or tensile strain, respectively). Application of linear elasticity theory gives the following value for the strain in the perpendicular direction:

$$e_\perp = e_{zz} = -2\nu e_{//}/(1-\nu) = -2e_{//}C_{12}/C_{11}$$

where ν is the Poisson's ratio and the C_{ij} are the elastic constants of the layer. If the deposited layer is an alloy, the lattice parameter and the elastic constants are calculated by means of Vegard's law.

Due to the elastic energy stored in the strained layer, elastic or plastic relaxation processes may occur after the deposition of a critical layer thickness. For high mismatch, the relaxation process is initially elastic (Figure 1b) and is characterized by surface buckling or formation of separated islands on the substrate [2]. The driving force for such a three-dimensional (3D) growth (Stranski–Krastanov growth mode) is the elastic relaxation occurring at the top of the islands which balances the increase of surface energy. Plastic relaxation can occur after the 3D growth during the islands coalescence or in two-dimensional (2D) layers for lower mismatch cases. This leads to the formation of misfit dislocations at the interface plane (Figure 1c). The full description of the various dislocation nucleation mechanisms involved in the 2D or 3D growth modes is beyond the scope of this paper and the interested reader is referred to the corresponding literature [3–7].

TEM Specimen Preparation

In order to get a full description of the strain state, thin foils for electron microscopy must be taken from different sections of the specimen. Horizontal and vertical sections are particularly useful and are usually referred as to plan-view and cross-sectional specimens (Figure 2).

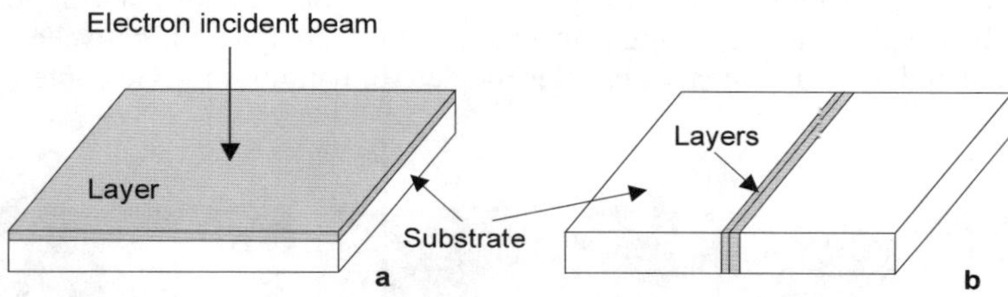

Figure 2. Plan-view (a) and cross-sectional (b) specimens in the simple case of a hetero-junction. Cross-sectional specimens are usually prepared by sticking two specimens together.

Several methods are available to prepare thin foils for electron microscopy. One has to be aware of the possible modifications of the strain state which could result from the creation of free surfaces or from modification of the geometry of the device. The preparation methods should also preserve a large observable area. The classic dimpler and ion-mill processes introduce severe thickness gradients (the thin-foil thickness is not constant, even at a local scale), and the softer chemical polishing method always reveals surface topography responsible for local strain heterogeneities. The newer FIB is good for parallelism, but introduces strong amorphization on each sidewall. In addition, the observable area is always limited to a few microns (20 • 10 μm^2 typical).

The tripod polishing appears as a very efficient method. Since its introduction in 1989 by Benedict *et al.* [8], tripod polishing has been applied to many tricky cases. It basically involves a mechanical polishing of the TEM thin foil down to electron transparency. The sidewalls are free of scratches, ions and chemicals, and the dead amorphous layer has a nanometric thickness. The observable area is wide (up to 1 mm) and the sidewalls are almost parallel (20' angle attainable). Finally, the process is amenable to the preparation of both plan-view and cross-sectional specimens.

Conventional Two-beam Image

Strain results in the bending and/or distortion of some lattice planes. Subsequently, local variations in diffraction conditions occur and these are associated with a change in the image contrast. The best contrast is observed in two-beam bright-field (BF) or dark field (DF) conditions (the specimen is tilted so that only one strong diffracted beam is observed; BF and DF images are formed with the transmitted beam and the diffracted beam, respectively). Interpretation of this contrast is generally possible by comparison with simulated images calculated by modelling the strain field associated with the studied structure [9, 10]. Finite element methods are now widely used to achieve these calculations. As an example, Figure 3 shows experimental and simulated DF images of a Ge_xSi_{1-x} island lying on a Si substrate observed in a plan-view specimen. In this case, the comparison between the experimental and simulated images enables the determination of the Ge content of the island.

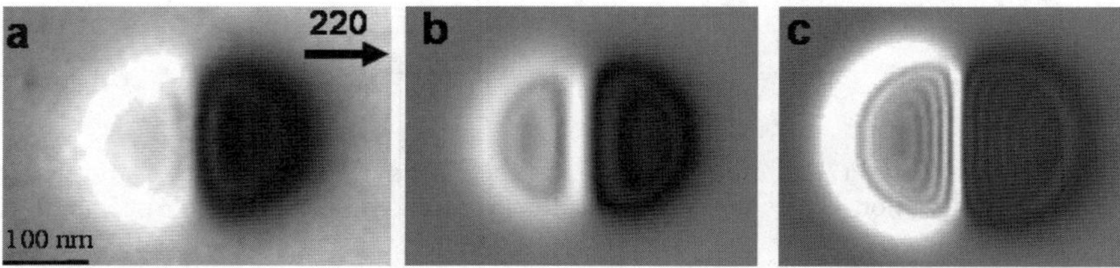

Figure 3. Two-beam dynamical dark-field images of a Ge_xSi_{1-x} island grown on a Si(001) substrate as observed in a plan-view specimen. a) Experimental image. b) and c) Simulated images calculated using the finite element analysis of the strain field associated with the islands for $x=0.4$ (b) and $x=1$ (c). The best match is obtained for $x=0.4$. Courtesy of Y. Androussi and T. Benabbas.

High-resolution Electron Microscopy (HREM)

In this method the crystal is aligned so that its atomic columns are parallel to the incident beam. The corresponding image results from interferences and, depending on the experimental conditions and the resolution of the microscope, the columns appear as white or dark dots. In the presence of local strains the atomic columns are slightly shifted as well as the corresponding dots on the HREM images. Two methods are used to measure the strains from HREM images.

The first one, called the peak finding method, consists of accurately measuring the shift of the atomic columns with respect to their ideal position in the unstrained materials. One example, taken from a $In_xGa_{1-x}As$ island on a GaAs substrate, is given in Figure 4 [11].

Another elegant method called the geometric phase analysis was proposed by M. Hÿtch [12]. In this method, the Fourier transform is used to determine the local phase of lattice fringes in the image, the phase being very sensitive to slight strain variations. An example of the strain field located around an edge dislocation is given in Figure 5 to illustrate this method.

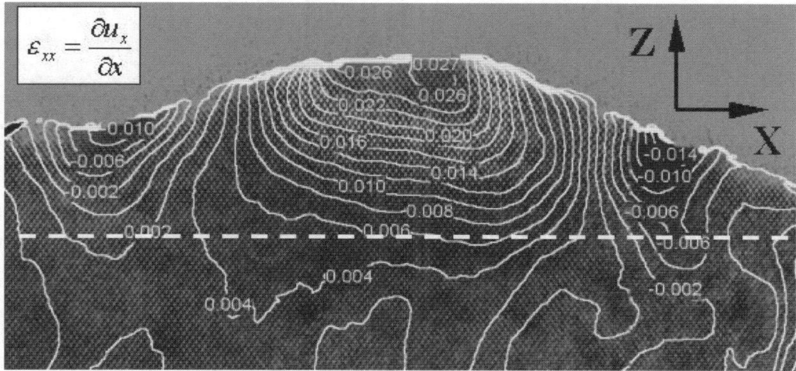

Figure 4. Strain characterization from a high resolution image of an $In_xGa_{1-x}As$ island deposited on a GaAs substrate. Strain map obtained by the Peak Finding Method. Courtesy of S. Kret, C.Delamarre and J.Y. Laval.

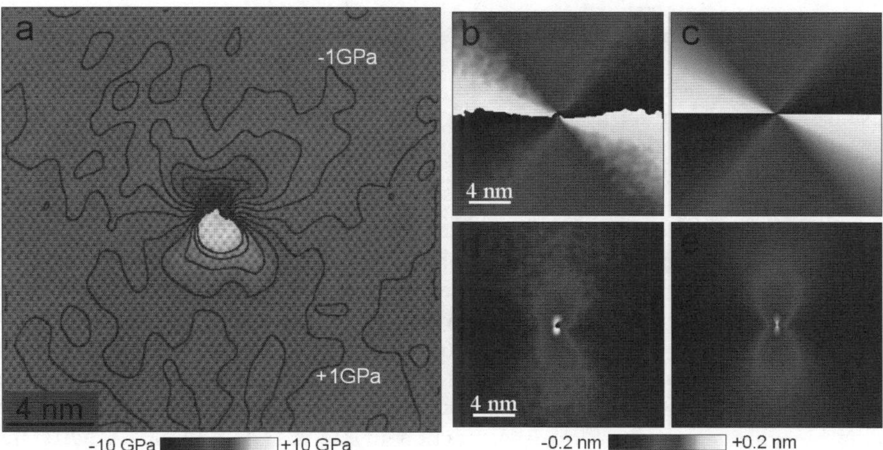

Figure 5. Geometric phase analysis (GPA) of local stresses and strain in silicon: a) high-resolution image of an edge dislocation viewed end on. The stress field measured by GPA is superimposed (stress contours for every 1 GPa up to ±5 GPa). b–e) Corresponding displacement field (u_x, u_y) measured experimentally (b,d) and from anisotropic elastic theory (c,e). The agreement with theory is ±3 pm. Courtesy of M. Hÿtch.

Electron Diffraction

Electron diffraction is a very valuable tool to characterize strained materials. Depending on the size and the convergence of the incident beam, selected-area electron diffraction (SAED), convergent-beam electron diffraction (CBED) and large-angle convergent beam electron diffraction (LACBED) [13] are available (Figure 6a).

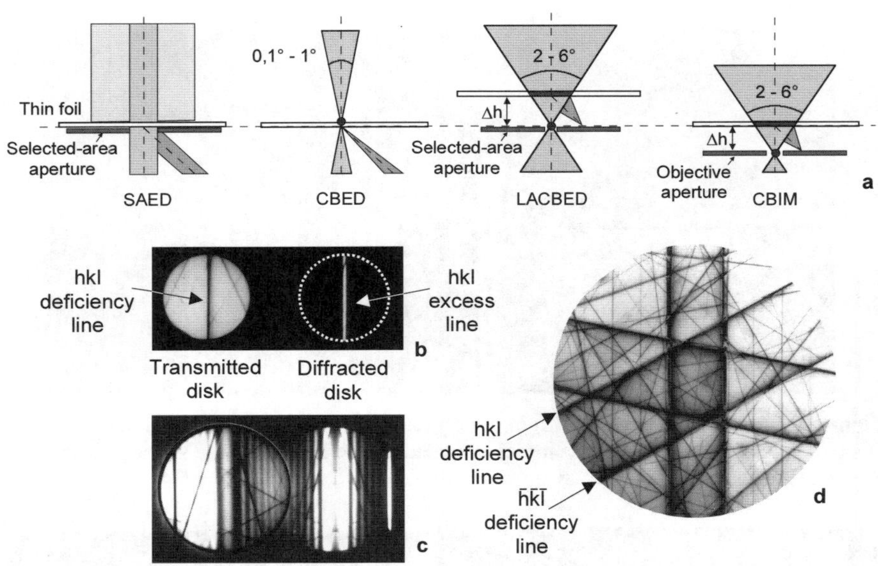

Figure 6. a) Description of the main electron diffraction techniques as a function of the size and the convergence of the incident electron beam. b) Two-beam CBED pattern displaying "pseudo-kinematical" hkl excess and deficiency lines. c) Two-beam CBED patterns displaying dynamical lines. d) LACBED pattern.

SAED produces spot or ring patterns. It is not really useful for local strain analyses since the pattern is an average pattern coming from a relatively large diffracted area. In addition, the spots or rings are not very sensitive to slight strain variations. CBED is a focused method. When performed with a very small spotsize (in the range of 2 to 100 nm with modern microscopes) it is perfectly well adapted to local strain analyses. LACBED is a defocused method displaying simultaneously the diffraction patterns and the shadow image of the diffracted area. It can thus be considered as an image-diffraction mapping. Local strains modify this mapping, allowing identification of the "strain map" of a relatively large specimen area (up to a few microns).

CBED and LACBED display disks containing lines [14]. As a matter of fact, each set of (hkl) lattice planes in diffraction position (vertical or nearly vertical planes) gives a set of parallel lines: an excess hkl line and a deficiency hkl line. In a two-beam CBED pattern (Figure 6b, c), the excess line is located in the hkl diffracted disk and the corresponding deficiency line is in the transmitted disk (bright-field disk). Weak "quasi-kinematical" lines are very sharp (Figure 6b) while strong "dynamical" lines are made of a set of black and white fringes (Figure 6c). In the multi-beam conditions, several deficiency lines are simultaneously observed in the transmitted disk (Figure 7a). In the LACBED method, the transmitted disk is very large and thanks to the selected-area aperture (Figure 6a), the patterns only display deficiency lines (called Bragg lines) which are arranged into pairs of parallel hkl and -h-k-l lines as shown on Figure 6d.

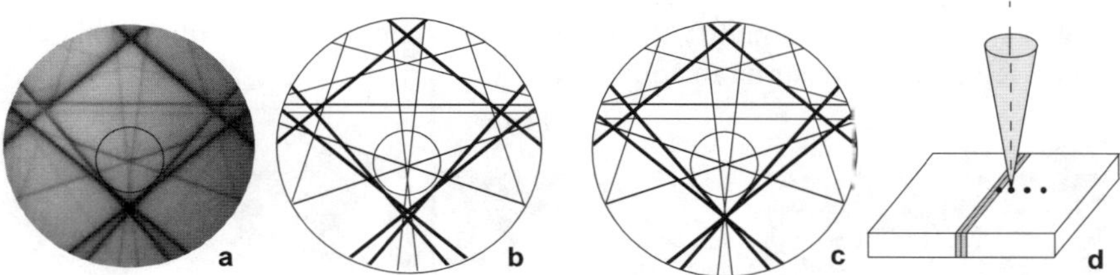

Figure 7. Effect of a variation of the lattice parameters on a Bright-Field CBED pattern. a) Experimental pattern. b) Theoretical pattern in agreement with a. c) Theoretical pattern after a slight change of the lattice parameter. The aspect of the circled intersection is modified. d) Experimental conditions used to observe a cross-sectional specimen. Simulations made with the "Electron Diffraction" software [15].

In strained materials, local atomic displacements modify the lattice parameters. This effect can also be considered in terms of variations of the interplanar distances and/or rotation of the lattice planes (see, for example, the rotation of the diagonal lattice planes which occurs between the substrate and the layer in Figure 1a). Then, the diffraction conditions are locally modified and a slight shift and/or a rotation of the lines occur on the patterns. Two main experimental solutions are used to probe the strained areas.

A first solution consists of observing the intersection of three or more lines whose aspect can be strongly changed even if the lines are very slightly shifted or rotated (see for example the circled intersection shown on Figure 7a). To obtain good accuracy, it is required to consider very sharp kinematical lines (called HOLZ lines) observed on zone-axis pattern (ZAP) and to calibrate the TEM voltage. Comparisons with simulated kinematical or dynamical patterns [15, 16] or interpretations from dedicated software [17] allow the identification of the local lattice parameters. CBED performed with a very small spotsize (down to a few nm) is well adapted to this approach especially when the experiments are performed on cross-sectional specimens (Figure 7d). The beam is then located on various positions on the substrate and layer for local analyses [18, 19].

A second solution involves the observation of the local displacement of a specific line in a strained area with respect to its normal position in the unstrained area. This approach is mainly used with LACBED patterns as illustrated in the examples given in Figure 8. In this case, the shadow image of the specimen superimposed on the diffraction patterns is very useful and allows a direct identification of the strained specimen areas [20-23]. Note that strain lying at the free surface of a sample can also strongly modify the intensity of some dynamical lines. This effect can be used to measure strain in epilayers [24].

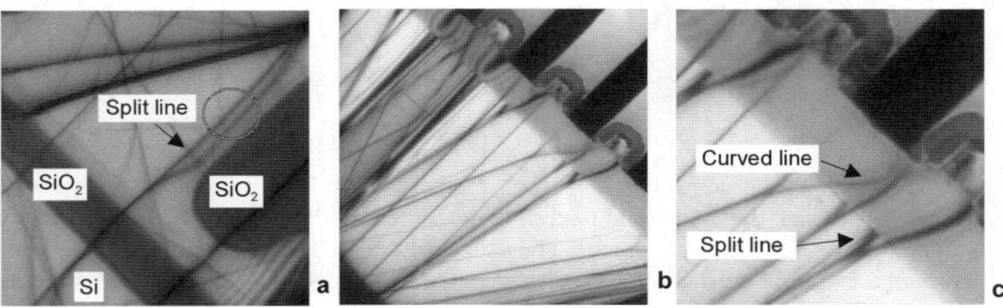

Figure 8. LACBED patterns of industrial specimens. a) Plan-view of a constraining isolation trench. The Bragg line is clearly split in the strained area. b) Cross-section through a line of transistors. c) Enlargement of pattern b. The splitting and curvature of some Bragg lines are progressive leading to a typical V-shaped aspect. The shadow image of the illuminated area, superimposed on the diffraction patterns, is clearly visible on these patterns.

Other TEM Methods

Two other TEM methods were also used to characterize strained materials:

- convergent-beam imaging (CBIM) (Figure 5a) which gives patterns similar to LACBED patterns but in image mode [25].
- bend contours observed in TEM images of buckled crystals as a transposition of the curvature method to TEM [26].

Concluding Remarks and Common Difficulties

Several key points must be taken into account for the characterization of strained silicon:

- these TEM methods measure the displacement field of the strained areas. The connection with strain requires knowledge of the chemical composition of these areas in the case of alloys and /or the lattice parameters of the unstrained areas.
- the displacement field observed and measured in the thin foil may not be exactly the one present in the bulk material. This is mainly due to stress relaxations which may occur during the preparation of the thin foil. In addition, local stress can also be introduced during this stage. Therefore, the choice of preparation method(s) is crucial.
- the connection between strain and stress can be made using Hooke's law, knowing the elastic constants of the materials. The finite element analysis is now widely used for these calculations.
- the various TEM methods are complementary and most of the time not self-sufficient.

There are, however, some experimental difficulties encountered when using TEM to characterize strains. HREM requires a dedicated microscope and very thin foils where relaxation phenomena can be very large. When performed with very small spotsizes, CBED can be sensible to specimen contamination. Electron-beam sensitive materials are also difficult to study with this method. Strongly strained areas located close to interfaces usually produce poor CBED patterns which are not useful for the analysis. LACBED has many advantages. Since it is a defocused method, the contamination is very low. The quality is excellent due to a strong filtering of the inelastic electrons by the selected-area aperture. The presence of the superimposed shadow image is very valuable. Nevertheless, LACBED requires relatively large crystal areas and is not well-suited to small crystals.

Other local methods that are now available to characterize strains, include X-ray microdiffraction and micro-Raman spectroscopy. A comparison of these two methods with the CBED method is given in reference [27].

Because of its versatility, high spatial resolution and great sensitivity, TEM is the only available tool to characterize strains at a nanometer scale and with accuracy in spite of the great care that is required to quantitatively interpret the results.

Acknowledgements

The authors kindly acknowledge Y. Androussi and T. Benabbas (LSPES, UMR CNRS 8008, Villeneuve d'Ascq, France), S. Kret, C. Delamarre and J.Y. Laval (LPS, ESPCI, Paris, France) and M. Hÿtch (CECM-CNRS, Vitry-sur-Seine, France) for providing some of the documents used to illustrate this chapter.

References

1.   F. Schaffler, Semicond. Sci. Technol. 12, 1515 (1997).

2.   A.G. Cullis, D. J. Robbins, A.J. Pidduck, P.W. Smith, Journal of Crystal Growth 123, 333, (1992).

3.   M. Albrecht, S. Christiansen, H.P. Strunk, P.O. Hansson, E. Bauser, Solid State Phenomena 32-33, 433 (1993).

4.   Y. Androussi, A. Lefebvre, T. Benabbas, P. François, C. Delamarre, J.Y. Laval *et al.*, J. Cryst. Growth, 169, 209 (1996).

5.   B. J. Spencer, J. Tersoff, Physical Review B 63, 205424 (2001).

6.   J. Zou, X.Z. Liao, D.J.H. Cockayne, Z.M. Jiang, Applied Physics Letters 81, 1996 (2002).

7.   I. Berbezier, A. Ronda, A. Portavoce, Journal of Physics: Condensed Matter 14, 8283 (2002).

8. J.P. Benedict, S.J. Klepeis, W.G. Vandygrift, R. Anderson, Electron Microsc. Soc. Amer. (EMSA) Bull. 19, 74 (1989).

9. T. Benabbas, P. François, Y. Androussi and A. Lefebvre, Journal of Applied Physics, 80, 2763 (1996).

10. X.Z. Liao, J. Zou; D.J.H. Cockayne, R. Leon and C. Lobo, Phys. Rev. Lett. 82, 5148 (1999).

11. S. Kret, C. Delamarre, J.Y. Laval, A. Dubon, Phil. Mag. Letters 77, 5, 249 (1998).

12. M.J. Hÿtch, J.L. Putaux, J.M. Penisson, Nature 423 (6937), 270-3 (2003).

13. M. Tanaka, R. Saito, K. Ueno, Y. Harada, J. Electron Microscopy 29, 408-412 (1980).

14. J.P. Morniroli, Large-Angle Convergent Beam Electron diffraction. Applications to crystal defects, Monograph of the SFμ, Paris 2002.

15. J.P. Morniroli, Electron Diffraction, Software to interpret electron diffraction patterns.

16. P. Stadelmann, JEMS, Software to interpret image and diffraction patterns.

17. European program STREAM; http://stream.bo.cnr.it.

18. Y. Tomokiyo, S. Matsumura, T. Okuyama, T. Yasunaga, N. Kuwano, K. Oki, Ultramicroscopy 54, 276-285 (1994).

19. R. Balboni, S. Frabonni and A. Armigliato, Philosophical Magazine A, 77, 1, p. 67-83 (1998).

20. D. Cherns, R. Touaitia, A.R. Preston, C.J. Rossouw, D.C. Houghton, Philosophical Magazine A, 64, 3, 597-612 (1991).

21. C.J. Rossouw, D. D. Perovic, Ultramicroscopy 48, p. 49-61 (1993).

22. X.F. Duan, D. Cherns, J.W. Steeds, Philosophical Magazine, A, 70, 6, 1091-1105 (1994).

23. F. Banhart, Ultramicroscopy, 56, 233-244 (1994).

24. R. Vincent, T.D. Walsh, M. Tozzi, Ultramicroscopy 76, 125, (1999).

25. C.J. Humphreys, D.M. Maher, H.L. Frazer, D.J. Eaglesham, Philosophical Magazine 58, 787-798 (1988).

26. M. Cabié, A. Rocher, A. Ponchet, J.L. Gauffier, Proc. ECM 2004, p. 147-148, Belgian Society for Microscopy, Liège (2004).

27. I. De Wolf, V. Senez, R. Balboni, A. Armigliato, S. Frabonni, A. Cedol et al., Microelectronic Engineering 70, 2-4, 425-435 (2003).

Part II

Material Aspects of Non-Volatile Memories

An Introduction to Nonvolatile Memory Technology

T. Mikolajick, C.-U. Pinnow

Infineon Technologies Dresden, Memory Development Center, Dresden / Germany

Introduction

Driven by an increasing demand for mobile devices, the market for nonvolatile memories is rapidly growing [1, 2]. Today the majority of nonvolatile memories is based on charge storage and is fabricated by materials available in complementary metal oxide semiconductor (CMOS) processes. These devices have some general shortcomings like slow programming, limited endurance as well as the need for high voltages during programming and erasing. The mentioned shortcomings imply severe restrictions on the system design side. A memory working like a random access memory — similar to static random access memory (SRAM) or dynamic random access memory (DRAM) — and being nonvolatile at the same time would therefore greatly simplify system design, since one universal memory could be used, where two or three memories are required today. Additionally, densities much higher than those possible today will be required for storing multimedia data like videos. This calls for a memory with an extremely small bit cell size. Since the realization of such a memory will require a switching between two stable states to achieve this goal, new materials that enable new switching mechanisms have to be introduced into the CMOS process flow. Figure 1 shows the hierarchy of possible CMOS memory material extensions used as a basis for the following discussion. The following articles of this chapter will introduce the material side of possible future nonvolatile memory options.

Nonvolatile Memories Based on Charge Storage

The vast majority of nonvolatile memories available on the market today is based on charge storage [3]. Two possibilities of storing charges are commonly used, namely floating gates and charge trapping layers. The dominant device today is the floating gate device. In charge storage devices the requirement for 10 years of data retention calls for a large energy barrier that has to be overcome during programming and erasing [4, 5]. This, on the other hand, leads to massive shortcomings. A high voltage (1020 V) is needed to pass the potential barrier during program and erase operation, write/erase speed is rather slow (from microseconds up to seconds) and write/erase endurance is limited (typically about $10^5 10^6$ cycles).

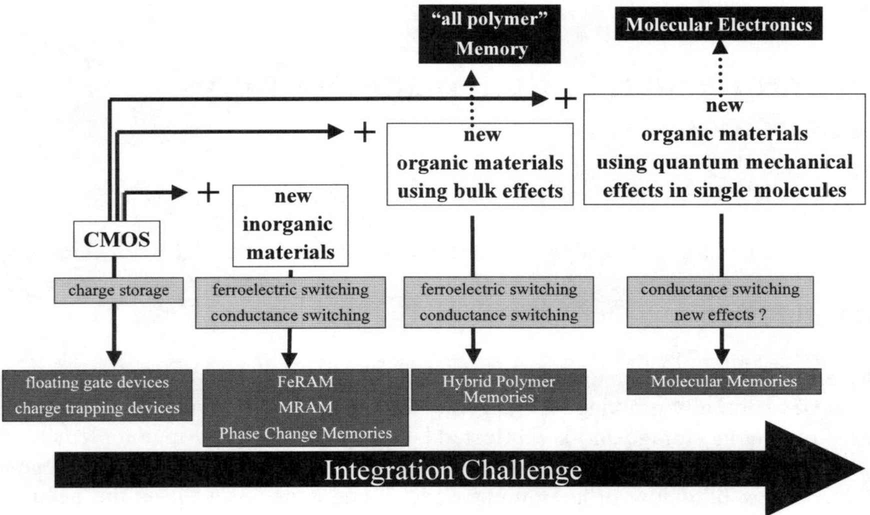

Figure 1. The hierarchy of nonvolatile memories from a material's perspective

The different options of nonvolatile memories on the market today can be mainly characterized by the amount of alterability and the cell size. A mask read only memory (ROM) can only be programmed during fabrication but has a very small cell size. An electrically erasable programmable read only memory (EEPROM), on the other hand, is writeable and erasable on a byte level, but has a rather large cell size utilizing two transistors per cell. Flash memories, which are programmable on a byte or word basis and erasable on a block or sector basis, can be implemented using a single transistor as a memory cell and, therefore, have proven to be the best compromise for many applications.

Among the Flash memories there are two different array architectures that are commonly used. In the not or (NOR) architecture the cells on one bitline are arranged in parallel making a fast random access (well below 100 ns) possible. In the not and (NAND) architecture the cells connected to one bitline are arranged in series [6].

Figure 2 shows the schematic of a NOR array and a standard NOR Flash cell. The cell consists of a metal oxide semiconductor (MOS) transistor with an additional floating gate between the channel and the control gate. The cell is programmed using channel hot electrons leading to short programming times in the 1-10 μs range and a high programming current caused by the low efficiency of hot electron injection. The cells are erased by using Fowler Nordheim tunneling. Up to very recently it was common to erase by tunneling to the source junction of the transistor. Using a channel erase scheme leads to smaller possible channel lengths and therefore to a better scalability [7] of the cell. In 0.13-μm technology and below, channel erase is essential for cell size scaling [7].

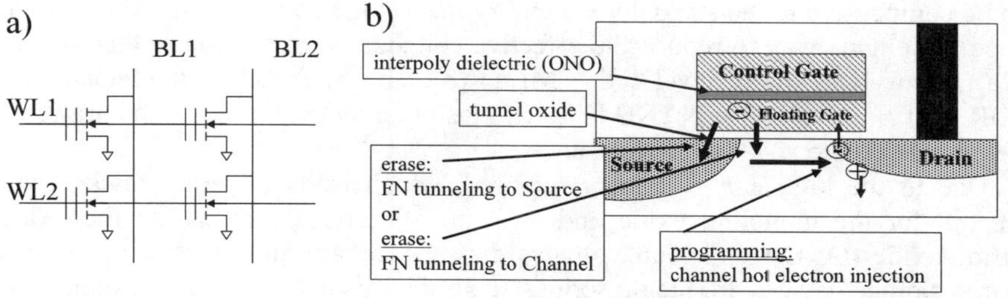

Figure 2. NOR Flash Array (a) and typical NOR Flash cell (b)

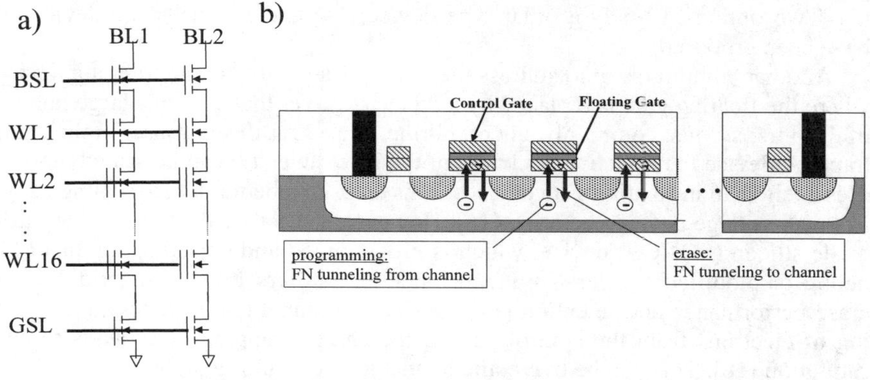

Figure 3. NAND Flash Array (a) and typical NAND Flash cell (b)

Figure 3 shows the schematic of a NAND array and the cross section of a few NAND cells. Again the cell consists of a MOS transistor with an additional floating gate in between the channel and the control gate. Programming and erasing is done using Fowler Nordheim tunneling from and to the channel. This leads to much longer programming times of typically a few hundreds of microseconds and requires higher voltages than in the NOR cell (up to 20 V). The programming current, on the other hand, is much lower than for channel hot electron programming, making it possible to program a significantly larger number of cells in parallel and achieve much faster effective programming speed if single Byte programming is not required.

The main advantage of the NAND architecture is the very small effective cell size due to the fact that no contacts are necessary between individual cells. This is paid for by a much larger random access time in the 10 μs range compared to the NOR Flash that enables sub-100-ns random access. Therefore, NAND Flash memories are used where large amounts of data that are accessed in larger portions (*e.g.* pictures, music, *etc.*) have to be stored. NOR Flash memories, in contrast, are primarily used for code storage.

To be competitive in the standalone memory market the cell size has to be as small as possible. One way to reduce the effective cell size is to store more than one bit in a memory cell (multi level cell = MLC) [8]. This approach is implemented in NOR [9, 10] as well as in NAND [11, 12] Flash memories, but it has the drawback of a reduced write and read performance.

Due to the high energy barrier required, which leads to rather thick oxides (>8 nm for the tunneling oxide and >15 nm effective thickness for the oxide-nitride-oxide (ONO)) and high voltages during program and erase, conventional charge storage devices are facing serious scaling limits below 100 nm ground rules [13]. To overcome these scaling limits and to address some of the other issues associated with charge storage devices, a number of alternative devices are investigated. One way to address the scaling limit is to use a device with a vertical channel. Two options, namely a pillar type device [14] and a trench type device [15] have been proposed.

Another common way to address the scaling issues are charge trapping devices, where the floating gate is replaced by a dielectric layer that contains large numbers of deep traps. Most commonly silicon nitride is used for this purpose. In the charge trapping device the effective thickness of the gate dielectric can be scaled more aggressively than in floating gate devices. This leads to a better device scaling as well as lower voltage device operation [16]. However, the classical silicon oxide nitride oxide silicon (SONOS) device, which is programmed and erased using direct tunneling or modified Fowler Nordheim tunneling, suffers from a tradeoff between erase performance and retention [17]. During the tunnel erase process, the tunneling of electrons from the control gate to the charge trapping layer leads to erase saturation [18]. This can be overcome by using a very thin tunneling oxide, a high-k top oxide [19], a control gate with higher electron work function [18], or hot hole erase. Using a very thin tunneling oxide leads to the aforementioned tradeoff between erase performance and retention. Both high-k top oxides and other control gates require further material optimization. One of these, or the combination of both approaches, could lead to a cell that makes the NAND architecture scalable into the deep sub 50-nm-range [20].

The hot hole erase is utilized in multi-bit charge trapping devices [21]. In this alternative way to store two bits in one memory cell one bit is stored on each side of a charge trapping device. The second bit in a cell can be accessed simply by reversing the direction of the cell and therefore this approach causes no performance drawback. The symmetric operation requires a virtual ground NOR array. This approach leads to effective bit sizes of about 3-4 F^2 (F is the minimum feature size of the manufacturing process), which is roughly 30% of the bit size of a conventional one bit NOR Flash.

Figure 4 shows a typical multi-bit charge trapping device. The device is programmed using channel hot electron injection and erased by hot hole injection. The hot holes are generated using band to band tunneling at the respective junction allowing fast erase for rather thick bottom oxides in the range of 5 nm. By reading the device in reverse mode compared to programming and using a sufficiently large drain voltage, the effect of the second bit can effectively be suppressed.

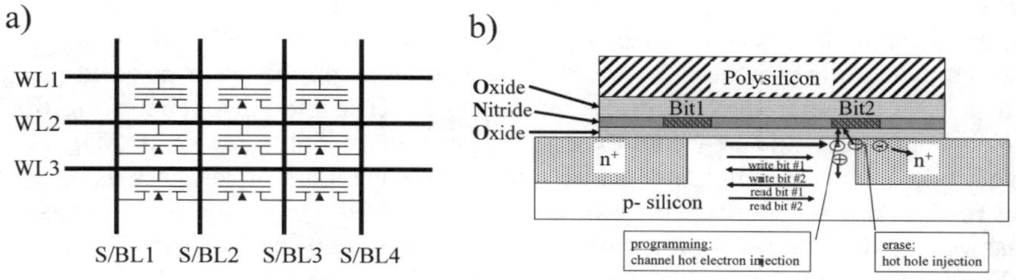

Figure 4. Virtual ground NOR array (a) and cross section of a multi-bit charge trapping cell

Many research and development activities are currently observed with multi-bit charge trapping devices [22, 23, 24] and high density products have appeared on the market [25, 26].

The scaling limit as well as the low write performance and limited endurance of both floating gate and charge trapping devices can be addressed by using thinner tunneling barriers. To overcome the retention problems of such an approach, a Coulomb blockade effect is utilized in silicon nanocrystal devices [27] and a combination of different materials as tunnel oxide is used in crested barrier devices [28]. Sufficient nonvolatile retention in combination with significantly better performance remains to be proven in approaches implementing a very thin tunneling oxide. On the other hand, nanocrystal devices with an increased bottom oxide thickness are an interesting alternative to charge trapping devices eliminating the erase saturation issue [17]. The challenge lies in the reproducible fabrication of the nanocrystals with respect to size and density and in maintaining the achieved electrical properties after the complete CMOS integration. The following two articles of this chapter will give a comprehensive overview on the two main fabrication methods for nanocrystals. The first article will cover nanocrystal devices fabricated using low pressure chemical vapor deposition (LPCVD) while the second one will focus on ion beam synthesis on nanocrystals.

The write/erase speed and endurance issue can also be overcome by using a device that is merged with the actual storage device allowing a lowering of the potential barrier during writing and erasing and maintaining the high barrier during storage. Memories called phase-state low electron drive memory (PLEDM) [29] and scalable two transistor memory (STTM) [30] by the respective authors are memory cells that apply this concept by using multiple tunnel junctions covered by the control gate and an additional signal line to supply the charge during write and erase. Again, the nonvolatile retention of these devices has not been demonstrated yet.

Nonvolatile Memories Utilizing Inorganic Switching Materials

To overcome the limitations of charge storage devices alternative switching mechanisms have to be applied. Since, up to now, integrated circuit technology deals mainly with inorganic materials, a lot of research and development activities have been focused towards novel inorganic materials [31, 32]. These memory mechanisms include ferroelectric and magnetic switching as well as material phase change to store information. Most of these memory options are aspired to merge into a CMOS technology platform by adding specific modules. Therefore, these concepts are generally expected to result in products within a few years. The additive modules may be implemented as switchable capacitors or resistors using various reading and writing schemes. In order to achieve the highest possible densities passive cross-point arrays rather than active transistor cells are desired. The smaller cell size, however, has to be paid for by a reduced performance due to a more complex readout.

Ferroelectric memory technology [33] is based on the electric dipole switching of a certain class of materials. These crystalline ferroelectric materials can be programmed in two different states by simply applying an electrical bias to the films. The electric field causes mobile ions to align along the applied field. Moreover, the individual unit cells of the crystal interact with their neighbouring cells to form ferroelectric domains. The nonvolatility is based on the fact that the majority of the polarized domains will retain their polarization state along the external field after the removal of the electric field.

Common ferroelectric materials are $PbZr_xTi_{1-x}O_3$ (PZT), $SrBi_2Ta_2O_9$ (SBT) and $(Bi,La)_4Ti_3O_{12}$ (BLT). The most common way to integrate a ferroelectric capacitor into a memory cell is the 1T/1C cell, which is similar to a DRAM cell. In order to sense the polarization state of the ferroelectric film a switching of the polarization is required. Depending on the initial polarization state of the film a small or a large amount of charge will flow to the circuit, when the capacitor is biased into one direction. The initial state can be identified by comparing this signal to a suitable reference. However, a write back is necessary to restore the initial information. In the chain FeRAM architecture one select transistor and one ferroelectric capacitor is connected in parallel rather than in series. Commonly, 8 or 16 elements of this kind are connected in series to form a chain. Since both contacts to the capacitor are shared between two cells, this leads to a very compact cell. The chain FeRAM architecture was used to demonstrate 8-Mb and 32-Mb devices [34, 35]. Combined with a capacitor that has its electrodes oriented perpendicularly to the semiconductor surface [36] this concept could improve the scalability of ferroelectric memories. In the 1T ferroelectric field effect transistor (FeFET) cell, the gate oxide of a MOS transistor is replaced by a ferroelectric film. The readout can be done – similar to a Flash cell – by sensing the drain current of this transistor and therefore it is nondestructive.

Ferroelectric materials show some specific reliability challenges. One is the decrease in signal with the number of switching cycles. This so-called fatigue phenomenon depends on the material, its deposition and the choice of electrodes. Additionally the capacitor has the tendency to prefer a state in which it has been stored for extended periods of time. This so-called imprint can lead to a failure, if the opposite state is written and subsequently read. Today ferroelectric memories for low densities are well established in the memory market [37]. In the third article of the chapter, the scaling issues that are associated with ferroelectric memories, and that have hindered the success of the technology for higher densities so far, will be illustrated.

Resistance switching in perovskite materials is another candidate for future nonvolatile memories [38, 39]. Certain perovskite materials show a pronounced change in electrical resistance, if voltage pulses are applied. These materials include the class of materials that show the colossal magnetoresistance effect, the best known is $(Pr,Ca)MnO_3$ (PCMO) [40] as well as Cr-doped $SrTiO_3$ and $SrZrO_3$ [41].

For the realization of magnetic random access memories (MRAM) three different physical effects, namely anisotropic magnetoresistance (AMR), giant magnetoresistance (GMR) and tunneling magnetoresistance (TMR) are potential candidates [42, 43]. AMR was used for the first MRAM products, but is obsolete today due to the very low resistance change. The GMR effect is based on the spin-dependent interface scattering of charge carriers in a sandwich structure of a conducting layer embedded between two magnetic layers. A resistance change of up to 6% is observed between the case, where the two magnetization states are parallel and the case, where they are anti parallel. If one magnetic layer is pinned and the other is free to switch, the structure is referred to as a spin valve (SV). By modifying the cell design and the reading sequence a pseudo spin valve (PSV) can be constructed leading to a doubling of the observed resistance change [44]. Like with AMR, GMR-based cells consist of all-metal structures leading to low resistance cells, which are not favourable for high density memories.

Finally, in the tunneling magnetoresistance concept a magnetic tunnel junction (MTJ) is used, where the tunneling current from one ferromagnetic layer to another one passing through a very thin insulating layer is dependent on the magnetic state of the two ferromagnetic films relative to each other. The resistance change between the case where the two magnetization states are parallel and the case where they are anti parallel is about 30–50% [45]. The MRAM cell is written by current pulses through the bitline and the wordline in order to generate a magnetic field, which is large enough to switch one of the ferromagnetic layers, but not the second one. The magnetic domain switching and thus writing time of the cell is in the nanosecond range. In contrast to ferroelectric memories the spin flipping mechanism does not degrade with cycling and therefore no endurance limitation is expected. Medium density MRAM memories based on MTJs are expected to hit the market very soon [46, 47]. In the fifth article of this chapter the different devices constructed from magnetic tunneling junctions will be covered.

Phase change memories are based on the reversible phase change between the amorphous and the crystalline phase of a chalcogenide glass [48]. These two physical states of matter differ in their resistivity. The transition from the crystalline to the amorphous state is performed by applying a very short electrical pulse to a resistive heater in contact with the phase change material, thereby melting the material and afterwards rapidly cooling it below the glass transition temperature in order to freeze the amorphous phase. The crystalline state is written by a little longer and lower pulse, enough to heat the material between the crystallization temperature and the melting temperature.

The most promising candidates for phase change memory applications are Ge-Sb-Te (GST) and Ag-In-Sb-Te alloys, which are also used in rewritable DVD optical memories [49]. The phase transition in these materials is very fast and can be repeated 10^{12} times [48]. Product demonstrators up to 64 Mb have been built based on this technology [50]. Articles number six and seven of this chapter will cover the current status of phase change memory technology as well as a detailed material study of the switching in GST.

Ionic conduction is used in a number of approaches to switch between a high and a low resistive state. An oxidizable anode and an inert cathode are used to contact a solid-state electrolyte in the conductive bridging RAM (CBRAM). In the erased state the resistance is very high, however, by applying an external voltage, metal cations formed at the anode diffuse into the electrolyte and thus metallic dendrites form to bridge the high resistive matrix. Typical metal ions used are Ag or Cu and the solid electrolytes can be glasses like As_2S_3, Ge_xSe_{1-x} [51], Cu_2S [52], or oxide layers like WO_x [53, 54]. By forming the metallic bridge, the resistance of the cell can be reduced by several orders of magnitude. The main advantage is the low and reproducible switching voltage defined by the electrochemical reaction involved.

Organic Memories

Instead of using an inorganic switching material it is also possible to implement organic materials. The advantage of organic materials lies in the fact that the properties of organic materials can easily be tailored. The main drawback is that usually the stability — especially the temperature stability – is limited. Therefore, organic materials are usually limited to memory concepts that are processed in the back end of line (BEoL). The processing in BEoL, on the other side, introduces the possibility to stack several memory layers on top of each other and to drastically increase the memory density. The CMOS circuits necessary to operate the array can be placed under the array resulting in very high cell efficiency [31, 32]. The organic memory layers are stacked on top of each other separated by insulators. Depending on the switching effect used, electrodes may also be shared among different memory layers reducing the processing steps for the memory array [55]. This approach is not limited to organic memories, in fact the first stacked OTP-type products use amorphous silicon antifuses [56], but organic materials have a large potential to combine all the necessary requirements to achieve a rewritable version of such a memory [55].

In this paper we discriminate between organic memories, which are memories using bulk switching effects in organic materials, and molecular memories which use the properties of a single (usually also organic) molecule. Figure 5 illustrates this concept showing an organic memory cell in Figure 5a and a molecular memory cell in Figure 5b. The latter will be covered in the following section of this article.

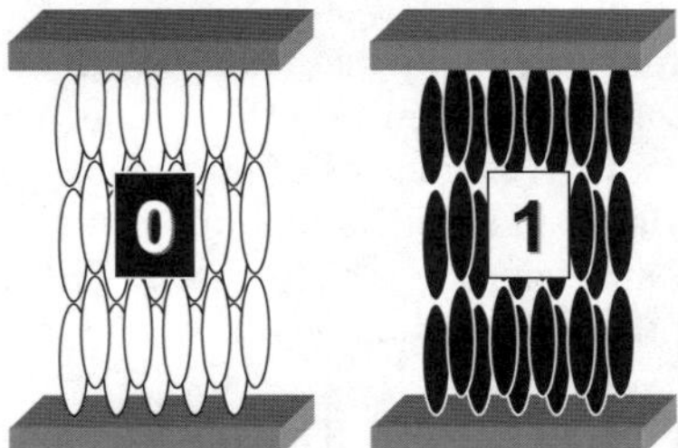

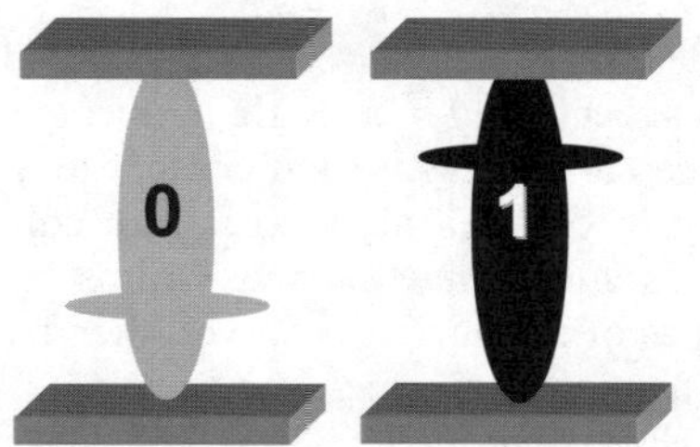

Figure 5. Schematic representations of an organic (a) and a molecular memory (b)

For organic memories either ferroelectric switching or conductance switching is used. Ferroelectric polymers have a much lower permittivity and a much slower switching speed than inorganic ferroelectrics [57, 58]. This may lead to the possibility of making the above mentioned pure cross point arrays, which were not successful with inorganic ferroelectrics [37]. The slow switching speed can be compensated for by massive parallelism in high density data storage applications like digital cameras, MP3 players or digital videos.

For resistance switching a number of different concepts have been published [54]. Krieger *et al.* have proposed the use of structural electronic instability in one-dimensional molecular systems [59]. In a concrete realization they used a polyconjugated compound (polyphenylacetylene) with 5–7% of potassium chloride. By applying an electric field the sodium and chloride ions are separated and form conducting complexes with the one-dimensional molecular system. Fast switching, a switching endurance of 10^8 cycles, data retention of 1.5 months as well as a multilevel cell storing four different resistance values were demonstrated [60, 61].

Potember *et al.* proposed an electrically induced phase transition in charge transfer complexes like silver and copper tetracyanoquinonodimethane (TCNQ) salts as another class of material showing promising switching behavior [62]. Charge transfer complexes recently also have demonstrated promising memory properties and integration results [63]. A large number of other materials showing switching effects have been reported in the literature [64, 65, 66, 67]. In most cases further work will be required to achieve stability and switching properties that are suitable for memory applications. The last article in this chapter will explain the main approaches under study today.

On top of the pure electrical approaches there are also approaches that use combinations of electrical and optical writing and reading effects [68]. The integration of these effects into high density memories still has to be proven.

Molecular Memories

The term molecular memory is used here for memory elements that make use of effects occurring in a single molecule involving quantum mechanical effects. It is not necessary to establish a contact to a single molecule. It is generally more convenient to create a monolayer film consisting of a large number of functional molecules and contacting these films by depositing an electrode layer on top. These types of memories, however, then have the potential to be scaled down to the scale of a single molecule.

Rotaxane molecules were investigated by Collier *et al*. [69]. Though a reasonable on/off ratio has been proven, it has also been found that these molecules were only programmable once. Using [2] catenane molecules the same researchers established reversible switching [70]. Recently, modified rotaxane molecules were shown to allow reversible switching and a very compact 64-bit memory was realized [71].

By attaching nitro-redox centers to suitable phenyl-based molecules Reed and Tour produced a large reversible switching effect [72, 73]. With self-assembled monolayers of these molecules embedded into an Au/molecular ML/Au electrode structure a fast and reversible resistance switching with a thermally activated decay time of about 10^3 s at room temperature was achieved. The retention is thought to depend on the energy barrier given by the thiol/Au contact. The current understanding is that the charge transfer to or from a molecule alters its geometric form and thereby its electronic orbital structure leading to the observed resistance change. Multiporphyrin nanostructures use a redox reaction to store charge in the molecule as the basic operational principle. The effect has been demonstrated in liquids [38] but no solid state realization of this concept has been shown.

Nanoelectromechanical nanotube memories are a crossbar arrangement of carbon nanotubes with well-defined on/off states [74]. The proposed memory array consists of single-walled nanotubes arranged in parallel rows lying on an insulating substrate. These rows are crossed by perpendicularly arranged nanotubes, which are suspended from periodic supporting pillars. Memory functionality is based on the electrostatic forces which act in between transiently charged nanotubes. Attractive or repulsive electrostatic forces bring the nanotubes in contact or separate them. The resistance at a matrix crosspoint is by some orders of magnitude lower if the two nanotubes are in contact. The degree of mechanical strain required for bistability is well below the elastic limit of the material. This type of memory could lead to ultrahigh density of 10^{12} elements per square centimeter [74]. However, no procedure for defined manufacturing of the single nanotube elements is known yet. The first realization, again, uses monolayer films of nanotubes [75].

Finding adequate deposition and structuring methods and tailoring the chemical properties of the materials is subject of the current research on molecular memories. The characterization and optimization is facing severe challenges and will certainly require much more time. Especially the thermodynamic stability of the different molecular states and the reaction kinetics have to be studied in more detail in order to give a deeper insight in the suitability of such molecules for their application in future nonvolatile memories.

Summary

In this article the main concepts for nonvolatile memories both in production and in the research stage have been summarized with the focus on material science aspects as well as basic concepts and a classification of the various approaches is given. Today's nonvolatile memories are based on charge storage with the majority of the products using charge storage devices. Charge trapping devices are becoming more important due to their superior scaling promises. To overcome the limitations of charge storage devices new switching materials have to be introduced in the CMOS process. Memories using inorganic switching materials, especially ferroelectric, magnetoresistive and phase change memories are in an advanced development stage. Ferroelectric memories are already established for lower density products. Organic memories hold the promise of low cost and very high densities. However, the material stability will need significant improvement. Molecular memories are scalable down to molecular dimensions if the basic fabrication issues as well as the contact formation challenges can be solved.

The following articles will give in-depth coverage on nanocrystal, ferroelectric, magnetoresistive and organic memories. All of the more advanced short to mid-term concepts are therefore covered in detail here, while the organic memory section gives an outlook into the mid and long-term future.

Acknowledgements

The authors thank Raimund Engl for Figure 5.

References

1.   A. Niebel, Proceedings of the Non-Volatile Semiconductor Memory Workshop, p. 9-13 (2000)

2.   A. Niebel, Proceedings of the Non-Volatile Semiconductor Memory Workshop, p. 14-19 (2004)

3.   W.D. Brown and J.E. Brewer (Ed.), Nonvolatile Semiconductor Memory Technology, IEEE Press, New York, (1998)

4.   S. Lai, Proceedings of the Seventh Biennial IEEE International Nonvolatile Memory Technology Conference, p. 6-7 (1998)

5.   S. Mori *et. al.*, IEEE Trans. Electron. Devices 38, 2, 386-391 (1991)

6.   P. Pavan *et al.*, Proc. IEEE 85, 1248 – 1271 (1997)

7.   S. N. Keeney, IEDM digest of technical papers p. 2.5.1-4 (2001)

8.   B. Eitan, G. Crisenza, P. Cappelletti, and A. Modelli, IEDM Digest of technical papers, p.169-172 (1996)

9. G. Atwood, A. Fazio, D. Mills and B. Reaves, Intel Technol. J. Q4, 1-8 (1997)

10. A. Modelli, A. Manstretta, and G. Torelli, IEEE Trans. Electron. Devices 48(9), 2032-2042 (2001)

11. T. Cho et al., ISSCC Digest of Technical papers, 28-29 (2001)

12. S. Lee, ISSCC Digest of Technical papers, p. 52-53 (2004)

13. A. Fazio, S. Keeney, S. Lai, Intel Technol. J. 6, No. 2, 23-30 (2002)

14. H. Pein and J.D. Plummer, IEEE Electron. Device Lett. 14, No. 8, 415-417 (1993)

15. D.S. Kuo, M. Simpson, L. Tsou and S. Mukherjee, Symposium on VLSI Technology Digest of Technical Papers 51- 52 (1994)

16. C. T. Swift *et al.*, IEDM Digest of technical papers, 927-930 (2002)

17. M. Sadd, Proceedings of the Non-Volatile Semiconductor Memory Workshop, 75-76 (2004)

18. H. Bachhofer *et al.* J. Appl. Phys. 89, 5, 2791-2800 (2001)

19. Y. Shin, Proceedings of the Non-Volatile Semiconductor Memory Workshop, 58-59 (2003)

20. K. Kim, Proc. 24th International Conference on Microelectronics 1, 377-384 (2004)

21. B. Eitan *et al.*, IEEE Electron. Device Lett. 21, No. 11 543-545 (2000)

22. W. J. Tsai *et al.*, IEDM Digest of technical papers, 32.6.1-4 (2001)

23. Y. Roizin, A. Yankelevich and Y. Netzer, AIP Conference Proceedings, No. 550, 181-185 (2001)

24. Y.K. Lee *et al.* Symposium on VLSI Technology Digest of Technical Papers, 208-209 (2002)

25. E. Maayan *et al.*, ISSCC digest of technical papers, 100-101 (2002)

26. B. Q. Le *at al.*, IEEE Journal of Solid State Circuits 39, No. 11, 2014-2023 (2004)

27. S. Tiwari *et al.*, IEDM Digest of technical papers 521-524 (1995)

28. K. Likharev, Appl. Phys. Lett. 73, No. 15, 2137-2139 (1998)

29. K. Nakazato *et al.*, ISSCC Digest of Technical Papers, 132-133 (2000)

30. J. H. Yi *et al.*, IEDM Digest of technical papers, 36.1.1-4 (2001)

31. T. Mikolajick and C.-U. Pinnow, Proceedings of the Non-Volatile Memory Technology Symposium, 3-12 (2002)

32. C.-U. Pinnow and T. Mikolajick, Journal of the Electrochemical Society 151, No. 6, K13-K19 (2004)

33. T. Mikolajick *et al.*, Microelectronics Reliability 41, No.7, 947-950 (2001)

34. D. Takashima *et al.*, ISSCC Digest of technical papers, 40-41 (2001)

35. S. Shiratake, IEEE Journal of Solid-State Circuits 38, No. 11, 1911-1919 (2003)

36. N. Nagel, VLSI Technology Digest of Technical Papers, 167-147 (2004)

37. D. Boundurant, Ferroelectrics 112, 273-282 (1990)

38. Y. Watanabe *et al.*, Appl. Phys. Lett. Vol. 78, No. 23, p. 3738-3740 (2000)

39. S.Q. Liu *et al.*, Appl. Phys. Lett. Vol. 78, No. 19,p. 2749-2751 (2000)

40. C. Pagagianni *et al.*, Proceedings of the Non-Volatile Memory Technology Symposium, 125-128 (2004)

41. G.I. Meijer, Proceedings of the Non-Volatile Semiconductor Memory Workshop, 13 (2004)

42. R.A. Sinclair *et al.*, IEEE International Nonvolatile Memory Technology Conference, 38-42 (1998)

43. W.J. Gallager *et al.*, J. Appl. Phys. Vol. 81, No. 8, p. 3741-3746 (1997)

44. S. Tehrani *et al.*, IEEE International Nonvolatile Memory Technology Conference, 43-46 (1998)

45. S. Parkin, J. Appl. Phys. 85, No. 8, 5828-5833

46. J. DeBrosse *at al.*, Symposium on VLSI Technology Digest of Technical Papers, 454-457 (2004)

47. M. Durlam *et al.*, IEEE International Conference on Integrated Circuit Design and Technology, 27-30 (2004)

48. S. Lai and T. Lowrey, IEDM Digest of technical papers 36.5.1-4 (2001)

49. S. Ogawa *et al.*, Proc. SPIE Vol. 3401, 272-276 (1998)

50. S. Lee *et al.*, ISSCC Digest of technical papers, 52-53 (2004)

51. M. Kozicki *et al.*, Electrochemical Society Proceedings 13, 298-309 (1999)

52. T. Sakamoto, Appl. Phys. Lett. 82, No. 18, 3032-3034 (2003)

53. M. Kozicki *et al.*, Proceedings of the Non-Volatile Memory Technology Symposium, 10-17 (2004)

54. Ju. H. Krieger and S. Spitzer, Proceedings of the Non-Volatile Memory Technology Symposium, 121-124 (2004)

55. S. Lai, Intel Developer Forum, 21-24 (2002) (http://www.intel.com/research/silicon/StefanLaiIDF0202.htm)

56. M. Johnson *et al.*, IEEE J. Solid State Circuits 38, No. 11, 1920-1928 (2003)

57. T. Wang, J. Herbert, and A. Glass (eds.), The Applications of Ferroelectric Polymers, Blackie&Son, Glasgow and London (1988)

58. L.-Jie, E. Schreck, and K. Dransfeld, Appl. Phys. A 53, p. 457-461(1991)

59. Yu. H. Krieger, J. Struct. Chem. 40, No. 4, 594-619 (1999)

60. Yu. H. Krieger, S.V. Trubin, S.B. Vaschenko, N.F. Yudanov, Synth. Metals 122 , 199-202 (2001)

61. Yu. H. Krieger, N.F. Yudanov , I.K. Igumenov S.B. Vaschenko, J. Struct. Chem. 34, No. 6, 966-970 (1993)

62. R. S. Potember, T. O. Poehler and D. O. Cowan, Appl. Phys. Lett. 34, 405 (1979)

63. R. Seczi *et al.*, IEDM Digest of technical papers 10.2.1-10.2.4 (2003)

64. D. Ma, M. Aguire, J.A Freire, and I.A. Huemmelgen, Adv. Mater. 12, No. 14, 1063-1066 (2000)

65. K. Sakai *et al.*, Appl. Phys. Lett. 53, No. 14, 1274-1276 (1988)

66. H.K. Henisch *et al.*, Thin Solid Films 51 265-274 (1978)

67. L. P. Ma, J. Liu, and Y. Yang, Organic electrical bistable devices and rewritable memory cells, Appl. Phys. Lett. 80, No. 16 (2002) 2997-2999

68. C.-Y. Liu and A.J. Bard, Acc. Chem. Res. 32, 235-245 (1999)

69. C.P. Collier *et al.*, Science 285, 391-394 (1999)

70. C. P. Collier *et al.*, Science 289, 1172-1175 (2000)

71. Y. Luo *et al.*, Chemphyschem, No.3, 519-525

72. M.A. Reed *et al.*, Appl. Phys. Lett. 78, No.23, 3735-3737 (2001)

73. M.A. Reed *et al.*, IEDM Digest of technical papers, 227-230 (1999)

74. K.M. Roth *et al.*, J. Vac. Sci. Technol. B 18, No.5, 2359-2364 (2000)

75. T. Rueckes *et al.*, Science 289, 94-97 (2000)

76. J. W. Ward *et al.*, Proceedings of the Non-Volatile Memory Technology Symposium, 34-38 (2004)

Floating-dot Memory Transistors on SOI Substrate

O. Winkler[a], M. Baus[a], M. C. Lemme[b], R. Rölver[a], B. Spangenberg[a], H. Kurz[a]

[a] Aachen University, Institute of Semiconductor Electronics, Aachen / Germany

[b] Advanced Microelectronic Center Aachen, Aachen / Germany

Introduction

Mobile, battery-powered electronics are becoming more and more common and indispensable [1]. Cellular phones, personal digital assistants, digital cameras and music players are widely spread and fuel the need for more semiconductor data storage devices that retain the data even when the supply voltage is disconnected. Due to the opening of this mobile electronics market non-volatile memories have gained huge attention and semiconductor market shares. The commonly used non-volatile memory is the so-called Flash memory. The Flash cell is a charge-storing memory cell which is based on a metal oxide semiconductor (MOS) field effect transistor (FET) with a charge-storing island embedded in the gate oxide. This type of transistor is most commonly used for digital semiconductor logic circuits and also acts as the basis of other non-volatile memories such as the charge-trapping NROM [2]. Like MOSFETs, Flash and floating-dot memories are compatible to complementary MOS (CMOS) technology. This is very important for the integration of non-volatile memory into logic circuits. However, conventional Flash memory cells suffer from limited write endurance due to degeneration of the isolation material that surrounds the charge-storing island. Floating-dot memories circumvent this handicap by employing a nano-dot layer that is insensitive towards local defects. To understand the details of these memory devices the next paragraph deals with the basic functional principle of Flash and floating-dot memories.

Fundamentals

The MOSFET transistor is the workhorse of data processing logic devices like computer processors and microcontrollers. In case of silicon MOSFETs the gate insulator is commonly fabricated from silicon dioxide (SiO_2). An introduction of charge carriers into this gate insulator leads to a change of the electric field within the gate insulator and thus to a change of the influenced carrier density inside the transistor channel. This leads to a threshold voltage shift ΔV_{th} and subsequently, the drain current I_D can be influenced by several orders of magnitude. This effect is exploited by charge-storing devices like FLASH and floating-dot memories.
As depicted in the cross-section schematic of a Flash memory cell (see Figure 1a) a second gate is embedded into the gate dielectric. Because this gate electrode is not

contacted and thus its electrical potential is not directly controlled it is called floating. This floating gate is electrically insulated from the gate electrode above and from the transistor channel beneath by two insulating layers. While the upper insulating film is used to prevent any electrical current between the floating gate and the gate electrode the task of the lower dielectric layer is more sophisticated. This layer is a tunnel dielectric layer. By applying appropriate voltages to the gate electrode, charge carriers can be forced to tunnel through this layer to and from the floating gate. After disconnecting the write or erase gate voltage these carriers remain on the floating gate and influence the transistor's threshold voltage V_{th}. This threshold voltage shift ΔV_{th} can be easily measured so that the stored information can be read from the memory transistor. The floating gate divides the gate capacity into a control oxide (cox) capacity C_{cox} and a tunnel oxide (tox) capacity C_{tox} (see schematic on the left in Figure 1).

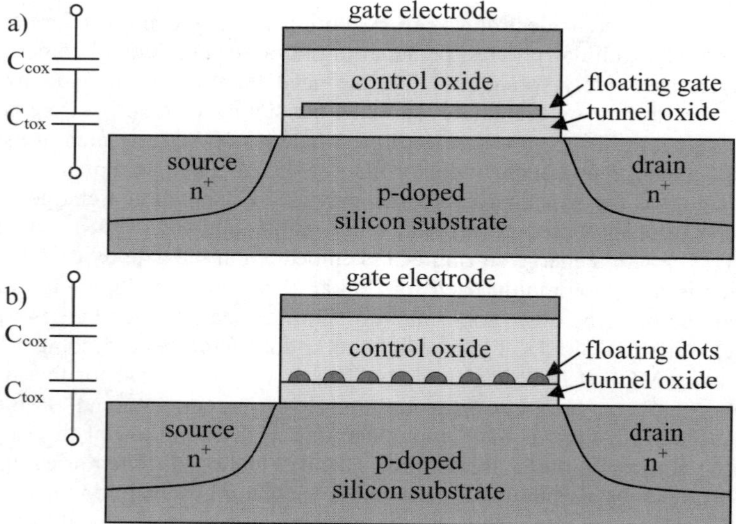

Figure 1. Cross-section schematic of a FLASH (a) and a floating-dot (b) memory cell transistor.

The threshold voltage shift ΔV_{th} that is induced by the charge on the floating gate only depends on the floating gate charge Q_{fg} and the control oxide capacitance C_{cox}:

$$\Delta V_{th} = -Q_{fg}/C_{cox}$$

The charge storage capabilities of the floating gate are directly governed by the electrically insulating properties of the oxide layers. The information storage time is determined by the leakage current through the oxide layers. Since the tunnel oxide is usually thinner than the control oxide, its insulating quality must be very high to effectively suppress leakage currents. On the other hand, the tunnel currents during write and erase processes damage the tunnel oxide locally and weaken its insulating properties. Finally, repeated write and erase procedures lead to a local conducting path through the tunnel oxide so that the charge can quickly flow back into the transistor channel after a charging process. After such a fatal oxide breakdown the memory cell is defective since its charge storing capability is destroyed. In order to enhance FLASH's oxide breakdown tolerance, floating-dot memories have been introduced [3-5]. Here, instead of one floating gate many separated floating dots are employed for charge storing. Thus, a local oxide defect only affects and discharges a locally limited area of one or a few dots but not all of them. Therefore, the oxide breakdown tolerance is improved and longer device lifetimes with more write/erase cycles can be achieved. Furthermore, a floating dot memory device is less sensitive to inhomogeneities of the tunnel oxide thickness which can lead to a premature oxide breakdown. Figure 1b) shows the cross section of such a floating-dot memory transistor's gate stack exhibiting the charge-storing dot layer embedded into the gate insulator.

Fabrication

Different approaches exist for the fabrication of floating-dot memory devices. The main challenge is the manufacture of the nano-dots with a density as high as possible while also retaining the electrical insulation from each other. These processes comprise the deposition and thermal annealing of amorphous silicon [6] or silicon rich oxide (SiO_x, $x<2$) [7] and the implantation of silicon into silicon dioxide [8, 9]. Another possibility is the self-organized deposition of silicon nano-dots by low pressure chemical vapor deposition (LPCVD) [10, 11] and the subsequent deposition of the control oxide. Devices based on this latter fabrication method have been manufactured and electrically characterized. Silicon on insulator (SOI) substrate was chosen since it offers unique advantages in terms of current control and scalability of these devices. SOI substrates consist of a thin (10–100 nm) monolithic silicon film that acts as the active semiconductor layer and that is electrically insulated from the bulk silicon material by a buried oxide layer in-between. By etching completely through this top silicon layer down to the silicon dioxide it is possible to fabricate insulated silicon structures and devices. SOI substrate offers the possibility of three-dimensional devices like triple-gate transistors [12, 13] and FinFETs [14, 15] which possess extremely high drain current controllability due to the improved gate influence.

For the fabrication of the memory devices slightly p-doped (boron, $2 \cdot 10^{15}$ cm^{-3}) SOI substrates with a top silicon layer thickness t_{top} of 100 nm were used. The transistor channel was lithographically defined and etched using a reactive ion etching (RIE) process. The 3-nm-thick tunnel oxide was thermally grown. Subsequently, the silicon nano-dots and the control oxide were deposited. These two key processes are described in detail in the following paragraphs. After the manufacturing of the memory gate stack the gate electrode is deposited by poly-silicon LPCVD, lithographically defined and RIE-etched. Finally, an arsenic implantation and annealing is used for n-doping the source and drain regions and the gate. Figure 2 depicts a simplifying perspective schematic of a floating-dot memory transistor on SOI substrate. The source, drain and channel regions fabricated of the top silicon can be recognized as well as the poly-silicon gate electrode on top and the gate insulator with the embedded floating silicon nano-dots in-between. The SOI top silicon thickness, the channel width W and the gate length L are also indicated.

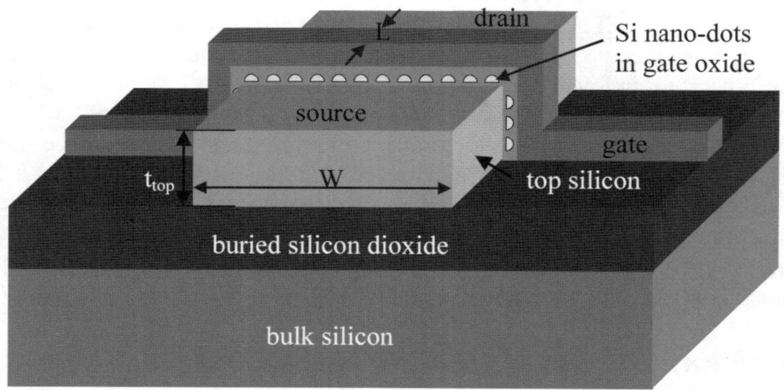

Figure 2. Schematic of a floating-dot memory transistor on SOI substrate.

Fabrication: Self-organized Silicon Nano-dot Deposition by LPCVD

For the memory effect of floating-dot storage devices the fabrication of the charge-storing islands is the key process. In contrast to silicon implantation and annealing the deposition of nano-dots by LPCVD on top of the tunnel oxide offers the advantage of a precise vertical localization of the nano-dots within the gate stack. Therefore, the tunnel oxide has got an exact and easily controllable thickness since it can be thermally grown. When developing an LPCVD process for the deposition of silicon nano-dots, one must consider that the amounts of deposited silicon are very small. In order to keep the deposition time from falling below an uncontrollable and unreproducible minimum the process parameters must be chosen carefully. Apart from the furnace geometry the parameters of the LPCVD process are the temperature, the gas mixture, the gas flux, the pressure, and the deposition time. In case of the applied process the used gas is silane (SiH_4). The temperature, pressure and silane flux are kept low in order to keep the density of reactants inside of the furnace and their reactivity low. Also, the pretreatment of the tunnel oxide surface influences the results of the nano-dot deposition. It has been reported that the density of the dots can be enhanced by an HF-treatment of the underlying silicon dioxide layer. This HF-treatment leads to an OH-bond termination of the silicon dioxide surface and to a decrease of the activation energy of the silane-based vapor phase deposition [16]. The deposition of nano-dots by LPCVD runs through three different phases as depicted in Figure 3. These phases are nucleation, dot growth and coalescence. During the nucleation the dot density increases because new nucleation sites are used for silicon accumulation out of the gas phase. Then, the dot growth sets in and the dot density remains almost constant. The deposited silicon is mainly integrated into existing dots which become larger and finally start to coalesce with each other in the last phase. Here, the dot density decreases until eventually there is only one contiguous silicon layer.

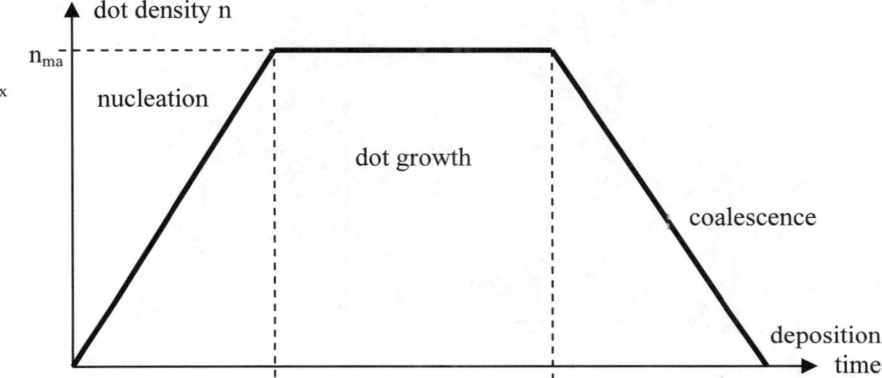

Figure 3. Idealized, temporal evolution during the deposition of silicon nano-dots by LPCVD exhibiting three phases: nucleation, dot growth and coalescence.

Coalescing silicon dots on top of a thin silicon dioxide layer are shown in the scanning electron microscopy (SEM) image of Figure 4. It can be seen that coalescing silicon dots form larger contiguous areas. The amount of deposited silicon is too high resulting in a non-optimal dot density.

By adjusting the deposition parameters appropriately it is possible to decrease the deposited silicon amount and to maximize the dot density. In Figure 5 a SEM image of silicon nano-dots on a 3-nm-thick tunnel oxide is shown. In comparison to Figure 4 the amount of deposited silicon material is smaller resulting in an optimal dot density of $4 \cdot 10^{11}$ cm^{-2}. The dot diameter is between 5 and 10 nm. This dot deposition process is used in the devices presented in the following paragraphs.

Fabrication: High Quality Silicon Dioxide Deposition by RPECVD

Silicon dioxide is one of the best electrical insulators for standard silicon process technology in terms of interface quality as well as electrical and mechanical stability. Furthermore, it can be very easily grown from silicon material by thermal oxidation as it is applied for the fabrication of the tunnel oxide. During this process silicon material is exposed to an oxygen atmosphere at high temperatures (800-1100°C) and reacts with it to silicon dioxide. 44% of the final silicon dioxide layer was formerly silicon material. So, during this process silicon is consumed. In the case of floating-dot memories it is necessary to deposit the control oxide instead of growing it. During a thermal oxidation process the silicon nano-dots would be completely consumed and thus, their charge-storing function eliminated.

Figure 4. SEM image of coalescing silicon dots.

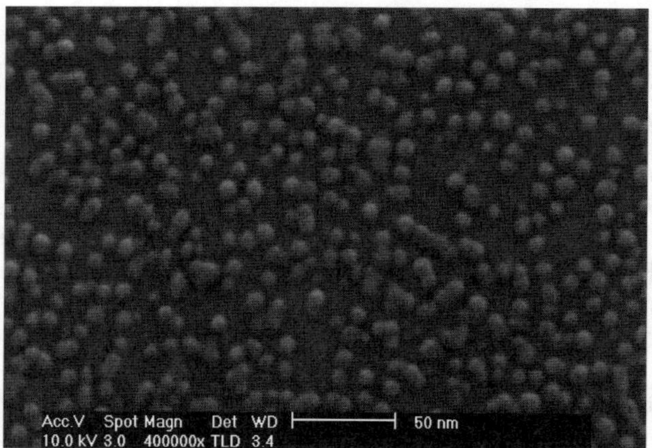

Figure 5. SEM image of LPCVD-deposited silicon dots with a density of $4\cdot10^{11}$ cm^{-2} and diameters of 5-10 nm.

The dielectric layers in FLASH and floating-dot memory devices play a key role in regard to the function and reliability of these devices. The deposited silicon dioxide must therefore be of high quality which can be recognized by a very low electrical conductivity, low oxide charge, low oxide trap and low interface state density as well as a high electrical breakdown field. To meet these requirements a remote plasma enhanced chemical vapor deposition (RPECVD) process is used.

The RPECVD chamber system is depicted in Figure 6. Due to its construction it offers several advantages allowing for the deposition of very high quality silicon dioxide. For the deposition the sample is heated up to a temperature of only 250°C. This low temperature is possible because of the additional application of plasma energy. The plasma is generated inductively (inductively coupled plasma, ICP) by a high frequency (HF) coil which is energized by an HF generator and a matching unit. The plasma region is spatially separated from the sample substrate and thus does not damage the sample surface by irradiation. Furthermore, it is possible to select the gases which pass the plasma region and are excited. In this case only oxygen and helium are let into the chamber passing the plasma region. The helium is used as an energy transfer medium transporting the plasma energy from the plasma region to the substrate surface. These selective gas inlets limit the possible chemical reactions and the resulting chemical products which improve the quality of the deposited oxide. Silane is introduced by a sample-near gas ring contributing the needed silicon. At the sample surface silane and oxygen form silicon dioxide assisted by the plasma energy delivered by the helium and oxygen molecules. Except for the silicon dioxide only volatile substances like O_2, He, H_2 and H_2O (gaseous) emerge which are pumped out of the system. With this process it is possible to deposit high-quality silicon dioxide at a constant deposition rate. After the deposition the sample is thermally annealed in an RTP chamber system at 1000°C for 30 s in order to further improve the silicon dioxide and its interface quality. The achievable quality of the deposited silicon dioxide is close to thermally grown SiO_2. The electrical properties are summarized in Table 1 [17, 18].

Table 1. The electrical properties of the RPECVD-deposited SiO_2.

Oxide charge density	Interface trap density	Breakdown field
$7 \cdot 10^{10}$ cm^{-2}	$3 \cdot 10^{10}$ cm^{-2}eV^{-1}	12 MV/cm

Floating-dot Memory Transistors

Applying the aforementioned processes, floating-dot memory transistors were fabricated on silicon on insulator substrate with a top silicon layer thickness of 100 nm. The transistor channel width W is 20 µm, the gate length L is 2 µm. All structures were defined by optical lithography. The tunnel oxide is 3-nm thick, the control oxide is 50-nm thick. Two SEM images of this device are shown in Figure 7.

The memory properties of these devices are characterized by electrical measurements. Figure 8 depicts the transfer characteristics of a floating-dot memory transistor which show the effect most clearly. The drain current I_D is plotted logarithmically versus the gate voltage V_G. The drain source voltage V_{DS} is 100 mV in this case. The first measurement runs from negative to positive gate voltages. The backward measurement was done in the opposite direction. The high negative and positive voltages at the beginning of the measurements led to discharged and negatively charged nano-dots, respectively. This charging effect can be observed by a

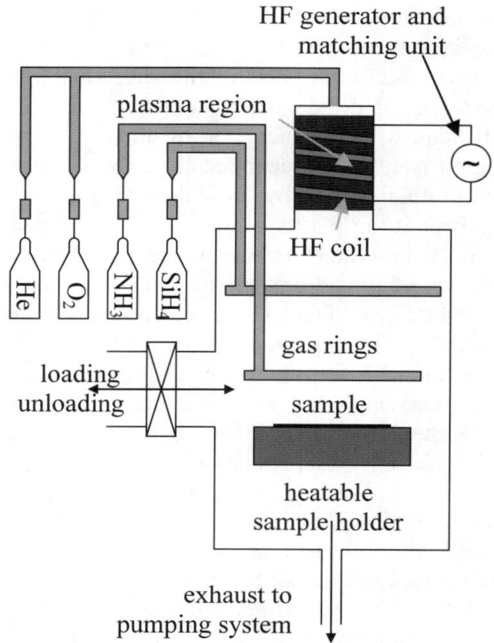

Figure 6. Schematic of an RPECVD chamber system.

shift of the I_D curves along the gate voltage axis. After applying a high positive voltage the silicon nano-dots were charged with electrons which led to a right shift of the second curve. The hysteresis between the two curves, i.e. the threshold voltage shift ΔV_{th}, is a direct indicator for the amount of charge stored inside of the nano-dots. A threshold voltage shift of 5.3 V is observed corresponding to a charge difference of about six electrons per silicon nano-dot between the first and second measurement.

The drain current ratio between these two curves is shown in Figure 9. It can be used to evaluate the memory signal strength when reading the information out of such a memory transistor. It can be seen that the drain current modulation exceeds 10^6 at a gate voltage of about ⁻5 V which allows a distinct read-out of the stored information.

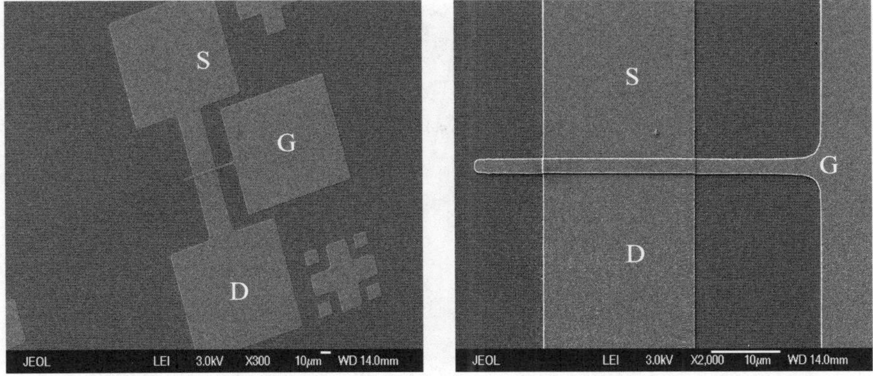

Figure 7. SEM images of a floating-dot memory transistor with W=20 µm and L=2 µm.

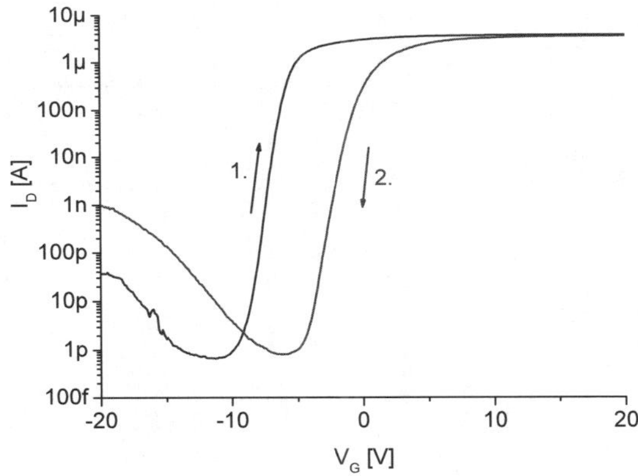

Figure 8. Transfer characteristics of a floating-dot memory transistor revealing a threshold voltage shift of 5.3 V due to charged nano-dots. V_{DS}=0.1 V.

In order to retain the stored charge for commercially relevant periods (~ 10 years) the tunnel SiO_2 must not be thinner than about 8 nm [19]. In this case the tunnel oxide is only 3-nm thick resulting in an exponentially shorter retention time. It can be observed within days and hours by monitoring the drain current I_D with fixed drain source and gate voltages. In Figure 10 such a retention time measurement can be seen. The drain source voltage V_{DS} is 100 mV, the gate voltage is kept at 0 V after charging and discharging the silicon nano-dots by applying +40 V and −42 V for 1 min respectively. The monitoring time period is 12 h. A worst-case linear approximation results in a storage time of three days. By using a thicker tunnel oxide drastically increased storage times will be achievable.

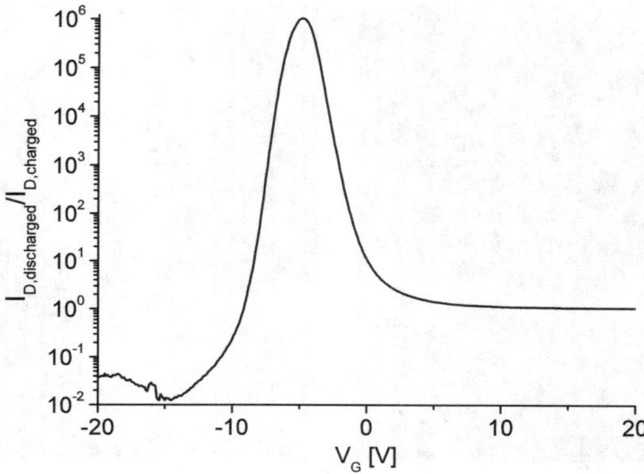

Figure 9. Drain current ratio $I_{D,discharged}/I_{D,charged}$ versus gate voltage V_G, t_{cox}=50 nm.

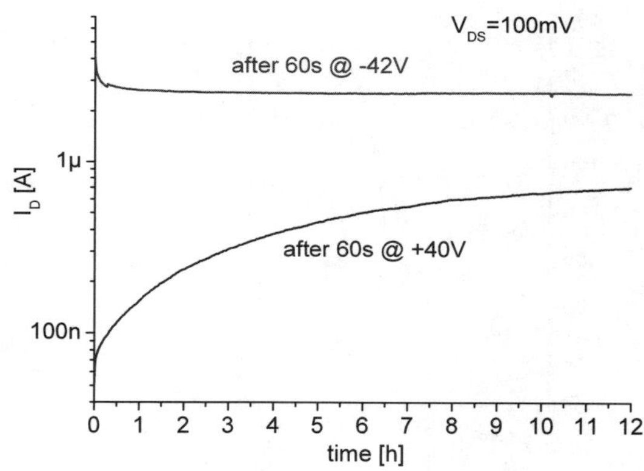

Figure 10. Storage retention time measurements after charging and discharging. V_{DS}=0.1 V.

In order to lower the programming voltages it is necessary to use a thinner control oxide on top of the nano-dots. Exploratory devices have been fabricated with a 20-nm-thick control oxide. This lowers the threshold voltage hysteresis as well as the write/erase voltages. In Figure 11 the transfer characteristics of one of these devices is shown.

The write and erase voltages are +18 V and –21.5 V, respectively. The channel width W of this exploratory memory transistor is 40 µm. As can be seen in Figure 11 the measured threshold voltage hysteresis is 1.5 V which indicates an average charge of 4.2 electrons per silicon nano-dot.

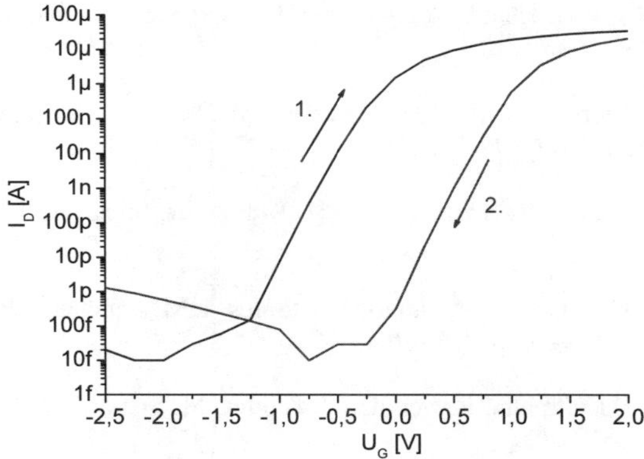

Figure 11. Transfer characteristics of a floating-dot memory transistor with only 20 nm control oxide exposing a threshold voltage shift of 1.5 V. V_{DS}=0.1 V.

Conclusion

Non-volatile memories have become an indispensable part of today's digital data processing. The quickly growing mobile electronics market especially fuels the demand for these devices. The presented floating-dot memory concept discloses a related and CMOS-compatible alternative with enhanced write/erase endurance compared to FLASH while not demanding severe changes of the manufacturing process at the same time. Here, the charge-storing silicon nano-dots are deposited by a self-organized LPCVD technique. The introduced concept is based on advanced SOI substrates, which exhibit fabrication as well as device advantages and offer higher scaling potential than conventional bulk silicon substrates. The electrical data of the presented examination devices prove the suitability of the floating-dot memory concept and pave the way for enhanced non-volatile memory devices.

Acknowledgements

The authors would like to thank the Bundesministerium für Bildung und Forschung (BMBF, under contract number 01M3101C), the Ministerium für Wissenschaft und Forschung des Landes Nordrhein-Westfalen (MWF-NRW) and the European Commission (network of excellence Silicon-based Nanodevices - SINANO, project reference 506844) for funding.

References

1.   S.M. Sze, Future Trends in Microelectronics, Wiley p. 291 (1999).

2.   B. Eitan, P. Pavan, I. Bloom, E. Aloni, A. Frommer, D. Finzi, IEEE Electron Device Lett. 21, pp. 542-545 (2000).

3.   S. Tiwari, F. Rana, H. Hanafi, A. Hartstein, E.F. Crabbé, K. Chan, Appl. Phys. Lett. 68 (10), 1377-1379 (1996).

4.   O. Winkler, F. Merget, M. Heuser, B. Hadam, M. Baus, B. Spangenberg *et al.*, Microelectron. Eng. 61-62, pp. 497-503 (2002).

5.   O. Winkler, M. Baus, B. Spangenberg, H. Kurz, Microelectron. Eng. 73-74, pp. 719-724 (2004).

6.   L. Tsybeskov, G.F. Grom, M. Jungo, L. Montes, P.M. Fauchet, J.P. McCaffrey *et al.*, Mater. Sci. Eng. B69-70, p. 303 (2000).

7.   M. Zacharias, J. Heitmann, R. Scholz, U. Kahler, M. Schmidt, J. Bläsing, Appl. Phys. Lett. 80 (4), p. 661 (2002).

8.   M. Hao, H. Hwang, J. C. Lee, Appl. Phys. Lett. 62 (13), p. 1530 (1993).

9.   T. Shimizu-Iwayama, N. Kurumado, D. E. Hole, P. D. Townsend, J. Appl. Phys. 83 (11), p. 6018 (1998).

10. Y. Shi, K. Saito, H. Ishikuro, T. Hiramoto, Japan J. Appl. Phys. 38, 2453-2456 (1999).

11. E. Nagata, N. Takahashi, Y. Yasuda, T. Inukai, H. Ishikuro, T. Hiramoto, Japan J. Appl. Phys. 38, pp. 7230-7232 (1999).

12. M.C. Lemme, T. Mollenhauer, W. Henschel, T. Wahlbrink, M. Baus, O. Winkler *et al.*, Solid State Electron. 48, 4, 529-534 (2004).

13. M. Lemme, T. Mollenhauer, W. Henschel, T. Wahlbrink, M. Heuser, M. Baus et al., Microelectron. Eng. 67-68,. 810-817 (2003).

14. Y. Choi *et al.*, IEEE Elec. Dev. Lett., 23, 1, pp. 25-27 (2002).

15. R. Chau *et al.*, 2002 Int. Conf. on Solid State Devices and Materials, Japan, (2002).

16. S. Miyazaki, Y. Hamamoto, E. Yoshida, M. Ikeda, M. Hirose, Thin Solid Films, Elsevier, 369, 55-59 (2000).

17. R. Rölver, O. Winkler, M. Först, B. Spangenberg, H. Kurz, 13th Workshop on Dielectrics in Microelectronics, 2004, Cork, Ireland, to be published in Microelectronics Reliability.

18. R. Rölver, O. Winkler, M. Först, B. Spangenberg, H. Kurz, 19th European Solar Energy Conference and Exhibition, 2004, Paris, France, to be published in the conference proceedings.

19. M. Specht, M. Städele, F. Hofmann, ESSDERC 2002, Proceedings of the 32nd European Solid-State Device Research Conference, Bologna, Italy, pp. 599-602 (2002).

Ion-beam Synthesis of Nanocrystals for Multidot Memory Structures

V. Beyer, J. von Borany

Forschungszentrum Rossendorf, Institute of Ion Beam Physics and Materials Research, Dresden / Germany

Introduction

According to the International Technology Roadmap for Semiconductors (ITRS) [1] multidot memory dominates the research activities of emerging memory devices. First published in the mid 1990s [2] this memory concept bases on a layer of well-separated Si or Ge nanocrystals (NCs) embedded in the transistor gate oxide substituting the floating gate of classical Flash-memory devices. Transferred through an injecting oxide, charges are captured or trapped in or at NC, which leads to a change of the transistor-characteristic. The separation of storage nodes allows further reduction of device dimensions, excess voltages and times with retaining non-volatility and improved endurance. Moreover, this concept supports the ability to store two bits in one transistor cell [3]. An overview of preparation methods of multidot memories is summarized in ref. [4]. The NC-formation by ion-beam synthesis (IBS) utilizes processes of self-organization and redistribution of supersaturated and perturbed matrices. Mainly small NCs ($d_{NC} \leq 3$ nm) are prepared with very high density of $N_{NC} \geq 10^{12} \text{cm}^{-2}$. Combined with cost-effective processing IBS is a promising method for NC formation.

The Principle of Ion-beam Synthesis

From the technological point of view IBS is a combination of high-dose ion implantation (II) and subsequent annealing. II leads to a supersaturation by impurity atoms (Si, Ge) inside the gate oxide. Generally speaking, their distribution, which is of gaussian-type by implantation, is adjusted by the mass and energy (in keV) of the implanted ions. Their average position (projected range R_p) is determined by energy, the achieved NC-size and NC-density by II-dose (in ions/cm^2). The NC-formation is described in terms of self-organization, nucleation and growth (Figure 1) [6]. Their mean size, density and spatial distribution changes during Ostwald ripening. These more general considerations have to be extended for high dose ion implantation in thin gate oxides taking "swelling" and "interface(IF)-mixing" into account. Furthermore, for Ge-II a "redistribution" of impurity atoms was observed.

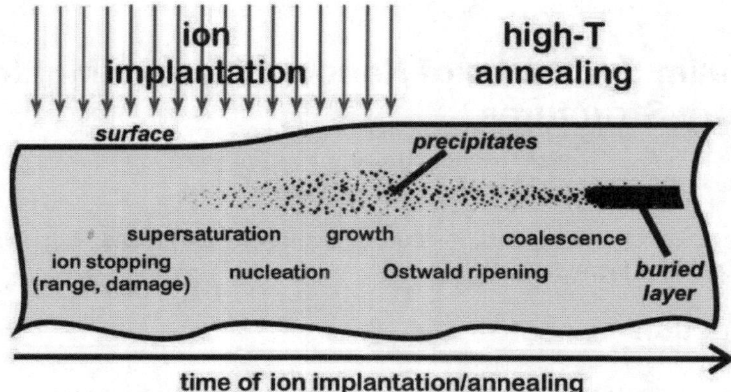

Figure 1. Scheme of ion-beam synthesis of nanostructures after [5]. By subsequent anneal-ing or even during ion implantation NCs nucleate inside the supersaturated region and grow in the following. Buried layers or wires can be formed for very high ion doses by a process of coalescence.

These aspects have a strong impact on the formation of NC-based memory devices by IBS.

Passing through the matrix the implanted ions lose energy by interactions with electrons and nuclei of matrix atoms within the SiO_2, whereas the latter one domi-nates for low ion energies. At high-dose II a cascade of displaced atoms leaves a strongly damaged gate oxide behind. As a consequence, the damaged oxide surface is sensitive to the local environment of cleaning and annealing conditions [7, 8]. The subsequent annealing is necessary to form NC and to restore the oxide as well.

Different Approaches of NC Formation by IBS

Figure 2 shows three main concepts of synthesis leading to different distributions of NC. (i) Ultra-low-energy (ULE) implantation of Si^+-ions in ≤ 10 nm SiO_2 using II-energies of ~1 keV and very high II-doses ($D \geq 1 \times 10^{16} cm^{-2}$) leads to a very shal-low Si-profile in a near-surface region, where Si-NCs form during annealing. (ii) A completely different concept is pursued by irradiation of a poly-Si(Gate)|SiO_2|Si(bulk) stack. Damage cascades of displaced atoms lead to a mixing of both formerly sharp Si/SiO_2-IFs. During subsequent annealing and IF-reconstruction a rather symmetric arrangement of Si-NCs forms in vicinity to the gate- and bulk electrode within the SiO_2. (iii) By low-energy (LE)-implantation both components – primary implanted Si ions and detached Si atoms by IF-mixing — contribute to the resulting NC distribution. A comparison of Si- and Ge-II shows that in the latter case additionally a redistribution of Ge during annealing has to be taken into account. Each of these concepts is discussed in the following exemples.

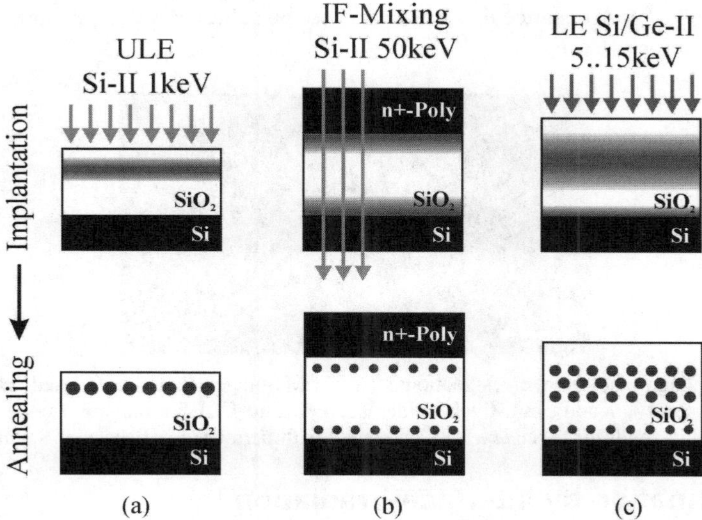

Figure 2. Three main concepts of NC-formation by means of IBS differing in the applied ion energy lead to different NC-distributions.

Ultra-low-energy Implantation

As mentioned before the ULE-IBS is carried out in thin gate oxides (~ 7nm) using a Si-ion energy of 1 keV and very high doses (1×10^{16}cm^{-2}) [8–10]. As a consequence, the resulting Si excess in the oxide reaches ~35 at.% [10]. This high amount of added atoms leads to a continuous swelling of the oxide during implantation and thus to a very shallow insertion of Si atoms. This high degree of perturbation requires annealing temperatures of T ≥ 950°C for at least 30 min in N_2 or Ar atmosphere. Kinetic Monte Carlo (KMC) simulations [11] show that in those cases the process of NC-formation has to be discussed in terms of "spinodal decomposition" and percolation. Transmission electron microscopy (TEM) investigations (Figure 3) indicate far more the formation of a structurally and/or electrically connected layer with crystalline reflexes than of isolated and well separated NCs [8, 12].

A memory window of ~2 V is reported with applied write/erase pulses of ±9 V for durations of ≥10 ms [8]. Neither degradation nor drift in memory window was detected after 10^6 write/erase cycles. Long-term extrapolation indicates a 10-year retention with a memory window of about 0.4 V. Here the control oxide is improved by dilution of the annealing gas with 1.5% O_2. The NC-layer is positioned more towards the gate than to the Si-bulk electrode (see Figure 3 for 7 nm SiO_2). In contradiction to that the charge exchange between Si-substrate and the NC-layer dominates [8, 9]. Thus, the mechanism of charge transfer through the oxide is supposed to be trap-assisted. The attempt to reduce the tunneling distance to <5 nm between the Si-NC layer and the Si substrate by increasing of II-energy failed (for details see [10]). A distance of ~2 nm can only be achieved by implantation in thinner gate oxides (~5 nm).

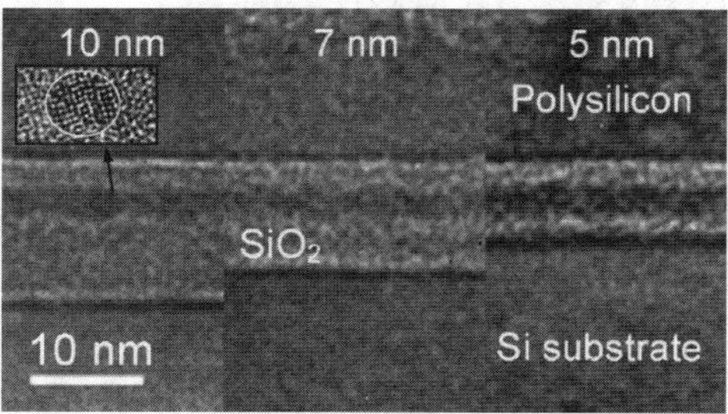

Figure 3. High-resolution cross-sectional (X-)TEM images under defocused bright field conditions for 10, 7 and 5-nm-thick oxides according to ULE implantation under same II- and annealing conditions (see text) [8], presented with permission of Elsevier Science B.V.

NC Formation by Interface-irradiation

Here the supersaturation of Si in the oxide is not directly related to the implanted species of Si. The Si-excess near the Si/SiO_2-IFs is achieved just by irradiation of a stack as presented in Figure 2b, in principle more or less independent on type of ion. In this case a 50 nm n^+-poly-Si gate covers 15 nm SiO_2 grown on p-Si [13]. This stack is irradiated by 50 keV 1×10^{16} Si^+-ions/cm². This concept of NC-formation by irradiation is protected by patents for device-relevant applications [14].

As a consequence of II, matrix atoms (Si and O) are displaced and mixed; both formerly sharp Si/SiO_2-IFs get smeared out. For the as-implanted state the expected Si-excess is shown in the inset of Figure 4a. The majority of implanted Si-ions is positioned deeply in the Si-bulk, thus a direct contribution of those atoms to the NC-formation is negligible.

During subsequent annealing (T ≥ 950°C) the instable mixture of Si and SiO_2 starts to separate. During reformation of the IFs some more strongly displaced Si atoms stay within the SiO_2, form precipitates and grow to NCs. The same happens to displaced O atoms within the Si-bulks forming SiO_2-clusters in the gate and substrate. The process of IF-reformation and NC-formation is confirmed by KMC-simulations (see Figure 4). The number of Monte Carlo (MC) steps corresponds to annealing temperature and time.

Aspiring towards equilibrium conditions the Si- and SiO_2-clusters disappear with on-going annealing time, but due to the higher mobility of O in Si than Si in SiO_2 the Si-NC are far more stable. The process of dissolution and growth can be described by means of Gibbs–Thompson relation and Ostwald ripening [6].

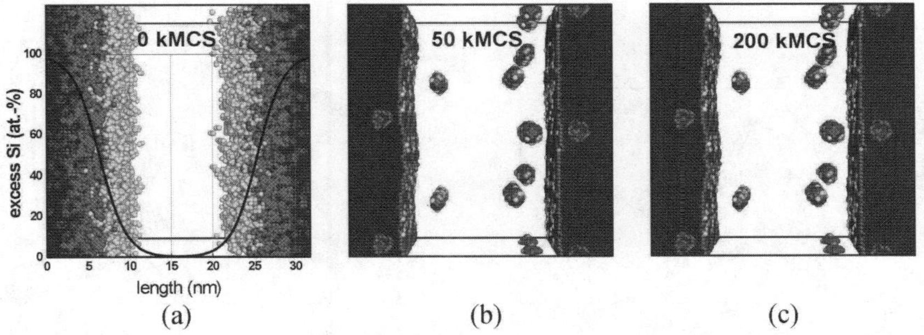

Figure 4. A 3D-KMC simulation of an irradiated stack as presented in Figure 2b. The gray-scales reflect the number of diatomic Si-Si bonds. After 50000 MC steps Si-NC and SiO_2-clusters are formed in the SiO_2 and in the Si-bulk, respectively. With proceeding simulation/annealing time the Si/SiO_2-IF gets smoother and the NCs dissolve until an equilibrium of totally separated regions is reached. The KMC-simulation was performed by K.-H. Heinig, Forschungszentrum Rossendorf 2003, and is presented with permission (see also [5, 11, 16]).

As indicated in the Figures 2b and 4c, after annealing two layers of ~1-2 nm small NCs form almost symmetrically in vicinity to each Si/SiO_2-IF. They are self-aligned by a denuded zone of 1-3 nm SiO_2 separated from the Si-gate and substrate electrode. The distance depends on the degree of IF-mixing, annealing temperature and time. Although so far the direct detection of these Si-NCs by fringes in conventional high resolution (HR)X-TEM failed, decoration experiments prove their existence [15]. This concept has the distinct advantage that the sensitive SiO_2 region is covered and thus unaffected by cleaning and annealing ambient.

First electrical investigations of samples annealed at 1050° C for 30 s in N_2 show clear memory behaviour [13]. Write/erase-pulses of 10 ms at voltages of ~±6 V reach programming windows of ≤0.6V with a data retention of hours at 85°C. A trade-off between the degree of mixing and NC-size, NC-density and their distance to the Si-bulk limits the application for memory devices. A stronger perturbation for larger injection distances needs a higher temperature budget for reconstruction. As a consequence larger NCs ripen in much lower density.

Low-energy Implantation and Redistribution

The third concept deals with low-energy (LE) implantation of Si or Ge-ions in gate oxides of 10-30 nm thickness (Figure 2c). With respect to the different mass of Si and Ge the parameters of II (energy and dose) has to be adjusted to achieve similar ion distributions in SiO_2 and a comparable and moderate damage at the Si/SiO_2-IF. For d_{ox}=20 nm the implantation is carried out for $^{74}Ge^+$-ions at 12 keV with 5.0×10^{15} cm^{-2} (LD) and 1.5×10^{16} cm^{-2} (HD) and for $^{28}Si^+$-ions at 6 keV with 7.0×10^{15} cm^{-2} (LD) and 2.0×10^{16} cm^{-2} (HD), respectively. The samples are subsequently annealed at 950°C (Ge) and 1050°C (Si) for 30 s in Ar.

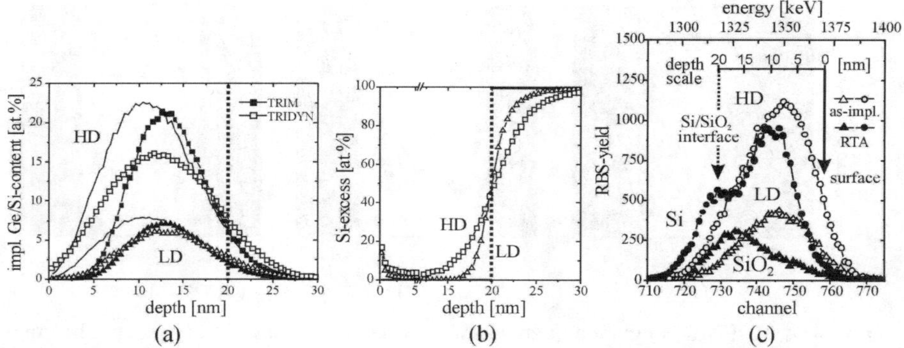

Figure 5. Effects of sputtering, swelling, IF-mixing and a redistribution have to be taken into account discussing IBS. (a) Implanted content of Ge and Si. Obtained Ge-II-profiles calculated by TRIM and TRIDYN result in considerable differences by swelling, especially for excess-concentrations ≥ 10 at.%. Si-TRIM profiles are referenced by the straight line. (b) A Si-excess by IF-mixing has to be taken into account, here shown by Ge-LE-II (TRIDYN). (c) After subsequent annealing a significant redistribution of Ge is observed by Rutherford backscattering spectrometry (RBS).

The distributions of implanted Ge and Si in SiO_2 are calculated by TRIM [17] (Figure 5a). For excess concentrations of ≥ 10 at.% dynamic effects of swelling and sputtering during implantation have to be taken into account, here calculated for Ge using TRIDYN [18]. A fraction of implanted ions and recoils reaches the Si/SiO_2-IF leading to mixing effects (see Figure 5b for Ge-II), which results in an additional Si-excess within the SiO_2.

During annealing Ge starts to redistribute (Figure 5c) [19, 20]; a fraction diffuses towards the Si/SiO_2-IF. Moreover, a loss of Ge-content is observed reaching about 30% of the implanted amount. In contrast implanted Si can be obtained as locally stable proved by isotope experiments [21]. Solely oxidation reduces the amount of free Si, whereas this fraction gets lost for the process of Si-NC-formation.

(HR)X-TEM investigations (Figure 6a) reveal that for Ge-IBS (LD) the NC-distribution differs significantly from the as-implanted Ge-profile. For the higher dose larger NCs form in the middle of the SiO_2 (Figure 6b). The position of Si-NCs (Si-II (HD)) follows the implantation profile (Figure 6c) as predicted by TRIM-simulation (Figure 5a). For the lower dose no fringes were detected. The evidence of Ge-NCs is much clearer than for Si-NCs due to an additional Z-contrast imaging. Similar to the IF-irradiation Si-NCs form in vicinity to the Si/SiO_2-IF as discussed before. Diffusing Ge atoms decorate the Si-NCs, thus compensating the tendency to dissolve during annealing.

Electrical analysis for Ge-IBS (LD) gives the expectation of programming times of ~100 ns for a memory window of ≥ 1 V at pulses of –8 V [22]. With write/erase pulses of ±8 V for 100 ms a large memory window of ~8 V (Ge-II) is achieved that lasts on the order of seconds with ~0.2 V remaining after days. For Si-LE-II(HD) the memory window saturates at ~2 V with a retention of 0.5 V for weeks with applied programming pulses of $\geq\pm8$ V for a duration of ≥100 ms. For similar structures a 256K-nvSRAM [23] is demonstrated proving long endurance capability.

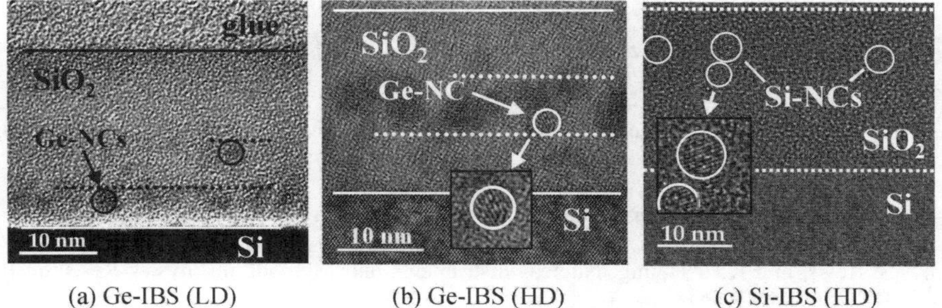

(a) Ge-IBS (LD)   (b) Ge-IBS (HD)   (c) Si-IBS (HD)

Figure 6. (HR)X-TEM investigations according to LE-IBS. Redistribution effects changes the position of NCs significantly (a). For high-dose II the position of NCs follows the impurity profiles of implantation (b,c).

Conclusion

Three different approaches for the IBS of nanocrystals in the gate oxide of MOS structures were discussed. In all cases clear memory behaviour was observed. A variance of multidot devices from DRAM-like to non-volatile-semiconductor-memory-like applications was shown. For high-dose IBS swelling, sputtering and IF-mixing have to be taken into account, which have a strong influence on the memory capabilities. It has been found that for Si-IBS or Ge-IBS, annealing influences the elemental depth profiles and the corresponding NC-distribution in a very different way. As a consequence, samples prepared by Si-IBS tend more to NVRAM-like behaviour, whereas the redistribution of Ge observed in Ge-LE-IBS provides DRAM-like properties. The latter combines the formation of NCs in a short distance to the Si substrate caused by IF-irradiation with a high density of small, well-separated NCs, which are fundamental requirements of common and future memory devices. Thus, IBS has the potential for use in cost-effective multidot memory applications, for example, in a multidot DRAM with prolonged retention time.

Acknowledgements

The work on LE-II was performed in cooperation with Infineon Technologies Dresden supported by the Federal Ministry of Education and Research of Germany, which is gratefully acknowledged. The work on the other concepts was sponsored by the European Community under the auspices of GROWTH Project No. GRD1-2000-25619 (NEON). Special thanks to R. Grötzschel and A. Mücklich for RBS and TEM investigations and T. Müller, K.-H. Heinig and B. Schmidt for their support, *e.g.* with TRIDYN and KMC simulations.

References

1. International Technology Roadmap for Semiconductors (ITRS), Semiconductor International Association (2003), http://public.itrs.net/

2. S. Tiwari, F. Rana, H. Hanafi, A. Hartstein, E.F. Crabbe, and K. Chan. Applied Physics Letters, 68(10):1377-1379 (1996)

3. B. Eitan, P. Pavan, I. Bloom, E. Aloni, A. Frommer, and D. Finzi. IEEE Electron Device Letters, 21(11):543-545 (2000)

4. J. De Blauwe. IEEE Transactions on Nanotechnology, 1(1):72-77, (2002)

5. S. Reiss, and K.H. Heinig. Nuclear Instruments and Methods in Physics Research B, 102:256-260 (1995)

6. K.H. Heinig, T. Müller, B. Schmidt, M. Strobel, and W. Möller. Applied Physics A, 77:17-25 (2003)

7. B. Schmidt, D. Grambole, and F. Herrmann. Nuclear Instruments and Methods in Physics Research B, 191:482-486 (2002)

8. P. Normand, E. Kapetanakis, P. Dimitrakis, D. Skarlatos, K. Beltsios, D. Tsoukalas, *et al.* Nuclear Instruments and Methods in Physics Research B, 216:228-238 (2004)

9. P. Normand, E. Kapetanakis, P. Dimitrakis, D. Tsoukalas, K. Beltsios, N. Cherkashin, *et al.* Applied Physics Letters, 83(1):168-170 (2003)

10. C. Bonafos, M. Carrada, N. Cherkashin, H. Coffin, D. Chassaing, G. Ben Assayag, *et al.* Journal of Applied Physics, 95 10:5696-5702 (2004)

11. T. Müller, K.H. Heinig, and W. Möller. Applied Physics Letters, 81(16):3049-3051 (2002)

12. M. Perego, S. Ferrari, M. Franciulli, G. BenAssayag, C. Bonafos, M. Carrada *et al.* Journal of Applied Physics, 95(1):257-262 (2004)

13. B. Schmidt, K.H. Heinig, L. Röntzsch, T. Müller, K.H. Stegemann, and E. Votintseva. Nuclear Instruments and Methods in Physics Research B, (to be published in 2005)

14. B. Schmidt and K.H. Heinig. Patent: EP 1 070 768 A1 (1999)

15. L. Röntzsch, K.H. Heinig, and B. Schmidt. Materials Science in Semiconductor Processing, (to be published in 2004)

16. M. Strobel, K.H. Heinig, W. Möller, A. Meldrum, D.S. Zhou, C.W. White, *et al.* Nuclear Instruments and Methods in Physics Research B, 147:343-349 (1999)

17. TRIM program is free for download. http://www.srim.org

18. W. Möller and W. Eckstein. Nuclear Instruments and Methods in Physics Research B, 2:814-818 (1984)

19. K.H. Heinig, B. Schmidt, A. Markwitz, R. Grötzschel, M. Strobel, and S. Oswald. Nuclear Instruments and Methods in Physics Research B, 148:969-974 (1999)

20. M. Klimenkov, J. von Borany, W. Matz, R. Grötzschel, and F. Hermann. Journal of Applied Physics, 91(12):10062-10067 (2002)

21. D. Tsoukalas, C. Tsamis, and P. Normand. Journal of Applied Physics, 89(12):7809-7813 (2001)

22. V. Beyer and J. von Borany. (to be published)

23. J. von Borany, T. Gebel, K.-H. Stegemann, H.-J. Thees, and M. Wittmaack. Solid-State Electronics, 46:1729-1737 (2002)

Scaling of Ferroelectric-based Memory Concepts

R. Waser

Institut für Festkörperforschung and Center of Nanoelectronic Systems in Information Technology - CNI, Forschungszentrum Jülich, Jülich / Germany and Institut für Werkstoffe der Elektrotechnik, RWTH Aachen, Aachen / Germany

Ferroelectric-based Memories within the RAM Family

For all matrix-based memories, the storage principle is based on physical states that can be read electrically by addressing the matrix element. Matrix-based memories (Figure 1) are based today on semiconductor chips and can be grouped into read and write random access memory (RAM), typically used for data storage, and read only memory (ROM), typically used for instruction storage.

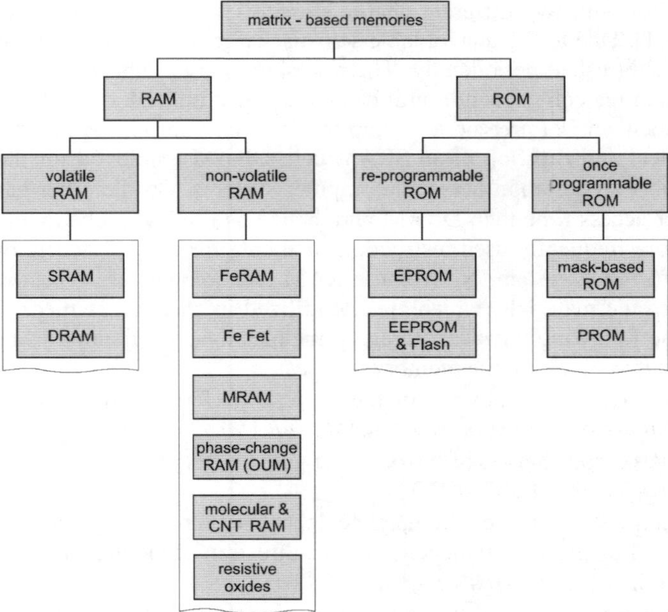

Figure 1. Categories of matrix-based memory chips showing the most relevant types. The abbreviations are explained in the text. The boundary between non-volatile RAM and reprogrammable ROM is not sharp. Today, it is given by the write times, which are in the ns to µs range for non-volatile RAM and in the ms to s range for reprogrammable ROM [1].

The latter group can be further divided into once-programmable ROM and reprogrammable ROM. The programming of the first group is either performed during fabrication of the chip by an appropriate layout of the last metallization mask (mask-based ROM), which is cost-effective only in very large numbers (> 10 000 pcs.) or, in the case of programmable ROM (PROM), by the customer in a first programming step in which, for example, tiny metal bridges at the matrix nodes are either fused or left intact to represent the binary information. In the case of reprogrammable ROM, MOSFETs with an additional *floating gate* are used as storage elements. The floating gate is charged to different voltages during the programming sequence to open or close the MOSFET channel in a non-volatile fashion. Before a new programming sequence, the information must be erased by discharging the floating gates of all memory cells. For electrically erasable PROM (EEPROM and Flash) reprogramming is conducted by an enhanced programming voltage, typically generated on the chip. In all types of reprogrammable ROM, the re-writing is far more time-consuming than the reading. Hence, they are used only when rewriting is rarely required.

RAM devices are classified into volatile RAM and non-volatile RAM. The volatile RAM types comprise static RAM (SRAM) and employ flip-flop-based latches as storage elements, while the so-called dynamic RAM (DRAM) uses a tiny capacitor with two different charge storage levels to represent the binary information [1]. Due to the unavoidable self-discharge of capacitors, the information needs to be refreshed periodically. The refreshing period is a fraction of a second. Since the storage cell capacitor and the single select transistor of a DRAM cell requires a much smaller area on a Si chip than the six transistors needed to make up the cell select and flip-flop of an SRAM cell, DRAM employed for the main primary memory in common personal computer systems. On the other hand, SRAM has a faster access time than DRAM and, hence, is used as cache memory to temporarily store frequently used instructions and data for quicker access by the processing unit(s) of a system. Nonvolatile RAM technologies are typically based on novel electronic materials, which are not utilized in classical semiconductor technology. The following physical storage principles have been employed or are suggested for future non-volatile memory devices:

- A capacitor with a hysteretic dielectric, *i.e.* a ferroelectric, is employed in the non-volatile *ferroelectric RAM (FeRAM)*.
- Polarization charges of ferroelectric gate oxides of a FET are employed in *ferroelectric FET (FeFET)*.
- Electrical resistance of a magnetoelectronic tunneling junction in which the tunneling probability depends on the direction of the magnetization is utilized in *magnetoresistive RAM (MRAM)*.
- Electrical resistance of a phase-change material that can be thermally switched between a crystalline and an amorphous phase is used by a novel non-volatile RAM concept called ovonic unified memory (OUM) [2], based on the same writing principle as the rewritable DVDs.
- Electrical resistance at a crosspoint of carbon nanotubes switchable by external electrostatic potentials.

- Electrical resistance of organic molecules, that depends on a switchable polarization or configuration state.

Other variants and combinations have indeed been reported in the literature. Still, this list provides the relevant devices currently in use or shows a high potential for future systems. In most cases, the physical storage principles for RAM are based either on the charge stored at the node (charge-based RAM) or the node shows a different resistance (resistance-based RAM) for the logic states. In some cases, the physical storage principle is combined in the gate of the select transistor, for which the channel resistor is readout (*e.g.* in the case of EEPROM or FeFET).

This chapter will cover the operation principles and the scaling properties of non-volatile memories based on ferroelectric oxides. It will comprise FeRAM, FeFET, as well as concepts based on the switchable resistance of ferroelectric oxides. Prior to this specific description, we will consider some general issues of the scaling of matrix-based integrated memories. More detailed information on the entire topic is given in [1].

General Scaling Trends for Future Memory Generations

The future scaling of matrix-based memory chips is determined by economic, technological, and physical boundary conditions. The historical development of memories in the past decades and the International Technology Roadmap for Semiconductors (ITRS) [3] provide reasonable guidelines for further evolution. Here, we will only consider general aspects while specific topics are discussed in the following sections. In contrast to the ITRS, we will not give any information on the estimated year of introduction but only on trends concerning the geometrical and electrical specifications. The data for the predictions are mainly taken from [3–5].

First, we introduce definitions to describe the geometry scaling shown in Figure 2 vs. the minimum feature size, F. The die size, A_{DS}, is the total area of a memory chip grown in the past, but is expected to grow only slightly during the coming memory generations for economic reasons. The fraction of the storage matrix area on the chip is given by X_{Matrix}. For the 256 Mb DRAM generation, the storage matrix area is approximately 55% of the total die area, *i.e.* $X_{Matrix} = 0.55$. During the next generations, this value may increase to 0.75 … 0.85 (Figure 2). The area of an individual storage cell at a node of the matrix is denoted A_{CA}, which is given by the square of the feature size, F^2, times the cell area factor, X_{CA}, which describes how many F^2 are needed to realize the cell:

$$A_{CA} = X_{CA}F^2 \tag{1}$$

For example, on a 256 Mb DRAM chip a cell area factor $X_{CA} = 8$ is required to realize a cell. According to the ITRS, it is expected that it will be possible to reduce X_{CA} to 4 in the long run (Figure 2).

From these considerations one can calculate the total storage capacity of a chip, *i.e.* the RAM capacity (bits/chip) by

$$\text{RAM Capacity} = \frac{X_{\text{Matrix}} A_{\text{DS}}}{X_{\text{CA}} F^2} \tag{2}$$

Figure 2 indicates, for example, that a possible future 1 terabit (Tb) memory generation is expected to require a feature size in the range of 10 to 12 nm together with a die area of about 600 mm². This is true of all types of matrix-based memories able to follow the extrapolations for X_{Matrix} and X_{CA} sketched in the diagram. It should be noted that the storage density [(bits/cm²) = RAM capacity / A_{DS}] shows the progress in technology more clearly than the total RAM capacity. In fact, in the long run the ITRS predicts a decreasing chip size for economic reasons, which will delay the introduction of RAM generations on a single chip while the functional density is scaled aggressively.

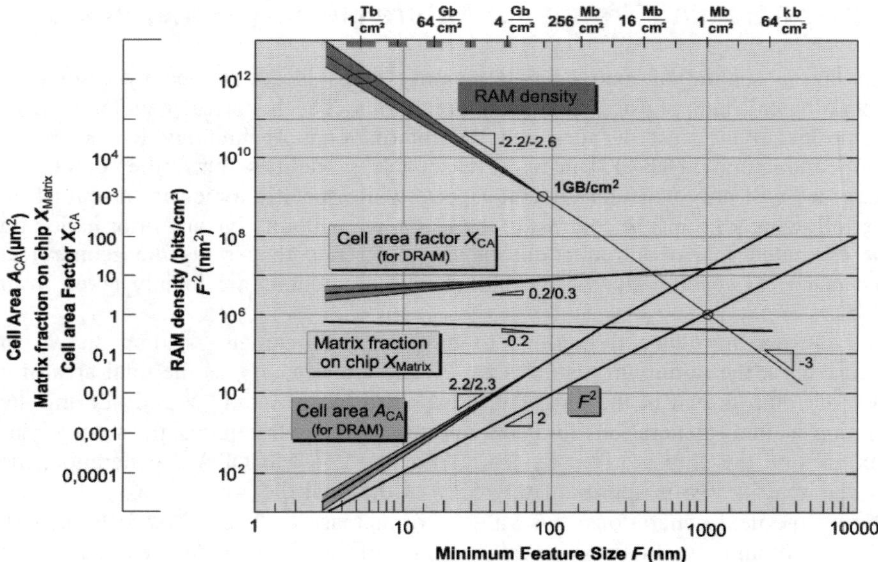

Figure 2. Roadmap of the chip and cell geometry. The symbols are explained in the text [1].

In addition to the geometrical aspects, there are general trends for electrical specifications that are independent of the specific type of memory (Figure 3).

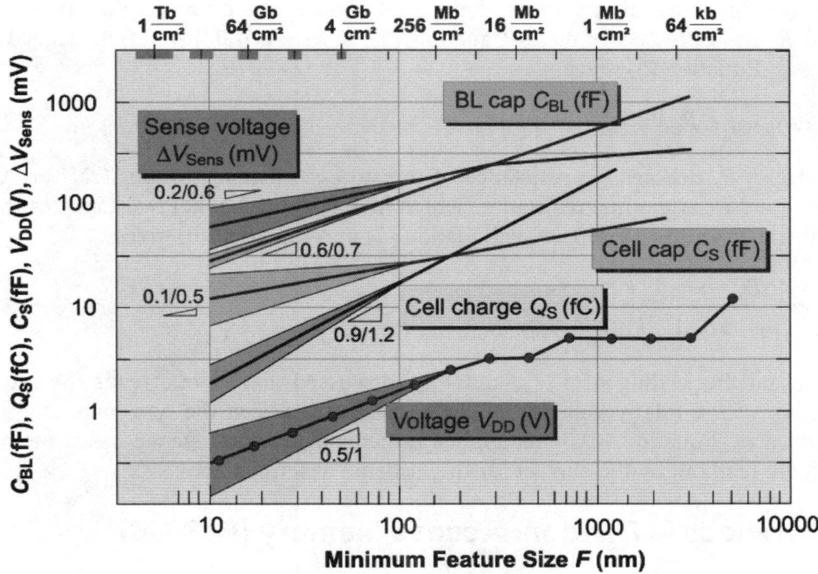

Figure 3. Roadmap trends of electrical specifications such as the BL capacitance, the minimum signal voltage and the operating voltage. The BL capacitance is shown for 512 storage cells per BL, which is typical of a 64 Mb DRAM. The cell capacitance is relevant for DRAM. The symbols are explained in the text [1].

The sense amplifier margin $V_{S,min}$ is expected to decrease slightly. This is also true of the bit line capacity C_{BL} due to the shorter line for connecting the same number of cells. The product reveals the minimum charge Q_S needed to be stored in a cell:

$$Q_S = V_{S,min}(C_{BL} + C_S) \tag{3}$$

In order to reduce Q_S significantly, the BL may be divided into sub-BLs to reduce C_{BL}. This may be very helpful since the major challenge for future charge-based RAM lies in the area needed for the storage capacitor, which is proportional to Q_S. This topic is discussed briefly in [1]. In addition, the operating voltage V_{DD} will continue to decrease (Figure 3). According to the ITRS roadmap, V_{DD} will scale in the range between F^1 and $F^{1/2}$, depending on an optimization for minimum power or for maximum speed. In conjunction with Q_S, this gives the scaling of the storage cell capacitor C_S for DRAM

$$C_S = \frac{Q_S}{V_{DD}} \tag{4}$$

while for FeRAM a different relationship applies.

For resistance-based memories, there is a boundary condition on the product of I_0 and $R_{S,OFF} + R_L$ according to Equation (5), since $V_{BL}(\text{"0"})$ may not exceed V_{DD}. Hence, Equation (5) reveals

$$I_0 \cdot \left(R_{S,OFF} + R_L \right) = V_{BL}(\text{"0"}) \leq V_{DD} \tag{5}$$

Here, R_S denotes the resistance of the storage element (in the "ON" or "OFF" state) and R_L is the line resistance, which includes the channel resistance of the access transistor, too. Insertion of Equation (5) into Equation (6) gives

$$\frac{\Delta R_S}{2\left(R_{S,OFF} + R_L \right)} = \frac{V_S}{V_{DD}} > \frac{V_{S,min}}{V_{DD}} \tag{6}$$

From the scaling of $V_{S,min}$ and V_{DD}, discussed above, one reads the required scaling of the relative change of the resistance between the two logic states, expressed by Equation (6). This applies in general to all resistance-based memories that are read out by the current sensing scheme described above.

Ferroelectric Random Access Memory (FeRAM)

Modern ferroelectric random access memory (FeRAM) using one transistor and one capacitor per cell node (1T-1C concept) is built similarly to DRAM, with the difference that it incorporates a ferroelectric film as a capacitor to hold data, instead of a simple dielectric film. The ferroelectric film has the characteristic of a remanent polarization, which can be reversed by an applied electric field and gives rise to a hysteretic P-E loop. By using thin film technologies, capacitors of submicron thickness can be prepared so that operation voltages are reduced to a level below standard chip supply voltages. FeRAM uses the P-E characteristic to hold data in a non-volatile state and allows data to be rewritten fast and frequently. In other words, an FeRAM, as all non-volatile RAM, has the advantageous features of both RAM and ROM.

Voltage pulses are used to write and read the digital information. If the electrical field of the applied pulse is in the same direction as the remanent polarization, no switching occurs. The change of polarization ΔP_{NS} is due to the dielectric response of the ferroelectric material. If the initial polarization is in the opposite state of the field, the polarization reverses giving rise to an increased switching polarization change ΔP_S.

The different states of the remanent polarization ($+P_r$ and $-P_r$) cause a different transient current behavior of the ferroelectric capacitor to an applied voltage pulse, as shown in [6]. By integrating the current, the switched charge ΔQ_S and the non-switched charge ΔQ_{NS} can be determined. The difference in charge $\Delta Q = A\,\Delta P$ enables one to distinguish between the two logic states. Scaling is characterized, in general, by the shrinking of the feature size, the reduction of the operation voltage and the expected enhancement of the voltage sensing. For the future scaling of FeRAM, a possible scenario is shown in Figure 4. Today, the development has reached a 64-Mb chip with a feature size of 0.13 μm.

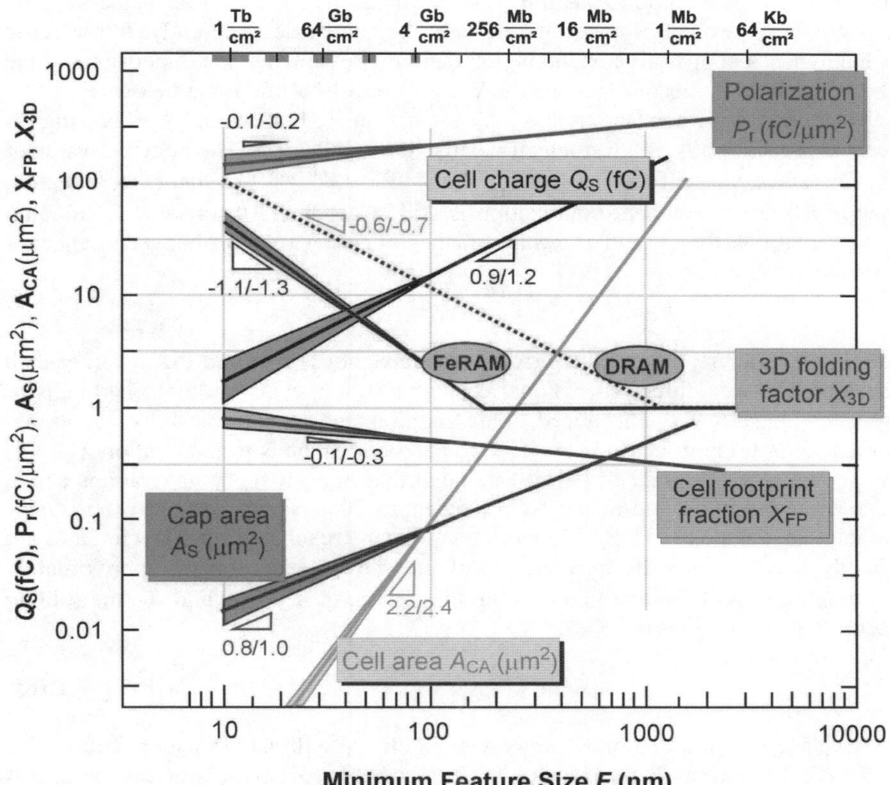

Figure 4. Roadmap trends of FeRAM. The symbols are explained in the text.

The area of an individual storage cell A_{CA} is given by the square of the feature size F and an area factor X_{CA} describing how many F^2 are needed to realize the cell. Today, X_{CA} for FeRAM ranges between 15 and 30, while it is typically 8 for DRAM. Cell optimization could lead to a decrease of X_{CA} for FeRAM. This resulted in a scaling of the cell area, A_{CA}, which is slightly steeper than F^2. The capacitor footprint area is the projection of the capacitor on the cell area A_{CA}. The factor X_{FP} indicates how much of cell area is covered by the ferroelectric capacitor: $A_{FP} = X_{FP} A_{CA}$. This factor is expected to be slightly enhanced along further miniaturization ($X_{FP} \propto F^{-0.1...-0.3}$). The limitations of the actual area of the ferroelectric capacitor A_S are determined by the switched cell charge and the remanent polarization: $A_S = Q_S / 2 P_r$. A reduction of A_S is possible as far as the charge is high enough to be detected by the sense amplifier. As shown above, it is expected that Q_S decreases with $F^{0.9...1.2}$, *i.e.* at feature size of 70 nm, a charge of 10 fC should be detectable.

In addition, scaling effects on the material properties of the ferroelectric which are independent of all design aspects, should be taken into account. In principle, the remanent polarization P_r is independent of the area within a ferroelectric plate capacitor. However, this may not be true at the nanoscale. Inherently, ferroelectric behaviour is a cooperative phenomenon that will cease when the dimensions of the ferroelectric body become too small. We will return to this issue below. Extrinsically, there are further factors that may impact on P_r. For example, processing effects at the sidewalls of a ferroelectric capacitor will decrease the effective value of P_r. Therefore, we will assume that P_r follows: $P_r \propto F^{0.1...0.2}$, so that even at feature size of 10 nm a remanent polarization is still larger than 10 $\mu m/cm^2$. Combining the estimations for Q_s and P_r results in the following relation between capacitor area and feature size:

$$A_s \sim F^{0.8...1.0} \tag{7}$$

When an actual area of the ferroelectric capacitor is required that is larger than the footprint area, three-dimensional capacitor structures as multi-stacked cells or trenches must to be introduced. This requirement is expressed by the three-dimensional folding factor $X_{3D} = A_S / A_{FP}$ and corresponds to the relation $X_{3D} > 1$. As shown in Chapter 22 of [1], it estimated that only memory generations above 256 Mb will require non-planar technologies. In this respect, FeRAM offers a significant advantage over DRAM where non-planar (trench and stack) techniques are already needed since the introduction of the 4 Mb generation (at approximately 1 μm feature size). Taking into account the scaling of A_{CA}, X_{FP} and A_S, the folding factor depends on F with:

$$X_{3D} \sim F^{-1.0...-1.5} \tag{8}$$

Long-term extrapolation to high density FeRAM with 64 Gb with a feature size of 30 nm the folding factor is about 10. This value is conceivable by the extrapolation of current techniques, but it requires a production at low tolerances.

The final memory density for FeRAM at the end of the roadmap depends on inherent physical properties, technological limitations, and economic aspects. Concerning the inherent physical issues, the potential limitation by the cooperative of the ferroelectricity has been mentioned already. This is called the superparaelectric limit. However, it has been shown theoretically and experimentally that films made from ferroelectric oxides keep their ferroelectric properties to thicknesses as low as 2 to 3 nm [7]. Lateral scaling seems to be limited to size above approximately 20 nm [8] by the superparaelectric limit. This is no immediate limitation, since the FE capacitor is highly folded anyway. Still, it is not clear yet how the superparaelectric limit is linked to aspect ratio of the ferroelectric body.

Another more simple physical limitation is given by the geometry. As described, the 3D folding helps to circumvent the discrepancy between the strongly decreasing cell area and the more moderate decrease of the required capacitor area. However, this does not immediately take into account that the capacitor has a finite thickness. If the size of the cell comes close to the thickness of a (folded, *i.e.* upright oriented) capacitor, scaling will come to its end. Suppose that ferroelectric capacitors of only 3-nm thickness can be fabricated. To this thickness we add electrodes of *e.g.* 3-nm thickness. If this stack is oriented perpendicularly, it requires 9 nm. The side of a square 4 F^2 cell for the 1 Tb/cm^2 generation shows approximately this length as can be read from (Figure 4). Hence, this will be the ultimate scaling limitation for FeRAM. Most probably, the technology and economy limits will come much earlier. Still, it is interesting to note that the limitation just discussed is slightly in favour of FeRAM compared to DRAM. In the multi-Gb era, DRAM will require high-k dielectrics. On the other hand, the leakage in DRAM cells is a very critical parameter because it leads to self-discharge and, hence, affects the refresh scheme. Because of the leakage behaviour of high-k dielectrics, dielectric thicknesses well below 10 nm are not conceivable. Leakage is much less an issue in FeRAM because of its storage and read-out principle. Therefore, we used the smaller thickness for our estimation.

A final consideration has to be given to the scaling of the coercitive voltage of ferroelectric capacitors. For proper read and write operations, the coercitive voltage, V_C, must always stay below approx. half the operation voltage, V_{DD}. Since V_{DD} decreases with the reduction of F, as explained above, V_C needs to decrease as well. However, in the sub-micrometer thickness regime, V_C does not scale proportional to thickness but much less (see Chapter 22 in [1]). The technological and inherent physical contributions to this relationship are not yet completely clear.

Ferroelectric Field-Effect Transistors (FeFET)

The concept of a ferroelectric field-effect transistor (FeFET) exploits the combination of a ferroelectric material as a gate insulator in a MOS type FET. The ferroelectric polarization charge adds to the charge at the gate dielectric, which controls the channel conductivity (Figure 5). By switching the direction of the ferroelectric polarization, the threshold voltage of the transistor can be shifted. This leads to a device that can be used for logic gates and still offers a non-volatile memory function. A detailed survey on FeFETs is provided by Chapter 14 in [1].

Although the concept appears simple, many problems and challenges are encountered in the realization. First of all, there is no useful ferroelectric oxide material that forms a "clean" interface with Si on the atomic scale. Typically, there must be an optimized non-ferroelectric buffer layer (I) between the Si (S) and the ferroelectrics (F), giving rise to a MFIS stack structure (M = metal electrode). This buffer has to be highly insulating to suppress injection and leakage currents and should have a relatively high permittivity in order to reduce the voltage drop across layer I. Buffer materials such as ZrO_2, HfO_2, or CeO_2 are reported. At the interface with Si the formation of few atomic layers of SiO_2 can not be avoided due to the oxygen affinity of Si. Still, charge injection often cannot be completely avoided, leading to apparent hysteresis effects without any ferroelectric origin. In order to optimize the processing of the buffer layer I and the ferroelectric layer F independent of each other, a MFMIS is built. The center metal M acts as a floating electrode in this structure. A second challenge arises from the fact that the remanent polarization P_r of a typical ferroelectric oxide exceeds 10 µC/cm^2. An analysis of the potential and charge distribution shows, however, that large P_r values lead to huge electric fields in layer I which may far exceed the breakdown voltage. Within a

MFMIS structure, this problem can be avoided by reducing the area of the F layer on top of the inner layer M. A third, interrelated, challenge is the limited retention time of all FeFET structures built up to now. Due to small leakage currents, the polarization charges of the ferroelectrics are screened within a period that is in the range of days, at most. The requirement for a universal non-volatile memory is a retention time of at least 10 years. A fourth challenge arises from the application of FeFETs in high-density 1T non-volatile memory circuits (1T = one transistor per cell). The NAND matrix offers (for ideal FeFETs) a layout in which X_{CA} can be reduced to a value as small as 4, similar to the situation in NAND Flash memories. However, the disturb pulses to non-addressed cells is so high that it cannot be tolerated under real operation conditions. The disturb scenario is significantly improved in the AND matrix scheme, at the expense of a larger cell. For details see [9].

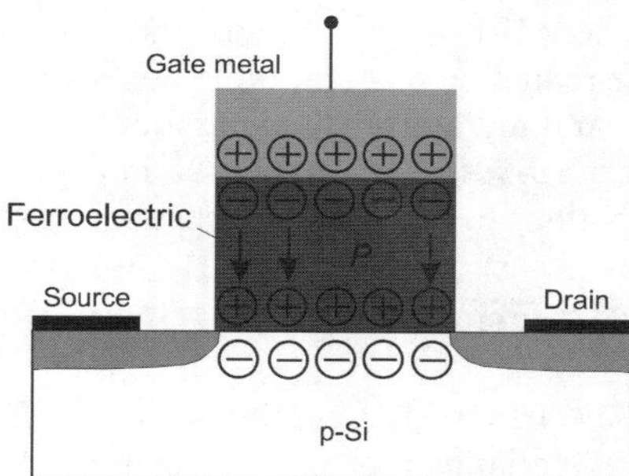

Figure 5. Schematic cross-section of a FeFET. The gate dielectric is a ferroelectric material. As an example, a polarization state including surface charge layers is shown.

With regard to scaling, the *ideal* FeFET can be treated very similarly to MOSFET and can follow the same scaling rules [3]. A precondition is that the properties of the ferroelectric materials can be adjusted accordingly. Especially, there is the need to reduce the coercitive voltage (similar to the scaling requirements of the FeRAM, see above). In addition, the remanent polarization needs to be kept at an adequately *low* level while the hysteresis loop must remain square-shaped. An ultimate limitation to scaling of ideal FeFETs is the superparaelectric limit below a critical lateral size of approximately 20 nm [8].

However, the scaling of real FeFET devices faces more immediate limitations, which are deduced from the challenges mentioned above. First of all, a reduction of the films' thicknesses leads to charge injection and leakage currents, which introduce additional, undesired hysteresis effects and a screening of the ferroelectric polarization. If the thicknesses are kept constant during the lateral scaling of the device, the operation voltage can not be reduced as required for the scaling of the standard MOSFET. Details of two scaling approaches using a constant gate stack thickness and a variable gate stack thickness are reported in [10].

Ferroelectric Resistive Switching Effects

There are various conceivable interactions between polarization charges and the resistance of passive layer structures in which the direction of the polarization affects the resistance value.

One such interaction is expected to occur in ultrathin ferroelectric films, the thickness of which is low enough to allow for tunnelling currents (ferroelectric tunnel junction, FTJ). The direction of the electron tunnelling current through such an MFM structure can be either parallel or antiparallel to the direction of ferroelectric polarization. There are different mechanisms through which polarity of the ferroelectric polarization may impact on the barrier and, hence, the tunnelling currents. As a consequence, the current-voltage (I–V) characteristics are expected to show dedicated features at the position of the coercitive voltage and a hysteretic behaviour. There are several theoretical and experimental papers that deal with this phenomenon [1–13].

The interaction mechanism in an FTJ may be, for example, the piezoelectric response (the so-called butterfly curve), which changes the width of the barrier and, because of the strong thickness dependence of the tunnelling currents, causes a detectable change in the I–V curve. Furthermore, an interaction of the ferroelectric polarization on the band offset and/or the effective mobility of the charge carriers is conceivable [13].

Another interaction of the polarization charge directly affects the shape of the potential barrier and, as a consequence, the charge transport by any transport mechanism (thermoionic emission, diffusive transport, *etc.*). Different variants are reported in the literature. The ferroelectric Schottky diode assumes a conductive, ferroelectric film with an ohmic contact on one side and a Schottky contact on the other [14]. Depending on the direction of the ferroelectric polarization, the depletion layer is either strengthened or compensated, leading to a low or high resistance of the structure. A related concept is based on a two-layer stack of a ferroelectric film and a (moderately) conducting, non-ferroelectric layer [15]. The direction of the ferroelectric polarization determines whether a potential barrier or a potential valley is formed at the ferroectric layer/non-ferroelectric layer interface, leading to a corresponding modulation of the resistance (Figure 6). In fact, it can be shown that the ferroelectric Schottky diode and this two-layer system are the two limiting cases of the same concept.

It is worthwhile to mention that further resistive switching effects have been reported for non-ferroelectric oxides (*e.g.*, [16]). The scaling potential of all resistive switching concepts cannot be estimated as long as the mechanism is not clarified. One important aspect relates to the effective area of the phenomena. The concepts mentioned above assume that the resistive switching takes place uniformly over the entire electrode area. Alternatively, however, the effect may take place at local "hot spots", representing channels of significantly enhanced conductivity through which most of the current flows. The extent of this localization determines the scaling potential for future resistively switching memory devices. If the hot spot mechanism holds, it must be guaranteed by the material and the technology that one (or defined number of) hot spot exists per memory element even when the area becomes very small. If the uniform conduction and switching mechanism holds, geometrical scaling is possible down to very small feature sizes. In this case, the challenge is given by a sufficiently high conductivity of the stack, that is still switching.

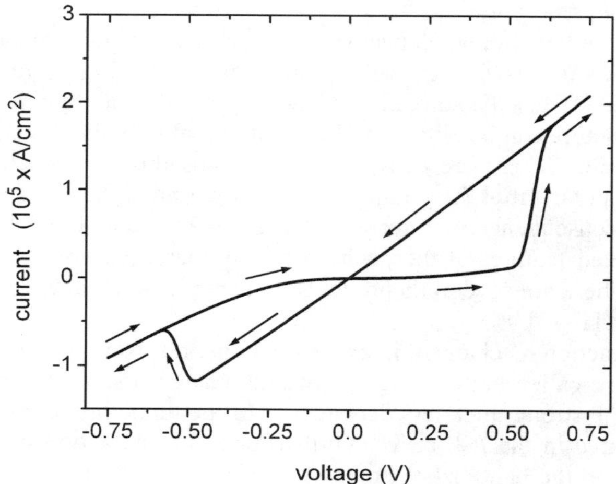

Figure 6. Calculated *I–V* curve for a two-layer stack consisting of a ferroelectric and a non-ferroelectric layer. Carrier concentration at both electrodes: 10^{19} cm^{-3}, thickness of the ferroelectric layer: 50 nm, spontaneous polarization: 10 µC/cm^2, thickness of the dielectric layer: 10 nm. Calculation performed by R. Meyer [15].

Acknowledgements

The author gratefully acknowledges G. Bednorz, U. Böttger, A. Gerber, H. Kohlstedt, R. Meyer, Y. Mustafa, J. Rickes, K. Szot and S. Tiedke for valuable discussions.

References

1.   R. Waser (ed.), Nanoelectronics and Information Technology, Wiley, 2003.

2.   S.R. Ovshinsky, Phys. Rev. Lett. 36, 1469 (1968). RAM based on phase-change materials have been introduced by Ovonyx Inc., http://www.ovonyx.com , http://www.ovonic.com ; IEDM 2001 Techn. Digest, Session 36, (2001); G. Wicker, SPIE vol. 3891, 2-9, (1999).

3.   International Technology Roadmap for Semiconductors (ITRS), Semiconductor International Association (2003), http://public.itrs.net/

4.   K. Itoh, *VLSI Memory Chip Design*, Springer series in advanced microelectronics, Springer, Berlin Heidelberg New York, (2001).

5.   A. Nitayama, Y. Kohyama, and K. Hieda, IEDM 1998 Techn. Digest, (1998).

6.   O. Lohse, M. Grossmann, D. Bolten, U. Boettger, R. Waser, MRS Symp. 655, CC 7.6.1 (2001).

7.   J. Junquera and Ph. Ghosez, Nature 422, 508 (2003).

8.   A. Roelofs, T. Schneller, K. Szot, and R. Waser, Nanotechnology 14, 250-253 (2003).

9.   M. Ullmann, PhD thesis, Hamburg (2002).

10.   M. Fitsilis, Y. Mustafa, R. Waser, to be published.

11.   L.Esaki, R.B. Laibowitz and P.J. Stiles, IBM Tech. Disc. Bull. 13 (1971) 2161; L.L. Chang, L.Esaki L, IBM Tech. Disc. Bull. 14 (1971) 1250.

12.   R. Wolf, P.W.M. Blom, and M.P.U. Krijn, U. S. Patent Nr. 5,541,422 (1996).

13.   J. Rodriguez Contreras, H. Kohlstedt, U. Poppe, C. Buchal, N.A. Pertsev and R. Waser, Appl. Phys. Lett. 83 (2003) 495; J. Rodriguez-Contreras, PhD thesis, ISBN 3-89336-368-8 (in English), University of Cologne (2003).

14.   P.W.M. Blom, R.M. Wolf, J.F.M. Cillessen, and M.P.C.M. Krijn, Phys. Rev. Lett., 73, 2107 (1994).

15.   R. Meyer, J. Rodriguez Contreras, A. Petraru, and H. Kohlstedt, accepted by Integrated Ferroelectrics.

16.   Y. Watanabe, J.G. Bednorz, A. Bietsch, Ch. Gerber, D. Widmer, A. Beck, Appl. Phys. Lett. 78 (2001) 3738; C. Rossel, G.I. Meijer, D. Brémaud, D. Widmer, J. Appl. Phys. 90 2892 (2001).

Device Concepts with Magnetic Tunnel Junctions

H. Brückl, J. Bornemeier, A. Niemeyer, K. Rott

University of Bielefeld, Nano Device Group, Bielefeld / Germany

Introduction

The joint venture of IBM and Infineon Technologies unveiled a prototype 16-Mbit magnetoresistive RAM (MRAM) in June 2004. This 16-Mbit MRAM chip features read and write cycles of around 30 ns, which makes it competitive with established DRAM memory chips. The prototype with a cell size of 1.4 μm^2 is manufactured with a 0.18 µm CMOS process used for standard logic chips. A 4-Mbit MRAM of Freescale Semiconductor Inc. possesses a similar performance and is a standard product, that is already available to customers. The next development steps include transferring the design to a 0.13 micron process.

The fast read and write features give a huge advantage over other non-volatile memory devices like flash cards. Furthermore, MRAM is less demanding with respect to power consumption than comparable DRAM chips. The first projected markets are therefore the replacement of flash memories in RF-ID, memory cards, or mobile phones.

Besides automotive and sensor applications, magnetic logic devices are said to be another lucrative market by arguments resting on the same advantages as MRAM technology. The non-volatile character allows a reprogrammable and re-configurable logic. Especially tempting is the fact that memory and logic are then based on the same technology platform. This opens the unique opportunity to build up a unified system on a single chip.

The core of this progress is the magnetic tunnel junction (MTJ), which basically consists of two ferromagnetic (fm) layers separated by a nanometre-thick insulating barrier that electrons are able to tunnel through. In the following, the concept and the material aspects are illustrated. New developments with double barriers and more sophisticated three-terminal devices will be introduced. And finally, concepts are proposed for a full integration of magneto-electronic devices in Si technology.

Material Aspects of Single MTJ

Although different companies follow different routes to implement magnetic tunnel junctions in the common Si environment, lower process temperatures and contamination reasons favor the preparation of junction stacks in the backend-of-line (BEoL) process, *i.e.*, after the transistor levels are finished. A write line and a bit line generate the local magnetic fields for bit writing (Figure 1). Selective readout is managed by one or two additional transistors per cell (1Tr1MTJ by Freescale, or

2Tr1MTJ) or by a cross-point architecture where several junctions are connected in series (NEC).

The serial circuitry with the transistor has the consequence that the absolute junction resistance has to be much larger than the transistor resistance. Therefore, depending on the junction size, the area resistance of the tunnel stack should be about 0.1 to 10 $k\Omega\mu m^2$, which corresponds to a starting Al thickness of 0.6 to 0.9 nm plus oxidation (described later). Depending on the technology process, the discrimination of the high and low level additionally requires a voltage gap of 100 to 400 mV. Thus, a resistance change as large as possible is desirable. Following the simple model of Julliere [1], the relative resistance change, called tunnel magnetoresistance (TMR) amplitude, is directly connected to the spin polarization P of the ferromagnetic material next to the barrier. We have tested a variety of sputtered polycrystalline materials. The best values are yielded from Fe-based alloys ($P_{Co70Fe30} = 50\%$, $P_{Co50Fe50} = 48\%$, $P_{Ni80Fe20} = 53\%$) whereas simple transition ferromagnets give lower values ($P_{Fe} = 43\%$, $P_{Co} = 38\%$). Thus, TMR amplitudes of 50 % could be easily reached by proper choice of ferromagnetic materials. Even larger amplitudes can be achieved by half-metallic Heusler alloys like Co_2MnSi (P = 60%) [2] or $La_{0.67}Sr_{0.33}MnO_3$ (P = 78%) [3], although their fabrication is complicated. Astonishingly, the alloy CoFeB recently showed much larger TMR amplitudes of 70% at room temperature [4]. The reason for this is unclear, and it is speculated that its amorphous state leads to a smoother surface, and hence, a smoother Al growth for the barrier. Consequently, the oxidized barrier is more homogeneous and defect-free. Although not contained in the original Julliere theory, the defect-free quality of the barrier is also decisive for large TMR amplitudes. The importance of the barrier was best recognized by a Japanese group who measured 230% TMR with epitaxial Mg oxide barriers [5]. The deposition process, however, cannot easily be transferred to replace the standard Al oxide barrier.

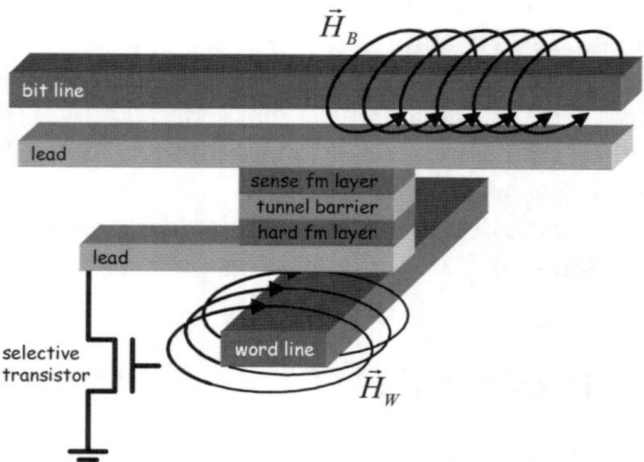

Figure 1. Principle layout of a 1Tr1MTJ cell by Freescale Semiconductor Inc.

Since MRAM chips are processed on Si wafers with a diameter of at least eight inches, the layer homogeneity is a further challenge in order to achieve a uniform area resistance across the whole wafer. Especially, the exponential behavior of the tunnel current requires a film thickness uniformity of ±0.1 Å. In order to avoid roughness-induced magnetic interactions [6] of the hard and sense magnetic layers, the interface roughness should be minimized, too. This can be achieved by optimizing the deposition parameters and choosing suitable materials, *e.g.*, amorphous or nanocrystalline instead of polycrystalline layers. Further issues like reliable magnetic switching [7], dielectric stability [8], and dielectric stability (discussed later) are of immense importance.

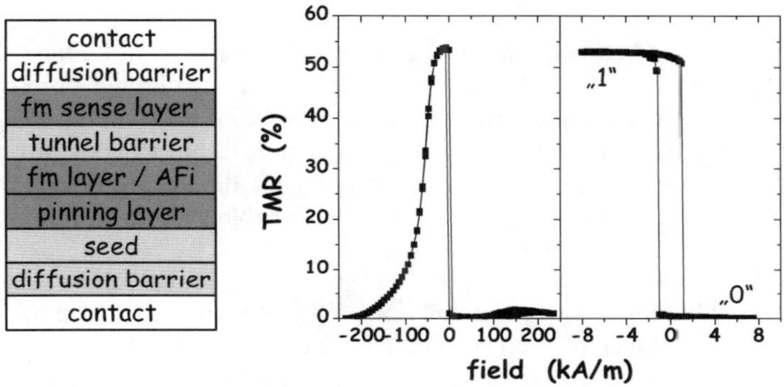

Figure 2. Left: setup of a layer stack with contacts and diffusion barriers. Right: major and minor loop of a TMR hysteresis. The two states of a bit are available at zero field.

The realization of MTJs comprises three main challenges in fabrication: the stack deposition, the barrier oxidation and the patterning. The latter wasn't an issue for the prototypes of the 0.18-µm technology, but isn't solved if the cell size enters the 100-nm limit. The tolerable standard deviation of the area resistance across an eight inch wafer is 2%, *i.e.*, the cell radii have to be kept constant within ±0.5 nm. On the other hand, the resistance must not change as well. Since the tunnel current grows exponentially with the thickness of the insulating barrier layer, the maximum allowable thickness variation is ±0.1 Å. Not only due to these criteria, the appropriate choice of material is essential for reliable MRAM fabrication. Figure 2 shows a typical layer stack which is suitable for MRAM application.

The bottom and top contact layers (Cu, Al, W) serve as an electrical connector to the enclosed functional magnetic tunnel cell. Diffusion barriers prevent the intermixing of layers, which would disable correct operation. The optimization of its functionality requires a more complicated layer stack than simple pictures implicate. The most important requirements are smooth growth, small crystallites or an amorphous state, hard/soft architecture of the magnetic properties, vanishing magnetic coupling between the layers, a perfect barrier without any defects, thickness homogeneity, and more. The hardmagnetic electrode typically consists of an antiferromagnetic (AF) pinning layer, which supports a ferromagnetic layer or an artificial ferrimagnet (AFi). Usually, proper growth of the AF layer requires an additional seed layer. An insulating tunnel barrier separates the hardmagnetic and sense electrode. The latter softmagnetic electrode is again usually capped by a diffusion inhibitor. The material aspects and functionality of the different layers are explained in the following sections.

Exchange Biasing

The hardmagnetic electrode should be as insensitive as possible to external magnetic fields, which are generated, for example, by the switching lines or by stray fields of neighboring layers. Furthermore, it should be thermally stable and simple to fabricate.

The simplest way to realize a hardmagnetic electrode seems to be the use of a single hardmagnetic layer, *e.g.,* CoCr [9] or CoCrPt [10]. Although this solution provides a high thermal stability, the magnetic interaction with the softmagnetic layer hinders independent switching. A different route is to pin a magnetic layer to a natural antiferromagnet (AF) by exchange biasing via direct magnetic exchange coupling. The pinning effect of the exchange bias is visible as a shift of the switching to large fields in TMR major loops (Figure 2).

Usually, the pinning effect is activated after film deposition by an annealing step at 300°C in a homogeneous magnetic field. While the magnetization of the adjacent ferromagnetic layer is aligned along an external field, the AF is heated above its Néel temperature. The direct exchange orients the AF interface spins along this direction and "freezes" them during the subsequent cooling. The strength of this coupling (typically several 10 kA/m) can be explained by a model of uncompensated spins; only a small fraction of interface spins contributes to the pinning (see the review by Berkowitz [11]). Typical AF materials are FeMn, IrMn, and PtMn. The two latter excel due to their corrosion resistance and high exchange field values [11–15]. For example, the composition $Ir_{17}Mn_{83}$ gives a relatively high Néel temperature and, hence, a high blocking temperature [16].

Thus, it is possible to completely separate the hysteresis curves of the hardmagnetic and softmagnetic electrodes, which is essential for MTJs.

The Artificial Ferrimagnet (AFi)

The AFi consists of two ferromagnetic layers separated by a nonmagnetic interlayer. A multitude of material combinations shows an indirect exchange coupling via conduction electrons in the interlayer. This forces an antiparallel alignment of the ferromagnetic layers at certain interlayer thicknesses [17, 18]. This interlayer coupling is an oscillating function with thickness, similar to the RKKY spin polarization in the neighborhood of a magnetic defect in a nonmagnetic host.

The most relevant AFi stack for MRAM application seems to be a FM/Ru/FM trilayer with CoFe, CoFeB, or similar as the ferromagnetic material [19, 20]. Ru is reported to have the strongest interlayer coupling with a second maximum at about 8 Å. A reduced net magnetic moment due to the antiparallel orientation of the ferromagnetic layers reduces the torque of an external field, and, therefore, enhances the stiffness of the hardmagnetic electrode. The AFi provides a clear improvement of the magnetic rigidity [21, 22]. In addition, the antiparallel alignment of the FM layers leads to a flux closure in laterally patterned junctions, which reduces the stray field interaction with the softmagnetic sense layer.

Alternately, trilayers consisting of ferromagnetic rare earth and transition metals show an antiparallel exchange coupling of the individual layers. Thus, they can be exploited as hardmagnetic electrodes as well. MTJs consisting of IrMn exchange-biased NiFe/Gd/NiFe trilayers show a compensation of the magnetic moment and a sixfold enhancement of the exchange-bias field [23]. Recent investigations show that this compensation even works at room temperature if Gd is replaced by GdCo alloy [24].

AFi stacks are also promisingly employed as softmagnetic sense layers due to their tendency to minimize stray fields and due to the possibility of spin flop switching [25], which provides a soft stepwise (by 90° angle) rotation of the magnetization in an otherwise compensated AFi.

Barrier Formation by Al Oxidation

The formation of the tunnel barrier is a challenging preparation step for the MTJs [26]. The simplest approach is direct Al oxide deposition, but better results are yielded by a post oxidation of thin Al films. This oxidizing mechanism, however, is discussed controversially in the literature [27, 28]. Recent high-resolution transmission electron images reported by a Japanese group visualized each oxidation stage including annealing and give a feeling of what happens [29]. If some metallic Al is left due to underoxidation, this leads to smaller tunnel resistances and TMR amplitudes. By overoxidation, however, the adjacent ferromagnetic layer is oxidized more or less by oxygen diffusion through the grain boundaries of the polycrystalline Al film, which generally also leads to a deterioration of the TMR and certainly to a much larger tunnel resistance.

Both the excellent wetting properties of Al on transition metals [30] and the apparently ideal oxidation without defects favor Al oxide barriers as the golden mean. Moreover, a working barrier seems to be possible with nearly all gentle oxidation methods as long as the Al film itself is continuous and smooth.

A lot of different oxidation techniques have been employed, *e.g.*, plasma [31], remote electron cyclotron resonance (ECR) plasma [32], ion beam [33], thermal [34], ultraviolet-light assisted [35], and natural oxidation [36]. The MTJs prepared by different oxidation methods show different barrier properties and TMR effects. For instance, natural und UV-assisted oxidation is better for low resistance MTJs in read heads, whereas plasma oxidation is favored for medium resistance memory cells in MRAM devices [37].

New Concepts by Magnetic Double Junctions

The development of magnetic double junctions (MDJs) that consist of either two stacked single MTJs or one MTJ in combination with a Schottky barrier is interesting from the physical and the application point of view [38]. On the one hand ballistic effects [39] and resonant states [40, 41] are predicted, leading to higher magnetoresistance values. On the other hand, it is possible to build a multi-valued logic and improve the bias voltage dependence in applications [42]. The essential electron transport regimes employed in MDJs are sketched in Figure 3 and described with respect to their advantages in the following.

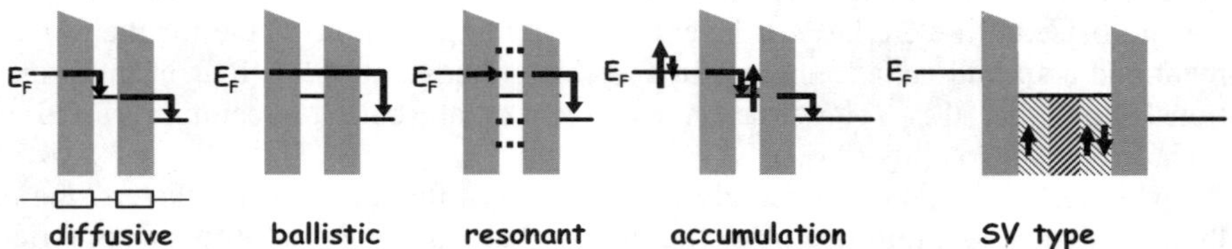

Figure 3. Comparison of different electron transfer mechanisms employable in MDJs.

Diffusive Electron Transport

Experiments with CoFe/Al oxide/NiFe/Al oxide/CoFe junction stacks demonstrate that a simple serial resistor model is sufficient to explain the electron transport at first glance [38]. The fast relaxation of electrons to the Fermi energy results in an overwhelming portion of diffusive transport. This means that the scattering time of tunnel electrons that cross the first barrier is smaller than the dwell time in the middle electrode. In this case, the Julliere model [1] predicts unchanged TMR amplitude, if (a) the spin polarization of all three electrodes is the same, (b) the tunnel electrons do not lose spin information during tunneling, and (c) there are no other contributions to the electron transport (see the following).

The serial circuit of two junctions considerably improves the bias stability, which is interesting for many applications including MRAM. The voltage drop at each junction is half of the bias voltage as shown in Figure 4. The benefits are larger TMR amplitudes at elevated bias voltages, improved electrical breakdown stability, and in addition, a multi-valued state: three different signal levels are addressable (Figure 4). This allows the storage of more than two states in a single elementary cell. The concept is expandable to four and more states for memory or logic applications [43, 44].

Resonant States

Resonant quantum well states, which could theoretically exhibit a large TMR amplitude, are not evident experimentally in MDJs consisting of metallic electrodes up to now. First hints to the presence of quantum well states are found in single junction experiments that show thickness oscillations with epitaxial Cu/Co bilayer electrodes [45].

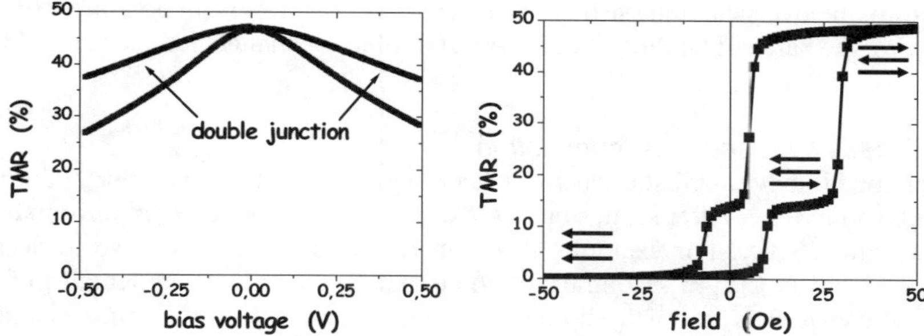

Figure 4. Left: bias dependence of a single and double junction. Right: triple state of a double junction. In case of equal TMR and area resistance of the single junctions, the levels would be equidistant.

Ballistic Electron Transport

All electrons crossing the barrier contribute to the tunnel current. These electrons are subject to a lot of scattering processes and relax to the Fermi level quite rapidly. The decay probability obeys an exponential law. Typical mean free paths before scattering are 1–10 nm at room temperature. The electrons that arrive at a certain distance without any energy loss are called ballistic or hot electrons.

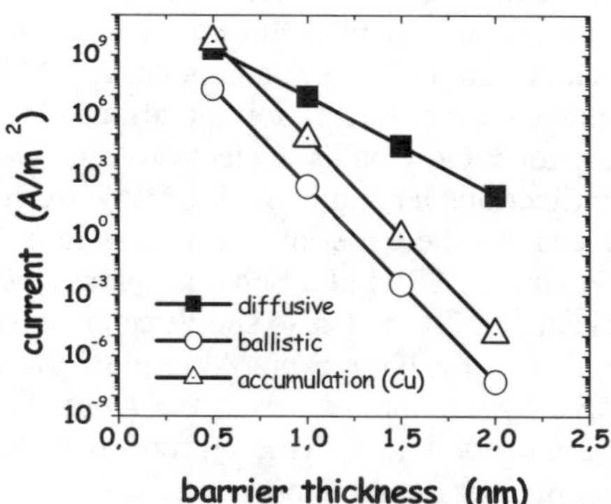

Figure 5. Calculated current density through a Co/Al oxide/Co (Cu)/Al oxide/Co MDJ for different transport regimes.

The absence of ballistic effects in MDJ can be understood by calculating the contribution of the ballistic to the total current. This calculation is carried out for Co/Al oxide 1.5 nm/Co 2 nm/Al oxide 1.5 nm/Co stacks at a bias voltage of 30 mV. The Schrödinger equation of free electrons is solved for the double junction potential in the anti- and parallel-magnetization state. Whereas the TMR amplitude rises from 44% for diffusive tunneling to 115% for ballistic transport (without any resonant state enhancement), the ballistic current is a factor of 10^7 smaller than the diffusive part. Thus, the larger ballistic TMR effect is always masked by the overwhelming diffusive current part. This ratio can be improved by decreasing the barrier thickness, but it is not possible to balance both currents (Figure 5).

Spin Excess by Electron Accumulation

As mentioned above, ballistic electrons keep their momentum and energy during their relaxation time. After scattering, the electrons can keep their spin information quite a while longer. For example, this spin relaxation time gives rise to decay length of 1 μm in Cu [46]. Contrary to noble metals, the spin decay length in ferromagnetic materials is much shorter (1–5 nm). Thus, a flow of spin polarized electrons across a barrier causes a local excess of one spin sort within the spin decay length, leading to spin accumulation (Figure 3). Still sustaining larger TMR amplitudes than single MTJ, contributions of accumulative currents can appreciably increase at thin barrier thicknesses (Figure 5, calculated with a Cu middle electrode).

The specified transport regimes are utilized in several new concepts of spin-dependent devices. They are quite promising with respect to their signal amplitudes. The next two sections are devoted to them.

Spin Filtering with Ballistic Electrons

Spin dependent measurements of ballistic electrons could distinguish between energy-dependent scattering events and filter out pure spin scattering. The sketch of Figure 6 shows the principle setup of a device containing a MTJ on top of a Schottky barrier. A ferromagnetic emitter injects spin-polarized electrons across a tunnel barrier to the second electrode. Only ballistic electrons are able to traverse the base to the Schottky barrier. Electrons having enough energy are collected in the semiconductor. This device enables the experimental access to ballistic electrons by a simple combination of a single MTJ and a Schottky barrier, which acts as a diode to collect the ballistic electrons. Figure 6 also shows an example of a measured ballistic magnetocurrent (BMC) of a NiFe 5 nm/Al oxide 1.8 nm/Co 5 nm stack on n-GaAs(100). The bias voltage is applied across the Al oxide barrier, whereas the lower Co electrode is grounded. The ballistic current over the Schottky barrier is measured by an ohmic contact at the semiconductor.

As usual, the TMR amplitude drops with increasing bias voltage. The BMC is zero up to the threshold of the Schottky barrier height of about 0.8 eV. Then, the BMC increases and reaches a maximum value of about 70% at 1.5 eV at a temperature of 50 K. This value increases at lower temperatures and is much larger than the usual TMR amplitudes in this voltage range.

Even larger BMC values were recently obtained by implementing a spin valve (SV) between two barriers (Figure 3). Spin valves consist of two ferromagnetic layers separated by a thin nonmagnetic layer, *e.g.,* Co/Cu/Co. Different spin decay lengths of majority and minority electrons cause spin filtering by traversing one of the ferromagnetic layers, and hence, a strong spin polarization of the ballistic current. Since the antiparallel magnetization state affects both channels of spin current, a very large magnetocurrent is available. The BMC depends exponentially on the thickness of the relevant magnetic layers. Therefore, it can reach infinity theoretically, but the total current also decreases exponentially. The balance of these opposite factors is a question of optimization of material, layer thickness, barrier transparency, and collection efficiency.

The so-called spin valve transistor (SVT) is a hybrid device, where a spin valve structure is embedded between two semiconductor layers exhibiting Schottky barriers. The collector current of such a SVT depends on the applied bias as well as the magnetic state of the base. A large BMC of typically 300% is obtained at room temperature. Although the fabrication is difficult for mass production, it clearly demonstrates the possibilities of such approaches. An extensive recent review on this topic can be found in [47].

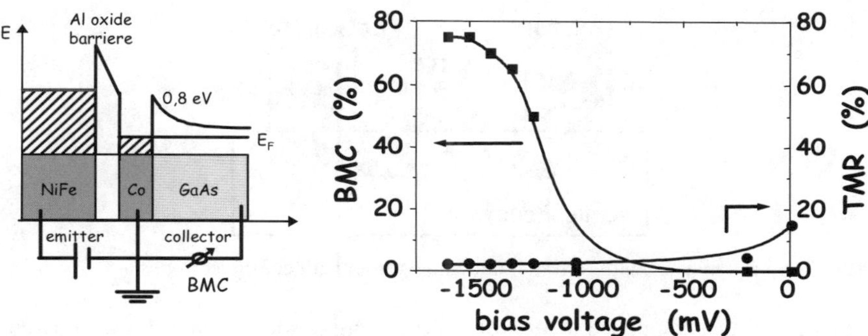

Figure 6. TMR and BMC amplitude vs. bias voltage in double junction stack consisting of a Co/GaAs (100) Schottky barrier and a single MTJ on top (80×80 μm^2) at a temperature of 50 K.

Magnetic tunnel transistors (MTTs) are similar to the SVTs, but inject hot electrons from a nonmagnetic metal emitter via a tunnel barrier instead of a Schottky barrier (Figure 4). BMC values exceeding 3400% are reported at a temperature of 77 K (with additional base voltage) [47]. Local studies of such spin valve structures (NiFe/Cu/Co/GaAs) by ballistic electron emission microscopy exhibit huge magnetocurrents of the order of 600% already at room temperature. BMC mappings show that the spin filtering effect of the spin valves is quite homogeneous on sub-μm scales [49].

MTT and SVT are suitable to inject a spin polarized current into the semiconductor, and to produce a spin accumulation zone near the contact. As shown, the degree of spin polarization can be quite large, if spin is also conserved during crossing the Schottky barrier. The principle has already been demonstrated by optical means [50, 51].

Spin Injection in and Ejection from Semiconductors

The implementation of MTJ or MDJ suffers from moderate temperature stability, which banishes the fabrication to the BEoL process. Typically, the TMR amplitude diminishes already above annealing temperatures of 300°C due to interdiffusion of the metallic species in the stack [52]. Most prominent is the diffusion of Cu to the barrier interface, which disables the spin polarization of the electrodes [53]. Thus, temperatures of up to 550°C in the FEoL process are far out of reach. The best solution would be the utilization of the semiconductor itself as spin host. This approach needs three crucial components: spin injection, spin transport, and spin ejection, *i.e.,* electrical detection (Figure 7).

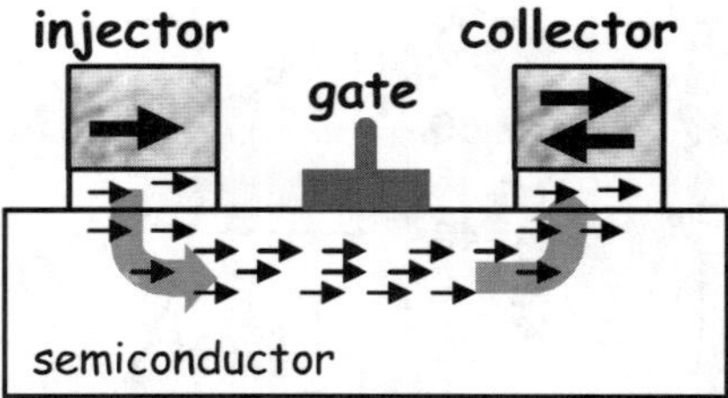

Figure 7. Spin transistor with ferromagnetic injector and collector electrodes.

Datta and Das supposed a so-called three-terminal device in 1990 [54]. It consists of a transistor with two ferromagnetic electrodes, the spin filter, and a third gate electrode, which steers the spin precession. The conductance of the device depends on the applied gate voltage, which allows switching between two states. Much simpler and more effective would be a direct detection of the injected spin-polarized electrons by a spin-conserving ejection to a ferromagnetic collector.

Precondition for the successful operation of such a spin transistor is a high spin injection rate into the semiconductor and a large spin diffusion length. The first issue proved to be difficult in praxis. In 1997, optical spin injection into a semiconductor was reported [55]. Further investigations of optical injection promised a long spin diffusion length corresponding to a decay time of at least 1 ms [56, 57]. The preceding chapter numerates several possibilities of injection. Other experiments used the injection of spin-polarized electrons from a ferromagnetic semiconductor, as discussed by Ohno [58]. The detection, however, is always achieved by optical means (*e.g.,* [58, 59, 60]). A spin polarization of about 20% at room temperature is shown by analyzing the polarization state of light emitted by a LED with NiFe contact leads [61]. Thus, both spin injection and transport in semiconductors are apparently feasible.

The opposite path, *i.e.,* the ejection from semiconductors, seems to be straightforward. Unfortunately, up to now there exists no report of a successful combination of electrical injection and detection of spin-polarized current in semiconductors. This experimental proof, however, would be the next step towards a spin transistor.

Acknowledgement

The authors are grateful for valuable discussions with G. Reiss and J. Schmalhorst and their support of this work. They also thank J. Schotter for careful reading of the manuscript. They acknowledge the collaboration with J. Langer and W. Maass from Singulus Technolgies AG, and the funding from the German Ministry of Education and Research (BMBF) in project #13N7989 and #13N8207.

References

1. M. Julliere, Phys. Lett. 54A, 225 (1975)

2. S. Kämmerer, A. Hütten, H. Brückl, to be published

3. D.C. Worledge, T.H. Geballe, Appl. Phys. Lett. 75, 900 (2000)

4. D. Wang, C. Nordman, J.M. Daughton, Z. Qian, J. Fink, IEEE Trans. Magn. 40, 2269 (2004)

5. http://www.aist.go.jp

6. L. Néel, C.R. Acad. Sci. 255, 1676 (1962)

7. H. Koop, H. Brückl, D. Meyners, G. Reiss, J. Magn. Magn. Mater. 272-276, e1475 (2004)

8. J. Schmalhorst, H. Brückl, M. Justus, A. Thomas, G. Reiss, M. Vieth *et al.*, J. Appl. Phys. 89, 586 (2001)

9. M. Justus, H. Brückl, G. Reiss, J. Magn. Magn. Mater. 240, 212 (2002)

10. S.S.P. Parkin, K.S. Moon, K.E. Pettit, D.J. Smith, R.E. Dunin-Borkowski, M.R. McCartney, Appl. Phys. Lett. 75, 543 (1999)

11. A.E. Berkowitz, K. Takano, J. Magn. Magn. Mater. 200, 552 (2000)

12. Y. Saito, M. Amano, K. Nakajima, S. Takahashi, M. Sagoi, J. Magn. Magn. Mater. 223, 293 (2001)

13. K. Yagami, M. Tsunoda, M. Takahashi, J. Appl. Phys. 89, 6609 (2001)

14. M. Hayashi, Y. Ando, T. Miyazaki, Jpn. J. Appl. Phys. 40, L1317 (2001)

15. M. Urbaniak, H. Brückl, F. Stobiecki, T. Lucinski, G. Reiss, Acta Phys. Pol. A 105 (3), 307 (2004)

16. G. Anderson, Y. Huai, L. Miloslawsky, J. Appl. Phys. 87, 6989 (2000)

17. M.N. Baibich, J.M. Broto, A. Fert, F. Hguyen van Dau, F. Petroff, P. Etienne *et al.*, Phys. Rev. Lett. 61, 2472 (1988)

18. G. Binasch, P. Grünberg, F. Saurenbach, Phys. Rev. B 39, 4282 (1989)

19. H. van den Berg, W. Clemens, G. Gieres, G. Rupp, M. Vieth, J. Wecker et al., J. Magn. Magn. Mater. 165, 524 (1997)

20. J. Schmalhorst, H. Brückl, G. Reiss, G. Gieres, and J. Wecker, J. Appl. Phys. 91, 6617 (2002)

21. H. Brückl, J. Schmalhorst, H. Boeve, G. Gieres, J. Wecker, J. Appl. Phys. 91, 7029 (2002)

22. H.A.M. Berg, W. Clemens, G. Gieres, G. Rupp, M. Vieth, J. Wecker *et al.*, J. Magn. Magn. Mater. 165, 524 (1997)

23. Y.-H. Fan, H. Brückl, Appl. Phys. Lett. 83, 3138 (2003)

24. A. Niemeyer, H. Brückl, to be published

25. D.C. Worledge, Appl. Phys. Lett. 84, 4559 (2004)

26. J.S. Moodera, G. Mathon, J. Magn. Magn. Mater. 200, 248 (1999)

27. N. Cabrera, N.F. Mott, Rep. Prog. Phys. 12, 163 (1948/49)

28. V. Kottler, M.F. Gillius, A.E.T. Kuiper, J. Appl. Phys. 89, 3301 (2001)

29. J.S. Bae, K.H. Shin, H.M. Lee, J. Appl. Phys. 91, 7947 (2002)

30. F.J. Himpsel, J.E. Ortega, G.J. Mankey, R.F. Willis, Adv. Phys. 47, 511 (1998)

31. X.-F. Han, M. Oogane, H. Kubota, Y. Ando, and T. Miyazaki, Appl. Phys. Lett. 77, 283 (2000)

32. A. Thomas, H. Brückl, M.D. Sacher, J. Schmalhorst, G. Reiss, J. Vac. Sci. Techn. B 21, 2120 (2003)

33. S. Cardoso, V. Gehanno, R. Ferreira, P.P. Freitas, IEEE Trans. Magn. 35, 2952 (1999)

34. S.S.P. Parkin, K.P. Roche, M.G. Samant, P.M. Rice, R.B. Beyers, R.E. Scheuerlein *et al.*, J. Appl. Phys. 85, 5828 (1999)

35. U. May, K. Samm, H. Kittur, J. Hauch, R. Calarco, U. Rüdiger et al., Appl. Phys. Lett. 78, 2026 (2001)

36. Z. G. Zhang, P. P. Freitas, A. R. Ramos, N. P. Barradas, and J. C. Soares, J. Appl. Phys. 91, 8786 (2002)

37. J. M. Daughton, J. Appl. Phys. 81, 3758 (1997)

38. A. Thomas, H. Brückl, J. Schmalhorst, G. Reiss, IEEE Trans. Magn. 39, 2821 (2003)

39. J.H. Lee, I.-W. Chang, S.J. Byun, T.K. Hong, K. Rhie, W.Y. Lee et al., J. Magn. Magn. Mater. 240, 137 (2002)

40. X. Zhang, B.-Z. Li, G. Su, F.-C. Pu, Phys. Rev. B 56, 5484 (1997)

41. L. Sheng, Y. Chen, H.Y. Teng, C.S. Ting, Phys. Rev. B 59, 480 (1999)

42. Y. Saito, M. Amano, K. Nakajima, S. Takahashi, M. Sagoi, J. Magn. Magn. Mater. 223, 293 (2001)

43. R. Richter, L. Bär, J. Wecker, G. Reiss, Appl. Phys. Lett. 80, 1291 (2002)

44. H. Brückl, M. Brzeska, D. Brinkmann, J. Schotter, G. Reiss, W. Schepper et al., J. Magn. Magn. Mater. 282, 219 (2004)

45. S. Yuasa, T. Nagahama, Y. Suzuki, Science 297 (2002)

46. F.J. Jedema, A.T. Filip, B.J. van Wees, Nature 410, 345 (2001)

47. R. Jansen, J. Phys. D 36, R289 (2003)

48. S. van Dijken, X. Jiang, S.S.P. Parkin, Phys. Rev. B 66, 94417 (2002)

49. R. Heer, J. Smoliner, J. Bornemeier, H. Brückl, Appl. Phys. Lett. 85, 4388 (2004)

50. A.T. Hanbicki, B.T. Jonker, G. Itskos, G. Kioseoglou, A. Petrou, Appl. Phys. Lett. 80, 1240 (2002)

51. H.J. Zhu, M. Ramsteiner, H. Kostial, X. Marie, T. Amand, C. Fontaine et al., Physica E17, 358 (2003)

52. J. Schmalhorst, H. Brückl, G. Reiss, J. Appl. Phys. 91, 6617 (2002)

53. H. Brückl, J. Schmalhorst, G. Reiss, G. Gieres, J. Wecker, Appl. Phys. Lett. 78, 1113 (2001)

54. S. Datta, B. Das, Appl. Phys. Lett. 56, 665 (1990)

55. J.M. Kikkawa, I.P. Smorchkova, N. Samarth, D.D. Awschalom, Science 277, 1284 (1997)

56. J.M. Kikkawa, D.D. Awschalom, Nature 397, 139 (1999)

57. I. Malajovich, J.M. Kikkawa, D.D. Awschalom, J.J. Berry, N. Samarth, Phys. Rev. Lett. 84, 1015 (2000)

58. Y. Ohno et al., Nature 402, 790 (1999)

59. R. Fiederling *et al.*, Nature 402, 787 (1999)

60. H.J. Zhu *et al.*, Phys. Rev. Lett. 87, 16601 (2001)

61. V.F. Motsnyi, J. DeBoeck, J. Das, W. van Roy, G. Borghs, E. Goovaerts *et al.*, Appl. Phys. Lett. 81, 265 (2002)

Phase-change Memories

R. Bez, A. Pirovano, F. Pellizzer

STMicroelectronics, Central R&D, Agrate Brianza / Italy

Introduction

Non-volatile memories (NVM) are playing a very important role in the semiconductor market, thanks in particular to Flash technology, that is used mainly in cellular phones and other types of electronic portable equipment (mobile PCs, mp3 players, digital cameras and so on). This trend will continue in the coming years and it is expected that portable systems will demand even more from NVM either with high density and very high writing throughput for data storage application, or with fast random access for code execution in place. Although Flash memory technology has been able to follow the evolution of the semiconductor roadmap, since its introduction in late 1980s, and further scaling down is forecasted to continue into the coming years, there are some physical limitations to be faced and the downscaling beyond the 45 nm technology node is still considered critical. Hence other technologies, alternative to floating gate devices, have been proposed and are under investigation. These new proposals exploit different physical mechanisms to store the information, like magnetism, ferroelectricity, solid electrolysis, and phase change. Among the different NVM based on these mechanisms, apart from the floating-gate concept, phase-change memories (PCM), also called ovonic unified memories (OUM), are one of the most promising candidates to become a mainstream NVM, having the potentiality to improve the performance compared to Flash – random access time, read throughput, direct write, bit granularity, endurance – as well as to be scalable beyond Flash technology.

Origins, Concept, and Development

Research on the electronic properties of disordered materials dates back to the 1950s with the original work of S. R. Ovshinsky that, in the years from 1958 to 1961, investigated the chemistry and the metallurgic properties of amorphous semiconductors and discovered two types of reversible switching phenomena [1]. The first mechanism, called ovonic threshold switching (OTS), is a field-assisted and reversible transition that makes an amorphous semiconductor switching from a highly resistive to a conductive state.

Once the amorphous resistivity drops, a second transformation may occur in several compounds. This mechanism, called ovonic memory switching (OMS), is a reversible phase-change from the amorphous to the crystalline state induced by Joule heating due to the current flow. From then on, threshold and memory switching were observed in many materials [2–5], and increasing efforts were devoted to exploiting both mechanisms in applications like xerography [6] and electronic memories [7].

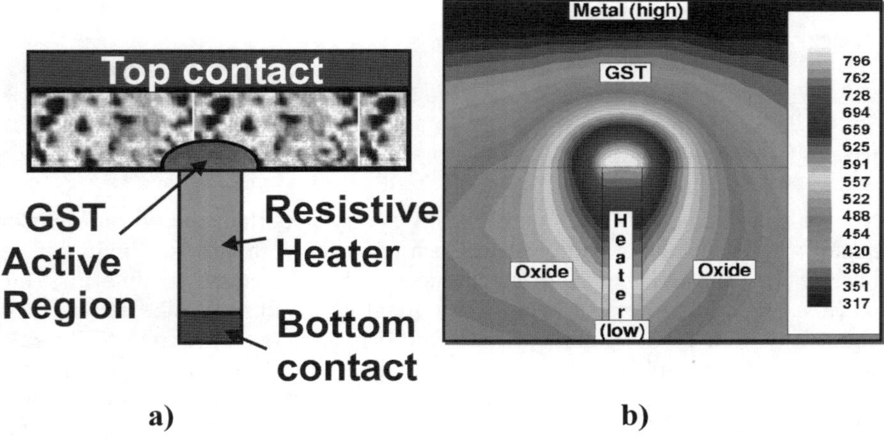

Figure 1. Conceptual PCM cell structure (a) and simulated thermal profile showing the Joule heating effect employed for electrical programming (b).

The use of phase change chalcogenide alloy films to store data electrically and optically was first reported in 1968 [1] and in 1972 [8], respectively. In the early phase of their development, phase change memory devices used tellurium-rich, multicomponent chalcogenide alloys with a typical composition of $Te_{81}Ge_{15}Sb_2S_2$. During the 1970s and 1980s, significant research efforts by many industrial and academic groups were focused on understanding the fundamental properties of chalcogenide alloy amorphous semiconductors [9], and prototype optical memory disks and electronic memory device arrays were also announced [10]. A major step for the development of chalcogenide-based applications was the introduction of rapidly crystallizing alloys by several optical memory research groups [11]. These new compounds, derived from the germanium-antimony-tellurium ternary system, did not phase segregate upon crystallization like the earlier Te-rich alloys, exhibiting congruent crystallization [12] with no large-scale atomic motion. Fast programming operations were thus achieved, allowing one to exploit the phase-change chalcogenide alloys for optical storage. Starting from the 1990s, Energy Conversion Devices, Inc., focused the research on the amorphous semiconductor alloy $Ge_2Sb_2Te_5$, also known as GST, developing the phase-change technology now used in rewritable DVDs [11, 13].

In optical disks the information storage relies on the OMS mechanism. In this application, a laser pulse focused onto the area corresponding to the bit size heats up the GST making it locally change between the amorphous and the crystalline phase. The different diffraction index of the two phases makes possible the optical read out. Since the phase change is also accompanied by a dramatic resistivity variation, the same material has been proposed for applications in semiconductor memories [7]. In this case the phase change is electrically induced and the stored information is associated to the corresponding high and low resistance values. Since both states are stable, no energy is required to keep the data stored, resulting in an inherently non-volatile memory technology.

The PCM cell is essentially a resistor of a thin-film chalcogenide material with a low-field resistance that changes by orders of magnitude, depending on the phase state of the GST in the active region (Figure 1a). The switch between the two states occurs by means of local temperature increase. Above the crystallization temperature, crystal nucleation and growth occur and the material becomes poly-crystalline (set operation). To bring the chalcogenide alloy back to the amorphous state (reset operation), the temperature must be increased above the melting point and then quenched down very quickly to preserve the disorder and not let the material crystallize. From an electrical point of view, it is possible to use the Joule effect to reach locally both critical temperatures (Figure 1b) using the current flow through the material by setting proper voltage pulses with durations in the range of tens of nanoseconds. The cell read out is instead performed at low bias.

Although the phase change concept has been well known for many years and the first studies date back to the 1970s, its application for NVM has known renewed efforts and in 1999 the idea to have a semiconductor NVM cell based on the phase change of a chalcogenide material has been presented again [14, 15]. Since then the effort to bring the PCM basic concept to a mature technology level has constantly increased and many groups have started to study, to develop, and to integrate in MultiMegabit arrays the memory cell. As a demonstration of the interest and of the activities on PCM, in 2003 six papers, related to the phase change memory technology, were presented at the International Electron Devices Meeting [14, 16–20]. Recently, PCM technology reached the phase of large array demonstrator (from 8 Mb to 64 Mb) showing a promising level of maturity for manufacturability and fast growing capabilities [21, 22].

Chalcogenide Compounds Physics

The information storage in PCM devices relies on the different resistivities of the GST alloy in the polycrystalline phase (from now on referred to as crystalline state) and in the amorphous one, varying from 25 mΩ·cm to values three orders of magnitude higher. In complete analogy with optical disk technology, data storage relies on the OMS phase transitions. However, PCM devices are programmed through electrical pulses. Therefore, the GST electrical properties are of fundamental importance for the device operation. Since the crystalline state is highly conductive, a sufficiently high current can easily flow through the device causing the melting of the GST and allowing one to reset the memory cell.

On the other hand, the high-field threshold switching phenomenon (OTS) is mandatory to allow current flowing through the device and to heat the amorphous chalcogenide. When a threshold voltage is reached, the PCM device passes into a highly conductive state, also referred to as ON states, that allows the flow of a large current density at moderate voltage drops. The GST is thus heated to about 350°C and the amorphous to crystalline transitions can be accomplished in hundreds of nanoseconds. The threshold switching in amorphous chalcogenide is thus essential for device functionality and can be regarded to be as fundamental to PCM as hot carrier injection or electron tunneling is to floating-gate memory technology.

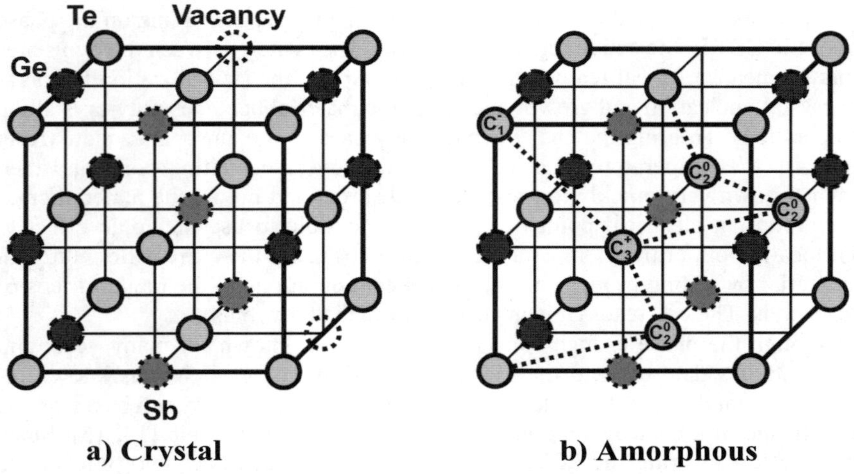

Figure 2. Crystallographic structures for the poly-crystalline (a) and amorphous (b) phase [26].

Despite the importance covered by the OTS mechanism for PCM functionality, the nature of the switching effect was debated for decades [23–25] and only recently its electronic nature was clearly demonstrated [26, 29]. Starting from the original Ovshinsky's study [1], several models have been proposed, with many researchers supporting the idea that OTS is essentially a thermal effect and that the current in an amorphous layer rises above the threshold voltage due to the creation of a hot filament [23]. Later Adler showed that, at least in thin chalcogenide films, the OTS effect is not thermal [24], in agreement with Ovshinsky's original picture. In his pioneering work [25], he demonstrated that a semiconductor resistor may feature OTS, without any thermal effect, provided that carrier generation, driven by field and carrier concentration, competes with a strong Shockley–Hall–Read recombination via localized states.

A complete description of the GST structure and of the physical mechanisms responsible for charge transport in chalcogenide alloys has been recently provided in [26]. Relying on a description of the GST microscopic structure in both the crystalline (Figure 2a) and amorphous phase (Figure 2b), it was argued that both structural states of the alloy could be treated as semiconductors with comparable energy bandgaps. From optical absorption spectra it was estimated that the bandgap is about 0.7 eV in the amorphous state and 0.5 eV in the polycrystalline state. Relying on low-field electrical measurements, a conductivity activation energy, E_a, of about 0.3 eV for the amorphous state and 0.02 eV for the crystalline state were estimated. In addition, the amorphous phase exhibits a very low, trap-limited hole mobility of 2×10^{-5} cm^2/V·s, while the crystalline phase shows band-type mobility of about 10–15 cm^2/V·s. These large differences are mainly due to the existence of disorder-induced localized electronic states inside the amorphous phase.

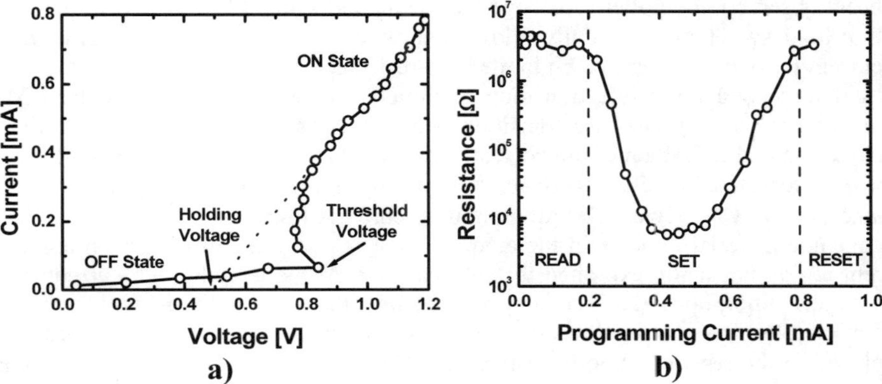

Figure 3. Amorphous GST I–V characteristic showing the electronic threshold switching effect (a) and corresponding programming curve, defined as the low-field resistance as a function of the programming current (b).

When chalcogenide alloy semiconductors are amorphized, electronic energy levels originating in the valence and conduction band are pushed into the original empty energy band gap of the crystalline material. These new gap states are spatially localized and do not extend throughout the material. Consequently, carriers move through the amorphous material either by hopping among the localized states or by being successively thermally excited to spatially extended band states and then being trapped into localized states. The experimental observation that E_a is about $E_g/2$ was then explained in terms of a large density of special negatively and positively charged traps that also result from structural disorder in amorphous chalcogenide alloys [25]. These charged traps (valence alternation pairs) act like compensating dopant levels in a conventional crystalline semiconductor, effectively forcing the Fermi level to lie near midgap between the energy levels of the two types of traps.

In the polycrystalline state, crystal vacancies are proposed [26] to give rise to acceptor-like states that move the Fermi level close to the valence band edge. This Fermi level position, plus the loss of the disorder-produced trapping states, gives rise to the nearly degenerate p-type high conductivity of the crystalline state. In the amorphous phase, the chalcogenide alloy material has a high electrical resistance at low electric fields. With increasing voltage, conductivity is initially ohmic, but it begins to grow exponentially when the field exceeds 10^5 V/cm. As shown in Figure 3a, when a particular "threshold voltage," V_{th}, is exceeded, the material switches rapidly into a highly conductive "dynamic on-state." The threshold switching results from the competition between a traps-based recombination mechanism and a concentration –assisted field-induced generating mechanism that fills the traps and inhibits the recombination efficiency [26]. The dynamic on-state is maintained as long as a sufficient "holding current" passes through the device. This transient high conductivity state is electronic in origin and does not involve a structural transformation from the amorphous to the low-resistance crystalline state since it has also been observed in molten chalcogenide semiconductors [27]. In PCM devices, threshold switching essentially allows one to deliver the required programming current needed to program a bit in the high-resistance state at low voltage [28].

The second reversible transition exploited for memory storage is the OMS mechanism. This is a phase transition between the glass and the polycrystalline condition and its dynamics are correctly explained by nucleation and growth kinetics, as reported in [29]. Figure 3b finally reports the programming curve of an amorphous PCM device. For programming currents lower than 200 µA, the chalcogenide material remains in the amorphous state and this low-programming current region is usually exploited for cell read out. At moderately high current density, the amorphous GST crystallize, reaching the set condition. Only when a sufficiently high current flows through the device is the GST melted and reamorphized in the reset condition. Despite the phase-change transition dynamics being analogous to the mechanism employed in rewritable optical disks, it has been shown that the existence of inhomogeneous electrical pattern due to the OTS effect leads to complicating phase distribution inside the PCM cell [30]. Even though the macroscopic properties of the intrinsic cell do not seem significantly impacted by this effect, the implication on the device properties and on the statistical behaviour are still largely unexplored and currently under investigation.

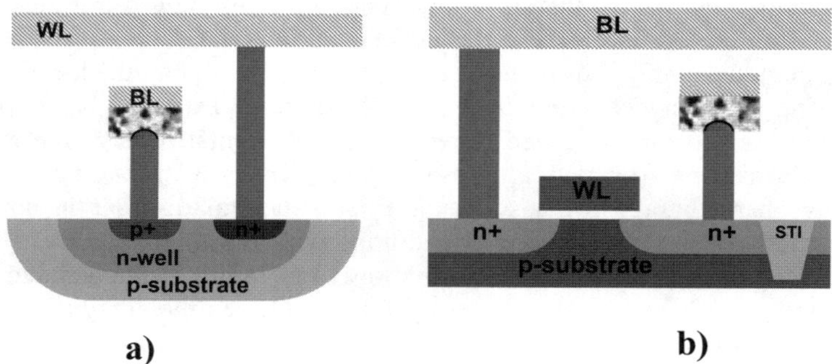

Figure 4. Schematic cross sections of the PCM cell with BJT (a) and MOS-selector (b).

Cell Architectures

Considering the electronic and transport properties of the GST alloy, either in the crystalline or in the amorphous state, in order to form a functional compact memory array, a PCM cell must be formed by the variable resistor with a selector device (transistor) in series. Hence, the basic PCM cell has a 1T/1R structure. The type of transistor and of data-storage varies respectively as a function of the application and of the process architecture strategy. For high-density memory, a more compact cell layout is achieved via the vertical integration of a *p-n-p* bipolar transistor (Figure 4a), while for embedded memory the transistor is a *n*-channel MOS (Figure 4b), where a larger cell size is balanced by a minimum process cost overhead with respect to the reference CMOS [31]. Only recently a MOS-selected PCM cell has been also proposed for high-density application, obtaining a compact cell layout thanks to a very narrow, but also very short, gate dimension [32]. A 3.5 nm gate oxide transistor should provide the necessary reset current with a supply voltage of 3V. In this approach, the process simplicity is balanced with a heavily stressed CMOS and the data-storage long endurance concern is translated into MOSFET reliability.

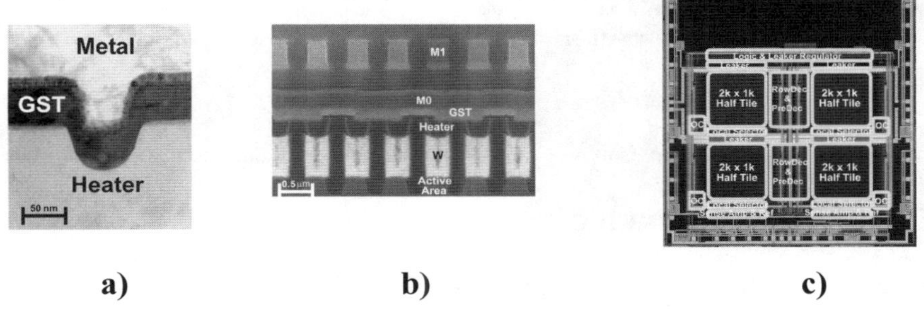

Figure 5. TEM picture showing the GST-heater interface in an optimised μ-trench cell with reduced programming currents (a), integration in an array with BJT selectors (b), and an 8 Mb demonstrator microphoto [22, 31].

The integration of the data-storage occurs between front-end and back-end modules of the CMOS process. The "simple" variable resistor, *i.e.*, the heater and GST system, may be obtained in different ways and the choice is a function of the understanding of process complexity, current performance, thermal properties and scaling perspective [14, 16]. Early phase change nonvolatile memory devices [10] were fabricated by sputtering the chalcogenide memory alloy into 5-μm pores etched in an oxide layer deposited on a silicon substrate containing the diode access devices. Memory operation was demonstrated, but the relatively large programming volume and poor thermal efficiency resulted in programming current too large for a competitive NVM technology. A possible approach to improve the memory element performance is to use a sub-litho contact heater with a planar GST or a modified version with a recession in the contact and GST confinement, which should improve the thermal properties and reduce the reset current [32]. A completely different approach relies on the definition of the contact area between the heater and the GST by the intersection of a thin vertical semi-metallic heater and a trench, called "μtrench" [31], in which the GST is deposited (Figure 5). The μtrench architecture keeps the programming current low and maintains a compact vertical integration. Since the μtrench can be defined by sub-litho techniques and the heater thickness by film deposition, the cell performance can be optimized by tuning the resulting contact area still maintaining a good dimensional control.

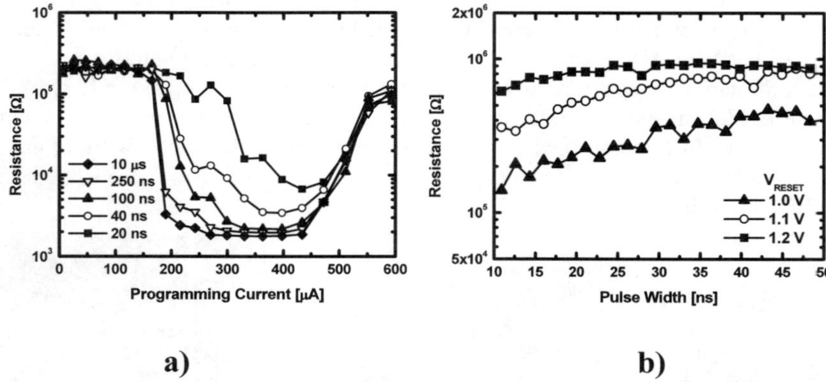

a)   b)

Figure 6. Programming curves of a PCM cell for different set (a) and reset (b) pulse widths.

Electrical Performance

To exploit the PCM technology for high-performance applications, fast write and read times are mandatory, while still preserving good data retention capabilities. One of the main concerns to speed up the writing operation is the trade-off between fast crystallization and good non-volatility properties. To assure 10 years data retention capabilities at temperatures in the range of 110–120°C [31], the GST compound with the stoichiometry 2-2-5 has been chosen. However, better retention capabilities imply longer programming pulses to crystallize the GST.

Figure 6a shows the programming curves of a MOSFET-selected PCM cell for several programming pulse widths. For very long pulses (10 μs), a complete crystallization is easily achieved with a resistance change of two orders of magnitude between the set (crystalline phase, low resistance) and the reset state (amorphous phase, high resistance). However, this programming time is unacceptable for real high-performance products. By reducing the pulse width, the GST is not able to fully crystallize, resulting in a higher set resistance. However, for pulses as short as 20 ns, a factor of 10 in the resistance change between the two programmed states is still achieved. Despite the read margin being reduced for very short pulses, Figure 6a clearly demonstrates a suitable working window for read out operations with a 20 ns set time.

The programming curve in Figure 6a also shows a reset programming current as low as 600 μA. This value, obtained with the μtrench approach, can be further reduced by tailoring the GST contact area and the heater resistance. Since the μtrench width scales with the lithography, a continuous improvement of power consumption is expected in the next generation devices. Moreover, optimizations of the heater material and thickness are proven to allow a reduction of the programming current, leaving room for further improvements.

Figure 6b reports the programmed amorphous resistance as a function of the reset pulse widths. The three curves correspond to different pulse amplitudes, *i.e.*, programmed reset states ranging from 10^5 to 10^6 Ω. By reducing the pulse width to as short as 10 ns, a small decrease of the programmed resistance can be appreciated. This effect could be related to the existence of a delay to reach the thermal steady state condition in the heated volume, mainly due to the high thermal resistivity of the GST film. In any case, the read margin is still granted and programming pulses of 20 ns and 10 ns for the set and reset operations, respectively, are suitable for a 10× resistance change of the PCM cell. These programming/erase capabilities, combined with a demonstrated read access time shorter than 50 ns in a multi-megabit demonstrator, clearly confirm that the PCM technology is suitable for high-performance embedded non-volatile memory products.

Endurance of phase change memory cells has been reported in several publications [28, 31, 32] to be between 10^9 and 10^{13} write/erase cycles – considerably in excess of the nominal 10^6 cycle endurance of Flash.

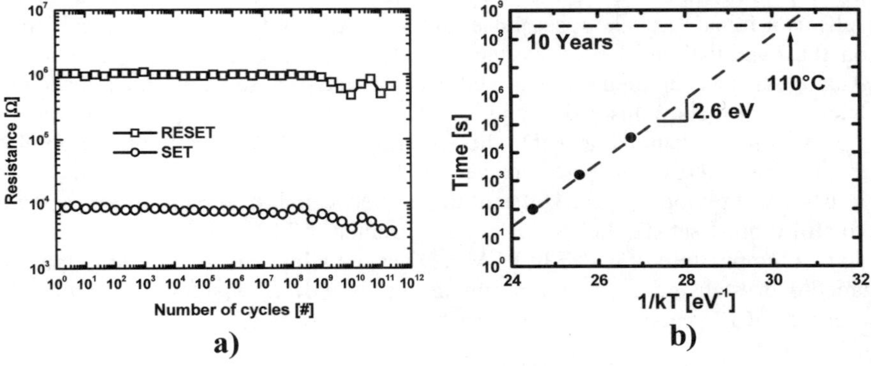

Figure 7. PCM cell endurance (a) and retention capabilities (b).

A typical endurance measurement is also reported in Figure 7a. A number of extrinsic mechanisms that limit cycle life have been identified for PCM cells [33]. Delamination of the chalcogenide semiconductor layer from the resistive electrode material can occur after repeated thermal cycles if proper attention has not been paid to surface preparation and film adhesion. Another mechanism results from changes in the GST composition due to interactions with electrode chemistry at elevated temperatures. As a result, the cycle life of the bit is found to be a power law function of the energy applied in the programming pulse.

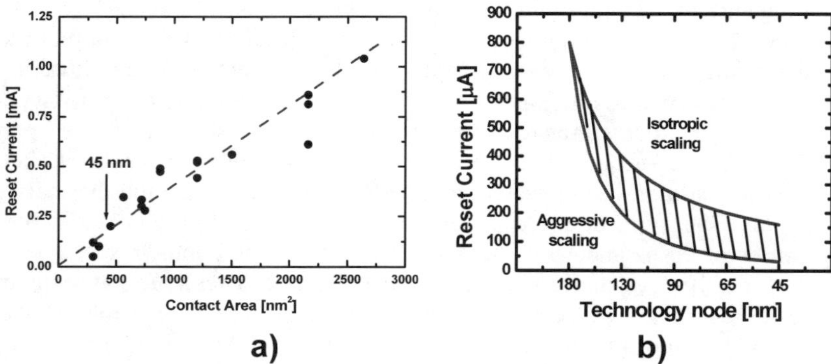

Figure 8. Experimental verification of programming current reduction with the decreasing of the contact area (a) and expected scaling in two different scenarios (b).

An advantage related to the reduction of the reset pulse width is the corresponding improvement of the PCM cell endurance. The cycle life measured on different cells as a function of the reset pulse width shows that the cell endurance depends on the overall time elapsed by the cell at higher temperature (reset operation). Analogous measurements performed with a variable set and fixed reset pulse widths do not show instead any cycle life dependence on the programming strategy. This result demonstrates that the actual limit for PCM cell endurance is related to the total energy dissipated inside the device, now being in the order of few Joules. As previously stated, the trade-off between fast crystallization and non-volatility must satisfy the mandatory requirement of 10 years storing capabilities up to a temperature of 85°C. In PCM devices, data loss can occur due to the spontaneous crystallization of an amorphous bit. However, as shown in Figure 7b, a lifetimes of 10 years at 110°C [31] has been demonstrated.

Future Challenges

Since PCM technology is based on the basic properties of the chalcogenide alloy, the integration of the material into a standard CMOS process still represents a challenging matter: not for the single cell concept, which is already proven to be very strong, but for the manufacturability of very high density NVMs, where the technology can be considered robust if demonstrated only over many billions of cells. Another critical topic is represented by the reset current: its controllability and reduction is fundamental to guarantee a compact and scalable cell size, a competitive writing power consumption and enhanced reliability. Hence the development efforts are mainly focused on the optimization of the data storage structure with respect to the reset current. Recent results, reported in Figure 8a, have shown significant improvements in the cell current reduction with the perspective of a continuous scaling according to the technology node (Figure 8b) [14, 16].

Considering its key features, PCM technology is a promising candidate as a mainstream NVM. Although today the technology is still in the early stages, the development results and the level of comprehension of the physical mechanisms obtained so far allow one to foresee a fast progress.

Acknowledgements

Thanks are extended to our colleagues at the STMicroelectronics MPG and Central R&D groups, and to the Ovonyx, Inc., team for contributions and valuable discussions on the various aspects of chalcogenide physics.

References

1.   S.R. Ovshinsky, Phys. Rev. Lett., 21, 1450-1452, (1968).

2.   J.F. Gibbons and W.E. Beadle, Solid State Electron., 7, 785–797, (1964).

3.   Argall, Solid State Electron., 11, 535–541, (1968).

4.   J.G. Simmons and A. El-Badry, Solid State Electron., 20, 12, 954–961, (1977).

5.   A.E. Owen and J.M. Robertson, IEEE Trans. Electron Devices, 20, 105–122, (1973).

6.   Fundamentals of Amorphous Semiconductors, National Academy of Sciences-Nat. Res. Council, Washington, DC, 1972.

7.   R.G. Neale and J.A. Aseltine, IEEE Trans. Electron Devices, 20, 195–205, (1973).

8.   J. Feinleib, J. de Neufville, S. C. Moss, and S. R. Ovshinsky, Appl. Phys. Lett., 18, 254-256, (1971).

9.   D. Adler, Scientific Amer., 236, 36–49, (1977).

10. R.G. Neale, D.L. Nelson, E. Moore, Electronics, p. 56, (1970).

11. M. Chen, K. Rubin, and R. Barton, Appl. Phys. Lett, 49, 502-505, (1986).

12. J. Gonzalez-Hernandez, B. S. Chao, D. Strand, S. R. Ovshinsky, D. Pawlik, P. Gasiorowski, Appl. Phys. Comm., vol. 11, n. 4, p. 557, (1992).

13. S. Tyson, G.Wicher, T. Lowrey, S. Hudgens, K. Hunt, in Proc. Aero Space Conf., 5, 385–390, (2000).

14. S. Lai, IEDM Tech. Dig., pp. 255-258, (2003).

15. G. Wicker, SPIE Conf. on Electron. and Struc. for MEMS, Australia, (1999).

16. A. Pirovano et al., IEDM Tech. Dig., 699-702, (2003).

17. Y. N. Hwang *et al.*, IEDM Tech. Dig., 893-896, (2003).

18. J. H. Yi *et al.*, IEDM Tech. Dig., 901-904, (2003).

19. N. Takaura *et al.*, IEDM Tech. Dig., 897-900, (2003).

20. Y. C. Chen *et al.*, IEDM Tech. Dig., 905-908, (2003).

21. W. Y. Cho *et al.*, ISSCC dig. Tech. Papers, (2004).

22. F. Bedeschi *et al.*, Symp. on VLSI Circ., (2004).

23. A.E. Owen, J.M. Robertson, C. Main, J. Non-Cryst. Solids, 32, 29–52, (1979).

24. D. Adler, H.K. Henisch, S.D. Mott, Rev. Mod. Phys., 50, 2, 209–220, (1978).

25. D. Adler, M. S. Shur, M. Silver, S.R. Ovshinsky, J. Appl. Phys., 51, 6, 3289–3309, (1980).

26. A. Pirovano, A. L. Lacaita, A. Benvenuti, F. Pellizzer, R. Bez, IEEE Trans. Electron Devices, 51, 3, 452-459, (2004).

27. A.R. Regel, A.A. Andreev, M. Mamadaliev, J. Non-Cryst. Solids, 8, 10, 455, (1972).

28. S. Lai, T. Lowrey, IEDM Tech. Dig., 803, (2001).

29. A. Redaelli *et al.*, IEEE Electron Device Lett., 25, 10, 684-686, (2004).

30. A.L. Lacaita, A. Redaelli, D. Ielmini, F. Pellizzer, A. Pirovano, A. Benvenuti, R. Bez, IEDM Tech. Dig., (2004).

31. F. Pellizzer *et al.*, Symp. On VLSI Tech., (2004).

32. S. H. Lee *et al.*, Symp. On VLSI Tech., (2004).

33. A. Pirovano *et al.*, IEEE Trans. Device Mater. Rel., (2004).

Amorphous-to-*fcc* Transition in GeSbTe Alloys

S. Privitera[a], C. Bongiorno[a], E. Rimini[a], R. Zonca[b]

[a] Istituto per la Microelettronica e Microsistemi (IMM), Catania / Italy

[b] STMicroelectronics, Central R&D, Agrate / Italy

Introduction

GeSbTe alloys exhibit a large variation of the optical and electrical properties when changing from the amorphous to the crystalline state, in the metastable rock-salt structure (*fcc*). The amorphous-to-*fcc* transition can be reversibly induced by laser irradiation or current pulses. For these reasons, these materials have been used as optical rewritable storage media and have also been recently proposed for phase change nonvolatile random access memories (PCRAM) [1]. The storage mechanism is the structural reversible phase change. The two logic states are represented by the crystalline (low resistivity) versus the amorphous phase (high resistivity).

GeSbTe alloys with compositions centering around the GeTe–Sb_2Te_3 pseudobinary line, in the ternary phase diagram, are very promising as data storage media. They exhibit high crystallization rate and good reversibility between amorphous and crystalline phases. Furthermore, the phase change characteristics may by controlled by selecting the film composition [2]. Among the others, films with $Ge_2Sb_2Te_5$ composition combine a high transition rate together with sufficiently large activation energy (more than 2 eV).

In view of the application as storage material in PCRAMs, we have studied the amorphous-to-*fcc* transition in $Ge_{2+x}Sb_2Te_5$ (x = 0, 0.5) thin films. We have followed, under isothermal annealing, the change of the resistivity, that characterizes the two memory logic states. The effects of annealing temperature and time on the structure and the electrical properties of films with different composition have been compared. Moreover, since the storage mechanism is based on a phase transition that proceeds via nucleation and growth, these two involved processes need to be studied in more detail.

Electrical [3], but also optical and calorimetric measurements [4–6], commonly employed to characterize the phase transition, are only sensitive to average variations of the physical properties. They do not allow one to separate the contribution of the nucleation and the growth and to follow the early stage of the transition, during which the nucleation process strongly depends on time. This transient regime is directly related to the grain size distribution, which can only be measured by using microscopic techniques. *In situ* transmission electron microscopy (TEM) analysis represents a very powerful technique to experimentally determine the parameters that are strictly related to the dynamical evolution of the population of crystalline grains, such as the incubation time, the growth velocity, and the driving force of the transformation. However, it has been shown that the electron beam may influence the nucleation process, making difficult the study of the kinetics of the amorphous-to-*fcc* transition in $Ge_2Sb_2Te_5$ films by *in situ* TEM analysis [7]. In this work, the effects of the electron beam have been suppressed by switching off the electron beam current during *in situ* thermal treatments, and by analyzing only fresh areas. In this way, the density of *fcc* grains and the size distribution have been determined for various temperatures and times. By this information, the contribution of nucleation and growth has been separated and the scalability of PCRAMs has been evaluated.

Effect of Composition on the Electrical Properties and Structure

The amorphous-to-polycrystal transformation can be followed *in situ* by sheet resistance measurements, performed by using the four point probe method. Samples have been prepared by sputtering at room temperature, from a single target, 50-nm thick $Ge_2Sb_2Te_5$ films on oxidized Si substrates. The obtained amorphous films were then isothermally heated in a thermal chuck, at temperatures in the range of 130–190°C, to obtain the conversion into the low resistivity *fcc* phase. Figure 1 reports as full and open symbols, the resistivity *vs.* temperature, as measured after 13-min. isothermal anneals, for x = 0.0 and x = 0.5, respectively. The uncertainty in the resistivity determination is lower than the symbol size. Films with x = 0.0 exhibit a continuous decrease of the resistivity as a function of temperature. In $Ge_{2.5}Sb_2Te_5$ films, the resistivity instead reaches a small plateau at temperatures between 145 and 160°C, and then it further decreases as the temperature increases. After annealing at 190°C the resistivity of the two samples becomes comparable.

In order to relate the electrical properties to the film structure, samples annealed at different temperatures were analysed by X-ray diffraction (XRD). We used a SIEMENS D5005 diffractometer operating with Cu K_α ($\lambda = 0.15406$ nm), in grazing angle configuration with an incident angle of 0.7°. The measured diffraction patterns in $Ge_2Sb_2Te_5$ samples, annealed at 130°C for 120 min or at higher temperature, correspond to an *fcc* structure. The lattice parameter is $a = 0.602 \pm 0.001$ nm, in agreement with the values reported in literature [3, 8].

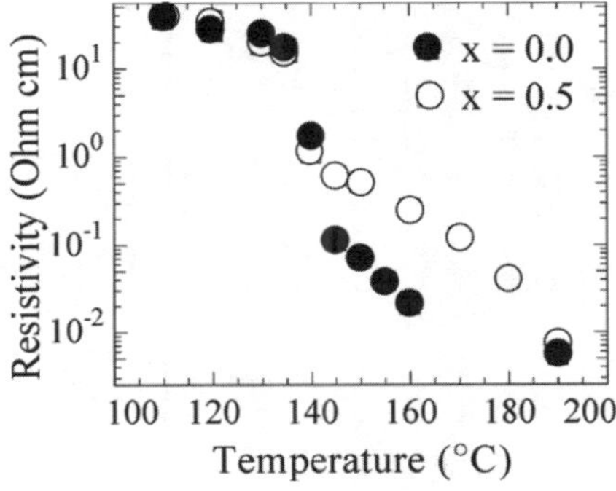

Figure 1. Resistivity as a function of annealing temperature, measured after 13 minutes isothermal anneals for samples with x = 0.0 and x = 0.5.

XRD spectra collected for films with x = 0.5, obtained in samples annealed at temperatures below 170°C, are instead the convolution of the spectrum of a *fcc* phase with that of the amorphous phase, even if the thermal treatment is protracted for a few hours. Therefore, in $Ge_{2.5}Sb_2Te_5$ films, the observed decrease of the resistivity at temperatures in the range 130–160°C, does not correspond to a full crystallization. XRD spectra show that the contribution of the residual amorphous material can be completely eliminated by annealing the samples at temperatures above 160°C.

TEM images and transmission electron diffraction patterns confirm the presence of residual amorphous material, uniformly distributed in the film, after annealing at temperatures in the range of 130–160°C.

Further information can be obtained by monitoring the position of the XRD peaks as well as the full-width at half maximum (FWHM) as a function of temperature. Data plotted in Figure 2a and b as full and open symbols have been obtained by profile fitting analysis using the pseudo-Voigt function for compositions x = 0.0 and x = 0.5, respectively. The width of 220 peaks in $Ge_2Sb_2Te_5$, shown as full squares in Figure 2a, slightly decreases as the temperature increases, while the peak position, plotted in Figure 2b, remains constant as a function of temperature. The FWHM and peak position *vs.* temperature measured for x = 0.5, plotted as open symbols, exhibit a different behavior. In the temperature range 130–160°C, the FWHM decreases as a function of annealing temperature, while the peak positions remain almost constant and close to those measured in $Ge_2Sb_2Te_5$. For temperatures higher than 160°C, the peak positions and widths instead gradually move to higher values with increasing temperature. In this temperature range (160–190°C) no residual amorphous material has been observed. Moreover, the FWHM measured in samples with x = 0.5 is always larger than that obtained for x = 0, indicating that in samples with an excess of Ge, the growth process is inhibited.

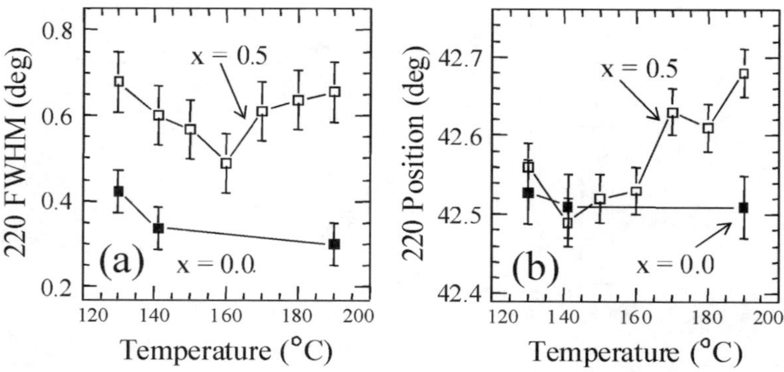

Figure 2. Width (a) and position (b) of 220 peaks, obtained by profile fitting analysis for samples with x = 0.0 (full symbols) and x = 0.5 (open symbols).

Phase separation has been reported in $Ge_2Sb_{2+x}Te_5$ films, with the formation of *fcc* crystals and a small amount of Sb, which remains amorphous at the grain boundaries [8]. The presence of a residual amorphous material in our samples, may be a confirmation that the structure tends to maintain a relatively large amount of vacancies. However, in the present work, the amorphous material, remaining by annealing samples in the temperature range 130–160°C, can be almost completely converted into a crystalline phase, by increasing the annealing temperature. The increase of the FWHM, as well as the change of the peak positions as the temperature increases, indicate that the residual amorphous material is converted into a different phase, with *fcc* structure, but with slightly lower lattice parameter. Reasonably, the crystalline phase formed under annealing at temperatures below 160°C, has the $Ge_2Sb_2Te_5$ composition. However, as the crystallization proceeds, the residual amorphous material becomes richer and richer with Ge, with a composition that depends on the crystalline fraction χ. As an example, for χ = 40%, the composition of the amorphous material is $Ge_3Sb_2Te_5$. A crystallization temperature above 160°C, as observed to completely crystallize samples with composition $Ge_{2.5}Sb_2Te_5$, is in quite good agreement with those reported in Ref. [2], for compositions away from the pseudobinary line, close to those that may form in our samples.

Transition Kinetics in Ge₂Sb₂Te₅

In order to separate the contribution of nucleation and growth, the two processes that govern the amorphous-to-crystal transition, X-ray diffraction and electrical measurements, which provide only average volume information, have been coupled with *in situ* TEM analysis, which allows one to also measure the grain density and size during the transition. The deposited $Ge_2Sb_2Te_5$ amorphous samples were made transparent to the electron beam for TEM analysis by grounding, dimpling, and etching. The chemical etch is preferred with respect to conventional Ar ion milling to avoid sample heating above the crystallization temperature, which may occur during ion milling. TEM analyses were performed using a JEOL 2010 TEM operating at 200 kV, with a sample holder equipped with a heating stage and a thermocouple. To obtain a good statistical evaluation of the grain density and size, for each time step, regions of about 20 μm^2 were analyzed. In order to minimize the effect of the electron beam on the nucleation process, the electron current was switched off during thermal heating. Moreover, after each annealing time, only fresh areas of the sample were observed. The transformation has been followed in the temperature range from 130 to 143°C, because outside of this range the transition becomes too slow or too fast for the TEM observation and for the ramp rate of the employed heater.

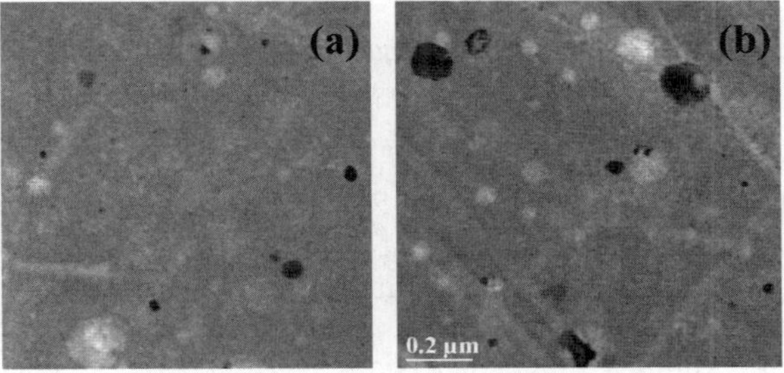

Figure 3. TEM images of a $Ge_2Sb_2Te_5$ film annealed at 137°C for 85 min (a) and 105 min (b).

Figures 3a and b show plan view images in bright field of a $Ge_2Sb_2Te_5$ sample annealed *in situ* at 137°C for 85 min and 105 min, respectively. Crystalline grains embedded in the amorphous matrix appear as dark or bright regions. From these images, the grain density and size can be evaluated as a function of annealing time, for all analyzed temperatures.

Figure 4 shows the grain density as a function of time for various annealing temperatures. The error bars are determined from the statistical fluctuations of the number of grains per unit area. After the incubation period, the grain density linearly increases with time, indicating that a constant nucleation rate is established. The nucleation rate J, defined as the average number of grains which appear per unit time and area, can therefore be evaluated from the slope of the linear fit to the data plotted in Figure 4. To take into account the reduction of the available volume, the obtained value has been divided by the amorphous fraction. However, since our analysis is limited to low transformed fractions, this correction result is negligible.

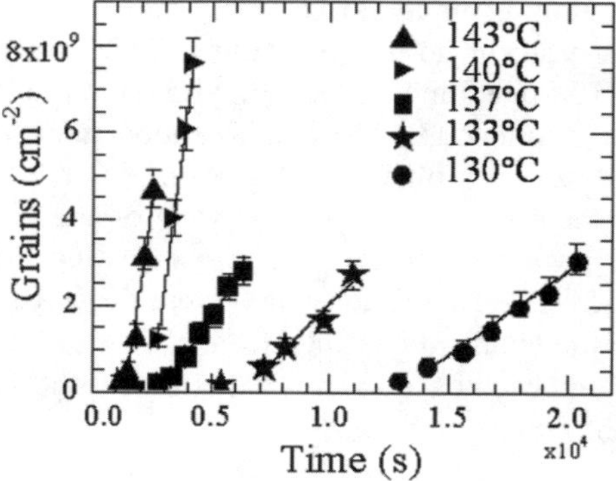

Figure 4. Number of grains per unit area, *vs.* annealing time at different temperatures. Solid lines are linear fits to data.

In Figure 5 the maximum grain radius has been plotted as a function of annealing time for different temperatures. Since these values could be affected by eventual pre-existing seeds, data in Figure 5 have been averaged between different images, collected after the same annealing time. The error bar length is the difference between the maximum and the minimum of the largest measured sizes. These, in general, exhibit a small dispersion. The grain radius linearly increases with time and the growth velocity can be evaluated by the linear fit to the data shown in Figure 5.

The nucleation rate and growth velocity can be plotted as a function of the reciprocal temperature, determining, from the linear fits, the activation energies for the involved processes. Values of $E_n = 2.9 \pm 0.5$ eV and $E_g = 2.3 \pm 0.4$ eV have been evaluated for the nucleation and growth process, respectively.

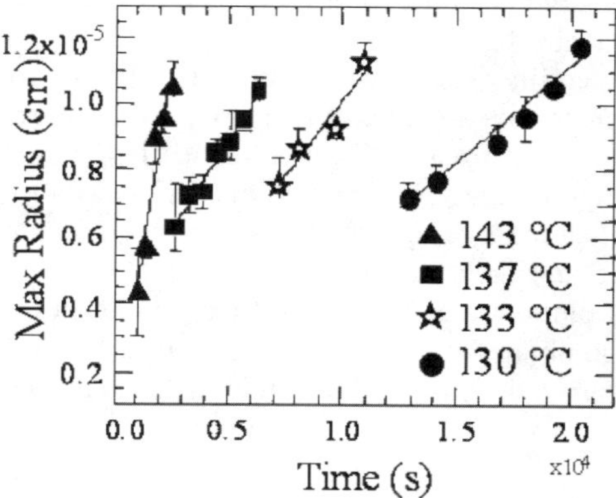

Figure 5. Maximum grain radius measured in $Ge_2Sb_2Te_5$ as a function of annealing time for various temperatures. Solid lines are linear fits to data.

In the classical theory of the nucleation and growth, the activation energies for the nucleation rate and for the growth velocity differs by the barrier energy ΔG^*, for the formation of a crystalline grain of critical size. Our determination, considering the experimental uncertainty, indicates a value of ΔG^* lower than 1 eV. This result suggests the occurrence of a heterogeneous nucleation phenomenon. The observed dependence of total activation energy for the crystallization on the materials between which the Ge–Sb–Te layer is sandwiched [3, 6, 9] confirms our results.

Moreover, it has been observed that as-deposited films crystallize with longer incubation times and higher total activation energy than melt-quenched amorphous films, because these materials contain quenched-in crystalline nuclei [4, 9, 10]. Following this explanation, the total activation energy for melt-quenched samples is only determined by the growth process. Differences of 0.2–0.3 eV [4, 10] have been reported in the literature between the total activation energies measured in as-deposited and melt-quenched films. It is straightforward to show that this difference is equal to $\Delta G^*/n$, where n is the Avrami exponent, which in thin films is expected to be between 2 and 4. Therefore, from these experiments [4, 10], ΔG^* can be estimated to be between 0.4 and 1.2 eV, in good agreement with our determination.

The separation between the nucleation and growth processes and the evaluation of their activation energies allow the estimation of the number of crystalline nuclei that form in a memory cell at the temperatures reached during programming through current pulses.
Let us consider a population of 10^{10} memory cells, each of volume V and with an average number of nuclei N, formed during programming. The probability to have one failing bit into a chip can be evaluated as the probability $P(0, N)$ that, due to statistical fluctuations, one cell out of 10^{10} does not form any crystalline nucleus during the programming pulse. Following the Poisson statistics, $P(0, N) = e^{-N}$, where $N=J\tau V$ is the average number of grains formed after the program time τ, with a nucleation rate J, that is assumed to have an Arrhenius behavior with a single activation energy. Since, according to these hypotheses $e^{-N} \approx 1/10^{10}$, this gives $N \approx 20$. For typical values of $\tau \approx 50$ ns and $T \approx 800$ K, we obtain a volume $V \approx 10^{-19}$ cm^3. This volume corresponds to a contact size of less than 8 nm.

References

1.   M. Gill, T. Lowrey, J. Park, IEEE International Solid State Circuit Conference, Proc. Vol., ISSCS 2002, p.158

2.   N. Yamada, E. Ohno, K. Nishiuchi, N. Akahira, and M. Takao, J. Appl. Phys. 69, 2849 (1991)

3.   I. Friedrich, V. Weidenhof, W. Njorge, P. Franz, and M. Wuttig, J. Appl. Phys. 87, 4130 (2000)

4.   V. Weidenhof, I. Friedrich, S. Ziegler, and M. Wuttig, J. Appl. Phys. 89, 3168 (2001)

5. T.H. Jeong, M.R. Kim, H. Seo, S.J. Kim, and S.Y. Kim, J. Appl. Phys. 86, 774 (1999)

6. N. Ohshima, J. Appl. Phys. 79, 8357 (1996)

7. B.J. Kooi, W. M. G. Groot, J. Th. M. De Hosson J. Appl. Phys. 95, 924 (2004)

8. N. Yamada, T. Matsunaga, J. Appl. Phys. 88, 7020 (2000)

9. P.K. Khulbe, E.M. Wright and M. Mansuripur, J. Appl. Phys. 88, 3926 (2000)

10. J. Park, M.R. Kim, W.S. Choi, H. Seo, and C. Yeon, Japan J. Appl. Phys. 38, 4775 (1999)

Organic Nonvolatile Memories

Y. Yang, L. Ma, J. Ouyang, J. He, H. M. Liem, C.-W. Chu, A. Prakash

University of California, Department of Materials Science and Engineering, Los Angeles, CA / USA

Introduction

Recently, organic memory devices, which use organics as the active materials, have gained much attention [1–21] due to the advantages of using organic materials, such as high flexibility, easy processing, low fabrication cost, and larger area fabrication by printing techniques. They can be used for nonvolatile memory device applications, such as Flash memory, random access memory, and can also be used as switches to drive organic light-emitting diodes for electronic paper applications. In this chapter, we review the development of two-terminal organic memories driven by an electric field. These devices exhibit a conductance change, where a low conductance state (off-state) may switch to a high conductance state (on-state) under a critical applied voltage. We hope this will provide an introduction to those who have interest in this field.

Organic Memories Using Charge-transfer Complex Materials

Dating back to the 1970s, the threshold switching phenomenon had been observed in metal–organic charge-transfer complex systems by Potember *et al.* [1, 2], where the transition between the high-conductance and the low-conductance states is realized by charge transfer between copper (Cu) and tetracyanoquinodimethane (TCNQ). It was explained that under an applied electric field, the charge-transfer complexes undergo a phase transition accompanied by a corresponding change in electrical resistivity, but the details about the phase transition were not clear [1]. The typical current-voltage characteristics for the charge transfer complex system are shown in Figure 1 [2]. The device switching speed was about 15 ns and a high current pulse was used for restoring the device to the off-state.

The memory effect based on the metal–organic charge transfer complex materials was very popular during the 1970s and 1980s. Recently, Oyammada *et al.* reported a switching and memory effect in Cu: TCNQ charge transfer–complex thin films, where a thin aluminum oxide (Al_2O_3) layer between the anode and the Cu:TCNQ layer was critical to the observed switching phenomenon [3]. The Al_2O_3 layer was obtained by exposing the Al (20 nm) film to a UV–ozone chamber (12 min, O_2 1 atm) or by sputtering Al_2O_3. The Cu:TCNQ film was deposited by the co-evaporation method. The typical I–V characteristics for the ITO/Al/(Al_2O_3)/Cu:TCNQ/Al device are shown in Figure 2. Figure 2 reveals the characteristic electrical switching of an ITO/Al/Cu:TCNQ/Al device. The device has a clear threshold from low impedance to high impedance at an applied voltage of 10.0±2.0 V and a reverse phenomenon at a negative bias of –9.5±2.0 V, when the authors progressively applied a voltage as indicated by the circled numbers. This characteristic switching phenomenon was observed more than 1000 times. However, they observed no switching effect in the ITO/Cu:TCNQ/Al device, suggesting that the Al layer plays an important role in switching (inset of Figure 2).

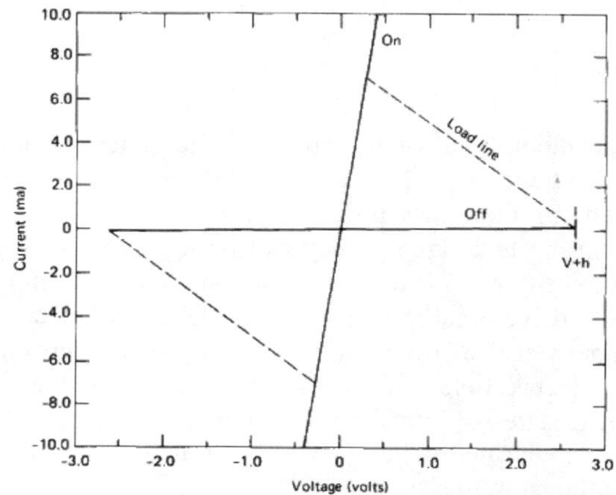

Figure 1. Typical dc current–voltage characteristics showing bistable switching in a 5-μm-thick Cu-TCNQ sandwich structure [2].

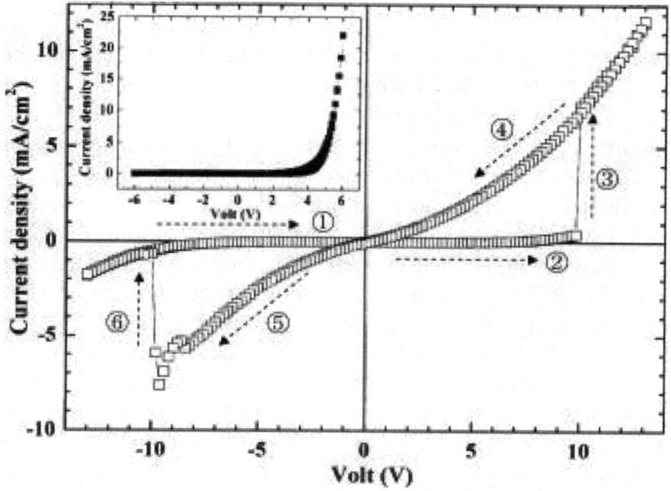

Figure 2. *J–V* characteristics of an ITO (300 nm)/Al (20 nm)/(Al$_2$O$_3$)/Cu:TCNQ (1:1, 100 nm)/Al (100 nm) device. (Inset) *J–V* characteristics of an ITO (300 nm)/Cu:TCNQ (1:1, 100 nm)/Al (100 nm) device [3].

Writing and erasing was realized by applying a positive and negative electric field (about 10^6 V/cm), respectively, to the films. The on/off ratio can reach as high as 10^4. No mention of write–erase cycle tests and retention tests were found in this report. They failed to observe the switching effect without the Al_2O_3 layer, and the device mechanism was not clear.

Organic Memories Based on Polymer Films

The switching phenomenon was also observed in some polymer films early in the 1970s [4, 5] and most of the earlier observed electrical memory effects turned out to be due to filament formation. Recently, at Princeton University, the Forrest research group reported a write-once-read-many times (WORM) memory effect by burning polymer fuses [6, 7].

The conductive polymer fuse consisted of polyethylene dioxythiophene:polystyrene sulfonic acid (PEDOT: PSS). Current transients were used to change the fuse from a conducting to a nonconducting state to record a logical "1" or "0". The burning resulted in a permanent reduction of forward-bias current by approximately five orders of magnitude. The conducting polymer film was obtained by spin coating technique. Figure 3 shows the I-V curves before and after 10 V voltage pulse was applied to an Au/PEDOT:PSS/n-Si device.

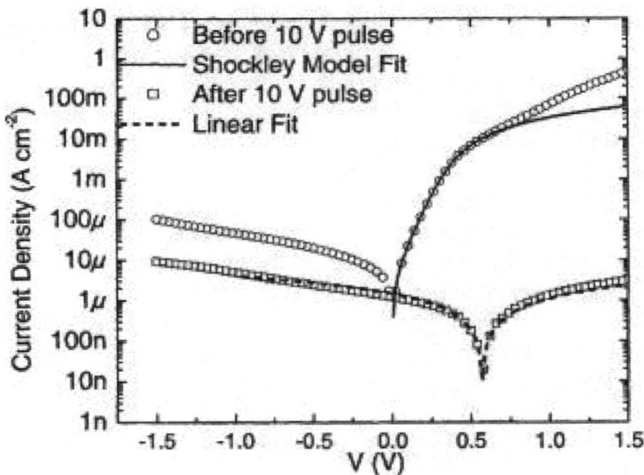

Figure 3. Current density *vs.* voltage characteristics of a Au/PEDOT:PSS/*n*-Si devices, after etching surrounding PEDOT:PSS, before (squares) and after (circles) switching with a 5-ms, 10-V pulse [7].

The Princeton group also demonstrated a simplified polymer/Si OI–HJ rectifying junction with potential applications for WORM memories. Polymer conductivity switching at high current densities was due to an oxidation/reduction reaction in the PEDOT:PSS film made possible by the injection of holes and electrons from the Au and Si contacts, respectively. Stabilization of the nonconducting state of

PEDOT: PSS occurs on shorter time scales and at significantly lower current densities than has been previously observed for Au/PEDOT/ITO double injection devices. They had confirmed that the thermal effects play a significant role in stabilizing the neutral PEDOT:PSS species formed by charge injection. The change from high to low conductivity state resulted in a permanent reduction in conductivity by a factor of $>10^4$ in a time period of ~300 ns. The switching process was found to be highly reliable and reproducible. It was suggested that this device structure had the potential for use in passive-matrix memory arrays in archival data storage.

Organic Memories Based on Functional Molecules

Electrical switching and memory effect were also reported for the devices using functional organic molecules as the active materials [8–12]. The basic idea is that organic molecules may have various conformations and states (reduction state or oxidation state), which may change under an applied electric field and cause the conductivity change [8]. The difficulty in this study is that it is not easy to verify that the observed switching and memory effect are caused by the real molecular conformation change or not.

Pal *et al.*, reported that conductance switching between three levels can be achieved in supramolecular structures of Rose Bengal [8], where electroreduction and conformational change of the molecules caused the conjugation modification, and as a result, conductance of the molecules was changed. At a negative voltage, Rose Bengal molecules are electroreduced to the high-conductivity state, while a positive voltage oxidized the molecules back to their low-conductivity state (Figure 4). The off-state is defined as state 1, while the on-state as state 2a. At a higher V_{Max}, the switching definition itself has been found to change. Figure 4 shows that with $V_{Max} > 4.5$ V, the material switched to a still higher conductivity state (state 3) at a positive V_{Max}. At a negative V_{Max} (< -4.5 V), Rose Bengal molecules switched off to a state (state 2b), which has a conductance around that of state 2a. They therefore had observed at least three conducting states in these devices, all of which can be achieved reversibly. The ratio between the two conductance states was as high as 2000. The ratio and V_{Max} at which the switching definition changed depends on the pH of PAH in the supramolecular matrix.

Organic Memories in Organics/Metal Nanocluster/Organic Structure

In the above-reported organic memories, metals already have been introduced in metal–organic charge-transfer complex systems. Recently, the UCLA research group demonstrated an organic electrical bistable device (OBD) by introducing metal nanocluster layers within an organic layer as the active medium [13–18], which shows rewritable nonvolatile memory behavior (Figure 5).

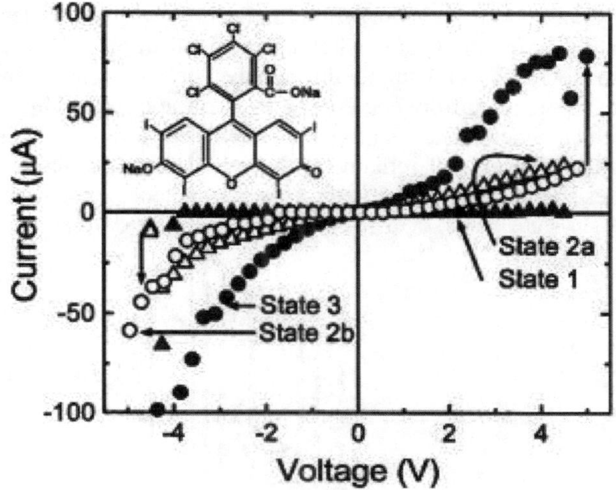

Figure 4. Current–voltage characteristics of a device based on ESA films of Rose Bengal (*p*H of PAH = 8.0) for two sweep directions and at two V_{Max}. Filled and open symbols represent sweeps from forward and reverse bias, respectively, in both the cases. V_{Max} was 4.5 (triangles) and 5.0 V (circles). The vertical arrows represent switching from state 1 to state 2a and from state 2b to state 3. The chemical structure of Rose Bengal in its ground state is shown in the inset [8].

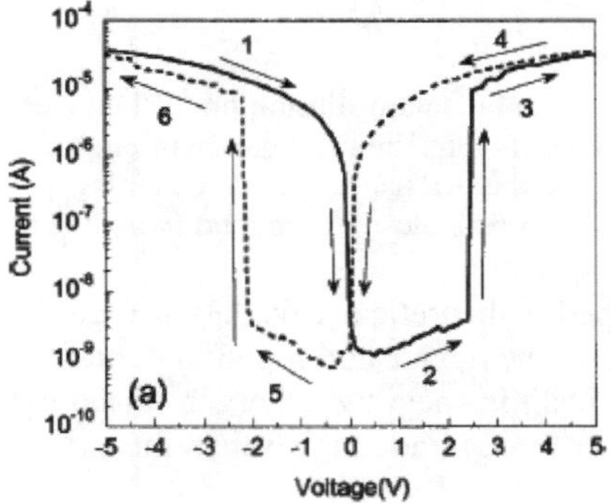

Figure 5. The *I–V* characteristics of an OBD device measured from –5 to +5 V (open circle) and +5 to –5 V (filled circle) [17].

The on-state and the off-state differ in their conductance by as much as 10^6 times and show a remarkable retention. In order to obtain these metal nanoclusters, a slow evaporation rate (~0.3 Å/s) for the metal vapor in a vacuum around 2×10^{-6} torr was utilized, so that the middle layer consisted of metal nanoclusters (metallic cores with insulating shells). The size of the nanoclusters was about 10 to 15 nm.

Alternatively, the metal nanoclusters can also be achieved by other means; for example, co-evaporation of Al vapor with organic insulating molecules.

Figure 6 illustrates the OBD switching mechanism [17]. Free electrons in the metallic Al-nanocluster cores tunnel through the energy barrier formed by the insulating shells of the cores, from one energy well to the other against the direction of the applied electric field. The negative charge will be stored on one side of the well and the positive charge will remain on the other side. The stored charge subsequently makes the organic layers undergo a conductance change and switches the device to the on-state.

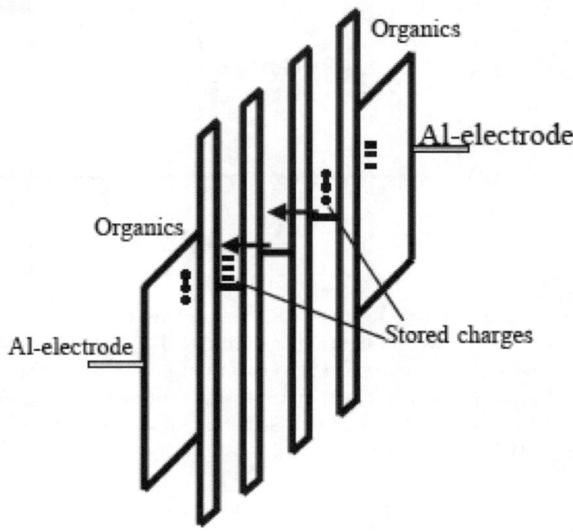

Figure 6. The schematic band diagram illustrating OBD under switching voltage electrons in the cores of Al nanoclusters tunneling resonantly through the energy barrier provided by the insulating coatings on the Al nanoclusters. Resulting positive–negative charges are stored on both sides of the Al-nanocluster layer, and switching the device to on-state [17].

Later we performed a theoretical calculation regarding the electron transmission probability through the device and found that indeed, the electron transmission probability with the positive–negative charges on both sides of the nanoclusters layer is several orders of magnitude higher than the case without charges within the nanoclusters layer [18].

Electrical bistable devices have been fabricated using the sandwiched structure of organic/metal nanocluster/organic as the active medium interposed between two Al electrodes. Experimental studies of the device operation mechanism have been carried out. It was found that the bistability is very sensitive to the nanostructure of the OBD. The UV–visible absorption spectrum, surface morphology, and electrical conductance studies suggested that the middle metal layer for the bistable device consisted mainly of partially oxidized, small metal nanoclusters, instead of pure metal as previously described [9]. An impedance study of the device indicated that more charge is stored for the device in a high-conductance state than in a low-conductance state. It is believed that the nanocluster layer is polarized after the applied bias is above a threshold voltage, consequently making the organic layers undergo a conductance change, switching the whole device to a high-conductance state. A proper reverse bias is needed to neutralize the stored charge and to restore the device back to a low-conductance state. It is also found that high-yield and high-performance OBDs must be fabricated with a slow deposition rate for the middle metal layer to allow the formation of very small-sized nanoclusters.

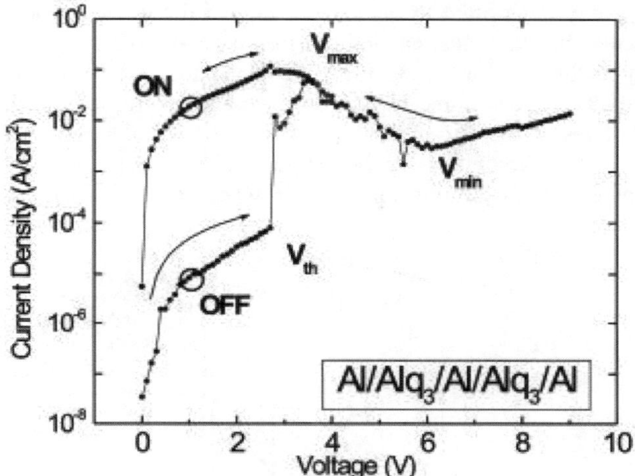

Figure 7. Current–voltage characteristic of an Al (50 nm)/Alq$_3$ (50 nm)/Al (5 nm)/Alq$_3$ (50 nm)/Al (50 nm) device [19].

The IBM research group at Almaden [19] also observed electrical bistability in this organic/metal nanocluster/organic system as shown in Figure 7, where the I–V characteristics were not the same as what we observed. The mechanism details are not clear, and need to be further verified. But one difference from our device was that the middle nanocluster layer was very thin (5 nm). They believed that the mechanism is a charge storage, as described by Simmons and Verderber [20].

Organic Memory by Controlling Cu-ion Concentration Within the Organic Layer

A high on-state current memory device was reported by our research group, by controlling the copper ions within the organic layer [21]. The device structure is copper anode/buffer layer/organic layer/cathode. The switching-on process was realized by applying a positive voltage pulse between the two electrodes, driving Cu^+ ions into the organic layer to transform the device into a high-conductance state.

Figure 8 shows the typical $I-V$ characteristics of a Cu–OBD. When the bias was ramped from 0 to 3.5 V, the current injection was initially very low. At a critical voltage (V_{c1}) of approximately 0.7 V the device switches from the low-conductance (off) state to a high-conductance (on) state. The device stays in the on state as the bias is increased until it reaches another critical voltage (V_{c2}) of around 2 V, where it reverts to the off state. Therefore, the transition from the off state to the on state (switch-on) can be achieved by applying a medium bias ranging from V_{c1} to V_{c2}, and the transition from on state to off state (switch-off) can be achieved by applying a voltage greater than V_{c2}. Interestingly, the switch-on and switch-off processes are nonvolatile processes, *i.e.*, once the device switches to either state it remains in that state for a prolonged period of time. A small probing (or reading) bias, less than the V_{c1}, may be applied to detect the state of the device. The bottom inset of Figure 8 shows the stress test carried out in the on and off states probed by 0.1 V. We also tested the retention time for both the states under no bias condition and found it to exceed six months. In addition, we also evaluated the thermal stability of both the states of the Cu–OBDs at 110°C for several hours in ambient conditions. The device remained in either state without degradation. Therefore, Cu–OBDs can be electrically programmed for rewritable nonvolatile memory applications.

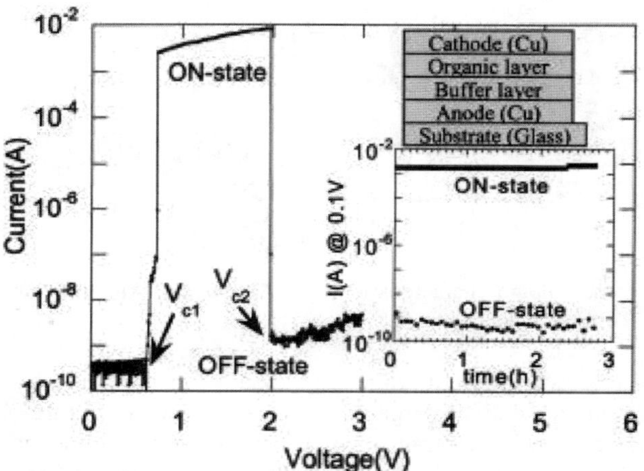

Figure 8. The typical *I–V* characteristics of Cu–OBDs. The inset shows the device structure and stress test for the device at both the on state and off state [21].

Organic Memory Based on Polymer Composite Containing Metal Nanoparticles and Small Organic Molecules

An electrical bistable device was recently developed by using a polymer film mixed with metal nanoparticles and conjugated small organic compounds [22, 23]. As shown in Figure 9, this device has a simple device structure. The active film between the two metal electrodes was formed by a solution process.

The materials in the device include an inert polymer such as polystyrene (PS) or poly(methyl methacrylate) (PMMA), metal nanoparticles, and conjugated organic compound such as 8-hydroxyquinoline (8HQ) or 9,10-dimethylanthracene (DMA). The Au nanoparticles have an average particle size of 2.8 nm in diameter. The inert polymer was used as the matrix. The device was fabricated through the following process. At first, the bottom Al electrodes (rows) were deposited in a high vacuum through a thermal evaporation. Then, the active film was formed by spin-coating a solution, such as 1,2-dichlorobenzenoic solution of 1.2% by weight PS, 0.4% by weight 1-dodecanethiol-protected Au nanoparticles (Au-DT), and 0.4% by weight DMA. Finally, the top Al electrodes (columns) were thermally deposited. The notation Al/PS+Au-DT+DMA/Al is used to represent this device.

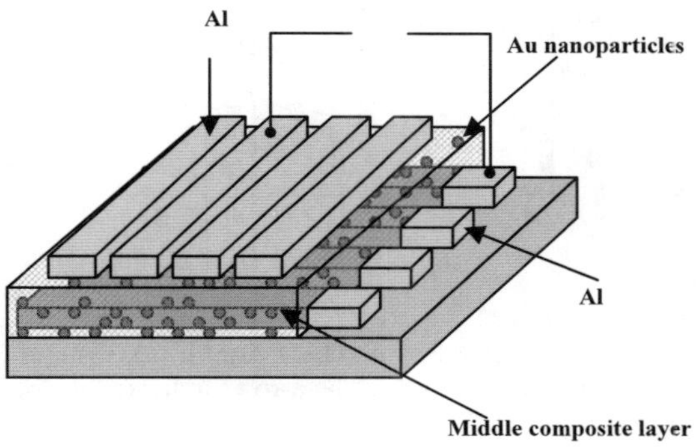

Figure 9. Device structure.

Figure 10 shows the I–V curves of the device Al/PS+Au–DT+DMA/Al tested in air. Initially, the device was in the off state, and the current was approximately 10^{-10} A at 1 V. The electrical transition from the off state to the on state took place at 6.1 V with an abrupt current increase from 10^{-10} A to 10^{-6} A (curve (a)). The device exhibited good stability in this high conductivity state during the subsequent voltage scan (curve (b)). The high conductivity state was able to return to the low conductivity state by applying negative bias as indicated in curve (c), where the current suddenly dropped to 10^{-10} A at -2.9 V.

This switching behavior can be repeated numerous times and can be driven by voltage pulses as well. Hence, this device can be used as a nonvolatile memory. The transition time from the off state to the on state is less than 25 ns.

Such electrical transition was observed for the device tested under a nitrogen atmosphere as well as in the air, which suggests that water and oxygen do not play any role in the electrical switching. The switching mechanism is not the filament formation between the metal electrodes, which was the reason for the electrical transition observed for some polymer films. The device Al/Au–DT+DMA+PS/Al in the on and off states was studied by ac impedance spectroscopy. No drop of the capacitance was observed after the device was electrically switched from the pristine state to the on state. Moreover, the electrical behavior of the device is strongly dependent on the nature of the metal nanoparticles and the conjugated small organic compound. The polymer may just play a role as a matrix for these two components.

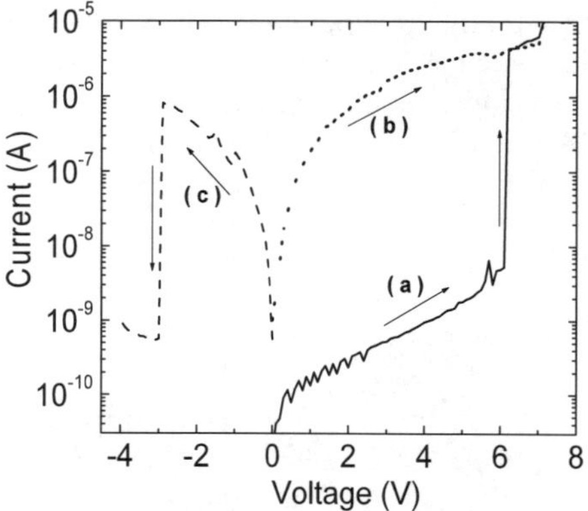

Figure 10. *I–V* curves of a device, Al/PS+Au-DT+DMA/Al. (a), (b) and (c) represent the first, second, and third bias scans, respectively. The arrows indicate the voltage-scanning directions.

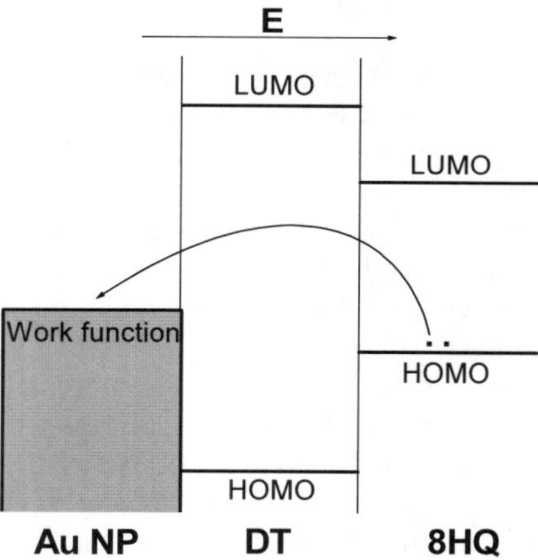

Figure 11. Energy diagram of Au nanoparticle (Au NP), 1-dodecanethiol (DT), and 8-hydroxyquinoline (8HQ). The two dots on the HOMO of 8HQ represent two electrons, **E** is the applied electric field, and the lower arrow indicates the electron transfer.

An electric-field induced charge transfer between the Au nanoparticle and 8HQ was proposed for the switching mechanism. It has already been demonstrated that 8HQ and Au nanoparticles can be an electron donor and acceptor, respectively [24–26]. When the electric field reaches a threshold, the electrons on the HOMO of 8HQ may gain enough energy and tunnel through the insulator coating on the Au–

DT nanoparticle into the metal core. This results in the generation of the charge carriers on 8HQ and the increase of the conductance of the film. It is wellknown that the conductivity of conjugated organic compounds will increase after their HOMO or LUMO become partially filled [27, 28]. The recombination of the positive charge on the 8HQ and Au nanoparticle was prevented by the insulator coating on the Au nanoparticle.

Figure 11 is a schematic energy diagram for such a change transfer between Au nanoparticles and 8HQ. The HOMO and LUMO levels of 8HQ, calculated using DFT B3LYP method with the 6-31+G(d,p) basis set, are 1.9 and 6.1 eV, respectively. The energy (E_c) to charge a Au nanoparticle with a diameter of 2.8 nm and capped with 1-dodecanethiol, is about 0.1 eV and calculated using [29]:

$$E_c = \frac{e^2}{2C}$$

$$\text{and } C = 4\pi\varepsilon_0\varepsilon_r \frac{r}{d}(r+d),$$

where C is the capacitance of the Au nanoparticle, ε_0 the permittivity of free space, ε_r the permittivity of the capped molecule on the Au nanoparticle, r the radius of the Au nanoparticle core, and d the length of the capped molecule. This charging energy is the energy to be overcome for the charge transfer to take place. It is possible for the electron to gain such energy under a high electric field.

This simple model can interpret the stability of the device in the high conductivity state as well as during the erasing process by applying a negative bias. Stability of the negative charge on Au nanoparticle is due to the insulator coating of 1-dodecanethiol on the Au nanoparticles, which prevents recombination of the charge after removal of the external electric field. Since the charge transfer is induced by an external electrical field, the film is polarized after the charge transfer. Only a reverse electric field can assist the tunneling of the electron from the Au nanoparticle back to the HOMO of 8HQ$^+$, resulting in a return to the low conductivity state.

Conclusion

We have briefly reviewed the development of organic memories. Due to the space limitation, there may be much more work on organic memories that has been omitted here. Organic memories are promising, considering their advantages of potential low cost, flexibility, and easy of fabrication, but there is still a distance to go before they are ready for real application. For most organic memories, the retention times and the device stability are the issues that need to be overcome before application. Organic materials incorporated with inorganic materials or metal nanoclusters may be a solution. We sincerely hope that these issues will be solved in the near future.

Acknowledgements

For financial support, the authors are indebted to the Air Force office of Scientific Research (FA9550-04-1-0215, Dr. Charles Lee), National Science Foundation (DMR-0305111, Dr. La Verne Hess), and the Office of Naval Research (N00014-04-1-0403, Dr. Paul Armistead). Thanks are due to our research group members for their help and support, particularly for Vishal Shrotriya's polishing of the manuscript.

References

1. R.S. Potember, T.O. Poehler, and R.O. Cowan, Appl. Phys. Lett. 34, 405 (1979).

2. R.S. Potember, T.O. Poehler, A. Rappa, D.O. Cowan, and A.N. Bloch, Synth. Met. 4, 371 (1982).

3. T. Oyamada, H. Tanaka, K. Matsushige, H. Sasabe and C. Adachi, Appl. Phys. Lett., 83, 1252 (2003).

4. H. Carchano, R. Lacoste, and Y. Segui, Y. Appl. Phys. Lett. 19 (1971) p. 414.

5. H.K. Henish, & W.R. Smith, Appl. Phys. Lett. 24 (1974) p. 589.

6. S. Moller, C. Perlov, W. Jackson, C. Taussig, S.R. Forrest, Nature 426 166(2003) .

7. S. Smith and S.R. Forrest, Appl. Phys. Lett., 84, 5019 (2004).

8. A. Bandyopadhyay and A.J. Pal, Appl. Phys. Lett., 84, 999 (2004).

9. C.P. Collier, G. Mattersteig, E.W. Wong, Y. Luo, K. Beverly, J. Sampaio *et al.,* Science 289 p.1172 (2000).

10. A.R. Pease, J.O. Jeppesen, J.F. Stoddart, Y. Luo, C.P. Collier, and J.R. Heath, Acc. Chem. Res. 34 p. 433 (2001).

11. Y. Chen, D. A.A. Ohlberg, X. Li, D.R. Stewart, R.S. Williams, J.O. Jeppesen *et al.,* Appl. Phys. Lett. 82 p. 1610 (2003).

12. M.A. Reed, J. Chen, A.M. Rawlett, D.W. Price, and J.M. Tour, Appl. Phys. Lett. 78 p. 3735 (2001).

13. Y. Yang, L.P. Ma, and J. Liu, US patent pending US 01/17206, 2001.

14. L.P. Ma, J. Liu, S.M. Pyo, and Y. Yang, Appl. Phys. Lett. 80 p. 362 (2002).

15. L.P. Ma, J. Liu, and Y. Yang, Appl. Phys. Lett. 80 p. 2997 (2002).

16. L.P. Ma, J. Liu, S.M. Pyo, Q.F. Xu, and Y. Yang, Mol. Cryst. Liq. Cryst. 378 185 (2002).

17. L.P. Ma, S.M. Pyo, J. Ouyang, Q.F. Xu, and Y. Yang, Appl. Phys. Lett. 82 p. 1419 (2003).

18. J.H. Wu, L.P. Ma, and Y. Yang, Phys. Rev. B 69 p. 11531 (2004).

19. L.D. Bozano, B.W. Kean, V.R. Deline, J.R. Salem, and J.C. Scott, Appl. Phys. Lett. 26 p. 607 (2004).

20. J.G. Simmons and R.R. Verderber, Proc. R. Soc. London, Ser. A 301, 77 (1967).

21. L.P. Ma, Q.F. Xu, and Y. Yang, Appl. Phys. Lett, 84 p. 4908 (2004).

22. Y. Yang, J. Ouyang, and C. Szmanda, US patent pending.

23. J. Ouyang, C.-W. Chu, C. Szmanda, L. Ma, and Y. Yang, Nat. Mater., to be published.

24. C.K Prout and A.G. Wheeler, J. Chem. Soc. A p. 469 (1967).

25. D.M. Adams *et al.* J. Phys. Chem. B 107 p. 6668 (2003).

26. B.I. Ipe, K.G. Thomas, S. Barazzouk, S. Hotchandani, and P. V. Kamat, J. Phys. Chem. B 106 p.18 (2002).

27. T. Oyamada, H. Tanaka, K. Matsushige, H. Sasabe, and C. Adachi, Appl. Phys. Lett. 83 p.1252 (2003).

28. X. L. Mo, *et al.* Thin Solid Films 436 p. 259 (2003).

29. J.F. Hicks, A.C. Templeton, S.W. Chen, K.M. Sheran, R. Jasti, R.W. Murray *et al.*, Anal. Chem., 71 p. 3703 (1999).

Part III

Materials for Interconnects

Interconnect Technology – Today, Recent Advances and a Look into the Future

M. Engelhardt, G. Schindler, W. Steinhögl, G. Steinlesberger, M. Traving

Infineon Technologies, Corporate Research, Munich / Germany

Introduction

In the table of the International Technology Roadmap for Semiconductors (ITRS) [1], displaying the MPU interconnect technology requirements for the long term, *i.e.* for the 2010 to 2018 timeframe, almost all fields are highlighted in red, standing for *"Manufacturable Solutions are NOT known"*. These so-called "red brick walls" illustrate that interconnect technology is becoming progressively demanding and that it is imperative to already be trying today to assess tomorrow's interconnect challenges. The manufacturability of potential solutions cannot be given today anyway, since tomorrow's manufacturing tools, in particular next generation lithography (NGL) are not yet available. Access to the critical dimensions (CDs) and fine pitches of these future semiconductor products is only possible by use of direct writing lithographic methods. To bypass the limitations imposed by these techniques on the generation of large numbers of test structures, alternate approaches based on removable spacers [2] and mask trims [3] were developed; these techniques allow, by use of standard manufacturing lithography, the fabrication of structures with tight CDs, however with the trade-off of relaxed pitches, reflecting lithography resolution. Recent results obtained with such an approach show that some red brick walls have cracks and suggest that Cu/low-k on-chip wiring may be extended evolutionarily rather than revolutionarily. Furthermore, in future chip generations, design will progressively become an enabler.

Contributors to RC Delay

The technological approach to reduce RC signal delay is to keep metal resistivity low and to use low-k materials for insulation to reduce inter- and intra line parasitic capacitance. RC signal delay can also be reduced by design and by architecture by keeping local interconnects short and utilizing hierarchical architecture, respectively. Design and architectural approaches can thus take some burden from technology to move to risky and cost intensive material innovations. Calculations have indeed shown [4] that SiO_2 could be used as an insulator even in 45 nm node products, without parasitic capacitances of the interconnect being the performance limiting element, if only design would keep local interconnect lengths shorter than 15 CDs.

Conductor Resistivity and Size Effects

Using SiO_2 as the intermetal dielectric (IMD), Cu-based nano-interconnects with CDs down to 30 nm were fabricated (Figure 1) with a spacer technique [2]. The electrical characterization yielded a strong increase of electrical resistivity with decreasing CD over a wide range of temperatures (Figure 2). The experimental data could be well described by a combined model [5]. This model takes into account charge carrier scattering at both external and internal interfaces, *i.e.* scattering effects resulting from the fact that at least one dimension of the conductor is no longer large compared to the mean free path of charge carriers [6, 7] and grain boundary scattering [8], respectively. The experimental data show that cooling becomes progressively inefficient to reduce resistivity with decreasing CDs [9], and that the resistivity value (2.2 $\mu\Omega$cm) displayed in the ITRS for future chip generations may only be achieved at very low temperatures. The lifetime of Cu lines (not connected to vias) with CD=65 nm was determined [10] with standard reliability methodology and yielded a value (120 years) comparable to values obtained for current products. In destructive tests the lines failed at current densities (60–100MA/cm^2) beyond those obtained for a wiring scheme with relaxed CDs [9].

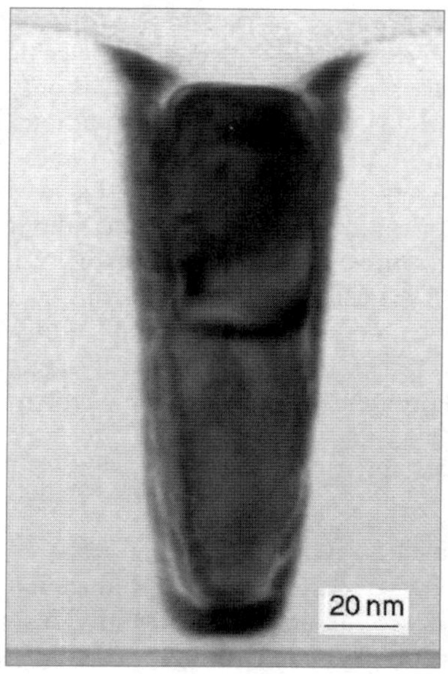

Figure 1. Metallized damascene trench meeting ITRS requirements regarding CD and barrier thickness for local interconnects in MPUs in 2013. The requirement for the aspect ratio (height:width) is even surpassed by about 100%

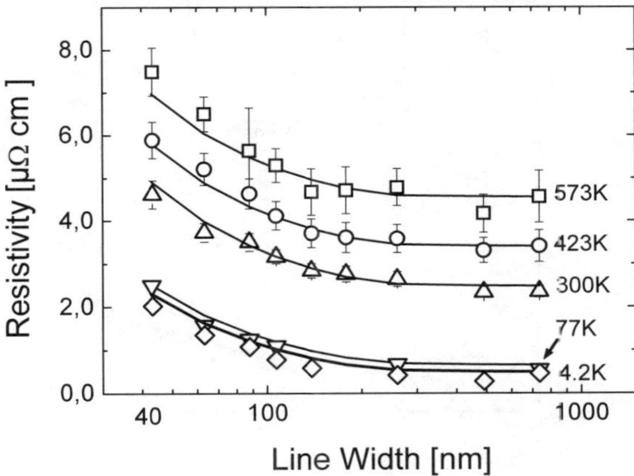

Figure 2. Dependence of the electrical resistivity of copper lines on line width measured at various temperatures

Diffusion Barrier Scaling

Barrier films have to be scaled down in thickness aggressively to leave maximum space for Cu as they exhibit electrical resistivities of at least an order of magnitude higher than that of Cu; this assertion holds for both lines and vias. For vias in particular, thin barrier films are mandatory as the current path leads right through the barrier film at the bottom of the structure. Ta-based barriers are most commonly used in Cu-metallization schemes. Direct deposition of Ta on the IMD results in tetragonal β-Ta, which exhibits high electrical resistivity (180 μΩ·cm). The desirable low resistivity modification (20–30 μΩ·cm) of cubic α-Ta is only obtained by deposition onto a TaN underlayer, which, in turn, exhibits high electrical resistivity (235 μΩ·cm). In minimizing the thickness of the TaN film in a Ta/TaN bi-layer barrier stack, α-Ta was obtained on TaN films as thin as 1 nm [11]. After annealing (450°C, 120 min) in forming gas, leakage current densities between neighboring damascene lines (Figure 3.) with effective thicknesses of Ta only barrier films down to the sub-1 nm regime were performed [12] and found to be as low as those obtained with a thick reference barrier used in a current RF product. The leakage current density obtained at an electric field strength of 0.4 MV/cm is still at noise level; under worst-case assumptions, this is expected to be the maximum field strength between neighboring conductor lines in end-of-roadmap (2018) high-performance products. In bias-temperature-stress (BTS) tests, barrier films with nanoscaled thicknesses were furthermore exposed to high electric fields (≈1 MV/cm) at elevated temperatures (200°C) for excessive durations (2000 h) and showed the same barrier integrity as the thick reference barrier [12]. These results suggest that fabrication of barrier films with end-of-roadmap thicknesses (2 nm for 18 nm node products in 2018) may not remain a blocking point.

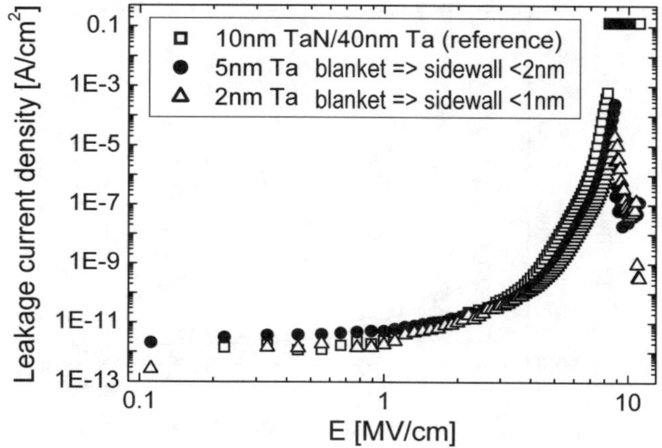

Figure 3. Dependence of leakage current density on electrical field for diffusion barriers with end-of-roadmap thicknesses; the result obtained for a thick barrier bi-layer in a current RF product is shown as a reference

Low-k Intermetal Dielectrics

Reduction of the contribution of the IMD to RC signal delay means to substitute SiO_2, the decade-long used insulator of choice with a material of lower electrical permittivity, *i.e.* a material with lower k-value than that of SiO_2 (k ≈ 4). A reduction of the k-value can be achieved by three approaches: by a reduction of polarity, by a reduction of density, and by leaving out the IMD. This is basically a pathway from SiO_2 (k ≈ 4), to doped oxides (k ≈ 3.5), to organic materials (k ≈ 2.5–3), to porous films (k ≈ 2.0–2.5), and finally to airgaps, featuring the ultimate k-value of vacuum (k = 1). This pathway to reduce the k-value is accompanied by an increasing need for auxiliary layers, used as hard mask, etch stop, *etc.* These auxiliary layers, in turn, contribute with their intrinsic k-value to the overall k-value, the effective k-value k_{eff}, of the metallization scheme; furthermore the reduction of the k-value is associated with an increase of the complexity of processing and of the complete integration scheme. Simulations of the k-value of porous films have shown [13] that a volume reduction of about 50% is required to reduce the bulk k-value of the IMD from 4 to 2. With focus only on k-value reduction, airgaps are the most promising approach. Recently it was shown [14] by an electrical characterization of airgaps (Figure 4) that a reduction of line-to-line capacitances in a SiO_2/SiN-based interconnect scheme of 50% can be achieved. Electrical reliability, as well as mechanical stability, in particular in packaging, are currently under investigation for both interconnect schemes with porous low-k films and with airgaps.

Figure 4. Airgaps between copper damascene lines in an SiO$_2$/SiN-based interconnect scheme [14]

Assessment of Novel Interconnect Schemes

The discussion of novel approaches in the backend-of-line (BEoL) has mainly focused on the red brick walls in the ITRS. As described above, recent investigations on the various aspects of deep sub-100 nm Cu-based interconnects have shown that some of the red brick walls in the interconnect chapter of the ITRS will probably not last in the future.

3D Architecture for an Advanced BEoL: Stacking of Device Layers

An advanced BEoL 3D architecture would mean to shift from today's metallization scheme (Figure 5) featuring multiple conductor levels extending over three dimensions on top of a single layer with active devices to a three-dimensional arrangement of layers with active devices, each of them having only a few wiring levels. 3D in the wiring scheme would be shifted to 3D for active devices. In such an arrangement resulting from the stacking of (rather similar) technology layers one could even imagine that within the individual device layers in the stack the interconnections could then be achieved with relaxed requirements regarding design, layout, and unit processes. Thus the global interconnects, the longest interconnections across a chip, could be kept shorter than with today's state-of-the-art approach with many levels of metal on a single device layer. Such a 3D approach is characterized by roughly requiring the same total area A of silicon as conventional technology, while featuring a reduced footprint (A/n) in the stack of n device layers. It has been reported [15] that stacking of five device layers may result in almost half the RC delay obtained for a single device layer not utilizing 3D architecture. Depending on the concept used, multilayer stacks imply the application of all the challenging process modules of thinning, gluing/bonding, vertical interconnecting repetitively whenever a further layer is added to the stack. Thus both the adhesion strength of the glue/bond and the integrity of the vertical interconnects through the layers already present on the chips may be repetitively subjected to mechanical stress, which may impact the overall yield and reliability. Therefore, the appearance of semiconductor products featuring stacks consisting of multiple active layers is more than questionable in the near future.

Figure 5. Interconnect scheme in an advanced MPU (130-nm CMOS) featuring nine levels of copper on a single layer of active devices (arrow). Photo courtesy: E. Zschech, AMD Saxony LLC & Co. KG

Optical Interconnects for an Advanced BEoL: III-V on Si

Beyond this BEoL-related 3D architecture for stacking of device layers, optical interconnects have been discussed as potential on-chip interconnect solutions for future chip wiring [1]. For the BEOL, the main application in mind is clock distribution. This is the second step of 3D integration, namely the stacking and vertical interconnection of (rather dissimilar) technologies. Regarding optical on-chip interconnects based on III–V materials there are some serious obstacles:

1.   The crossover of the optoelectronic conversion delay ($\tau \approx 10$–10 s) of optoelectronic components with the RC signal delay of a copper-based interconnect of the 50 nm technology node (assuming SiO_2 as intermetal dielectric for both cases) occurs at an interconnect length of about 300 μm (Figure 6.) and thus shows no need for optical wiring; the crossover with wider wires occurs at even longer interconnect lengths.

2.   Disparity of the technologies and materials: III–V materials required for photon generation are associated with a process technology that differs significantly from silicon process technology, which has no active photonic material in its portfolio of materials. The disparity of materials in the III–V

world and in the silicon world would require a 3D-integration scheme with, for example, III–V optoelectronics on Si-based electronics.

3. Disparity of supply voltages (Figure 7.): The energy gaps of III–V materials range from about 2.5 eV (AlP) down to 0.4 eV (InAs). Therefore, the minimum power supply voltages required to obtain emission of laser radiation ranges from 2.5 V (AlP) down to 0.4 V (InAs); in reality the threshold voltages are higher than these theoretical values. The ITRS predicts a continuous downscaling of the nominal power supply voltages for high-performance products with a decrease of about 30 mV/year for both the short and the long term. With Vdd=1.0 V for 90 nm node products expected to be in production in 2004, products with end-of-roadmap feature sizes (18 nm) are assumed to be operated at a power supply voltage of 0.5 V. To allow low power supply voltages, III–V materials with narrow band gaps must be used; InAs could therefore be the only candidate with a chance to be operated as active optoelectronic material at voltages comparable with the power supply voltage of end-of-roadmap products. Due to the steady decrease of supply voltages in the silicon world from technology node to technology node it is becoming increasingly challenging to operate both III–V materials and silicon with the same supply voltage.

4. Disparity of component sizes and CDs: The wavelength (in air) corresponding to energy gaps of III–V materials ranging from 2.5 eV (AlP) down to 0.5 eV (InAs) are in the 0.5–2.5-µm range. When using one supply voltage for both III–V materials and silicon, the downscaling of Vdd in silicon would require one to move to narrow band gap optoelectronic devices (such as InAs) and hence to long emission wavelengths. Since optical waveguides have lateral extensions comparable to the wavelength they have to guide, optical wiring would also become very area-intensive and require components with larger sizes, while feature sizes in silicon technology become smaller by following Moore's law for downscaling CDs.

5. High temperature operation of III–V material-based lasers such as vertical cavity surface emitting lasers (VCSEL) is problematic: Their operation temperature has to be kept well below 90°C. This is a serious obstacle for 3D of Si and III–V. The T_{use}=105°C in the silicon world is a temperature that is applied in the III–V world to promote accelerated degradation of optoelectronic components for reliability studies.

6. Packaging issues have not been solved: In hybrid integration the alignment of optical components has not been solved.

7. Lack of adequate design tools for integration of photonic components.

8. Optical wiring of silicon-based electronics would furthermore impose a testability issue as testing could only be done at the very end of the complete process flow.

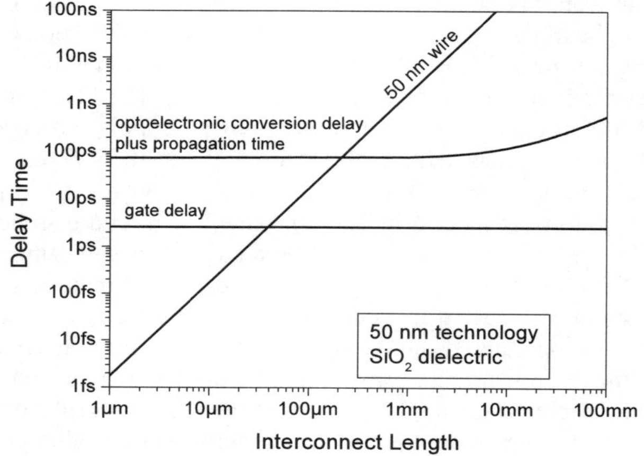

Figure 6. Comparison of conversion and propagation delay of optoelectronic devices with wire and gate delay for 50 nm node products assuming SiO$_2$ as intermetal dielectric

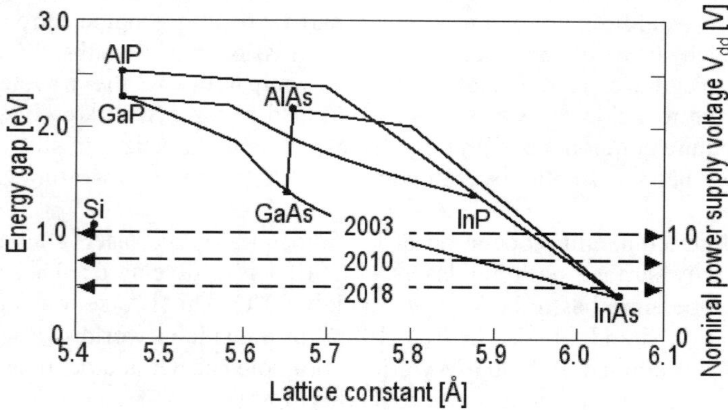

Figure 7. Comparison of the energy gaps and related minimum operation voltages of optoelectronic materials with the supply voltage requirements for semiconductor products for the 2003– 2018 timeframe (ITRS 2003)

RF Interconnects for an Advanced BEoL

Beyond 3D architecture for stacking of device layers and optical interconnects, RF/microwave-based approaches have been discussed for potential on-chip interconnect solutions for future chip wiring [1]. Like for optical interconnects, the main application in mind is clock distribution. RF/microwave-based clock distribution may, but must not necessarily utilize, 3D architecture. There is already RF/microwave on today's chips if one looks on the clock frequency of advanced MPUs! For the near future, the ITRS requirement for the clock frequency shows an increase from 3 GHz today up to 12 GHz in 2009 followed by an average long-term annual increase of 5–6 GHz up to about 50 GHz in high performance MPUs which are, according to the ITRS, expected to be in production in 2018. The wavelength (in air) for these frequencies today are 30 cm and will remain in the cm-range until 2015 and thus be comparable with chip extensions. In the literature [16], on-chip antennas have been discussed; the radiation patterns of such antennas compare well with those obtained for the same types of antennas in the VHF/UHF regime. The radiation patterns were, however, measured in an experimental off-chip setup. For on-chip applications, such as clock distribution, the Raleigh criterion for the minimum critical distance d that has to be realized between "transmitter" and "receiver" is given by

$$d > 2 \cdot D^2 / \lambda$$

with D and λ standing for the lateral extension of the antenna and the emission wavelength, respectively. For both dipole-type and loop antennas the lateral extension is typically on the order of half the wavelength; fractal element on-chip antennas feature a smaller footprint at the expense of higher area consumption, resulting in some decrease of the effective D due to their geometry; the Raleigh criterion, however, yields a critical distance comparable with wavelength and chip extensions for both types of antennas. Thus, the validity of the results obtained in off-chip experiments for on-chip applications is questionable. Furthermore, there are concerns for RF/microwave based on-chip interconnects regarding area and power consumption, cross-talk, interference, noise, and the component size of waveguides.

Generally, applying the Raleigh criterion for on-chip antennas requires transmitter-receiver distances comparable with chip extensions; an application of this technique to chip-to-chip data transfer is possible rather than an on-chip wireless LAN.

Carbon Nanotubes: Enablers for Future Interconnect Schemes?

The unique electrical, mechanical, and thermal properties of carbon nanotubes [17] outperform the corresponding properties of state-of-the-art electronic materials in silicon process technology. Furthermore, their diameters (1–30 nm) and lengths (μm–mm) are comparable with the CDs and geometrical length in future technology nodes. This consequently fuels discussions about the application and implementation of carbon nanotubes for on-chip interconnects (and active devices). Carbon-nanotube-based interconnects [18], FETs [19], and diodes [20] have been fabricated successfully with electrical characteristics exceeding those known in standard CMOS technology. This may give rise to a vision of a scenario of nanoelectronics (Figure 8) with carbon-nanotube-based elements for both electronic devices and wiring. The functionality of these building blocks has already been demonstrated. Outstanding material properties of carbon nanotubes have been obtained so far on handpicked samples; to compete with CMOS, one would already require today the realization of 10^8 functional, identical, interconnected active devices. In order to become a mainstream technology further progress is still required with a focus on selective growth with pre-defined properties, low contact resistance, and, in particular, self-organization.

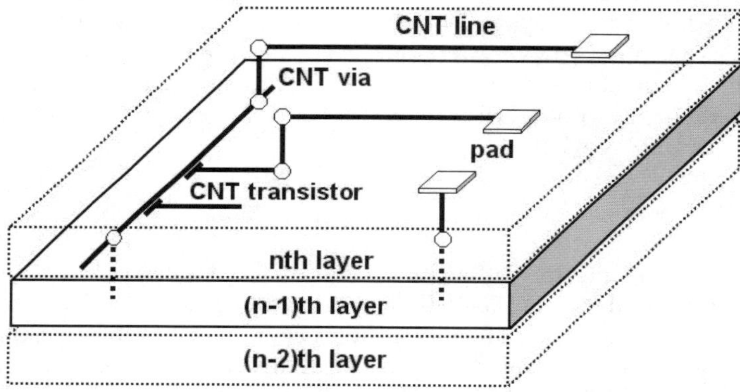

Figure 8. Vision of a scenario of nanoelectronics with carbon-nanotube-based elements for both electronic devices and interconnection

Conclusion

The results show that some red brick walls even for end-of-roadmap chip generations have cracks and that in the next 14 years until the end of the current edition of the ITRS the phrase "no known solutions" can be changed into "solutions known". The results also show that conductor resistivity (ρ_{Cu}=2.2 μΩ·cm) remains rated "no known solution". For the realization of future interconnect schemes the role of being enabler will, at least partially, shift from technology to design. Design has not yet been driven to its limits as it has been the case with technology. Encouraging examples for this assessment are the recent advances of design realizing hardware with X-architecture [21, 22] for on-chip wiring. Furthermore, an estimation shows [4] that if design can keep local interconnects short (<15 CDs), local interconnects might even be embedded in SiO_2 as intermetal dielectric for 45 nm node products without being primarily compromised by wire related RC signal delay. In the future (interconnect) technology *alone* will no longer be the enabler; further advances in performance and cost effectiveness will result from technology *and* design, with the latter discipline becoming progressively the enabler. As a result of interdisciplinary close concurrent efforts of technology *and* design, interconnect technology for future technology nodes will very probably be evolutionary rather than revolutionary.

Acknowledgements

Thanks are due to our colleagues at Corporate Research in particular to Z. Gabric, C. Hanke, W. Hönlein, D. Kehrer, W. Pamler, H. Riechert, W. Simbürger, and B. Stegmüller for contributions and valuable discussions on the various aspects of interconnect technology. Special thanks for continuous support are due to the process engineers of Infineon´s Munich cleanroom H84, the metallization group, the reliability methodology group, and the failure analysis group.

References

1. International Technology Roadmap for Semiconductors (ITRS), Semiconductor International Association, http://public.itrs.net/ (2003)

2. M. Engelhardt, G. Schindler, K. Mosig, G. Steinlesberger, W. Steinhögl, G. Gebara, Advanced Metallization Conference, Proc. AMC 2001, p. 11, (2001)

3. G. Steinlesberger, G. Schindler, M. Engelhardt, W. Steinhögl, M. Traving, International Interconnect Technology Conference, IITC 2004, Proceedings, p.51, (2004)

4. G. Schindler, W. Steinhögl, G. Steinlesberger, M. Traving, M. Engelhardt, Microelectron. Eng., 70/1, 7-12 (2003)

5. W. Steinhögl, G. Schindler, G. Steinlesberger, M. Engelhardt, Phys. Rev. B 66, p. 075414 (2002)

6. K. Fuchs, Proc. Cambridge Philos. Soc. 34, 100 (1938)

7. E.H. Sondheimer, Adv. Phys. 1, 1 (1952)

8. A.F. Mayadas and M. Shatzkes, Phys. Rev. B 1, 1382 (1970)

9. G. Schindler, W. Steinhögl, G. Steinlesberger, M. Traving, M. Engelhardt, Advanced Metallization Conference, Proc. Vol. , AMC 2002, p. 13 (2002)

10. G. Steinlesberger, A. von Glasow , M. Engelhardt, G. Schindler, W. Hönlein, M. Holz *et al.,* International Interconnect Technology Conference, IITC 2002, Proceedings, p. 265 (2002)

11. M. Traving, G. Schindler, G. Steinlesberger, W. Steinhögl, M. Engelhardt Advanced Metallization Conference, Proc. AMC 2002, p. 753, (2002)

12. M. Traving, W. Steinhögl, G. Schindler, G. Steinlesberger, M. Engelhardt, Advanced Metallization Conference, Proc. AMC 2003, p. 391, (2003)

13. W. Steinhögl, G. Schindler, Advanced Metallization Conference, Proc. AMC 2001, p. 393, (2001)

14. Z. Gabric , W. Pamler, G. Schindler, W. Steinhögl, M. Traving, International Interconnect Technology Conference, IITC 2004, Proc. p. 151, (2004),

15. P. Kapur, G. Chandra, S. Souri, K. Saraswat, SEMICON West 2002, SEMI Technical Symposium (STS), Semiconductor Equipment and Materials International, p. 159 (2002)

16. K.K. O, K. Kim, B. Floyd, J. Mehta, H. Yoon, C.-M. Hung *et al.*, International Interconnect Technology Conference, IITC 2003, Proc., (2003)

17. W. Hoenlein, F. Kreupl, G.S. Duesberg, A.P. Graham, M. Liebau, R. Seidel *et al.*, Mater. Sci. and Eng. C 23, 663, (2003)

18. A. M. Cassell *et al.*, J. Am. Chem. Soc. 121, p. 7975 (1999)

19. S. J. Tans *et al.*, Nature, 393, 49, May 1998

20. C. Zhou *et al.*, Nature, 290, 1552 (2000)

21. www.techonline.com/community/news/14441

22. www.golem.de/0402/29811.html

Dielectric and Scaling Effects on Electromigration for Cu Interconnects

P. S. Ho[a], K.-D. Lee[b], J. W. Pyun[a], X. Lu[a], S. Yoon[a]

[a]Laboratory for Interconnect and Packaging, University of Texas at Austin, Austin, TX / USA

[b]Silicon Technology Development, Texas Instruments, Inc., Dallas, TX / USA

Introduction

The fabrication of damascene structure for Cu interconnects requires novel materials and processes, including electroplating Cu, low-k dielectrics, chemical-mechanical polishing, ultra-thin barriers and passivation layers. The novel materials and processes give rise to distinct structure and defect characteristics raising reliability concerns for Cu/low-k interconnects. For the past several years, extensive efforts from the industry have successfully implemented the low-k interlayer dielectrics (ILD) into the 90 nm technology node and significant progress has been made to understand and improve Cu/low-k reliability. Of particular interest is electromigration (EM) reliability where the problem can be traced to basic issues relating to low-k dielectric materials, the damascene structure and interconnect scaling [1–3]. Compared with oxide, low-k dielectrics are softer, expand more and conduct less heat, which are material characteristics that can significantly degrade low-k interconnect reliability. The dual-damascene architecture gives rise to distinct EM characteristics for Cu interconnects regarding mass transport path, flux divergence and damage mechanisms. Statistical studies have revealed multi-mode failures in the Cu interconnects with early failures dominated by void formation at the via bottom surface [4, 5].

As the technology continues to advance, the implementation of porous low-k dielectrics beyond the 65-nm technology node will further degrade the thermomechanical properties of the low-k dielectrics. The porous dielectric plus continuing interconnect scaling will bring in new materials and processing issues to impact EM reliability. In this paper, we review recent results from our laboratory using a statistical approach to study the dielectric and scaling effects on EM for several low-k dielectrics including porous materials. We discuss first the effect of dielectric confinement on EM reliability of Cu line structures based on the concept of effective elastic modulus B [6]. This parameter B was evaluated using finite element analysis for Cu/oxide and Cu/low-k damascene structures and found to correlate well with the measured EM lifetime and the threshold current density-length $(jL)_c$ product. Then, we review results from EM studies of the scaling effect due to metal line width and barrier thickness on EM reliability for Cu/low-k interconnects. Three line widths: 0.25, 0.175, and 0.125 µm, were studied corresponding to the

180, 130, and 90 nm technology nodes. Results revealed an intrinsic scaling effect based on the observed line width dependence for the strong-mode EM statistics, which is as expected. However, process-related issues were observed leading to decreasing early failure lifetime and reliability degradation for the 0.125-μm interconnects.

Effect of Dielectric Confinement on EM

Consider the line/via element in a Cu dual-damascene structure shown in Figure 1. Under a current driving force, the drift velocity (v_d) of Cu ions induced by a direct current can be expressed as [7],

$$v_d = v_{EM} + v_{BF} = \mu \, (Z^* e \rho j - \Omega \Delta \sigma / L) \tag{1}$$

Here $Z^* e$ is the effective charge, j the current density, Ω the atomic volume, L the line length and $\Delta \sigma / L$ the EM induced stress gradient along the line. Equation 1 indicates that in a confined structure, the current induced mass transport v_{EM} is opposed by a back flow v_{BF} induced by the Blech stress gradient as a result of mass transport from the cathode to the anode. Accordingly, a threshold product $(jL)_c$ can be defined when $v_d = 0$ as:

$$(jL)_c = \Omega \Delta \sigma / Z^* e \rho \tag{2}$$

With Ω and $Z^* e \rho$ being material constants, $(jL)_c$ is proportional to $\Delta \sigma$. For a Cu damascene structure, the value of $\Delta \sigma$ that can be sustained depends on the confinement on the Cu structure imposed by the ILD and the surrounding barrier and cap layers. The barrier and cap layers although much thinner than the low k ILD, have much higher mechanical strengths and can make a significant contribution to confine the Cu lines in low k damascene structures. The $(jL)_c$ product provides an effective measure to evaluate the back-flow stress and the dielectric confinement effect on EM. In our study, the $(jL)_c$ product was measured for Cu damascene lines with oxide and low-k dielectrics.

The confinement effect can be quantified using an effective elastic modulus B following an approach first formulated by Korhonen *et al.* for Al interconnects [6]. Accordingly, B was defined to account for the stiffness of the interconnect structure responding to the stress generated by mass transport under confinement of the damascene structure as:

$$dC/C = - \, d\sigma / B \tag{3}$$

Here, dC/C is the volumetric strain induced by EM, and $d\sigma$ is the corresponding amount of stress generated as measured by an effective modulus B of the interconnect structure. Defined in this way, B depends on the elastic properties and geometry of the metal and dielectric structures and surrounding barrier and cap layer materials. It also depends on the mass transport mechanism because the mechanical response of the interconnect structure depends on how mass transport is distributed in different directions. In AlCu interconnects, mass transport is primarily through the grain boundaries, resulting in an isotropic mass distribution and an isotropic $d\sigma$ stress. In Cu interconnects, mass transport is dominated by diffusion at the cap layer interface [8], which results in an anisotropic mass distribution primarily along the normal to the line direction. Using a finite element analysis, Hau-Riege calculated B for AlCu interconnects assuming an isotropic mass transport [9]. He found that as the elastic modulus of ILD decreases from 72 GPa for oxide to below 10 GPa for low-k materials, the value of B is reduced but to a significantly less degree, from 25 GPa to about 10 GPa. This indicates that the Ti/TiN upper and lower layers in the AlCu line contribute substantially in additional to that from ILD to metal confinement. When the AlCu line was embedded in a damascene structure with additional side-wall barriers, B of low-k ILD was found to increase by about 50%. This demonstrated that the barrier and cap layers in the damascene structure make significant contributions to metal confinement, particularly for low-k dielectrics.

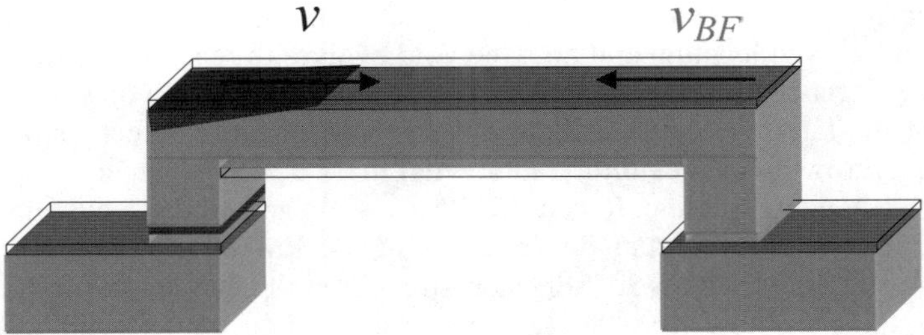

Figure 1. Mass transport under EM in a confined metal structure. The drift velocity driven by the current V_{EM} is opposed by a stress-induced back flow V_{BF}.

The confinement effect on EM reliability can be examined from Equation 1 where a weak confinement will decrease the back-flow stress $\Delta\sigma$ resulting in an increase of the net drift velocity and a reduction of the EM lifetime. To evaluate the effect on EM lifetime, not only the dielectric confinement on mass transport, the kinetics of stress evolution and void formation and their roles in controlling EM damage also have to be considered. Compared with Al interconnects, these parameters for Cu interconnects have distinct characteristics and impact EM reliability differently [10,11]. In Figure 2, we show the initial state of a Cu line with a uniform tensile stress at time t_0. Upon EM, mass transport builds up a small tensile stress at the cathode at t_1 with a corresponding compressive stress at the anode. With continuing EM to t_2, the tensile stress at the cathode can reach a critical value to induce void formation. For Al interconnects with a strong Al/oxide interface, void formation is not commonly observed. In contrast, for Cu interconnects, voids can form readily at the Cu interface near the cathode end requiring only a moderate tensile stress of about 100 MPa [11]. Upon void formation, the local tensile stress relaxes quickly to zero while the compressive stress in the Cu line away from the void continues to increase as shown at t_3 and finally reaches a steady state at t_4 as shown. This effectively increases the steady-state compressive stress at the anode end making Cu lines more prone to failure due to metal extrusion. This can degrade EM reliability, particularly for low-k interconnects with a weak interfacial adhesion in a damascene structure.

Based on void growth kinetics, Korhonen *et al.* deduced a critical void volume at the steady state as [10],

$$V_c = \frac{\sigma^T L}{B} + \frac{J\rho e Z^* L^2}{2\Omega B}$$

(4)

Here, σ^T is the residual thermal stress and L the length of the metal line where both the residual stress and EM mass transport contribute to void growth. The residual stress σ^T represents the stress existing in the line after processing. Assuming that the EM lifetime is determined by a critical void volume V_c, Equation 4 provides a relationship to correlate EM lifetime to the confinement effect. Accordingly, the lifetime is inversely proportional to the effective modulus B and dependent on the line length L under the current density j and the residual stress σ^T. However, the confinement effect on EM is generally more complicated, depending also on initial void location and how the void evolves to reach the critical void volume. The effect should be statistical in nature since the initial voids can occur randomly at the interface depending on the presence of local defects and the subsequent void growth depends on the interfacial mass transport which may vary from grain to grain along the line. Recent Cu EM results showed that void evolution and grain structures are important in determining the EM failure statistics [12]. For Cu/low-k structures with weak adhesion strength at the cap layer interface, the increase in the back-flow stress at the anode end can induce interfacial delamination to cause premature line failure by metal extrusion before the steady-state is reached. As a result, the EM lifetime will be lower than that estimated from B based on the confinement effect.

Based on these discussions, the confinement effect seems to be a complex statistical phenomenon depending on material properties, interconnect geometry and processing defects. A discussion of the statistical issues and their effects on EM reliability is beyond the scope of this paper and will not be included. The effect of dielectric confinement on thermal stress has been investigated using X-ray diffraction to measure Cu damascene line structures with oxide and low-k dielectrics [13]. The results confirmed that the cap and barrier layers are important and have to be considered in order to account for the observed thermal stress characteristics. In the following sections, we summarize first the results of EM studies on lifetime and $(jL)_c$ for 0.25 μm Cu damascene structures with oxide and low-k dielectrics. This will be followed by reviewing results on scaling effects on EM for Cu interconnects integrated with a porous MSQ low-k material with k ~ 2.3 as a function of line width and barrier thickness. EM results for three line widths: 0.25, 0.175, and 0.125 μm corresponding to the 180-, 130- and 90-nm nodes and for three Ta barrier thicknesses of 17.5, 12.5, and 7.5 nm will be discussed.

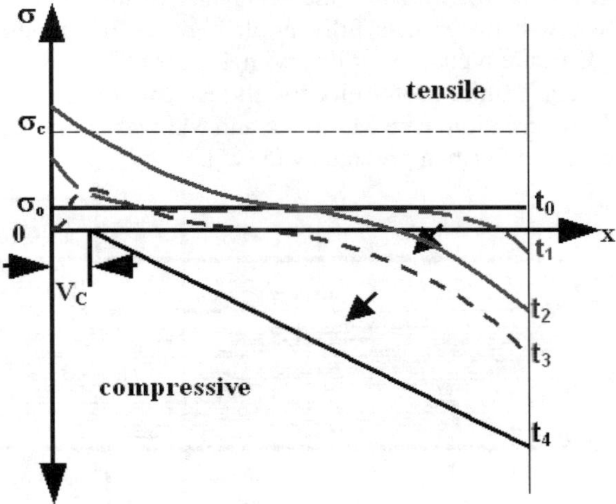

Figure 2. Stress evolution in confined Cu line under EM as a function of time. σ_0 is the initial thermal stress.

Experimental Details

Experiments were performed using statistical structures to examine the effect of low-k dielectrics on EM reliability of Cu interconnects. For this study, critical length (LC) and early failure (EF) test structures were used, which were designed for measuring the critical jL_c product and the early failure statistics, respectively. The test structures were fabricated by a dual-damascene process with two metal layers (M1/Via/M2). As shown in Figure 3a, the LC test structure consists of a collection of serially connected line/via interconnects arranged as six repeating sets of 14 interconnects where the M2 length varies from 10 to 300 μm and with a line width of 0.25 μm. The early failure test structures comprised of N=1, 10, and 100 of line/via multi-links with short M1 lines and long 300-μm M2 lines. The length of M1 remained constant at 10 μm in order to drive the failure to occur above the M1 level at either the via or the M2 trench to facilitate failure analysis. Designed in this way, the serial connection of test structures provides an ensemble of line/via elements with varying line lengths to measure the dielectric confinement effect on EM early failure statistics and threshold $(jL)_c$ product.

The low-k ILDs used included CVD low-k, porous MSQ and organic polymer and their EM behaviors are compared with that of oxide. The material properties of the dielectric materials are listed in Table 1. Test structures were designed at UT-Austin and fabricated at International Sematech and LSI Logic. The samples were prepared using 200 mm and 300 mm wafers and the two-level interconnect structures were based on a Ta/low temperature PVD seed Cu/electroplated (EP) Cu stack [14]. The metal lines in the test structures showed an apparent "near bamboo" microstructure with a significant amount of twinning that was associated with Cu film growth. In Figure 3b, we show the schematic of the LC Cu/low-k test structure, where the low-k material is fully implemented. EM experiments were performed in a high-vacuum chamber filled with 20 torr of N_2 gas to reduce oxidation and to improve temperature uniformity for all test samples. In all EM tests, a current in the up-flow direction from M1 to via and M2 was applied. More experimental details have been described previously [5, 15].

(a)

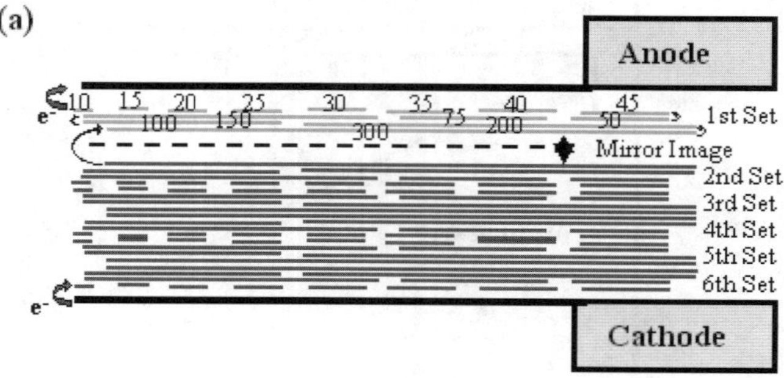

(b)

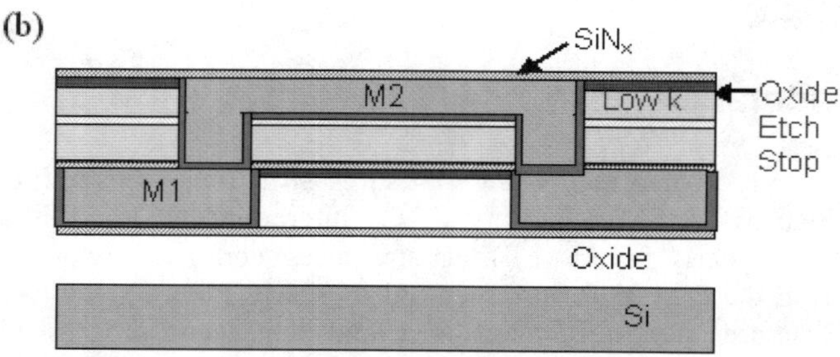

Figure 3. (a) Schematic overview of LC test structure. (b) Stacking layers of dual-damascene Cu/low-k interconnects.

Table 1. Dielectric constant (k), coefficient of temperature expansion (CTE) and Young's modulus (E) for different dielectric ILDs and simulated effective bulk modulus (B) and (jL)$_c$ for dual-damascene Cu interconnects are summarized.

	k	CTE (ppm/°C)	E (GPa)	B (GPa)	jL_c^* (A/cm)
Oxide	3.9	0.51	71.4	13.7	3700
CVD low-k	2.7	25	6	7.6	2000
Porous MSQ	2.3	7.3	3.6	7.3	2500
Org. polymer	2.7	66	2.5	7.2	1200

* These values were determined at 1.0 MA/cm^2.

Results and Discussion

EM Lifetime and Threshold $(jL)_c$ Product

EM experiments were performed on LC test structures at temperatures between 190 and 360°C. The first abrupt change or instability in resistance trace during EM stressing was considered as an "onset" EM failure and was used to determine EM lifetime. In Figure 4, Arrhenius plots of t_{50} *vs.* $1/kT$ obtained at 1.0 MA/cm^2 are plotted for Cu/oxide, Cu/CVD low-k, Cu/porous MSQ, and Cu/organic polymer structures integrated with Ta barrier and SiN$_x$ cap layer. The activation energies of Cu structures were found to be between 0.8 and 1.0 eV. This is commonly associated with mass transport at the Cu/SiN$_x$ interface, suggesting that interfacial diffusion dominates EM mass transport [15]. The test structures had the same Cu/Ta and Cu/SiN$_x$ interfaces, so EM occurred via similar interfacial diffusion paths independent of ILD.

In Figure 4, the EM lifetimes of Cu/low-k structures are generally shorter than that of Cu/oxide structures under a similar test condition, which can be attributed to the confinement effect from the low-k damascene structure. As shown in Equation 1, the weaker confinement by low-k ILD reduces the back-flow term v_{BF}, resulting in an increase in the net mass transport and a reduction of EM lifetime.

To estimate the confinement effect, B was calculated using finite element analysis for the Cu test structures used in this study assuming mass transport only at the cap layer interface. The results are listed in Table 1. The values of B for all low-k ILDs are very similar, equal to about half of that of oxide. The difference of B is significantly less than that of the elastic moduli E of the dielectric materials indicating that the barrier and cap layers are important in contributing to confinement of the Cu lines. Except for the organic polymer, the EM lifetime seems to be proportional to B, which is consistent with void-induced failure at the steady state, as described in Equation 4. For the organic polymer, the EM lifetime is lower than that estimated from B. Failure analysis showed that this is due to premature failures caused by interfacial delamination at the cap/etch stop interface and will be discussed later.

After a long EM stressing, individual lines in LC structures were examined using a focused-ion beam (FIB) microprobe and an optical microscope to identify EM induced damages. Most of the failures observed were caused by void formation near the cathode end. In this way, the failure statistics as a function of line length can be used to determine the jL_c product corresponding to the "strong-mode" due to void formation in the M2 lines. As shown in Figure 5, the failure probability drops sharply as the line length decreases. The critical length L_c for a given current density j can be determined from the intercept of the regression line generated by assuming that failure probability is proportional to the net drift velocity and thus the line length [16]. In Figure 5, L_c was found from the intercept to be 10 μm giving an $(jL)_c$ of 3000 A/cm.

The results of $(jL)_c$ measured for Cu/oxide and Cu/low-k structures are summarized in Figure 6. Compared with oxide, low-k ILDs have smaller threshold $(jL)_c$ products, corresponding to 3700, 2000, 2500, and 1200 A/cm with error limits of 250–500 A/cm for oxide, CVD low-k, porous MSQ and organic polymer, respectively. No temperature dependence for $(jL)_c$ was observed under our test conditions. There is a good correlation between the lifetimes in Figure 3 and $(jL)_c$ since both are correlated to $\Delta\sigma$, reflecting the confinement effect from the dielectric materials on EM characteristics. Similar to the lifetime, the extrinsic effect of interfacial delamination reduces $(jL)_c$ for the organic polymer material.

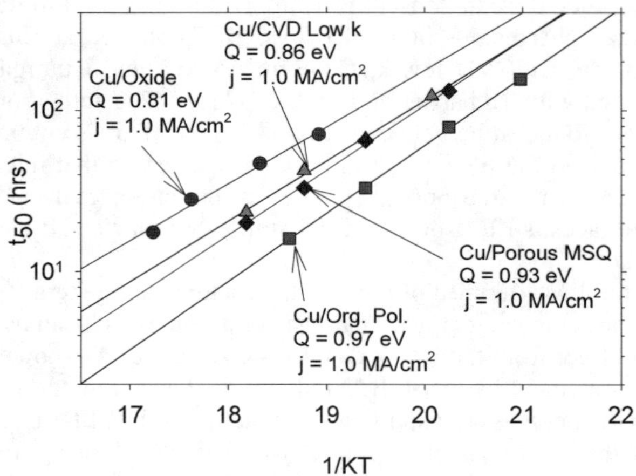

Figure 4. Graph shows t_{50} vs. $1/kT$ for Cu/oxide, Cu/CVD low-k, Cu/porous MSQ, and Cu/organic polymer. The activation energies determined from the slope of an Arrhenius plot were found to be between 0.8 and 1.0 eV.

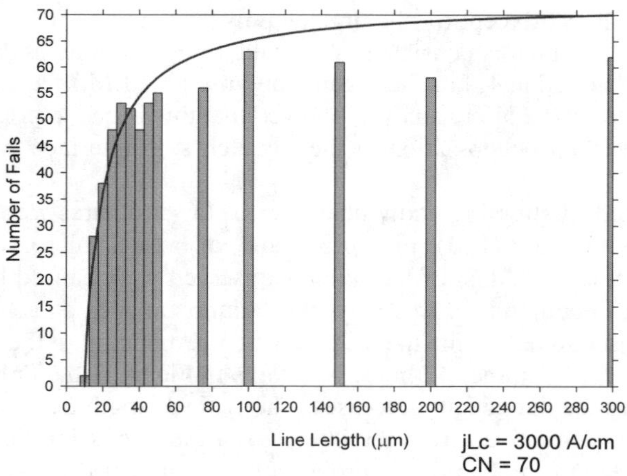

Figure 5. Failure distribution of Cu/porous MSQ LC test structures tested at 190°C and 3.0 MA/cm² is shown as a function of line length. Failure probability drops as line length decreases. $L_c = 10$ μm, $(jL)_c = 3000$ A/cm.

Failure Analysis

A focused-ion-beam (FIB) microprobe was used to identify the EM failure characteristics together with FIB-induced contrast (FIBIC) technique to locate the failure sites in the Cu lines as shown in Figure 7a for Cu/porous MSQ structure. Here we found that voiding at the cathode end was large enough to stop electric current flow causing a failure. In most cases, test structures were found to fail by cathode voiding in the Cu/porous low-k structures. Some of the Cu/porous MSQ structures showed lateral Cu extrusion near the anode under the SiN_x cap layer, followed by interfacial delamination, as shown in Figure 7b.

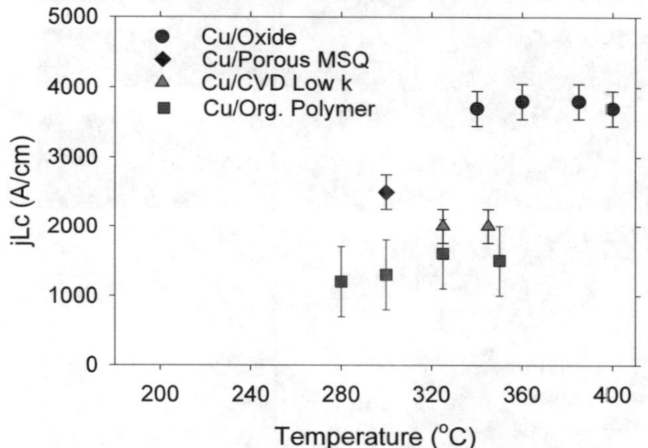

Figure 6. The graph shows $(jL)_c$ *vs.* T. There was no temperature dependence in our test conditions. $(jL)_c$ data were obtained from EM tests at 1.0 MA/cm^2, except for Cu/CVD low-k which was tested at 0.5 MA/cm^2.

While CVD low-k and porous MSQ structures failed mainly by cathode void formation, organic polymer structures failed by cathode void formation followed by anode extrusion [17]. In both cases, EM lifetime depends on the amount of void formation at the cathode, so regardless of anode extrusion, voiding seems to fail the line. If anode extrusion occurs prematurely due to low adhesion strength at the anode interfaces, such an extrinsic failure effectively reduces the back stress and accelerates void formation at the cathode. This mechanism can reduce the EM lifetime significantly as observed in the organic polymer structures.

FIB was used to examine the failure mode at the anode of organic polymer low-k interconnects and the results are shown Figure 8. Here the top view (Figure 8a) shows an extrusion failure near the anode end. The cross-sectional FIB micrograph in Figure 8b shows that the extrusion was initiated at the top corners of the Cu line beneath the oxide etch stop (see schematic drawing in Figure 8c). This suggests that the corner at the barrier, low k ILD and cap/etch stop layer intersection is a mechanical weak point where failure can occur by interfacial delamination induced by a compressive back-flow stress at the anode. This was confirmed by high-resolution SEM observations, which identified interfacial delamination and anode extrusion along the low-k and oxide etch stop interface. Additionally, atomic force microscopy (AFM) was used to examine the surface topology, revealing that SiN_x at the anode end was lifted by EM by about 180 nm [17]. Thus, the weak adhesion of low-k/etch stop interface causes the interfacial breakdown and premature EM failure. In this case, the Cu/low-k structure was not able to sustain the back-flow stress estimated from the effective elastic modulus.

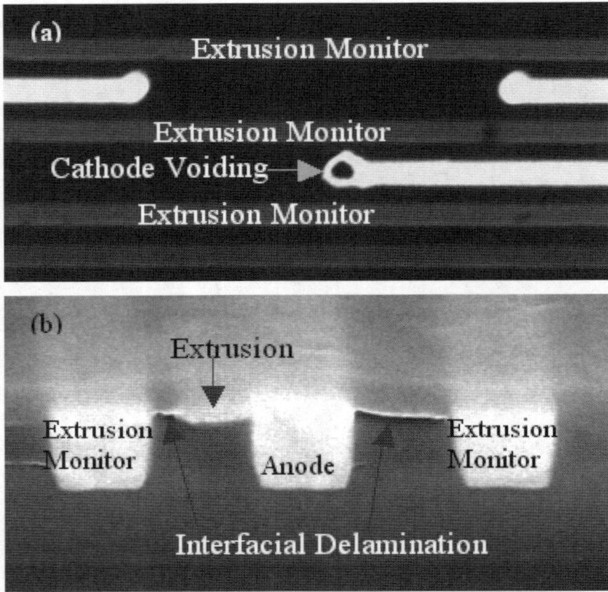

Figure 7. (a) Void formation at the cathode end and (b) extrusion at the anode end with interfacial delamination observed in Cu/porous MSQ structures.

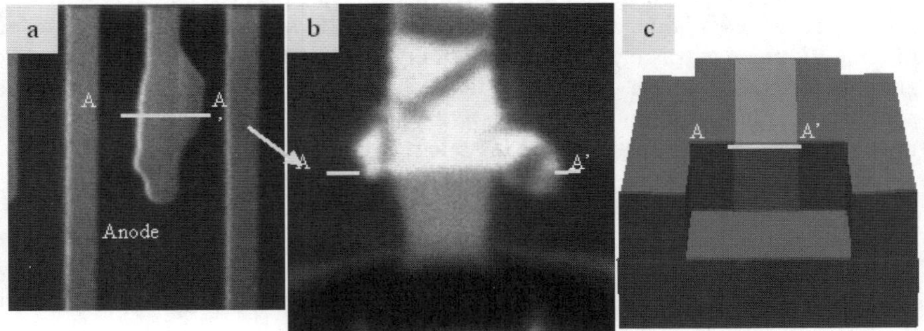

Figure 8. (a) Top-down view of anode extrusion and (b) A-A' cross section of anode of Cu/organic polymer test structures. (c) Schematic view of FIB cutting.

Line Width Scaling Effect

The EM results obtained at 330°C and 1.0 MA/cm^2 for the 0.25 μm test structures are shown in Figure 9. A Monte Carlo simulation was used to derive the lifetime and failure statistics for the strong mode and the weak mode as a function of N. The simulated results corresponding to an early failure of 0.32% are plotted for comparison with the data in Figure 9 and show a good agreement with the observed failure statistics for different N. In this way, we derived the early failure population and the statistics for the strong mode and the weak mode as a function of line width. The results are summarized in Table 2. The lifetime for the strong mode show a systematic decrease with scaling of the line width as shown in Figure 10.

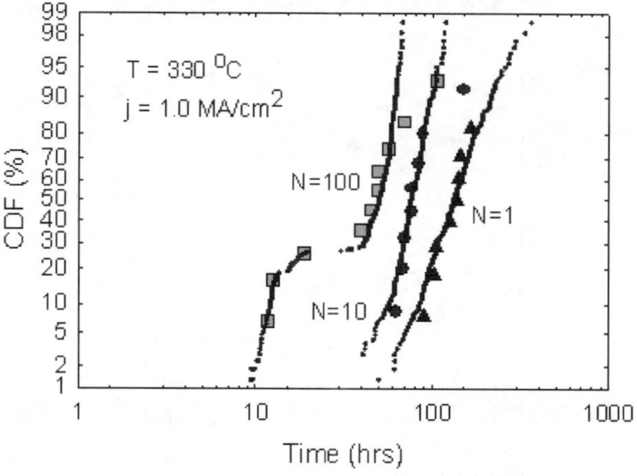

Figure 9. Experimental and Monte Carlo simulated cumulative failure distribution (CDF) plots.

Table 2. The portion of early failures and and the statistics for the strong mode and the weak mode for dual-damascene Cu/low-k interconnects with different line width.

	0.125 μm	0.175 μm	0.250 μm
EF population	0.40 %	0.40 %	0.45 %
t_{50} (strong mode)	60 h	100 h	170 h
σ (strong mode)	0.50	0.40	0.40
t_{50} (weak mode)	8 h	27 h	30 h
σ (weak mode)	0.30	0.35	0.35

In our test structure, the via diameter was scaled in the same way as the line width, so the amount of mass depletion at the cathode required to failure the line scaled also with the line width. On this basis and assuming that mass transport is dominated by diffusion at the cap layer interface [8], the observed scaling effect for the strong mode can be properly accounted for. Interestingly, the early failure population seemed to be relatively independent of the line width but the lifetime for the weak mode for the 0.125 µm line structures decreased significantly more than that expected from via size scaling. FIB analysis revealed that the early failures are mostly associated with via voiding, thus our result suggests defects related to via processing can induce additional early failure with scaling of line width to 0.125 µm.

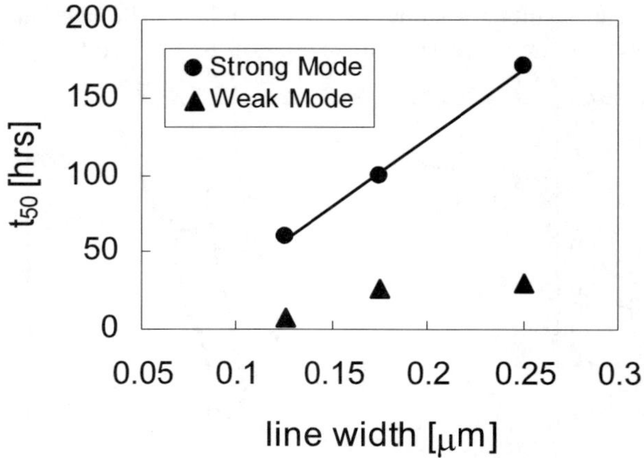

Figure 10. The strong mode and weak mode lifetime as a function of M2 line width.

Barrier Thickness Scaling Effect

The EM lifetime statistics obtained for the 0.175-µm LC test structures, as a function of Ta barrier thickness, are summarized in Figure 11. The strong mode electromigration lifetime data corresponding to failure statistics above 10% seem to be relatively independent of the Ta barrier thickness. There is a small increase in lifetime for the thinner Ta barrier structure, which can be attributed to a lower current density in the metal line due to a larger Cu cross section with a thinner barrier. In contrast, the early failure lifetime corresponding to failure statistics below 10% seems to decrease with decreasing barrier thickness. This result can be attributed to an increase in process-induced defects for thinner barriers. Thus, scaling in both line width and barrier thickness reduces the early failure lifetime, which is probably due to extrinsic process induced defects.

To determine the critical $(jL)_c$ product, we examined void formation used an FIB microprobe. Figure 12 shows that no void was found in a test structure of less than 20-µm line length, and voids were found in lines with longer length due to a lower back stress gradient. In this way, the $(jL)_c$ product was determined and the results for different Ta barrier thicknesses are shown in Table 3 together with the EM lifetime, and the calculated effective moduli for each Ta barrier thickness. The $(jL)_c$ value for the thinner Ta barrier is somewhat lower. To examine the confinement effect on the $(jL)_c$ product, we used finite element analysis to calculate the effective elastic modulus B and results are included in Table 3. Although the variations in B and the $(jL)_c$ product are consistent, the amount of change in the $(jL)_c$ product is smaller.

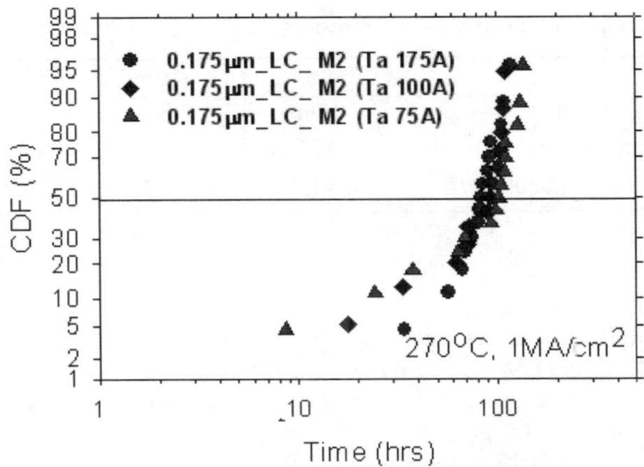

Figure 11. Experimental cumulative failure distribution (CDF) plots for different barrier thicknesses.

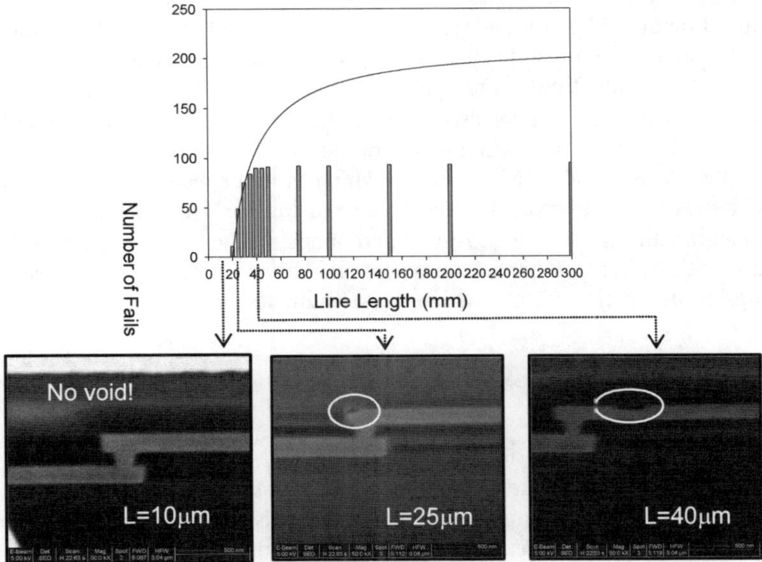

Figure 12. Failure distribution of LC structure as a function of line length, and cross-sectional FIB images.

This suggests that there are other factors making the confinement effect less effective. In our study, the $(jL)_c$ product is determined statistically, so this again points to the possibility of defect-induced line failures limiting the build-up of the back-flow stresses and thus the value of the $(jL)_c$ product.

Table 3. The summary of electromigration results for different barrier thicknesses.

	75 Å Ta	100 Å Ta	175 Å Ta
$(jL)_c$ [A/cm]	1700	1900	2000
t_{50} [h]	102	92	84
Effective modulus (B)	8.8 GPa	10.7 GPa	15.3 GPa

Summary

In summary, multi-link statistical test structures were used to study the dielectric and scaling effects on EM reliability for Cu interconnects. Experiments were performed on dual-damascene Cu interconnects integrated with oxide, CVD low-k, porous MSQ, and organic polymer ILD. The EM activation energy for Cu structures was found to be between 0.8 and 1.0 eV, indicating mass transport is dominated by diffusion at the Cu/SiN_x cap-layer interface, independent of ILD. Compared with oxide, the decrease in lifetime and $(jL)_c$ observed for low-k structures can be attributed to less dielectric confinement in the low-k structures. An effective modulus B obtained by finite element analysis was used to account for the dielectric confinement effect on EM. The scaling effect was investigated using a statistical approach as a function of line width and barrier thickness for three line widths: 0.25, 0.175, and 0.125 μm corresponding to the 180-, 130-, and 90 nm nodes. Results revealed an intrinsic scaling effect based on the observed line width dependence of the strong-mode EM statistics, which is as expected. However, process-related issues were observed leading to decreasing early failure lifetime and reliability degradation for the 0.125-μm interconnects. The jL_c product was found to decrease with decreasing barrier thickness and the trend can be attributed to a decreasing confinement effect, which was estimated using an effective elastic modulus.

Acknowledgement

The authors would like to thank the International SEMATECH and LSI Logic Corporation for providing test structures for this study. They also gratefully acknowledge the support from R. Augur and N. Henis from Sematech and the State of Texas, and the SRC Center for Advanced Interconnect Science and Technology.

References

1   C.K. Hu and B. Luther, Mater. Chem. Phys., 41, 1-7, (1995).

2   C.K. Hu and J.M.E. Harper, Mater. Chem. Phys., 52, 1998, 5-16, (1998).

3   E.T. Ogawa, V.A. Blaschke, A. Bierwag, K.-D. Lee, H. Matsuhashi, D. Griffiths *et al.*, Mater. Res. Soc. Symp. Proc., 612, D2.3.1-D.2.3.6, (2000).

4   E.T. Ogawa, K.-D. Lee, H. Matsuhashi, K.-S. Ko, P.R. Justison, A.N. Ramamurthi, *et al.*, 2001 IEEE International Reliability Physics Symposium Proceedings 39th Annual, 341-349, (2001).

5   K.-D. Lee, X. Lu, E.T. Ogawa, H. Matsuhashi, V.A. Blaschke, R. Augur *et al.*, 2002 IEEE International Reliability Physics Symposium Proceedings, Dallas, TX (April, 2002).

6   M.A. Korhonen, P. Borgesen, K.N. Tu, Che-Yu Li, J. Appl. Phys. 73, 3790, (1993).

7   I.A. Blech, J. Appl. Phys. 47, 1203 (1976).

8   C.K. Hu, R. Rosenberg, and K.Y. Lee, Appl. Phys. Lett. 74, 2945, (1999).

9   S.P. Hau-Riege and C.V. Thompson, Mat. Res. Soc. Symp. Proc., 612, D10.2.1, (2000).

10   M.A. Korhonen, P. Borgesen, D.D. Brown, C.-Y. Li, J. Appl. Phys. 74, 4995 (1993).

11   S.P. Hau-Riege, J. Appl. Phys. 91, 2014, (2002).

12   H. Kawasaki, Proc. Advanced Metall Conf, San Diego, CA (in press) Dec. (2002).

13   S.H. Rhee, Y. Du and P.S. Ho, J. Appl. Phys. (in press), (2003)

14   V. Blaschke, J. Mucha, B. Foran, Q.T. Jiang, K. Sidensol, A. Nelson, in 1998 Proceedings of the Advanced Metallization Conference, 43-49 (1998).

15   K.-D. Lee, E.T. Ogawa, H. Matsuhashi, P.R. Justison, K.-S. Ko, P.S. Ho *et al.*, Appl. Phys. Lett. 79, 3236 (2001).

16   E.T. Ogawa, A.J. Bierwag, K.-D. Lee, H. Matsuhashi, P.R. Justison, A.N. Ramamurthi *et al.*, Appl. Phys. Lett. 78, 2652 (2001).

17   K.-D. Lee, X. Lu, E.T. Ogawa, H. Matsuhashi, V.A. Blaschke, R. Augur *et al.*, 2002 IEEE International Reliability Physics Symposium Proceedings 40th Annual, 322-326, (2002).

Texture and Stress Study of Sub-Micron Copper Interconnect Lines Using X-ray Microdiffraction

I. Zienert, H. Prinz, H. Geisler, E. Zschech

AMD Saxony LLC & Co. KG, Materials Analysis Department, Dresden / Germany

Introduction

For high-performance microelectronic products like microprocessors, the continuous shrinking of the dimensions of on-chip interconnects, advanced backend-of-line (BEoL) manufacturing process steps and changed combinations of thin film materials result in new reliability challenges: different microstructure of the metal interconnects, other types of interfaces and so far unknown degradation phenomena. Electromigration (EM), stress-induced degradation and – in the case of low-k materials – mechanical weakness are reliability concerns for inlaid copper interconnects [1, 2].

Although a lot of theoretical and experimental work has been done on EM of inlaid copper interconnects, it is still not fully understood how the interconnect degradation takes place [3, 4]. However, numerous experimental studies have indicated that EM-induced degradation and eventually interconnect failure depend on both interface bonding and microstructure of the copper interconnect structures. Since copper interconnect microstructure becomes more and more important to ensure the required product reliability, a high BEoL process stability has to be guaranteed. One approach to control the stability of the on-chip interconnect manufacturing process is to monitor microstructure parameters like grain size, texture and stress for copper interconnects on a routine basis. According to our experience, stress and texture in copper lines are the most sensible microstructure parameters to control deposition processes in the interconnect technology [5]. X-ray microdiffraction (micro-XRD) is the preferred technique for the determination of texture and stress of inlaid copper lines, since statistically relevant data can be obtained in shorter periods of time than with orientation imaging microscopy (OIM) [6].

In this paper, the influence of the barrier material and of the insulating dielectrics between the metal lines on copper texture and stress is studied for several interconnect geometries using micro-XRD. Particularly, differently deposited Ta films and SiO_2 *vs.* low-k interlayer dielectric (low-k ILD) materials are compared.

Interconnect Microstructure and Reliability

Summarizing the experimental results that have been published on inlaid copper interconnect reliability, the EM-induced mass transport along copper interconnects seems to be dependent on competing activation energies for atomic transport processes along the interfaces and the grain boundaries [2]. As long as the activation energy for mass transport along at least one interface is smaller than along the grain boundaries, which is usually the case for the Cu/SiN_x interface in currently manufactured inlaid copper interconnect structures, the mass transport along the weakest interface dominates the EM-induced degradation process, i.e., the degradation of copper interconnects is a function of the bonding strength of the weakest interface [7]. For these cases, the copper microstructure is a second-order effect.

A physical model established by Sukharev [8] and the respective numerical simulation of EM-induced mass transport processes allow one to forecast degradation mechanisms and to make an estimate for the EM lifetime based on interface bonding strength and/or activation energies for atomic transport along interfaces. In the modelling of EM-induced transport, copper atom migration along interfaces (D_{int}), along copper grain boundaries (D_{GB}) and through "bulk" copper grains (D_B) have to be considered. As long as the interface diffusivity is enhanced ($D_{int} > D_{GB} \gg D_B$), i.e., as long as interfaces are considered as major channels for atom migration, the introduction of the copper microstructure into the simulation scheme does only slightly change the obtained picture of void evolution [9].

From our point of view, there are at least three reasons that current models and respective numerical simulations for EM degradation and failure do not consider copper microstructures:

- For most state-of-the-art interconnect systems, the microstructure is a second-order effect for EM-induced mass transport.
- The incorporation of copper microstructure into models is complex and increases the numerical effort.
- The copper microstructure parameters grain size, texture and stress are usually not known quantitatively.

However, the situation will change as soon as the interface bonding increases significantly, *i.e.* for $D_{int} \sim D_{GB}$. Then, the microstructure has to be considered in the simulation [10]. When the activation energy for interface diffusion along strengthened interfaces becomes comparable to the activation energy for copper grain boundary diffusion, mass transport along grain boundaries also has to be taken into consideration for interconnects with polycrystalline copper microstructures. Then, the determination of microstructure parameters like grain size, texture and stress for copper interconnects and their monitoring on a routine basis becomes even more important for BEoL process control [11]. The implementation of the copper microstructure into the numerical simulation will be the necessary next step to make models more precise, particularly in case of strengthened interfaces. These models will then help to provide a closer link between the microstructure data determined during process monitoring and EM lifetime.

Experimental

The comprehensive microstructure monitoring for inlaid copper structures is mainly focussed on grain orientation distribution (texture) and residual stress.

X-ray Microdiffraction

In contrast to unpatterned copper films, grains in inlaid structures are more or less effected by the sidewalls of the trenches, depending on the trench aspect ratio and process parameters. Therefore, arrays of parallel copper lines with 180 nm and 1.8 μm line width were used for microstructure studies. The test structures were manufactured using different barrier layers, capping layers and interlayer dielectrics (ILD).

X-ray microdiffraction analysis is the technique of choice for texture and stress analysis of arrays of parallel copper lines. For the investigations reported in this paper, a Bruker AXS D8 microdiffractometer system was used. This tool is equipped with a copper long fine focus X-ray tube, polycapillary X-ray optics, a Huber goniometer with a precise ¼ Eulerian cradle, video laser alignment, and a 2D detector. The smart combination of these components allows one to perform the measurements with a spot size of 100 μm [6].

Texture Analysis

For X-ray texture measurements, the sample is tilted/rotated to measure the diffracted X-ray intensity for the entire population of a particular family of crystallographic lattice planes, relative to the surface normal. Pole figures, *i.e.* two dimensional maps of a three-dimensional structure, which are typically represented as a pole plot with rotation in the *X-Y* plane, and the relative intensity of diffracted radiation in the *Z* direction are calculated from the raw data. There exist two ways to evaluate experimental pole figures. One method is to extract the full width at half maximum (FWHM) from Chi-cuts of the {111} pole figure. However, this method is only qualitatively, and provides limited information. For a quantitative analysis it is necessary to calculate the orientation distribution function (ODF). For this approach at least two experimental pole figures, in our case the {111} and {200}, have to be measured. The information that is extracted from ODFs is the inverse pole figure, the Eulerian cuts and the volume fraction of the respective orientations. There exist several commercially available software packages to evaluate pole figures. The software LaboTex was used for the investigations reported here.

In such a way, the X-ray measurements were performed with a step size Phi= 2° and with a measuring time of 60 s per frame. The samples were aligned in such a way that the line is oriented vertically in the plotted figures.

Stress Analysis

Until now, the most common method to determine the stress state in thin films was the $\sin^2\psi$ technique where the lattice strain is measured for several tilt angles ψ and plotted against $\sin^2\psi$. From the slope of this straight line the strain and stress values are extracted using the Young's modulus. The basic assumption of this technique is a biaxial stress condition with an out-of-plane stress. For the samples described, this assumption is not fulfilled. That means, it is necessary to evaluate the three-axial stress state. Complete parts of the Debye rings in every space direction can be measured using the 2D detector. From the Debye rings, the Bragg angles (and thus the lattice space d) are extracted for several ψ-tilts. Based on the measured 2Θ values, the strain and subsequently the three-axial stress state are calculated. With the determined stress values, a Debye ring is simulated and subsequently fitted to the experimental data. Since the error for the elastic lattice strain is smaller for larger Bragg angles, the Cu (311) diffraction line at $2\Theta = 89.4°$ for Cu-K_α radiation was used in this study.

Texture and Stress in Inlaid Copper Lines

Texture

Texture of Copper Interconnects with Different ILDs

Figure 1 shows the Cu {111} pole figures of arrays of copper lines (180-nm line width) embedded in standard SiO_2 and low-k ILD, respectively.

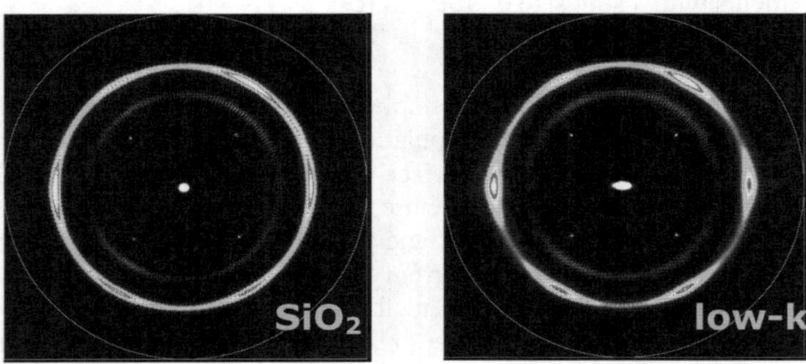

Figure 1. Cu {111} pole figures for Cu lines embedded in SiO_2 and low-k ILD.

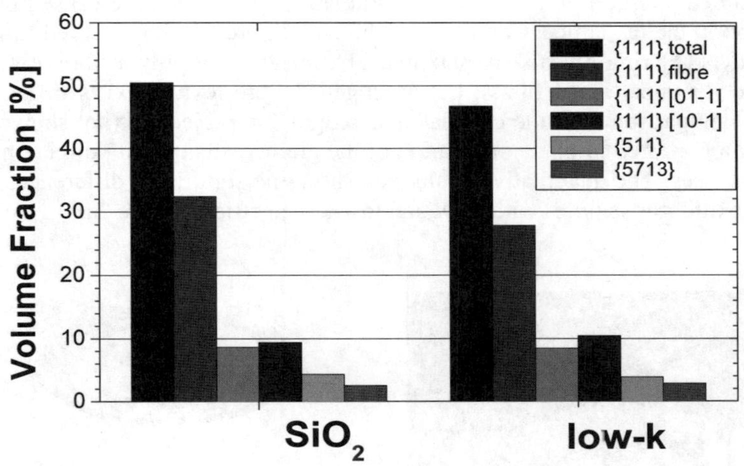

Figure 2a. Results of the quantitative texture analysis for different ILDs.

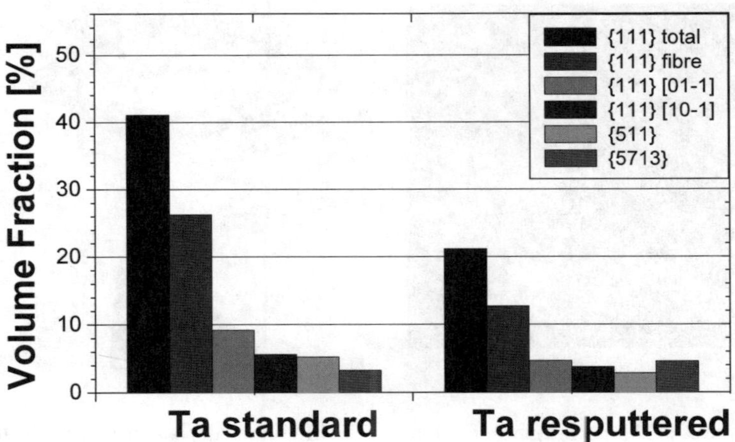

Figure 2b. Results of the quantitative analysis for different barrier layers.

At a first glance, there are no significant differences between the pole figures of these two samples. There is a {111} centre peak, a ring of symmetry-equivalent Cu {111} lattice planes at 70.5°, and an additional Cu {511} texture caused by twin formation in the fcc lattice. Compared to the pole figures for unpatterned films, the outer Cu {111} ring reveals six maxima. That means, not only a fibre texture (as observed for unpatterned films), but an engaged fibre texture exists in the metal lines. A comparison with the calculated stereographic projection (not shown here) revealed a preferred in-plane orientation of the grains with a <110> direction along the metal lines. The quantitative evaluation shows no significant differences of the copper texture for samples with SiO_2 and low-k dielectrics (Figure 2a).

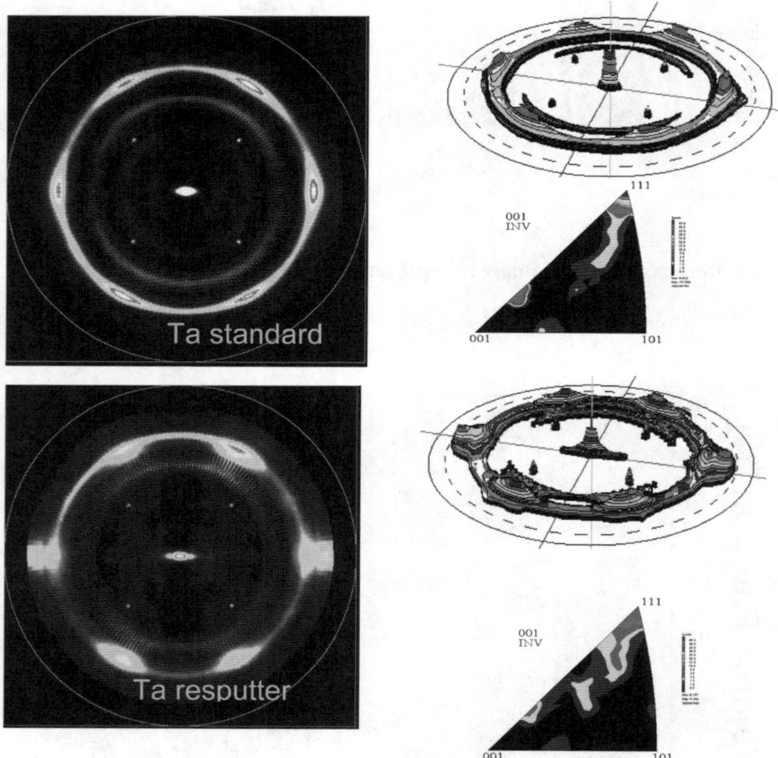

Figure 3. Cu {111} pole figures for lines with differently deposited barrier layers. The three-dimensional plots of the Cu {111} pole figures and the inverse pole figures are also shown.

Texture of Copper Interconnects with Different Barrier Layer

The Cu {111} pole figures of copper lines with standard and resputtered Ta liners are shown in Figure 3.

For the resputtered Ta liner, the Cu (111) centre peak of the pole figure is emerged horizontally, and additional peaks at the outer fiber ring are visible. A possible explanation is that copper grains are growing from slightly sloped side-walls near the trench bottom, *i.e.* from sidewall regions which are not perpendicular to the trench bottom (Figure 4). The quantitative texture analysis shows a higher fraction of {111} oriented copper grains for the standard Ta liner compared to the resputter Ta liner (Figure 2b).

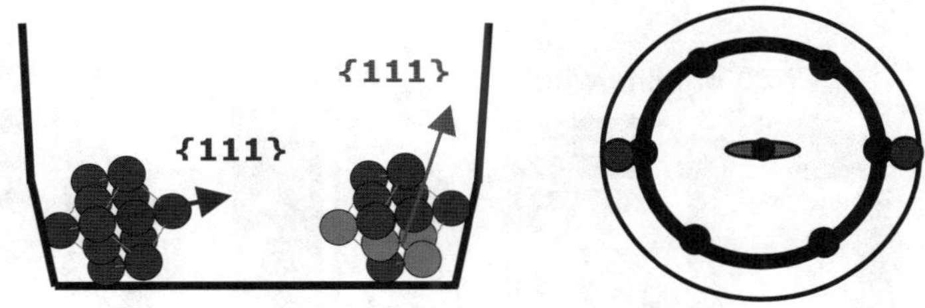

Figure 4. Schematic cross section through an interconnect, and a model of the appendant Cu {111} pole figure.

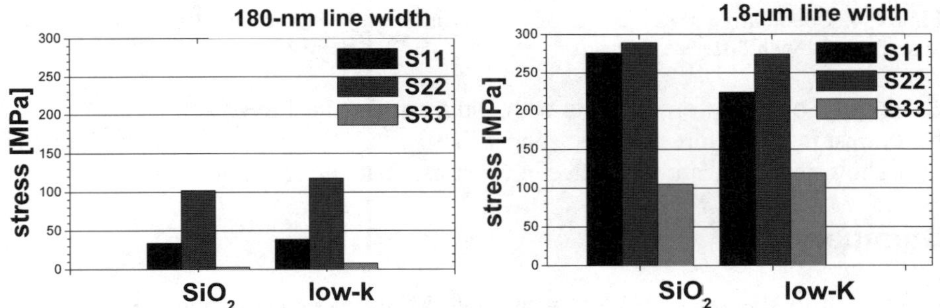

Figure 5. Quantitative stress results of copper interconnects with 180-nm and 1.8-µm line width encapsulated with different ILDs (SiO$_2$ and low-k material). S11: across the line; S22: along the line; S33: out-of-plane.

Stress

Stress of Copper Interconnects with Different ILDs

In order to investigate the three-dimensional stress state of copper interconnects with different ILDs, narrow (180 nm) and wide (1.8 µm) metal lines were analyzed using micro-XRD.

Figure 5 reveals a significantly higher stress state for the wider lines compared to the narrow lines. For both line widths, the stress along the lines showed the highest values. Across and along the narrow lines the stress is nearly zero, the stress in the wider copper interconnects is much higher in these directions. The fact that the differences of the stress state for the samples with different ILDs are marginal, is important. That means, the stress state in copper lines is not strongly influenced by the ILD material, even if the elastic properties (stiffness) differs by about a factor of 10.

One way to understand the experimental results better are with FEM simulations. Valeriy Sukharev provides very detailed insight into physical-based numerical simulations [12].

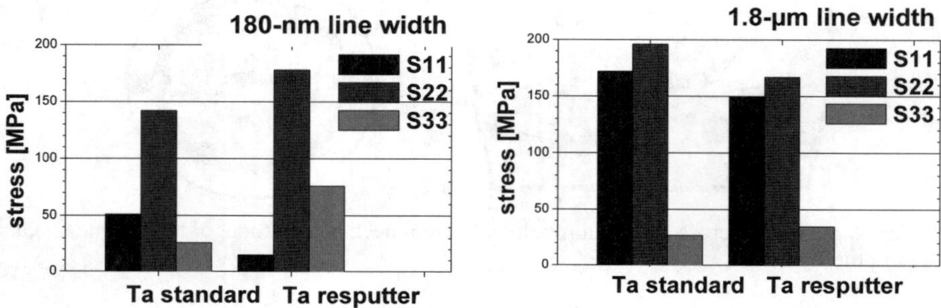

Figure 6. Quantitative stress results of copper lines with 180-nm and 1.8-µm line width manufactured with different Ta liners.

Stress of Copper Interconnects with Different Barrier Layer

In contrast to the texture analysis, copper lines with Ta resputtered and standard Ta lines show no significant difference in the stress state of the metal lines.

Summary

In addition to standard reliability tests, both a careful process control based on a large number of data to reach statistically relevant conclusions and the study of solid-state physical degradation mechanisms at representative samples are needed to understand weaknesses in the interconnect technology and to exclude reliability-related failures in copper interconnects. In addition, numerical simulation will help forecast the effect of process and material changes on interconnect reliability.

Microstructure information like the texture and stress of copper interconnects become more important for EM degradation if material transport along interfaces is reduced by interface strengthening.

Acknowledgement

The authors would like to thank Paul S. Ho, University of Texas, Austin/TX, and Valeriy Sukharev, LSI Logic, Milpitas/CA, for useful discussions.

References

1. P.S. Ho, K.D. Lee, E.T. Ogawa, S. Yoon, X. Lu, in Characterization and Metrology for ULSI Technology, ed. D. G. Seiler *et al.*, AIP Conf. Proc. (AIP: Melville, NY, 2003), 533 (2003).

2. E. Zschech, M. A. Meyer, E. Langer, Proc. MRS Spring 2004, in press (2004)

3. C.K. Hu, S. G. Malhotra, L. Gignac, Electrochem. Soc. Proc. 31, 206 (1999)

4. E.T. Ogawa, K.D. Lee, V.A. Blaschke, P.S. Ho, IEEE Trans. Reliab. 51, 403 (2002)

5. E. Zschech, W. Blum, I. Zienert, P.R. Besser, Z. Metallkd. 92, 320 (2001)

6. H. Geisler, I. Zienert, H. Prinz, M. A. Meyer, E. Zschech, in Characterization and Metrology for ULSI Technology, ed. D. G. Seiler *et al.*, AIP Conf. Proc. (AIP: Melville, NY), 469 (2003)

7. V. Sukharev, E. Zschech, J. Appl. Phys. 96, 6337 (2004)

8. V. Sukharev, 2003 IEEE Int. Integrated Reliability Workshop Final Report, Lake Tahoe, 80 (2003)

9. E. Zschech, V. Sukharev, Proc. of the Int. Conf. of Advanced Metallization (MAM), Dresden, in press (2005)

10. E. Zschech, M.A. Meyer, H. Prinz, I. Zienert, E. Langer, H. Geisler, Proc. of the 7th Int. Workshop on Stress Induced Phenomena in Metallization, Austin, TX (2004)

11. E. Zschech, H. Geisler, I. Zienert, H. Prinz, E. Langer, M.A. Meyer *et al.*, Proc. of the Advanced Metallization Conference (AMC), San Diego, 305 (2002)

12. V. Sukharev, this book, p. 251-263 (2005)

Stress Modeling for Copper Interconnect Structures

V. Sukharev

LSI Logic Corporation, Milpitas, CA / USA

Introduction

Implementation of copper and low-k materials as the major components of interconnect structures results in new reliability challenges: different microstructure of the metal interconnects, other types of interfaces and so-far unknown degradation phenomena. Electromigration (EM), stress-induced degradation and — in case of low-k materials — mechanical weakness are reliability concerns for inlaid copper interconnects [1]. All these degradation phenomena are driven more or less by the mechanical stress generated by differences between the thermal expansions of the component materials. Despite the smaller coefficient of thermal expansion of copper compared with aluminum, the higher stiffness of copper results in slightly greater biaxial thermal stress in copper films than in aluminum films subject to identical temperature changes [2]. This fact, taken together with the high yield strength of copper, led to the conclusion that the stress in copper metallization should exceed the stress in aluminum one. Furthermore, compared to aluminum, copper reveals a high degree of elastic anisotropy that should also contribute to the stress behavior. The multilevel metallization architecture of the chip interconnect assumes a need for the multiple thermal cycles with a maximal temperature of 350-400°C. As a result, a high tensile stresses evolve in the metal patterned structures upon cooling down from the anneal temperature. The triaxial tensile stresses that develop in metal structures as a result of the thermal contraction mismatch with the surrounding liners, dielectric barrier, inter-level dielectric (ILD) and silicon substrate, can initiate void nucleation and effect its growth [3, 4]. Electomigration-induced voiding is also effected by stress existing in the patterned metal structures [5]. The stress-dependent vacancy distribution along the conductor lines, as well as stress gradient-induced atom flux, contributes to the dynamics of void nucleation and growth [6]. A deep understanding of interconnect stress evolution is very important for robust chip design, advanced process development and reliability analysis.

Numerical modeling with the finite element method is a proven approach to the study of thermal stress evolution in interconnect structures [7-9]. This approach allows one to simulate a static stress distribution in interconnect structures characterized by complicated geometry and architecture as well as stress relaxation caused by different driving forces such as grain boundary diffusion, plasticity and creep [3]. This paper focuses on modeling stress distribution in copper lines and dual-inlaid copper interconnect structures when all essential details of the segment architecture such as metal liners, dielectric barriers and ILD are taken into consideration. The paper is organized as follows. Section "Passivated Copper Lines" considers copper lines of different widths embedded into either silicon dioxide or low-k ILDs. Thermal stress, caused by the metal lines cooling down from the zero-stress temperature to test temperature, are simulated as a function of metal line width and material properties of the intermetal dielectric. The influence of copper plasticity and copper elastic anisotropy on generated thermal stress is captured by implementing the elasto-plastic, elasto-plastic with hardening, and elastic orthotropic material models. Section "FEM Simulation-based Analysis" provides results of the simulation of thermal stresses in interconnect segments characterized by the presence of a wide metal line above or below via. Both architectures are analyzed regarding a role of the metal line width in stress-induced voiding. Section "Stress Gradient-Driven Atom Flux in Electromigration-Induced Degradation" considers a role of atom diffusion driven by stress gradient in electromigration-induced voiding.

Passivated Copper Lines

Stress migration in copper interconnects has been a topic of research since the implementation of the copper inlaid process in the semiconductor industry [10]. In the past it was not considered as a major reliability concern since the stiff silicon dioxide ILD fixed the relatively ductile copper in the line structures and almost no effect was observed in stress migration tests. Implementation of the low-k ILDs into the copper interconnect has significantly influenced its mechanical stability. Stress-induced degradation and mechanical weakness have become the challenging reliability concerns for inlaid copper interconnects. Experimentally determined thermal stress behavior in copper lines embedded into silicon dioxide and low-k ILDs mainly corresponds to preliminary information obtained from the studies of material properties of continuous copper films and stress behavior in patterned aluminum lines. But some experimental results have revealed a different stress behavior compared with an expected one on the basis of the previous experience. One of these "strange" results is related to a dependence of thermal stress in a copper line on the line width. Instead of having an expected decrease in the stress with increasing line width, the experimental data demonstrate an opposite behavior. Wider lines are characterized by greater thermal stress [11]. Data presented in [11] shows the longitudinal (S22-along line) and transverse (S11-in plane and S33-vertical directions) stresses in 180 nm and 1.8 μm copper lines embedded in silicon dioxide and low-k ILDs, measured with X-ray diffraction method. In both cases, the pitch is equal to the doubled line width. These data indicate the significantly higher stress state in the wide copper lines compared to the narrow lines. It is also shown that different ILDs do not produce significantly different stress in the copper line characterized by the same width. In order to understand this type of thermal stress behavior we have performed a 2-D plain stress simulation analysis based on the finite element method. Figure 1 shows a 180-nm copper line width interconnect segment where thermal stresses, developed due to thermal expansion mismatch during the cooling from 350°C (initial stress-free state) down to 20°C, have been simulated. The simulations were performed using commercially available FEMLAB finite element software.

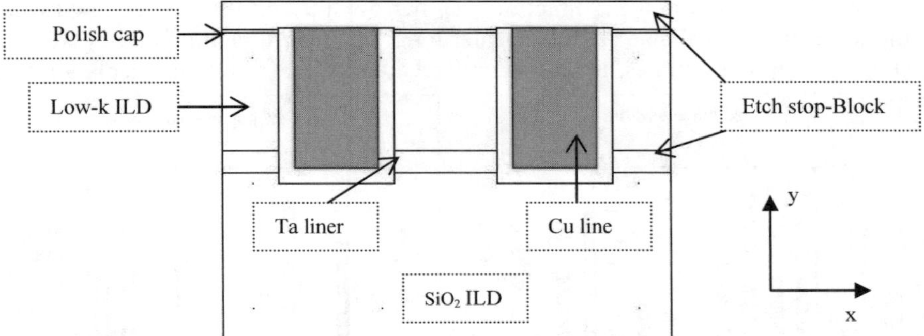

Figure 1. A representative section of the parallel copper lines embedded in ILD.

The base of the simulated segment was constrained from vertical displacement but free to experience a lateral displacement. The left face of the model was con-strained from displacement in the lateral direction but was allowed to move verti-cally. It was assumed that all components of the interconnect structure were per-fectly bonded. The silicon dictates the lateral expansion of the film since the substrate is many times thicker than the film. When the temperature is decreased, the substrate would experience thermal contraction in the lateral direction by an amount given by the expression:

$$\Delta L_{right} = \alpha_{Si} L \Delta T \tag{1}$$

where α_{Si} is the thermal expansion coefficient of the silicon, L is the total width of the model, and ΔT is the temperature change. The horizontal displacement given by Equation 1 was applied to the entire right edge of the model with no constraint in vertical direction. The Cartesian coordinate axes x, y, and z refer to directions transverse to the lines, normal to the substrate, and along the lines, respectively.

In the first run, all the materials used in the finite element simulation were as-sumed linear elastic and isotropic. The elastic properties of all materials are given in Table 1.

Table 1. Elastic properties of materials used in the finite element model.

Material	Elastic modulus, E, (GPa)	Poisson's ratio, ν	CTE, ($\times 10^{-6}$/K)
Copper	100	0.3	17
Silicon	165	0.22	2.3
Silicon dioxide	71	0.16	0.6
Silicon carbide	460	0.21	4
Tantalum	185	0.3	6.5

Figure 2 shows the volume-averaged components of the simulated triaxial and hydrostatic stresses in copper lines with 180-nm and 1.8-µm widths. It can be seen that wide lines in both cases of SiO_2 and low-k ILD materials are characterized by higher tensile stress in the x direction than narrow lines. In the case of low-k ILD, the wide lines reveal also slightly greater longitudinal (z direction) and hydrostatic stresses than narrow lines. If SiO_2 is used as a dielectric for interline isolation then

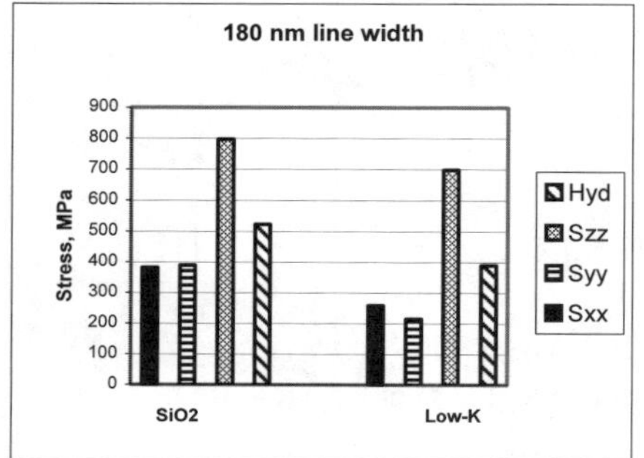

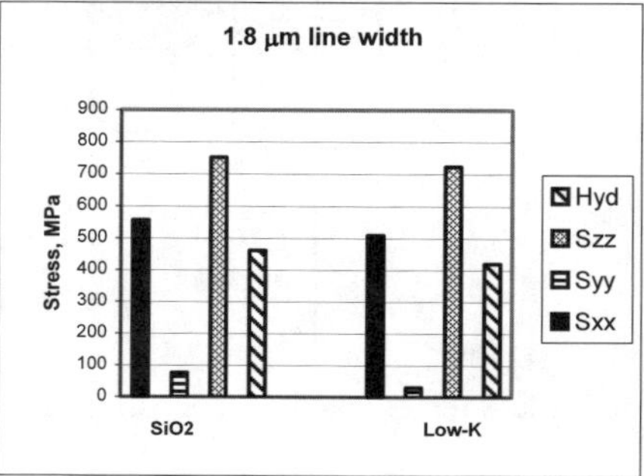

Figure 2. Room-temperature stress state of copper lines for different line widths and ILD materials, which was simulated with the linear elastic and isotropic material model.

the longitudinal and hydrostatic stresses are greater in narrow lines. Simulation predicts that, for both considered ILDs, the stress in the y direction is much smaller for wide lines. Comparison between the simulation results and experimental data [11] indicates a noticeable discrepancy. Simulated thermal stresses are two times greater than the measured stresses. All measured stress components are greater for wider lines for both ILDs. To address this misfit the finite element model should be modified in order to include some properties unique to copper. Introduction of the elasto-plastic material model for copper should reduce the maximal stress when overall yielding has happened. Taking into consideration the mechanical anisotropy of copper could modify the thermal stress dependence on the line width. Figure 3 shows the simulated stress states of 180-nm and 1.8-µm copper lines with SiO_2 and low-k ILDs when copper plasticity was included in the model. We assume that copper is nonhardening, and that yield strength is independent of the temperature. In our simulation we use 300 MPa as a representative value of the yield strength. Figure 3 clearly demonstrates that taking into consideration the copper plasticity indeed decreases the stress in copper lines but does not change the stress dependence on the line width. Indeed, wider copper lines are characterized by smaller stress than narrow ones for both considered ILDs. Introduction of isotropic hardening does not change this dependence, slightly increasing all stress components.

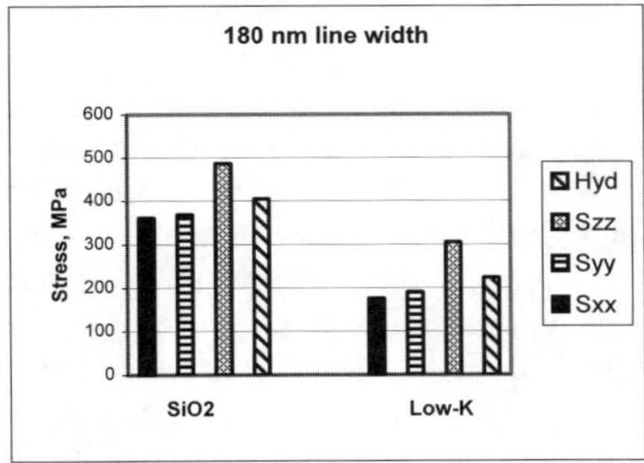

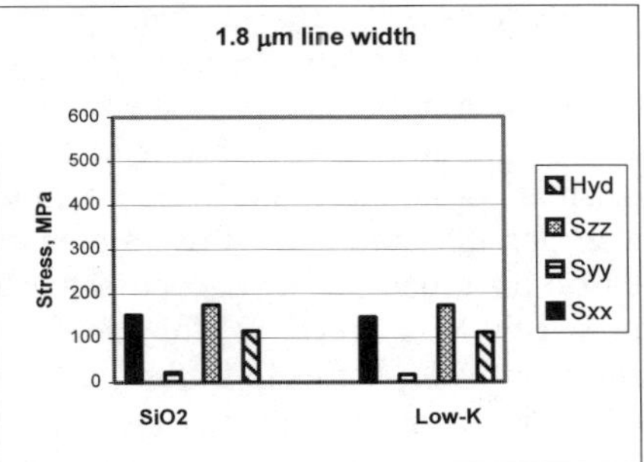

Figure 3. Room-temperature stress state of copper lines for different line widths and ILD materials, which was simulated with implementation of the elastio-plastic material model for copper.

In order to be able to capture the elastic anisotropy of copper, we have modified its material model by introducing the different material properties for the out-of-plane and in-plane directions. For out-of-plane we use $E_{(111)} = 191.1$ GPa and for in-plane $E_{(100)} = 66.7$ GPa as the elasticity modulus. All other material properties are the same for both directions. Figure 4 shows the simulated stress components for both line width and ILD materials. It can be seen that introduction of the copper elastic anisotropy does not essentially change the character of the stress dependency on line width, but almost proportionally reduces the value of all stress compo-

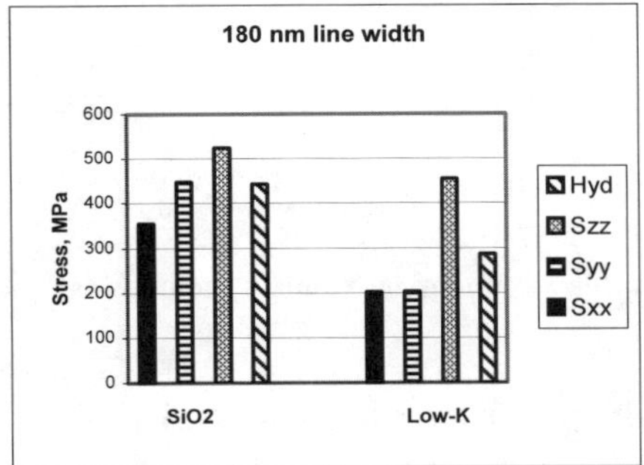

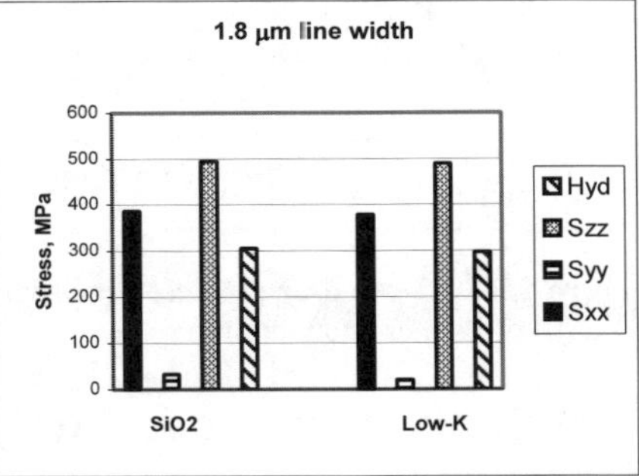

Figure 4. Room-temperature stress state of copper lines for different line widths and ILD materials, which was simulated with implementation of the orthotropic material model for copper.

nents. It is also worth noticing that the introduced copper anisotropy has destroyed the almost perfect hydrostatic character of the transverse stresses in the 180-nm line width for the SiO_2 ILD.

The scope of all employed modeling approaches was not able to unambiguously explain the stress reduction in all stress components when the line width was decreased, which was observed in experiment. It seems that implementation of the temperature-dependent yield stress as well as the kinematic hardening could not provide this explanation. It could be possible to obtain a stress dependency similar to the experimentally observed one by taking into account possible variations in major material properties of the interconnect components. Figure 5 shows the results of the performed analysis of the copper line hydrostatic stress sensitivity to the small variations in the elasticity modulus and CTE of copper and tantalum. Another possible solution to this problem could be in consideration of the major strain relaxation processes in copper patterned structures, similar to the analysis performed by Vinci *et al.* [2] for continuous copper films. The nonuniformity of the stress motivates atoms to diffuse to the region of large tension. These regions arise in the vicinity of the interfaces with the etch stop top layer and metal liners – diffusion barriers. The ratio of the total line surface to a line volume increases with the decrease in the line width. Hence, to accommodate the existing stress gradients a larger fraction of the volume atoms should be relocated and platted on the grain boundaries in the case of narrow line compared with the wide line. This fact should explain the greater uniform hydrostatic tension, which should be achieved in a state of equilibrium in the wide copper lines compared with the equilibrium stress in narrow lines.

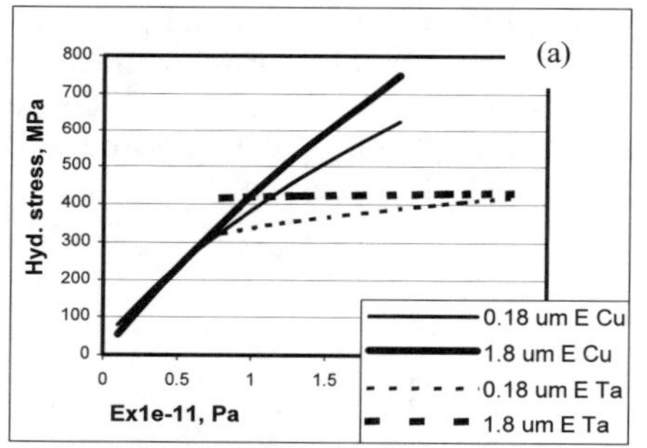

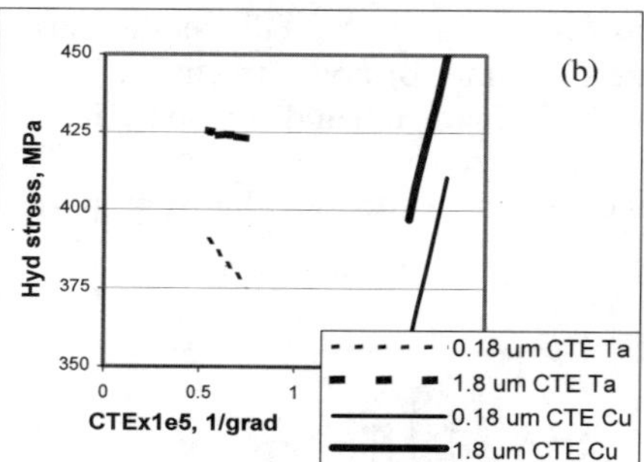

Figure 5. Copper hydrogen stress sensitivity to the variations in Young's modulus (a) and CTE (b) of copper and tantalum.

FEM Simulation-based Analysis of Thermal Stress in Interconnect Segment with Via Connected to Wide Metal Line

Severe voiding was observed in copper interconnects characterized by the presence of a wide metal line above or below via. First mode is accompanied by void generation inside via and the second mode is characterized by voiding in the wide metal line below via bottom. In both cases a specific size effect was observed. Voiding takes place when the size of the wide metal line exceeds some minimal size. To understand the role played by thermal stress in this voiding and to provide recommendations on how to minimize this voiding, a set of thermal elastic stress simulations have been performed.

We have found that the stress inside via has almost no defendence on the size of the wide upper metal line and on length of the narrow line, which connects the via with the wide line. We suggest that this type of voiding has resulted from a poor optimized processing. The second type of voiding has a stress-induced nature. We have simulated thermal stress in different segments characterized by different metal width and proposed a mechanism of the size effect. Finally, we have explored the possible benefits from the introduction of the dual via interconnect segments.

Top Wide Metal Line Connected to Via by the Variable Length Narrow Line
To evaluate an effect of the length of the upper narrow line (L) on the hydrostatic stress inside via, which was caused by the cooling down from the stress-free temperature $T_{ref}=430°C$ to $T=25°C$, we have simulated the stress distribution in the identical structures characterized by different L: 0.45 μm, 1 μm, and 2 μm. In all structures the geometry of the wide metal line was the same: $H = 2$ μm, W $= 2$ μm.

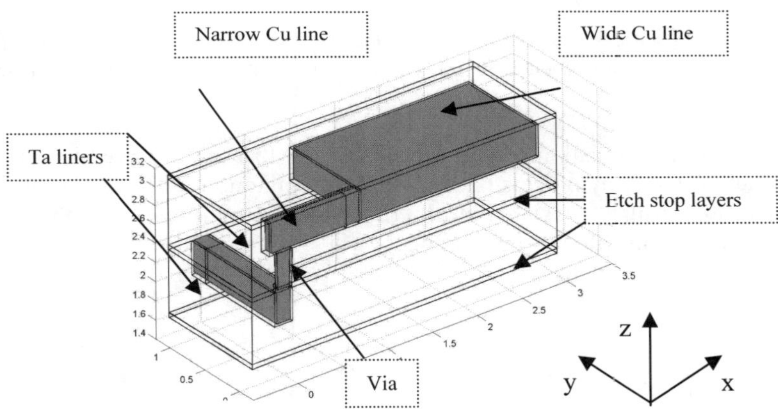

Figure 6. Simulated dual-inlaid two-layer structure.

Figure 6 shows the simulated dual-inlaid structure. As before, the base of the simulated segment was constrained from vertical displacement but free to experience a lateral displacement. The left and front faces of the model were constrained from displacements in the lateral directions but were allowed to move vertically. It was assumed that all components of the interconnect structure were perfectly bonded. The horizontal displacement given by Equation 1 was applied to the entire right and back edges of the model with no constraint in vertical direction. The Cartesian coordinate axes x, y, and z refer to directions along the lines, transverse to the lines, and normal to the substrate, respectively.

Figure 7 shows hydrostatic stress distributions along the via vertical axis of symmetry for three different lengths of the connecting narrow line. All considered segments reveal almost equal hydrostatic stress inside the via bulk. Reduction of the area of the wide metal plate (more than three times) as well as modification of the boundary conditions, from the segment sidewall displacements driven by silicon expansion to a driven by the etch stop (silicon carbide) expansion, do not provide any noticeable changes in the hydrostatic stresses inside the via.

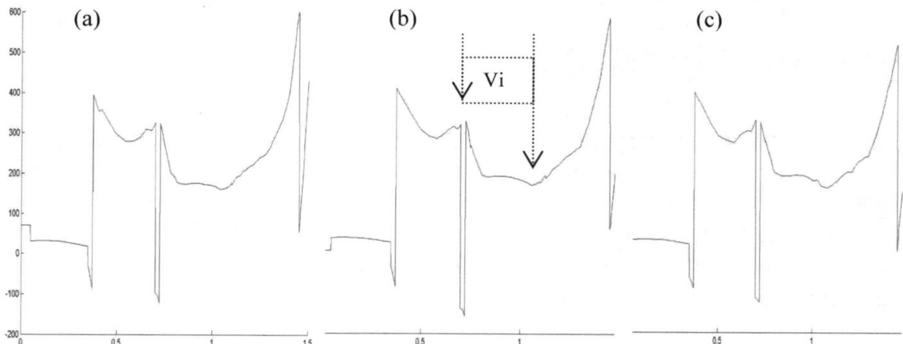

Figure 7. Hydrostatic stress distributions along the via axis of symmetry for different lengths of the narrow line connecting the via with wide metal line: 0.45 μm (a), 1 μm (b), and 2 μm (c).

Based on the simulation results we can suggest that the observed void nucleation is caused rather by processing than by stress migration. This conclusion corresponds to the results of Oshima *et al.* [12], which demonstrated that this type of voiding can be eliminated by optimizing the via cleaning process.

Wide Metal Line Below Via

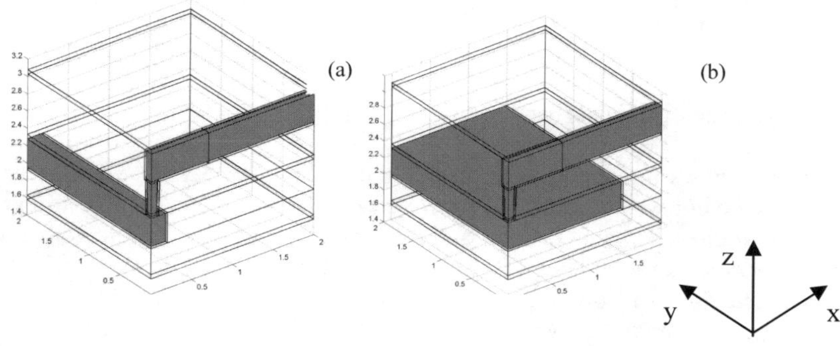

Figure 8. Simulated structures with different width metal lines located below via: 0.4 μm (a) and 3 μm (b).

Typical simulated structures that represent the interconnect segment with the wide metal line located below via are shown in Figure 8. Material properties and boundary conditions are the same as in the previous sections. Figures 9, 10, and 11 show the hydrostatic, normal vertical (σ_{zz}), and shear (τ_{yz}) stress distributions in the center cut yz of the bottom metal line in the vicinity of the copper-etch stop interface. It is clear from Figure 9 that for a wider metal line smaller hydrostatic stress is developed in the area, which is underneath the via bottom. It should be noted that we have resolved a local reduction in tensile stress in the area under the edge of the via for the widest of the considered metal lines (Figure 9c).

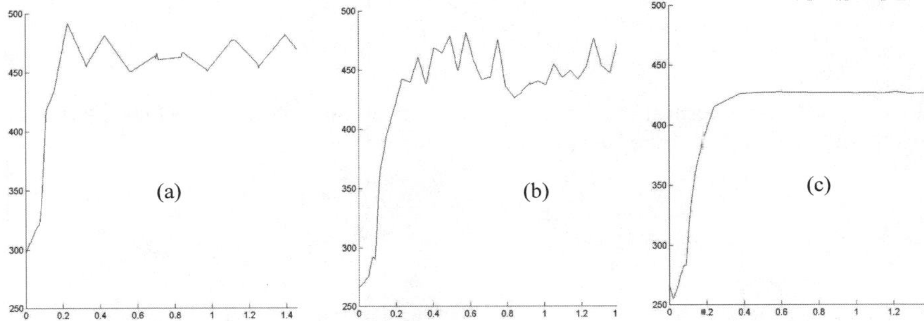

Figure 9. Hydrostatic stress distributions in the center cuts (yz) of the bottom metal lines in y-direction in the vicinity of the copper-etch stop interface. Metal line width: (a) 0.2 μm, (b) 0.4 μm, and (c) 3.0 μm

It means that in the wide metal line we can expect a stress-gradient-induced vacancy flow toward the area located beneath the via perimeter. This result corresponds to the prediction made by Ogawa *et al.* in [13].

A more pronounced size effect was observed in the dependence of the σ_{zz} stress on the line width (Figure 10). In all cases there is a compression in the vertical (z) direction in the portion of metal located below the via bottom. The maximal compression locates exactly below the via edge. Experiment demonstrates preferential but isotropic voiding at the perimeter of the via bottom rather than directly below the via center [12, 13]. Obtained dependence of the σ_{zz} on the line width can be explained by the difference in resistance of the different lines, characterized by different width, to the bending, caused by the via-induced pressure on the underlying metal line. The narrow line embedded in a soft low-k confinement, being more flexible to the bending, is characterized by lower stress than the wide line. As a confirmation we can referr to the results of the performed simulation that demonstrate that the σ_{zz} stresses developed in lines with low-k underneath layer and positioned on the hard base (zero vertical displacement) are quite different in narrow lines and almost the same in wide lines (Figure 12). This result clearly demonstrates that strain created by the via-line interaction, caused by the thermal expansion mismatch, is significantly accommodated by the wide line. Simulation results show that maximal bending of the via-underneath metal line, achieved in the case of the narrow line, is 0.75 nm and is 0.37 nm in the case of the wide line.

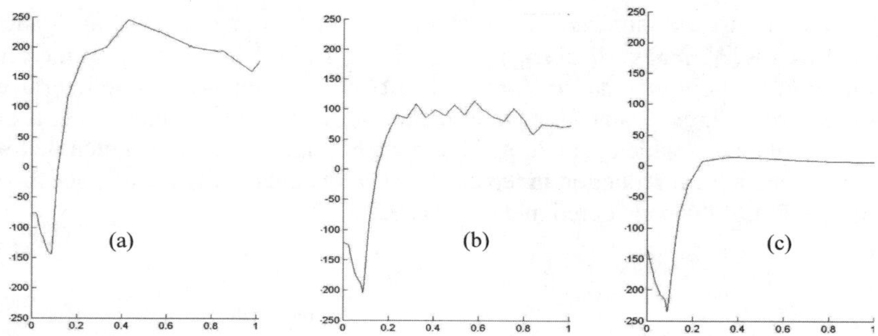

Figure 10. Normal vertical (σ_{zz}) stress distributions in the same cross sections. Metal line width: (a) 0.2 μm, (b) 0.4 μm, and (c) 3.0 μm.

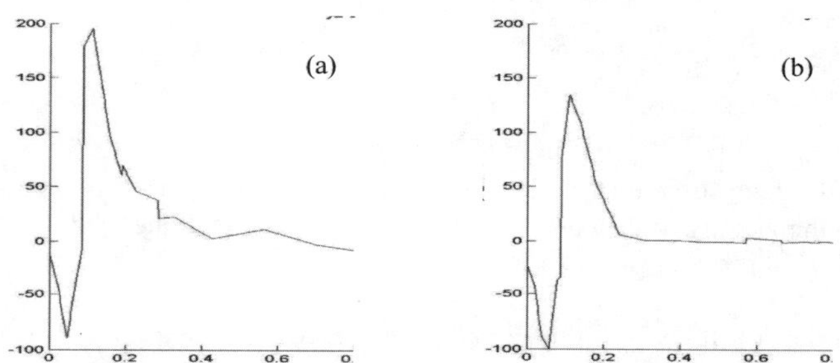

Figure 11. Shear stress (τ_{yz}) distributions in the same cross sections. Metal line width: (a) 0.2 μm and (b) 3.0 μm.

Based on these results we can suggest that a primary cause of the voiding, which takes place in the area of a wide line below the via edge, is a combination of the σ_{zz}, generated by the via pressure on the underlying metal line, and the shear stress at the copper-etch stop interface (τ_{yz}). When stress exceeds the critical level it can initiate wedge crack formation due to grain sliding or partial delamination of the copper-etch stop interface. The former mechanism will work if a grain boundary is available near the via edge. Existing gradient of the hydrostatic stress results in vacancy migration toward the nucleated crack. Grain boundaries and the copper-etch stop interface serve as the main channels for the vacancy diffusion. High resistance to bending, which is valid for a wide metal line, results in a higher level of internal stress (σ_{zz}) compared with the narrow line. It can be responsible for the higher probability of crack initiation and further void development in a wide line compared with a narrow line.

Implementation of a dual via provides a better resistance to EM-induced degradation [14]. Our simulations demonstrate that it can also be beneficial with regard to protection against stress-induced failure. Vertical stress σ_{zz} in the Cu line area located underneath the via bottom is smaller for dual via compared with the single via. For our simulation setup this difference is on the order of 50 MPa. If the developed stress is enough to provide grain sliding or interface delamination in the single via segment, then introduction of the dual via can suppress this failure. If one via of the dual via segments has failed due to stress-induced voiding in the wide metal line below the via, then the formed void, playing the role of vacancy scavenger, would reduce the probability of the second via failure.

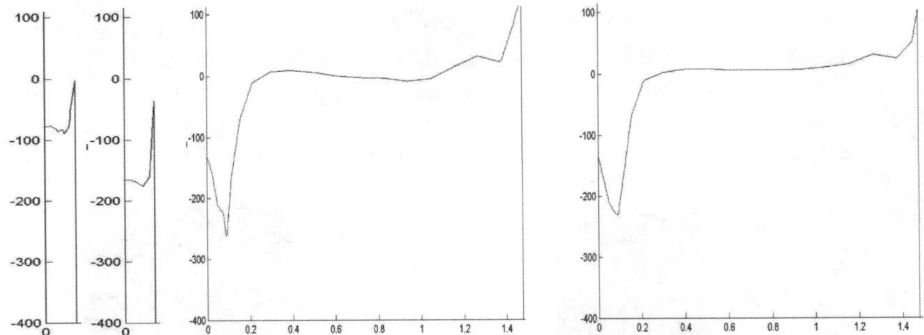

Figure 12. σ_{zz} stress distributions in the via underneath metal lines near the copper-etch stop interface in the xz cross-section. Low-k layer exists below metal lines of 0.2 mm (a) and 3.0 mm width (b); there is no low-k layer below metal lines, 0.2 mm (b) and 3.0 mm (d).

Stress Gradient-Driven Atom Flux in Electromigration-Induced Degradation

Electomigration-induced voiding is effected by stress existing in the patterned metal structures. A full EM model should take into account the stress-dependent vacancy distribution in interconnect segments as well as the stress-gradient-induced atom flux, which contributes to the dynamics of void nucleation and growth [4-6]. As an example we consider the simulation results of a material edge displacement in the Blech short strip test structure of electromigration [15]. In this structure a short Al segment is deposited on a strip of higher resistivity material such as TiN. In the region where Al is present the electrical current passes mainly through the Al because of its much lower resistivity. Otherwise, the current should pass through the TiN. Flux divergences take place at both ends of the Al segment and cause depletion of material at the cathode end and accumulation at the anode end. Figure 13 shows the initial hydrostatic stress distribution and the directions of the total atomic flux and current flow near the cathode end of the Blech structure. Development of thermal elastic tensile stress takes place everywhere in the Al strip except for the top corners where the compressive stress is developed. This stress distribution results in a net atomic flux out of the top corners. Before the electrical current has been applied these fluxes are balanced by the counter-fluxes caused by the developing concentration gradients. Applied current removes the concentration gradient at the cathode area by forcing the excessive atoms toward the anode area by means of electron momentum transfer. Figure 14a shows simulated void growth, taking place when all atom migration causes are taken into consideration. This picture fits well with the experimental data, which demonstrate that voids initially nucleate at the upper left corner and, propagating to the right and down, deplete the cathode completely [16]. An absolutely different picture of the void development can be obtained if the stress gradient-induced atomic flux is not taken into consideration or if the applied current density is high enough to control the atom migration. Figure 14b demonstrates the simulated void growth for this case. Atomic flux caused by the electron momentum transfer results in void nucleation at the lower left corner, where electron flux enters the Al strip, and further void extension occurs along the Al–TiN interface. This example demonstrates the importance of including all atom-migration driving forces in the prediction of voiding-induced failure.

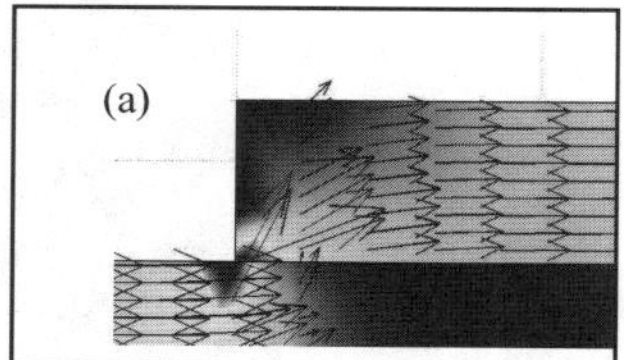 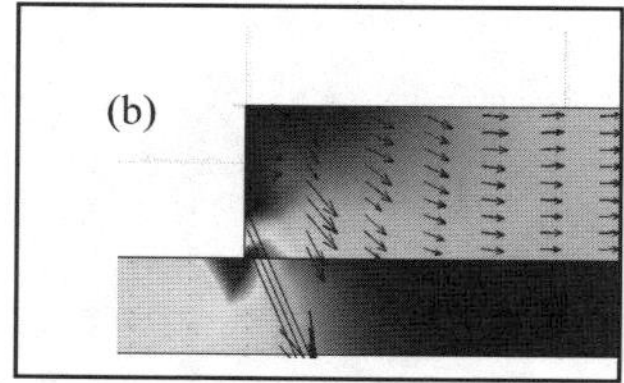

Figure 13. Initial distributions of the hydrostatic stress, current density (a) and total atomic flux (b) in the Al/TiN Blech structure of EM.

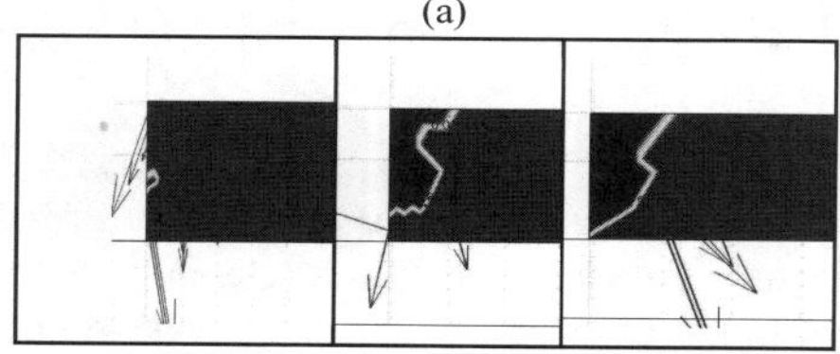 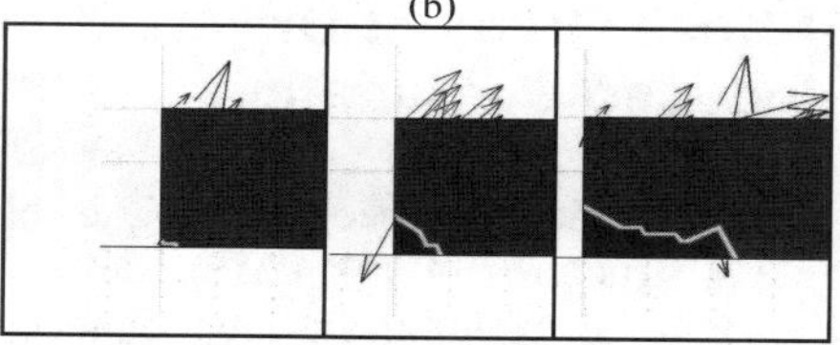

Figure 14. Simulated void growth when all atom migration causes are taken into consideration: atom fluxes induced by current and gradients of concentration, temperature and stress (a); simulated void growth when stress gradient induced flux is not included (b).

References

1. P.S. Ho, K.D. Lee, E.T. Ogawa, S. Yoon, X. Lu, in Characterization and Metrology for ULSI Technology, ed. D. G. Seiler *et al.*, AIP Conf. Proc. (AIP: Melville, NY), 533 – 539 (2003).

2. R.P. Vinci, E.M. Zeilinski, J.C. Bravman, Thin Solid Films 262, 142-153, (1995)

3. T.D. Sullivan, Ann. Rev. Mater. Sci. 26, 333-364, (1996).

4. Z. Suo, In Volume 8: Interfacial and Nanoscale Failure (W. Gerberich, W. Yang, Editors), Comprehensive Structural Integrity (I. Milne, R.O. Ritchie, B. Karihaloo, eds., Elsevier, Amsterdam, 265-324, (2003).

5. M.A. Korhonen, P. Borgesen, K.N. Tu, C.-Y. Li, J. Appl. Phys., 73, 3790, (1993).

6.   Y-J. Park, V.K. Andleigh, C.V. Thompson, J. Appl. Phys., 85, 3546, (1999).

7.   A.I. Sauter, W.D. Nix, IEEE Trans. Comp. Hybrids Manufact. Technol. 15, 594, (1991).

8.   Gouldstone, Y.-L. Shen, S. Suresh, C.V. Thompson, J. Mater. Res. 13, 1956-1966, (1998).

9.   J.A. Sethian, J. Wilkening, J. Comput. Phys. 193, 275-305, (2003).

10. M. Fayolle, G. Passemard, O. Louveau, F. Fusalba, J. Cluzel, J. Microelectr. Eng. 70, 255-266, (2003).

11. I. Zienert *et al.* , this book, p. 241-249 (2005).

12. T. Oshima, K. Hinode, H. Yamaguchi, *et al.,* Proc. IEDM, 757-760, (2002).

13. E.T. Ogawa, J.W. McPherson, J.A. Rosal, *et. al.*, Proc. IRPS, 312-321, (2002).

14. T.C. Huang, C.H. Yao, W.K. Wan, *et al.*, Proc. IITC, 207-209, (2003).

15. V. Sukharev, R. Choudhury, and C.W. Park, in 2003 IEEE IRW, pp. 80 – 85

16. I.A. Blech, J. Appl. Phys., 47, 1203-1208, (1976).

Conductivity Enhancement in Metallization Structures of Regular Grains

G. D. Knight

Ottawa-Carleton Institute of Physics, Ottawa, Ontario / Canada

Introduction

Metallization structures are challenged as the interconnect paradigm for electronic devices, when these and their separations are on the order of a quarter micron. The interconnect cross section is on the order of 0.01 $(\mu m)^2$ where electron trajectories near room temperature are curtailed by scattering potentials on bounding surfaces; and these dominate the transport dynamics. The search is on for specular surfaces and purified grains capable of supporting high current density even at reduced temperature so as to reduce noise and better facilitate cryogenic elements.

Not to be overlooked are surface texture and periodic geometries that may enhance conductivity by tapping the wave nature of electrons. Surface texture ensues from grain morphology, which can be engineered. The wave mechanics is more classical than for mesoscopic quantum dots or contacts, analogous to geometric optics with diffractive effects added in.

Geometrical effects are not unfamiliar: consider intrinsic anisotropy. Highly ordered pyrolytic graphite (HOPG), for example, has "easy" conduction directions along carbon slip-planes where electron mobility is high, whose conductivity is easily augmented by adding carriers. Thus an intercalated few sheets of HOPG may be as good or better a conductor than pure metal, especially in lateral confinement where it is less subject to the adverse scattering effects of surface roughness. Experiments could well be pursued with such sheets laid on a side in damascene trenches and around serpentines.

HOPG is not technologically realized as an interconnect material for reasons of handling, chemistry or deposition, but it is the base material that in cylinders of quantized diameter and torsional pitch makes carbon nanotubes. Structural autocorrelation, period and symmetry, selected by roll-and-seam physical chemistry, engender nondispersive wavefunctions or solitons, which are not soon scattered and support virtually ballistic axial transport. The clue for what in the long-range order "insulates" carrier wavefunctions from extraneous scatter effects lies both in the symmetries and the highly anisotropic mobility or reduced mass tensor — both energy-geometric effects.

Taking by contrast, an ideally pure dendritic metal line of similar dimension, a little disruption at its surface will ruin the prospect of good ohmic, let alone ballistic, transport. As such systems approach uni-dimensionality – when that represents a significant dimensional reduction – the presence of small disorder readily induces carrier localization.[1] Highly anisotropic mobility defers that effect by compressing the dimensional metric.

In well-known Nordheim behaviour, metal conduction is deeply modulated by sub-monolayer deposition of adatoms. This shows that especially isotropic mobility is sensitive to conditions at grain or bounding surfaces. Absence of this sensitivity in HOPG along easy directions is a result of highly anisotropic conductivity or mobility tensors, naturally paraxial carrier wave vectors having insignificant transverse momenta.

A useful question is, how transverse momenta get curbed without either dissipation or Anderson-like localization, which would have reduced transport. The clue for HOPG lies in the combination of potential surfaces and long-range (periodic) correlations of carbon-bond ring structures. In energy-geometric language, available geodesics are innately squeezed into paraxial flowlines, while in isotropic media they indicate expanding spherical wavefronts. Besides surface "roughness" being less intrusive for paraxial rays[2] surface scatter probability drops with smallness of transverse momentum as determined by the band structure. In other words, funneling carriers into paraxial bundles reduces surface encounters within the coherence length.

Repetition geometry shapes normative free carrier wavefunctions by affecting the space in which carriers move. In isotropic crystals the spherical Bloch function is embedded with crystal symmetry, and in HOPG it must be a stratified paraxial Bloch function with the periodicity of the graphite structure. Thus the *metric tensor* in the local geometry experienced by carriers is informed by the extended structure with its *repeating* patterns.

The sensitivity of isotropic metal to surface scattering in confined dimensions lies in metric tensors being largely *uninformed* of crystal truncation at a non-repeating surface, despite some alteration in the few reconstruction layers near it. There is little to limit carrier momentum from coming at high-angle to a bounding surface. It is more likely to experience non-specular or randomizing reflection. No flow-bundles curb the frequency of such encounters, and the surface has opportunity to reshape the mean free volume.

Mean free volume depicts local geometry and boundary specularity. Were every surface specular, being an elastic scatterer, it could not shorten coherence length. In the case of flat film or wire the effective mean free volume is then a full-sized sphere in isotropic metal or a very flat crêpe in HOPG. Approaching a sudden film or wire expansion from the wider side produces a kink in the free volume where propagators are negated at exterior steps in the field direction, an anisotropy that was noted in [1]. Space-charge should be acquired as narrowed structures reduce screening by inhibiting mobility, and the local potential may be useful to modulate as in a gated device.

Conductivity is a moment of the mean free volume in a field direction. Hence it's clear how shaping this volume affects conductivity and other material charac-

---

[1]In fact, Anderson localization is inevitable in the 1-D limit.
[2]Glancing rays scatter little in a Rayleigh approximation with roughness features foreshortened relative to the projected wavelength washing over them.

teristics such as the thermopower, TCR, and magnetoresistance. But the idea of kinds of geometric roughness that can usefully shape carrier momentum has barely begun to attract interest [2-4]. Ordered roughness of specular surface could mean symmetric structures, as in oriented mirrors[3] and diffraction surfaces, *or* strong length correlations impressed upon random roughness. By contrast, non-geometric roughness is randomness with no length correlation, or a diffusive surface at any scale[4].

While randomization of momentum by rough surfaces may reduce axial carrier group velocity, correlations impressed on the roughness characteristics can feasibly *frustrate* localization. Precedent for this behaviour exists in the optical analog of a step-index fibre or waveguide, where core surface roughness couples to lossy radiation modes. Change to surface roughness correlation length [5] can greatly reduce the coupling or coherent loss.

A further analogy in the waveguiding of electrons is the long-period grating. Grains might be engineered to have lengths a near multiple of carrier wavelength ~ 0.6 nm or thermal modulation (packet) length ~ 12 nm – though not greater than coherence length $L_\phi \geq 2\pi\ell \sim 250$ nm (at room temperature). As wavefunctions are just probability densities that obey laws of wave mechanics, a modulation wavelength can also be taken as the centre frequency modulated at a high-frequency. Features of wave mechanics can therefore be expected at the molecular scale of Fermi wavelength and at the "decanano" wavepacket size, involving resonances so far as a beam span of L_ϕ covers an extent of diffractive structure.

An electron traversing two crystallites undergoes refraction from difference in bravais lattice constant and tensor transformation of material anisotropy. Conservation of in-plane momentum yields Snell's law in so-called N-processes – while surprise coherent directions can also exist in U-processes depending on opposing irregular Fermi surfaces. Generally, the electron refractive index in any direction is determined by the coupling wavevector as a function of Fermi surface curvatures.

Forward coupling of momentum and energy in LPG fibres occurs through a phase-matched condition between core and low-order cladding modes. In a chain of metallic grains a 'cladding' equivalent is unobvious, but the atomic reconstruction layers near a bounding surface may serve this function. The additive role of a *spatial* wavevector to make-up axial phase for non-paraxial electron wavevectors is also analogous to quasi-phase-matching (QPM) gratings in nonlinear optical ampli-

---

[3]Oriented mirror surfaces would include light-gatherers and intensifier structures employed in space and in astronomy.

[4]Small-scale roughness without correlation randomizes wavevector orientation, while inelastic scattering "thermalizes" wavefunctions. The first leads to loss of coherence *in practice*, while the second – directly dephasing electrons – is a loss of coherence *in principle*. The effective loss of flux in the first case is by Anderson-like confinement from self-interference of coherent wavefunctions; but for convenience, both losses of coherent momentum flux are called incoherent.

fiers.[5] For electrons there is also the more direct interaction with polaritons in metallized gratings whose spatial period over electron velocity matches an optical period.[6]

Various further analogies in wave-mechanics motivate the search for modulated crystalline potentials through surface texture or grain size regularities to enhance coherence-preservation in momentum transport through a participating phonon medium. Simulation of such effects with salient parameters ought to be feasible through the use of beam-propagation methods (BPM) as employed in scalar models of photonic waveguides. Alternatively, Monte Carlo (MC) models for momentum flux may be pursued, but these need augmenting with diffractive effects. Straightforward in that goal is to beam-trace the lowest order diffraction peaks, though computational parallelism seems mandated by path multiplicity at every diffractive surface.

Knight & Smy [1, 6] solved the semiclassic BTE formulation of conductivity using 3-D charged particle optics (3CPO) to integrate along beam paths launched over the Fermi surface. This quasi-MC billiard method relegates randomness to the Beer's $\exp(-d(\mu)/\ell)$ relaxation kernel (μ being the beam cosine to a surface normal). The surface integral over the reduced mean-free volume is found, whose projective moments (in $\cos^2 \phi$) along field directions provides a tensor of conductivity components. The 3CPO integrator can accumulate and average across a via or interconnect cross section, or down its axis and throughout a multi-tiered metallization structure including sudden constrictions (induced by voids from electromigration). *Without* diffractive effects it obtains the Namba [7] result for corrugated surfaces, but also in serpentines and dual damascenes not amenable to Namba.

Subsequently, Knight [8-10] investigated a predicted all-metal diode induced by an asymmetric ramp-embossed texture (*cf.* [3] Figure 1) of the bounding surface on a monocrystal metal dentrite. For right-angled rect-echellia the axial conductivity component differed between forward and reverse current senses, and a soft-diode effect of more 6% appeared for a louvre angle of about 11°. This angle was shown [10] to maximize the difference between fore and abaft billiard scatter in a plot called the blue-bird (Figure 1).

Broken symmetry in axial conductivity thus predicted by 3CPO seems surprising for a billiard even with incidence-sensitive specularity. At first sight the optical reciprocity theorem seems violated; but as with light-gathering structures[7] increases the paraxial coherence to one direction: what is paraxial survives further, before a surface encounter. That diffuse scattering at transverse grain boundaries would vitiate this benefit is ample reason to look for the effect in well-annealed

---

[5]Whether there are material-dependent nonlinearities in electron transport analogous to optical Kerr or Pockels effects is an open question.

[6]The inverse Smith–Purcell effect couples input laser light via a grating of chosen pitch to propel electrons: the efficiency is sharply maximum when the electron velocity is such that it traverses a grating period in the time of one oscillation.

[7] Wind-driven seas [9] have similar anisotropic radar backscatter.

and pure lines. When the louvred surface is completely diffusive no anisotropic effect is to be found.

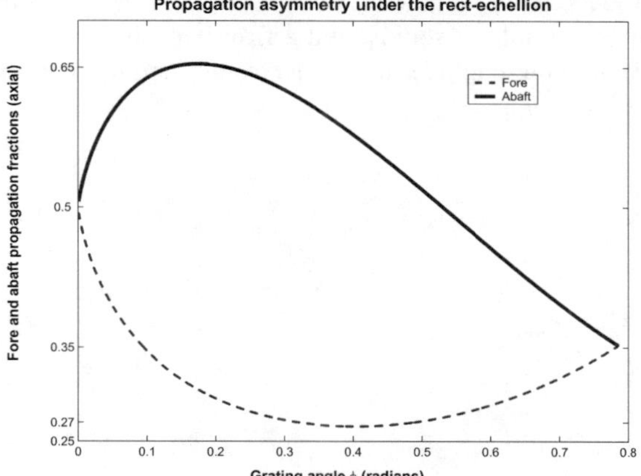

Figure 1. Blue-bird plot for rect-echellion [3] (Figure 1) showing axial moment of propagated flux of uniform spanning-beam in opposite components. Maximal difference occurs at about $\phi = 11°$ (non-unitarity at $\phi = \pi / 4$ is due to shadowing).

To further study the diffractive effect [8-10] scalar wave-scatter from an echellette surface was integrated in closed form following Beckmann [11]. Diffractive lobes for axis-in-plane incidence were divided fore and abaft to weight integration along the specular direction as if reflected at a planar surface (*cf.* [3] Figure 2). Coherence in fore lobes was taken as preserved, abaft as lost. Low-order lobes abaft of specular should partly fore scatter again, but the neglect was offset by over-counting fore lobes to be scattered abaft. As in a surface approximation of binary specularity at critical angle, results differed little from a billiard: definite asymmetry at the ~ 11° grating angle.

A similar "ratchet" effect [2] was found in 2DEG in a *nano-fabric* of oriented triangular pores. The structure efficiently rectifies alternating lateral current by preferential scatter either towards the unit triangle apex or off the base. A ribbon of fabric is similar to anti-phase ramped sidewall modelled above (Figure 2). Unexpectedly the rectifying direction switches between north and south depending on voltage or carrier density (Fermi energy), and possibly on coherence length (*e.g.* via carrier mean-free path and surface specularity – which may change with wavelength). Explanations are controversial; but modelling by 3CPO found similar switching when the mean-free path (hence coherence length) are varied over wide temperature range and when extinguishing vectors *had changed direction* at a proximal surface. As surface orientations change within reach of L_ϕ, the statistics of momentum reversals appears to follow a phase transition.

High-angle grain boundaries were added by Knight & Smy [3] for greater realism. Long grains of high transmissivity retained a sizeable effect; but the expected spread in uncertainty may obscure it in unfavourable experimental conditions ([3],

Figure 4b), unless sufficient material purity and anneals reduce grain impact. Regular grain sequence remains to be investigated extensively for electron-optic waveguiding effects, where challenges to grain engineering are daunting. This is a juncture at which technology stands, and a firm handshake between the physics of wave mechanics and the engineering of electronic (and optical) materials is hereby encouraged.

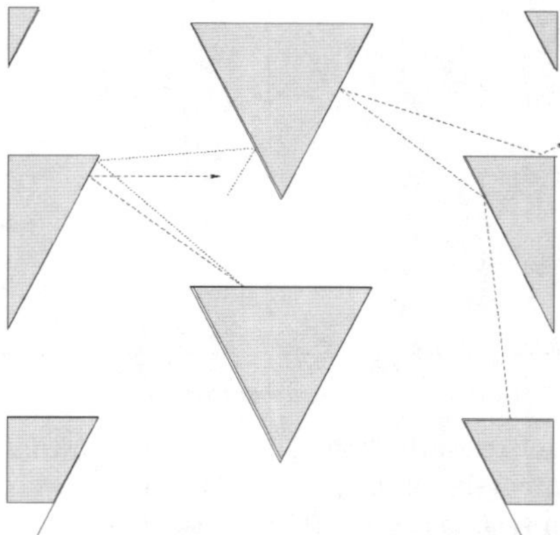

Figure 2. Nanofabric [2]. Rectification changes with mean-free path and can switch direction [12] with temperature.

Anderson-like Localization

We may compare the coherence effects of geometric roughness on conduction electrons to Anderson-localization in reduced dimensions, the latter is normally observed in disordered metals below liquid helium temperature. Anderson localization arises in conduction phenomenology for metal alloyed with impurities or heavily sintered with lattice defects, at very small mean-free path due to amorphous or sub-decananometer granularity. Below a critical temperature depending on disorder the resistivity begins to increase in an inverse T power law whose exponent is tied to so-called universal scaling or localization of the Anderson kind. It is inviting to relate this by analogy to the increase of resistivity – which also fits universal curves in thickness d – for very thin films and wires of relatively *pure* composition.[8]

A parallel analog for light exists. Light beams in fog produce from their coherent part a prominent albedo or backscatter of intensity at least four times that in any other direction of similar narrow solid angle. Coherent sources also produce coherent speckle backscattered in other random directions, from water droplets

---

[8]A suitable term for such coherent effects may be *Anderson-like* localization. At any rate, this is the term used here.

suspended in fog.[9] Prevalent in random media this *coherent* phenomenon is inevitable in strictly 2-D sheets with sufficient randomized scatterers and in 1-D lines with *any* non-periodicity.

Twenty years ago, experiments on amorphous Cu well below $10°K$ with an mfp of $\ell \sim 1nm$ showed a resistance increase of about $T^{-1.3}$ at sub-He temperature, where conduction channels become energetically less and less accessible at decreasing temperature. In very thin films the benefit to overall conductance of adding more material laterally is progressively lost to this energetic localization.

Energy restriction of carriers is a norm for pure crystalline metal; but in 3-D any *direction* restriction on momenta $\mathbf{k}$ (not counting the open Brillouin zone-crossing necks typical of noble metals) comes from geometric features including roughness. Macroscopically there is little scope for coherent interference of single electron wavelets, but as dimensions decrease in a thin film or wire surface, texture and degree of specularity play an increasing role in redirecting and limiting the available momenta.

Localization from increased self-coherent backscatter [9] via predominantly elastic scattering from nearby dislocations and other grainy imperfections may be *frustrated* or even undone by still other mechanisms that depend on coherence.[10] Paraxial electron gatherer/intensifiers or forward-coupled waveguide modes may be sought to frustrate that part of resistance induced by geometric (including randomizing) boundary roughness.

It is propitious to start with specular exterior surfaces, even if their relative orientations are random. So long as the size of facet is larger than the fermi wavelength and largely specular, it is feasible to model radiant momentum transfer by geometric optics. Diffractive effects most promising to counter localization are required to be added in the presence of texture periodicities or pseudoperiodicities. Even if such ordered facets far exceed λ_F what matters is that they be seen within reach of L_ϕ.

The approach is non-trivial since even simple optical waveguides are challenging to model when featuring irregular or asymmetrical corrugations. But the phenomenonology is ballistic in the sense of a large extent of coherence or correlation: even if the density of structural scatterers is high, they are largely *elastic*.[11] To this elastic feature of momentum transport is added a second: the exchangability of indistinguishable carriers. By this feature, perfectly elastic backscatter at a grain bound-

---

[9]This is localization of light radiance by mutual interference of coherent multiply-scattered rays from various droplets, each of which may be a "whispering gallery" to a single ray multiply scattered inside with phase preserving precision.

[10]Spin-orbit coupling is one mechanism altering the assumed independence of scatterers, most often probed using strong magnetic fields and the magneto-resistance.

[11]A dampening of the effects is, of course, also observed, due to a degree of inelastic scatter, but the inelastic relaxation time is considerably larger than the elastic, particularly for a cool temperature model – at about liquid nitrogen.

ary (rather than an external one) is conductive.[12] Of course, what subtracts from this conductivity is the loss of coherence by diffuse scatterers at boundaries or in the lattice.

Grain-Boundary Scattering

Grain boundaries can dominate metallic resistivity. Deposition conditions and anneals affect it, and specular though a surface might be, major grain effects remain. By 1965 Anderson [13] noted that despite extensive studies only monocrystal gold films showed a constant resistivity independent of thickness. By 1968 Mayadas [14] found the erstwhile constancy of $\rho_\infty \ell$ illusory, ρ_∞ decreasing with increased thickness. Mayadas *et al.* [15] found defect-free Al grains deposited at 200°C with implicitly specular bounding surfaces to exhibit this shift, and "since there is no compelling reason to believe that chemical purity is a function of thickness" [16] related the effect to increased grain size and developed with Shatzkes the MS model for columnar grains. MS, established 33 years after Fuchs [17] has widely been used for another 33; but the assumed constant impurity of growing grains has simultaneously been changing with awareness that the boundaries sweep up impurities.[13]

Intergrain potentials assumed in MS are zero-width delta functions of strength S^2 represented by a reflectivity $R/(1-R)$. Idealized smooth potentials would scatter bidirectionally were they all transverse to field and current, an idealization that masks deeper assumptions of random polar angles, and specularity of the boundaries nearly parallel to the current flow. Indeed coherent backscatter is tacitly assumed, and the R parameter may be reinterpreted to mean subtraction from the forward transported component. Figure 3 depicts the MS grain effect for nanowire.

Although R might approach 1 for highly correlated grain widths Mayades and Shatzkes note that "a periodic array of planes gives no resistance" (presuming there are no point scatterers in those planes). In a fuller account of diffractive effects, periods at a Bragg condition for deBroglie *or* group wavelenth may be refractive and others transmissive, with dependence on carrier density (Fermi energy) or temperature (modulating uncertainty of carrier wavelength). However, carriers that are elastically backscattered can be considered to contribute to momentum transport.[14] drawing analogy more closely between R and $q = 1 - p$, representing a lost specular part as in surface scattering. Thus Mayades and Shatzkes' statement should stand, mitigated by the known effect of impurities. This quasi-periodic hypothesis

---

[12]Labels are switched for carriers imaged in the boundary, a step that depends on a form of ergodicity and indistinguishability in quantum mechanics.

[13][18]: Impurities swept up by regrowing grains accumulate at boundaries where surface energy builds, eventually pinning competing boundaries to their final positions cementing grain morphology.

[14]This results from the continuation of phase memory and deceleration against the driving field, with an assumed background of indistinguishable electrons on the opposite side of a diaphanous boundary.

is still under experimental investigation for sidewall-textured Cu damascene nanowires.

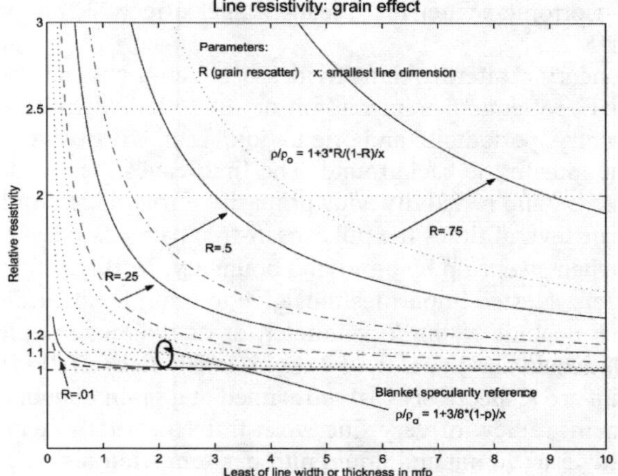

Figure 3. Increasing dominance of nanowire grain effect ($d \geq \ell$).

Loss parameter R is not incidence-sensitive in MS, an aspect that suits it to diffuse boundary scatter, even if not Mayades and Shatzkes' intent. The assumed specularity of parallel boundaries has support when these are geometrically smooth and lack effective coulomb scatterers. Ionized scatterers may be unimportant in macroscopic metal, but screening electrons confined by narrow bounding and rough surfaces are less mobile. Accordingly, DMR is expected (and found) in thin film and wire, being coupled to grain size and impurity levels.

If the scatter probability at a boundary is uniform with angle it must exist at low-angle grain boundaries found within the metal volume, not comprising the bounding surface treated separately by a Fuchs p or Soffer r parameter. MS was intended for no such cases, but it owes extended success to a second parameter, mean grain size D, and to anneals or high-temperature depositions that maximize this value – resulting in grains that do span a film or nanowire.

To progress, incidence-sensitive variable specularity similar to Soffer's is needed. A binary specularity could be applied, with angles of incidence below a critical angle θ_c diffusive and more grazing ones specular (R=0). This approach can be used to roughly approximate incidence at external surfaces, but before 3CPO it was not applied to interior grains. In 3CPO R is readily replaced by θ_c below which R=1, R=0 otherwise but it is as easy to treat angle-dependence with the more optical approach of Ziman and Soffer.

Such versions of parameter R may serve for as-grown films and deposited wires, but they make little difference to the annealed case where it is sufficient to treat incidence-sensitivity at exterior surfaces [1, 6]. This suggests that the diffuse scattering mechanisms at interior grain boundaries is isotropic, as with impurity scattering. It is even likely that the underlying grain scatter does not arise from boundaries *per se*, but from their accumulated impurities as noted in independent studies, a view that is not refuted by the known importance of the grain size since separation of isotropic scatter loci means longer effective free path in the separation direction.

Grain-boundary scattering thus divides between a coherent part from specular (and diffractive) reflection, transmission or refraction – whose directions depend on grain geometry, periodicity and orientation,[15] and a diffusive part from impurity scattering that adds to the background. The first leaves local conductivity substantially unaffected,[16] and resistivity adds primarily through the latter part when grains are shorter than several times the bulk mean-free path ($D < 2\pi\ell$). Even low impurity density, when swept-up by a moving boundary, readily produces a sheet of effective scatterers. Nested impurities thus give to grain boundaries diffusive scattering power and meaning to the R parameter, or its indirect complement t in the so-called multidimensional approach of Tessier and Toller (TTP) [19]. Fractions R or t of momentum are respectively lost or retained at a grain boundary.

In a landmark review of very fine work from the 1950s (with a film-thickness range at already 5 to 90 nm and liquid nitrogen temperatures) to studies in 1981 on epitaxially evaporated, $10^{-5}\%$ impure gold at <111> on mica and <100> on KBr, Sambles, Elsom and Jarvis [20] found that their excellent numerical model could explain high resistivity-ratios (between room and cryogenic temperature) of the large-grained films *only* by setting R=0.56 at the well-separated grain boundaries. This value, distinctly above the values 0.1 an 0.15 in small-grained films, led the authors to consider the alternative of inflated surface scattering effects. But even with Soffer's [22] superior incidence-sensitive specularity model, surface roughness had to be unreasonably high, and temperature dependence failed to represent data from over 500 temperature points (yet matched nicely with the R=0.56 hypothesis).

Such results on blanket Au films mean that while annealed grains may grow large (in this case several microns, much wider than the film is thick), reducing resistivity according to a grain-size denominator, the swept-up impurities counteract by increasing R or decreasing the t transmission parameter. In relation to film thickness, *cf.* [20] Figure 5, residual resistivity at liquid helium is seen to drop systematically as annealed grain size increases, consistent with reduced impurity density *in the lattice*.

Large grains cleaned of impurities can expose subtle surface character. In [20], increasing DMR (seen at variance with a Bloch–Gruneisen norm of TCR, *cf.* [20], Figures 4c, 5, 8c) in the case of large grains occurs as film thickness decreases.

---

[15]Grain orientation will have a certain refractive effect via changes in projected bravais lattice. This effect on resistivity is expected to be small unless it couples to another feature such as spin-selection in Cu/Co multilayers. Such alternating nanowires were found to be significantly resistive [21], but this may have been dominted by interface roughness.

[16]Although reflected, diffracted or refracted directions couple propagators to scattering surfaces and may cause deviations from Matthiessen's rule (DMR).

This behaviour, more visible where exterior surfaces are specular,[17] signifies that narrow films alter the phonon density-of-states distribution. Thin film and nanowire resistivities are generally non-parallel over a wide temperature range in their Bloch–Gruneisen (BG) plots (*e.g.* [20], Figure 3 shown to advantage in log–log form), with thinner film or wire declining in temperature to more elevated residuals and less kurtosis in the region of 20–80°K, where the TCR changes most sensitively for noble metals.

Non-parallelism and TCR variance share an explanation: the effect on phonon LDOS by reduced dimension or degrees of freedom. From cold temperature T up, the effect on TCR is to alter progression through *powers* of T: thinner films at 20°K tending from a lower power (than thick ones), which drops more quickly towards the asymptote of one. Thicker film dimensionality enables a high power law dependence on supply energy to continue into the upper energy regime (*cf.* [20], Figure 4).

The model that Sambles *et al.* [20] solved numerically for MS (high angle) grain boundaries and Soffer's surface specularity follows the above temperature dependencies remarkably, through effective mean-free path and competing grain size. Where grains span the film thickness as in most cases after anneal, this combined model serves well, as it also does in multilayers with planar interfaces (Sambles' *et al.* large-grain case, with an overlayer of micrograins), where separate conductive layers are amenable as an additive parallel network.

There are important deviations, however, in residual and kurtosis – where film thickness is not corroborated with the same precision as the functional form of temperature dependent resistivity.

Thickness determination by *in situ* resistance measurements at comparator temperatures, as used by Sambles *et al.* following von Bassewitz and von Minnigerode [23], has reappeared recently [24] to estimate nanowire width from the ratio:

$$\frac{L}{h}\Delta\rho_\infty / \Delta R$$

of difference in reference-point resistivities and measured resistances. It depends on reference temperatures spanning a linear region of the BG plot for a *width-unlimited* structure over a decade of resistance. With a temperature difference or resistivity ratio much lower, measured resistance may seem linear with temperature, but likely *not* at the linearity of the ∞ idealized structure. There may be "a temperature-dependent term arising from either surface or grain-boundary scattering [that] will make $\Delta\rho$ different from $\Delta\rho_\infty$"[18] and undermine applicability of the above ratio [20].

For film and wire with extensive TCR variation at cool temperature, the certainty of inferred thickness may drop with loss in linearity. Generally, films and wire of low resistivity ratio have less certain thickness in this method, while higher

---

[17]With Soffer's r or relative asperity of 0.1, an equivalent p is at least 0.65 in [20].
[18]The latter is the calibration basis of an inferrable reduced width.

ratio films (of large grain and surface size-effect) may deviate from Matthiessen's rule by coupled scatter processes. It is therefore worth adhering to independent verification as measurement temperatures drop. A method that successfully measured thin metal films as early as 1970 is the Fabry–Perot interference of X-rays [25] from the two surfaces. With pencil X-rays available and test damascene interconnects patterned in dense regularity, it is feasible to use low-angle diffracted beams to infer several widths at once: line, separation, barrier and seed widths.

In situ resistance measurement at temperatures is convenient, and will tend to cooler ranges. As efforts progress to enlarge and purify annealed grains in damascene trenches, it will be useful – in addition to high-temperature (as high as possible without activating further morphology change) – to measure resistance at a reference well below room or ice temperature. As thickness or width decreases a cool reference at about double the liquid nitrogen point would need raising, to avoid the encroaching nonlinear region from below. For thicker films of larger grain the trend is to overstate the true resistivity ratio and understate residual resistivity, exaggerating thickness – although the error is less marked across <100>-oriented grains. Overstating effective thickness is to overestimate room-temperature resistive contribution of grain boundaries and/or surface roughness.

Temperatures that affect TCR most dramatically (shown recently [26] to be a predictor of film quality) thus prove a sensitive indicator of model fitness. Elsom and Sambles [27] showed that the FS model fails in this regime by an unphysical p, but also in neglecting grain effects. The MS grain model goes far in redress, but even with a Soffer surface model deficiencies remain, including disparity of temperature-dependent data over grain-size and thickness interactions (DMR). The question is whether a model such as TTP [19] for grains in orthogonal (multidimensional) directions is needed. 3CPO offers a testbed capable of handling incidence-sensitivity consistently at *different* planes or low and high-angled boundaries. As a model 3CPO allows unlimited boundary conditions including interior boundaries. It avoids specifying integrals appearing in Sambles *et al.* such as:

$$\left\langle \int_0^1 K(\mu,\phi)/H(\mu,\phi)d\mu \right\rangle$$

where $K(\mu,\phi)$ is the Fuchs kernel with a $p(\mu)$ dependent on incidence following Soffer, and $\kappa = d/\ell$ replaced by $\kappa H(\mu,\phi)$, with H being $1+\alpha D/\chi(\mu,\phi), D$ the grain width in the (μ,ϕ) direction, χ its projection in the axial direction, and where $\langle Q \rangle$ means a moment in polar angle ϕ into the direction of momentum flow:

$$\langle Q \rangle = \frac{1}{\pi}\int_{-\frac{\pi}{2}}^{\frac{\pi}{2}} Q\cos^2\phi d\phi \cdot$$

In Sambles *et al.* this integral extends the MS grain scattering model to incidence-sensitive specularity, offering no such sensitivity at grain boundaries or difference at distinct surfaces; yet it was complicated enough to require numerical methods applied for the first time by Sambles *et al.*

3CPO, by contrast, allows reflection conditions of all kinds – including main diffraction orders – at each surface or boundary encountered along arbitrary momentum propagators from which an integral of attenuated coherence is constituted. Grain boundary angles are unlimited, inviting a comparison between isotropic scattering by impurities and coherent processes like inter-grain refraction. Geometric textures induced by exterior grain facets enter as boundary conditions that would be intractable for closed integrals.

Without diffractive effects beyond incidence-sensitivity, 3CPO shows that high-angle grain boundaries alone add little character to resistivity except to raise it *and* the bars of uncertainty [3]. More specular boundaries mean less resistive and variable grain effects. An apparent conflict with MS when R is interpreted as bare reflectivity disappears by considering R a measure of diffuse scatter at a boundary (whether forward- or backscatter) and its effective complement t (as in TTP) a measure of preserved coherence. Whether carriers are reflected or transmitted, their momentum is coherently transported: were it not, high-angle grain boundaries would always migrate in a field.

Conclusion

3CPO and beam-propagation methods offer unique means to model, characterize and design film or line surface with nano-engineered anisotropic texture, periodic grain morphology and purified boundaries.

Grain periodicity is amenable to the MS and TTP treatments and other legacy models. Periodic surface undulation without diffraction was also treated by Namba [7], directly applying Fuchs–Sondheimer [17, 28], and it would be useful to include Soffer's [22] incidence-sensitivity when zero correlation-length is separable from non-zero spatial period. But, even with that addition to the MS model – as should be done in extending Sambles' *et al.* approach – no model has built-in awareness of coherent effects. Without them, nothing can be said about novel device properties expected from marshalling the wave-nature of conduction electrons in radiant momentum transport.

Resistivity reduction techniques likely to be fruitful will include: specular barriers for damascene trenches (likely omitting Ta or other metals[19] over TaN), enlarging grains *e.g.* by laser-guided ablation shown to create bamboo structures, stress-release of grains with bossed sidewalls and removal of impurities from grain-boundaries in annealing – by gettering with volatile agents and outgassing from hot-anneal cycles. When techniques for purifying grain boundaries develop, grain-size goals may be relaxed, as specular grains are highly conductive (*e.g.*

---

[19] Ernur [31] appears to have obtained quite specular sidewalls by using TaN barriers without a Ta overlayer.

Sambles *et al.* [20]: even the small-grained Au films at <100> orientation had small *R* values).

In addition to these challenges, the possibilities of frequency-selective surfaces and grain periods that may enhance conduction of momentum, as outlined in the introduction. A particular test of periodic nature was the anisotropic surface texturing of damascene sidewalls, which 3CPO forecast to affect conduction even without a *full* diffraction model.

A closed diffractive solution for corrugated surface with a canted louvre angle was obtained [8], indicating anisotropy in the lowest orders fore and abaft. Non-diffractive numerical analysis established asymmetry under right-angled wedges (or rect-echellia) in a Littrow configuration, and found a grating angle of maximum effect (*cf.* Figure 1) to be ~11° or a ramp of 1:4. A virtual surface of repeating wedges was then simulated [3], finding 10% discrimination between forward and reverse conductivity when incidence-sensitivity as ersatz for diffraction was included. Experimental structure was proposed to verify a magnitude for such an effect.

A recent e-beam experiment [29] at KU Leuven imprinted SiO sidewalls with nanoscale texture of opposing mirror-image surface anisotropy in the same direction. The Ta barriers used in a standard process are unfortunately not specular [30], and yielded high, isotropic resistivities of 2.5–3.2 $\mu\Omega$•cm. However, together with notable grain effects there *was* evidence of period-sensitivity. In 10 nm textures on test-widths 70 and 100 nm in high-aspect (375 nm) trenches, a trend of conductance increase with spatial frequency suddenly dips for 70 nm and jumps for 100 nm lines at the intermediate pitch of 40 nm. Thus the increased mean Cu section is pointedly overwhelmed by a preferred size of grain, more of them pinning at 40 nm within 70 nm lines. In 100 nm lines 40 nm boss gaps may encourage three would-be grains of 100 nm to form 140-nm twins, reversing the resistance trend.

Proximity of 80 and 100 nm in the latter lines may also explain conductivity increase in 20 nm textures when pitch is 80 nm, versus symmetric double-scoop control textures: non-controls of smaller mean section being less resistive in 100 nm lines (but not less resistive than controls in 70 nm lines). The enhancement may evidence grains more regularly sized at 80 nm than they are distributed about 100 nm. To be validated this possible finding of periodicity effect requires further inspection in destructive sections (after planned low-temperature measurements). Clearly many more experiments of this kind are needed, and above all, specular improvements to Cu-diffusion barriers (*e.g.* with TaN alone) would clarify effects.

Experimental development of e-beam textured sidewalls supports several ends. It will perfect this lithographic method, determine suitable boss height and cant to keep the seed layer intact while reasonably faithful to boss profile, and identify geometry and resolution for best effect. Texturing is worth pursuit to pin grains at expansion joints for stress-relief on anneals – improving laminar integrity or sidewall adhesion – and provide vertical channels for outgassing or gettering impurities.

The *raison-d'être* of texture and periodic grain is to employ select periods and sidewall (boss) profiles to harness diffractive effects on carrier wavefunctions. Even if asymmetric conductivity is a challenging goal, this approach to quasiperiodicity of surface roughness, interacting as suggested in the above experiment with grain-size, should affect resistivity markedly.

For optical waves, correlation-length effects of random roughness have been modeled [5] in waveguide or fibre microbends. These indicate that texture correlation on a scale with mean-free path or coherence length may cause significant differences in scatter intensities. Provided sufficient surface specularity to attain substantial electron coherence length at certain periods and accessible cool temperatures, resisitivity both raised and lowered should therefore be looked-for in future.

References

1. GD Knight, T Smy, C Crisan: Calculating current flow in deep-submicron and nano-scale metal structures Proc AMC 2001 371-8 (Montreal) (2001)

2. AM Song: Electron ratchet effect in semiconductor devices and artificial materials with broken centrosymmetry Appl Phys A 75 229-235 (2002); AM Song *et al.* Appl Phys Lett 79, 1357-59 (2001)

3. GD Knight: Boundary texture and scattering: practical directions to a soft diode. Microelec Eng 76, 137-145 (2004)

4. Wen Wu: Fabrication and transport properties of ultra-fine copper interconnects Ph.D. dissertation, KU Leuven, Leuven (2004-2005)

5. KK Lee *et al.*: Effect of size and roughness on light transmission in a Si/SiO_2 waveguide Appl Phys Lett 77, 1617-1619; 2258 (2000)

6. GD Knight, T Smy: Modelling roughness, grain and confinement effects on transport in embedded metallic films. Microelec Eng 64, 417-428 (2002)

7. Y Namba: Electrical conduction of thin metallic films with rough surface. J Appl Phys 396117-18 (1968); Japan J Appl Phys 91326 (1970)

8. GD Knight: An all-metal diode Poster competition, TEXPO 2002 (Micronet-CMC Ottawa) (2002)

9. GD Knight, T Smy: 3CPO and carrier backscatter in periodic Cu grainsIn – Ming Yang (Ed.): Electronic proceedings ISTC 2002 (ECS Tokyo), p. 295-304 (2002)

10. GD Knight, T Smy: Electron wave diffractive transport in Cu interconnects: deep sub-micron model EUROMAT 2003 A2 (Lausanne, unpublished) (2003)

11. P Beckmann and A Spizzichino: The Scattering of Electromagnetic Waves from Rough Surfaces (MacMillan, New York) pp 58–67 (1963)

12. A Logfren *et al.*: Quantum behaviour in nanoscale ballistic rectifiers and artificial materials. Phys Rev B 67 195309 1-7 (2003)

13. JC Anderson: The use of epitaxial films in physical investigations In:The Use of Thin Films in Physical Investigations, NATO A.S.I: London 1965 (Academic Press, London New York) pp 1–7 (1966)

14. AF Mayadas: Intrinsic resistivity and electron mean free path in aluminum films J. Appl Phys 39, 4241-45 (1968)

15. AF Mayadas, R Feder, R Rosenberg: Resistivity and structure of evaporated aluminum films. J. Vac Sci Technol 6, 690-93 (1969)

16. AF Mayadas, M Shatzkes: Electrical-resistivity model for polycrystalline films. Phys Rev B 1, 1382-1389 (1970)

17. K Fuchs: The conductivity of thin metallic films according to the electron theory of metals. Proc Camb Phil Soc 34, 100-108 (1938)

18. W Wu *et al.*: Influence of surface and grain-boundary scattering on the resistivity of copper in reduced dimensions Appl Phys Lett 84 3838-40 (2004) S Brongersma *et al.* Copper grain growth in reduced dimensions IITC 2003 Proc (ieee) 3.10, 48-50 (2004)

19. CR Tellier, CR Pitchard and AJ Tosser: Statistical model of electrical conduction in polycrystalline metals thin. Solid Films 61349-54(1979)

20. JR Sambles, KC Elsom and DJ Jarvis: The electrical resistivity of gold films. Phil Trans R Soc Lond A 304365-96(1982)

21. PR Evans *et al.* Appl Phys Lett 76, 481-3 (2000)

22. SB Soffer: Statistical model for the size effect in electrical conduction. J Appl Phys 38, 1710 (1967)

23. A v.Bassewitz and G v.Minnigerode: Zeits f Phys 181368-90(1964)

24. G Steinlesberger *et al.* Impact of annealing on the resistivity of ultrafine Cu damascene interconnects. Mat Res Soc Symp Proc 766: E4.2 379-84 (2003)

25. F Abelès and V Van Nguyen: J Physiol, Paris C 179-84 (1970)

26. A v.Glasow *et al.* Using the temperature coefficient of the resistance (TCR) as early reliability indicator for stressvoiding risks in Cu interconnects IEEE Proc IRPS 41, 126-31 (2003)

27. KC Elsom and JR Sambles: Macroscopic surface roughness and the resistivity of thin metal films. J Phys F: Metal Phys 11, 647-56 (1981)

28. EH Sondheimer: The mean free path of electrons in metals. Adv Phys (suppl to Phil Mag) 1, 1–18ff (1950)

29. W Zhang, W Wu, GD Knight *et al.* (IMEC, unpublished) (2004)

30. SM Rossnagel: Characteristics of ultrathin ta and TaN films. J Vac Sci Technol B 20, 2328-36 (2002)

31. D Ernur: Narrow trench corrosion of Copper damascene interconnects. Japan J Appl Phys 417338-44(2002); resistivity data to be published (private communication) (2004)

Advanced Barriers for Copper Interconnects

M. Hecker[a], R. Hübner[a], J. Acker[a], V. Hoffmann[a], N. Mattern[a], R. Ecke[b], S.E. Schulz[b], H. Heuer[c], C. Wenzel[c], H.-J. Engelmann[d], E. Zschech[d]

[a] Leibniz-Institute for Solid State and Materials Research, Dresden / Germany

[b] Chemnitz University of Technology, Center of Microtechnologies, Chemnitz / Germany

[c] Dresden University of Technology, Dresden / Germany

[d] AMD Saxony LLC & Co KG, Materials Analysis Department, Dresden / Germany

Introduction

Performance and reliability of leading-edge microelectronic products are increasingly determined by design, technology and materials for on-chip interconnects. Inlaid copper interconnects, embedded into insulating material with continuously decreasing permittivity, will be used at least for the next decade to fulfill the fundamental requirement to transmit signals with high speed. Within the inlaid interconnect process sequence, a liner is deposited into the etched trenches and contact holes (vias), and subsequently, the structures are filled with copper. This liner protects the interlayer dielectrics and active transistor regions from copper atoms; *i.e.* it acts as a diffusion barrier.

To increase the performance of microelectronic devices like high-performance microprocessors, not only the sizes of transistors and interconnects have to shrink but also the barrier films must scale down. For the introduction of new, porous dielectrics and structures of high aspect ratios, improved diffusion barriers with thicknesses reducing from presently ~10 nm down to 2 nm in 2018 [1] and high conformality are mandatory. With reduced sizes the interfaces of interconnects also become increasingly essential for their electrical and reliability properties. High-quality smooth interfaces are prerequisites both for reducing electron scattering which limits the conductivity, and for minimizing diffusion at the interfaces, which is deleterious for electromigration lifetime of interconnects.

One approach to guarantee that these ultra-thin films with sub-10-nm thickness act reliably as diffusion barriers is to optimize their constitution and microstructure as well as their deposition and anneal parameters. The microstructure of the barrier layers becomes more and more critical with continuous scaling-down of the film thicknesses. Structural defects in barriers like grain boundaries or voids can cause short circuit diffusion and give rise to barrier inefficacy. Therefore, knowledge about the *mechanisms* leading to barrier failure and their correlation to the microstructure is crucial for the progress in interconnect technology. In this paper, the microstructure and degradation mechanisms of Ta- and W-based barrier systems are studied by complementary methods.

Demands on Diffusion Barriers

Apart from the obvious task of preventing the diffusion of mainly Cu, but also other elements like Si or process gases, there are additional requirements for diffusion barriers in Cu metallization. In particular, they should [2]:

- Possess high electrical and thermal conductivity,
- Act as nucleation layer for subsequent Cu deposition,
- Provide good adhesion to Cu and dielectric layers with smooth interfaces,
- Exhibit high mechanical and thermo-mechanical stability, in particular for porous dielectrics, at low internal stresses.

Therefore defect-free, homogenous thin barrier films are needed, which are stable under thermal stressing, form high-quality interfaces to the Cu and yield no detrimental diffusion paths for electromigration processes. Furthermore, the barriers should not, or as little as possible, contribute to an increase of the effective permittivity k of the dielectrics, which becomes increasingly important with the future introduction of low-k materials.

Barrier Materials

Diffusion barriers have been widely prepared on the basis of refractory metals like Ti, Ta or W [3], which are operating at low homologous temperatures leading to high activation energies for diffusion, and which do not react with Cu. Whereas in a single crystal a low diffusion coefficient is achieved, single-crystalline growth is difficult to obtain in real barrier films. In polycrystalline films grain boundaries can provide fast diffusion paths. One way to improve the barrier performance is to add light elements like N to the barrier, which can reduce the density of fast diffusion paths by microstructural changes like "stuffing" of the grain boundaries [4]. An elimination of the polycrystalline structure in favor of an amorphous one can be achieved by choosing proper process parameters and alloy compositions [5], *e.g.* by adding Si to the barrier material. Such amorphous barrier structures can prevent diffusion very effectively and provide smooth interfaces to the adjacent films. One issue, however, is their thermal stability. Heat treatment can again lead to grain nucleation and growth. Furthermore, the various properties required for diffusion barriers also have different, sometimes opposite dependencies on their composition. For example, the electrical conductivity of thin barrier films decreases with increasing nitrogen content, whereas their thermal stability increases. Thus it is necessary to find a compromise between highly stable amorphous structures and low electrical resistivity of the barriers.

One of the presently predominant industrial solutions for diffusion barriers in copper metallization is the bilayer system TaN/Ta. On top of a TaN underlayer providing good adhesion to SiO_2, Ta grows in the low resistivity b.c.c. phase (α-Ta, $\rho \sim 25$ $\mu\Omega$cm), whereas directly on the dielectric the high resistivity tetragonal β-Ta phase ($\rho \sim 170$ $\mu\Omega$cm [6]) would grow. In present technology, this system has proven excellent barrier properties. However, its adequacy in technology nodes requiring barrier thicknesses of 5 nm and below is questionable [1]. For example, the α-Ta has a polycrystalline structure containing undesired grain boundaries. Under thermal stressing, a redistribution of N towards a homogeneous chemical depth profile can occur in such bilayers, leading to a significant loss of thermal stability [7]. Furthermore, with decreasing size of the TaN underlayer, at the sidewalls of thin trenches a tendency of growing the unfavorable β-Ta phase has been observed [8]. Thus, there is ongoing research for alternative barrier systems.

Deposition of the Ta- and W-based Barriers

Widely-used techniques of barrier deposition are different variants of physical vapour deposition (PVD) as in particular modified magnetron sputtering techniques, but also arc deposition, and of chemical vapour deposition (CVD), for example, metal organic (MO) CVD or plasma-enhanced (PE) CVD. As an emerging technique appropriate to deposition into very narrow trenches with high-aspect ratios, the atomic layer deposition (ALD) is under development.

In the present study long throw magnetron sputter deposition was utilized to obtain Ta-Si-N barriers of different composition, and a PE-CVD process was developed for the preparation of W–(Si–)N barrier layers. By tuning the composition of the barriers, their performance was optimized.

The Ta–Si–N barriers were deposited onto thermally oxidized (100) Si wafers containing an about 140-nm-thick SiO_2 film. The deposition of the 10-nm-thick barriers was performed by rf magnetron sputtering, with a N_2 flow adjusted between 1 and 4 sccm, resulting in compositions of the barrier films according to Table 1. Without interrupting the vacuum, a 50-nm-thick Cu film was dc magnetron sputtered on top of the barriers.

Table 1. Designations, compositions as determined by RBS, and resistivities ρ of the Ta–Si–N barriers.

Sample	N_2 flow [sccm]	Composition [at.%]	ρ [$\mu\Omega$ cm]
TaSiN0	0	$Ta_{73}Si_{27}$	215
TaSiN1	1	$Ta_{62}Si_{20}N_{18}$	215
TaSiN2	2	$Ta_{56}Si_{19}N_{25}$	220
TaSiN3	3	$Ta_{41}Si_{20}N_{39}$	370
TaSiN4	4	$Ta_{30}Si_{18}N_{52}$	1700

One drawback of sputter deposition is the difficulty to obtain highly conformal films also in narrow trenches. Though the conformality of PVD techniques can be improved, *e.g.* by long-throw ionised sputter deposition, an alternative approach is achieved by CVD techniques. Whereas thermal CVD suffers from relatively high process temperatures (~750°C), lower deposition temperatures are possible by plasma enhancement. For deposition of thin W-based barriers, in this study a PECVD process based on $WF_6/H_2/N_2$ gases was developed. The WF_6 gas, which is reduced by H_2, acts as tungsten source, and N_2 delivers the nitrogen incorporated into the barrier. W–(Si–)N films were deposited with 10-nm thickness onto SiO_2 and covered with a Cu cap layer (80-nm nominal thickness). The process was first optimised for the binary W–N barriers (deposition temperature 400°C, gas flow ratios H_2/WF_6: 80, N_2/WF_6: 80), yielding homogeneous films with a resistivity of 210 $\mu\Omega$·cm and a composition of 80 at% W and 20 at% N. Subsequently, by a process modification using SiH_4 as Si supplier, the process was advanced to deposit ternary W–Si–N barrier films (*cf.* [9]).

Detection of Barrier Degradation

Though the barriers for Cu interconnects become thinner and thinner, their diffusion-preventing properties must be even better than in the former Al technology. Cu is an extremely fast diffuser in Si with an diffusion coefficient about 15 orders of magnitude higher than that of Al for a temperature of 500°C [10]. Since Cu also remains mobile at room temperature, it can diffuse over large distances from a local source, whereas Al accumulates at the barrier-substrate interface close to a barrier defect. Therefore, not only the structural characterization of ultrathin barrier films but also the detection of Cu that has diffused through a barrier poses a challenge for analytics.

To accelerate thermally activated processes and to investigate the mechanisms of barrier degradation, the dielectric/barrier/copper film stacks were annealed at several temperatures between 400°C and 700°C in vacuum ($p \approx 10^{-4}$ Pa). As an alternative atmosphere, H_2 gas was tested. For anneals of uncapped W–N samples at 450°C in H_2 gas, a reduction to pure W films was observed, accompanied by delivery of N_2. Also, a Cu cap could not prevent this process satisfactorily. Therefore, the thermal treatments were performed in vacuum. For each anneal a dedicated sample was used.

To investigate the microstructure and their annealing-induced changes which could lead to barrier failure, a series of different and also complementary methods was utilized. By X-ray reflectometry (XRR) and glancing-angle X-ray diffraction (XRD), the structural features of layers and their interfaces as well as the amorphous and crystalline order in the films, respectively, were analyzed on a macroscopic scale using a Philips X'Pert diffractometer with Cu-K_α radiation and thin film equipment. Elemental depth distributions of the layer stacks were obtained by glow discharge optical emission spectroscopy (GDOES) with a modified LECO-GD750 tool yielding a resolution of about 2.5 mm according to the crater diameter resulting from Ar ion sputtering. For high spatial resolution, cross-sectional transmission electron microscopy (TEM) was performed with a FEI Tecnai F30 microscope. To determine the onset of Cu diffusion through the barrier, different techniques were applied additionally (see below). In particular, by using atomic absorption spectrometry (AAS) not only a high sensitivity but also a quantification of the Cu traces in the substrate was obtained. For that, the Cu cap had to be removed by etching, and barrier, dielectric and substrate were dissolved in concentrated acids and analyzed separately.

Ta–Si–N Barriers

The amorphous structure of a TaSiN0 barrier is illustrated by the broad scattering maximum at $2\theta \approx 38°$ in the lower curve of Figure 1. At increasing N content in the TaSiN films, the amorphous character of this peak was retained, and a shift of its position towards lower angles indicates a modified short range order, particularly of the Ta atoms which are the strongest scatterers. Additional crystalline reflections in the diffraction pattern are due to the Cu cap film. According to the applied diffraction geometry with grazing incidence angle ($\alpha_i = 2°$), the intensive 220 reflection at $2\theta \sim 74°$ corresponds to a predominant <111> copper texture component, which was found to reduce slightly with increasing N content in the TaSiN films. A quantitative estimation on the basis of pole figure cuts yielded volume fractions of 68% and 16% for the preferred Cu <111> component in the cap layers of the TaSiN0 and TaSiN4 barrier systems, respectively. Additionally, slightly enhanced <511> twin components were observed, and the distribution of the remnant fraction was statistically.

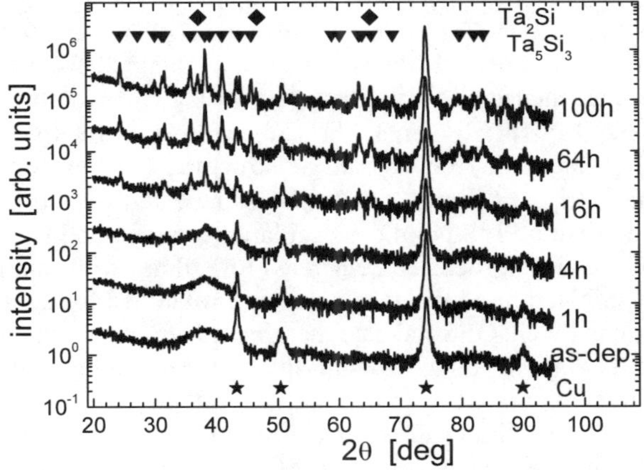

Figure 1. X-ray diffraction patterns of a 10-nm-thick $Ta_{73}Si_{27}$ barrier (TaSiN0) between Cu and SiO_2 after anneals at 600°C for different annealing times.

To analyze the thermal stability of the barriers, the samples were measured after anneals in vacuum performed at a temperature of 600°C for different times t_{an}. Figure 1 shows the stability of the amorphous state of the TaSiN0 barrier after a 1 h anneal, the onset of the crystallization at annealing for 4 h, and the subsequent increase of the crystalline fraction (Ta_5Si_3) for annealing up to $t_{an}=100$ h. Furthermore, a second phase (Ta_2Si) begins to crystallize during the 100 h anneals.

In analogy to the case of binary Ta–N barriers, where a nitrogen addition increases the thermal stability of the barriers [11], also the ternary Ta–Si–N barriers can withstand longer anneals without crystallization compared to the TaSiN0 sample. As the first crystalline phase in Ta–Si–N films, nitrides are obtained (Ta_2N for TaSiN1 and TaSiN2; Ta_5N_6 for TaSiN3), which are in the case of TaSiN1 followed by a silicide (Ta_5Si_3 firstly observed after t_{an} = 64 h). For the barrier system with the highest N content (TaSiN4), no signs of crystallization were observed for anneals at 600°C up to t_{an} = 100 h. An overview about the onset of phase formation is given in Table 2.

It should be noted, that without a Cu cap significantly higher stabilities against barrier crystallization were observed. One explanation is the formation of an oxidized sublayer at the surface of uncapped barriers. The oxygen can stabilize the amorphous barrier structure. Additionally, an acceleration of the crystallization by an influence of Cu atoms in case of the Cu-capped barriers cannot be excluded.

Whereas the wide-angle X-ray diffraction is sensitive on a scale corresponding to the crystal lattice, properties of the film stacks like layer homogeneity, thickness and interface morphology are revealed by XRR techniques in the low-angle region. Figure 2 compares XRR measurements of the barrier systems TaSiN0 and TaSiN4. In the as-deposited state, all Ta–Si–N films show pronounced oscillations with a wavelength of about 0.4° corresponding to the strong contrast between the barrier and the surrounding films. Additionally, short-period oscillations arise from the Cu layer. From fitting of calculated to the measured XRR curves, the thicknesses of the individual layers, their densities and r.m.s. interface roughnesses were determined. The thicknesses of the barrier films ranging between 10.4 nm and 10.8 nm were close to the nominal thickness of 10 nm obtained from the deposition parameters. The mass densities varied between 14.1 g/cm³ (sample TaSiN0) and 10.0 g/cm³ (sample TaSiN4). Whereas the Cu surface was relatively rough (σ_{rms} =2.4 nm) probably due to its polycrystalline structure, the low roughness of the barrier-Cu interface (σ_{rms} =0.1 nm ... 0.3 nm) is indicated by the intensive XRR oscillations. In the case of TaSiN0 these oscillations correspond to a homogenous barrier film with constant density, whereas for films with high N content, particularly for TaSiN4, a bimodal layer structure is obtained for the barrier. Depth profile investigations by GDOES indicate that this substructure can be related to an accumulation of N and Si in a ~0.7-nm-thick upper sublayer of the barrier, accompanied by a reduction of the Ta content.

After the anneals, a striking difference between the XRR curves of the two samples occurs. The barrier oscillations of the TaSiN0 system diminish at annealing corresponding to a changing layer set-up, whereas the remaining high contrast in the case of TaSiN4 indicates a stable layer stacking. Formally, the contrast reduction of the TaSiN0 sample can be related to a change of the r.m.s. interface roughness from σ_{rms} = 0.3 nm for the as-deposited film to σ_{rms} =0.5 nm after annealing for 1 h and to σ_{rms} = 0.7 nm after annealing for 4 h.

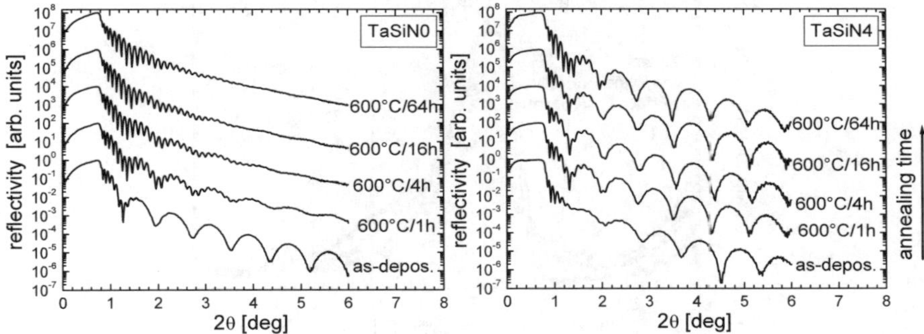

Figure 2. XRR measurements of the barrier systems TaSiN0 (left) and TaSiN4 (right).

More details of this different development of the interface morphology are revealed by local TEM analysis (Figure 3). For the nitrogen-free barrier system, the Ta_5Si_3 crystallites grow far into the Cu cap, leading to dissolution of the former barrier layer in agreement with the XRR results (Figure 2a) and thus enabling the Cu to come into direct contact with the SiO_2. In contrast, the Ta_2N crystallites, which grow as the first phase in the N-containing barrier layers, are extended within the barrier region. For the TaSiN4 sample neither changes of the barrier morphology nor crystallite formation are observed for heat treatments up to 100 h at 600°C (Figure 3b).

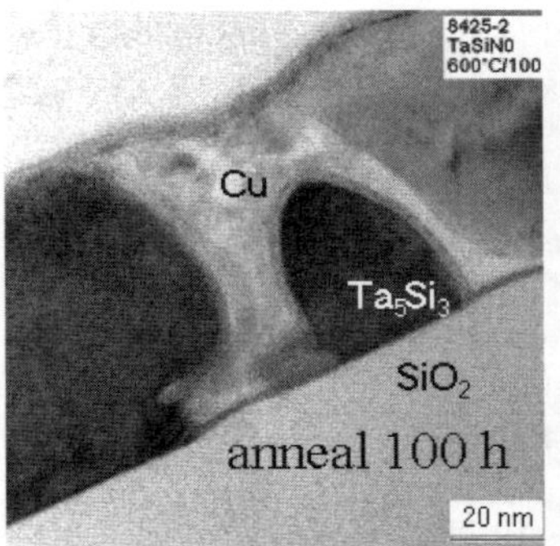

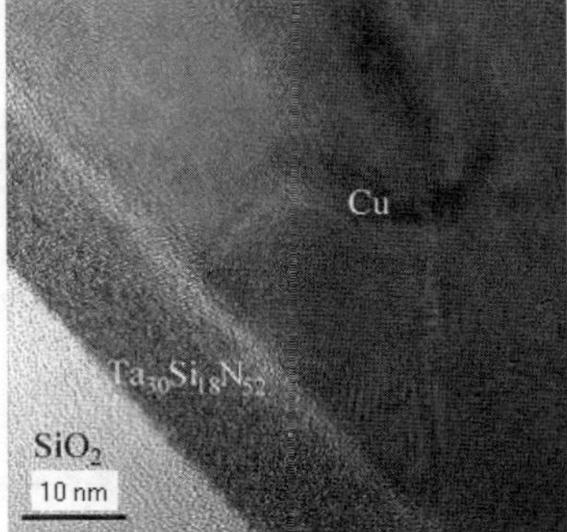

Figure 3. TEM cross sections of the barrier systems TaSiN0 (0 at% nitrogen, left) and TaSiN4 (52 at% nitrogen, right) after annealing at 600°C/100 h.

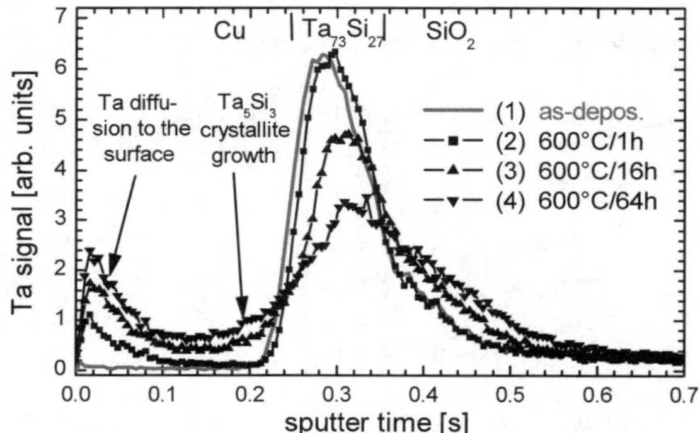

Figure 4. GDOES Ta depth profile of the TaSiN0 barrier system after different anneals at 600°C.

The anomalous Ta_5Si_3 crystallite growth is also striking since this phase does not reflect the composition of the nitrogen-pure barrier layers, which contain much more Ta. Using GDOES depth profiling, the traces of the surplus Ta can be revealed (Figure 4). In particular, the composition of such barriers changes by diffusion of Ta from the barrier through the Cu layer to the sample surface already prior to the onset of crystallization. Since an additional passivation layer on top of the Cu layer reduces the Ta diffusion [12], this process could be explained by the strong affinity between Ta and O_2 occurring in traces in the furnace. With increasing N content in the barriers, the tendency of the Ta outdiffusion decreases, as was also observed for binary Ta–N barriers [11]. Taking into account the steep increase of the resistivity of Ta–Si–N barriers with nitrogen contents above 25 at%, thus a suitable barrier system is represented by the TaSiN2 composition with still very high thermal stability and moderate resistivity properties.

Table 2. Annealing times for the onset of first phase formation and of diffusion towards the surface for barrier systems annealed at 600°C.

Sample	N-content [at%]	Phase formation [h]		Diffusion [h]	
		Ta nitrides	Ta silicides	Ta ↑	Si ↑
TaSiN0	0	-	4 / Ta_5Si_3	1	-
TaSiN1	14	16 / Ta_2N	64 / Ta_5Si_3	1	32
TaSiN2	26	8 / Ta_2N	-	8	8
TaSiN3	38	32 / Ta_5N_6	-	32	32
TaSiN4	51	-	-	-	-

W–N and W–Si–N Barriers

By optimisation of a PECVD process with the chemistry $WF_6(Ar)/N_2/H_2$, homogeneous binary W–N films with 10-nm thickness, a nitrogen content of ~ 20 at% and an amorphous microstructure were deposited onto SiO_2 and capped by a 80-nm-thick Cu film. The Cu cap was deposited either by CVD without interrupting the vacuum or by PVD, whereby the CVD deposited Cu films showed a higher content of statistically orientated grains and a weaker preference of the <111> texture component. From XRR investigations, a density of the barrier films of ρ = 16 g/cm³ was obtained, which is larger than that of the Ta–N films corresponding to a higher density of bulk W (19.3 g/cm³) compared to Ta (16.6 g/cm³). The XRR curves also show that the contrast of the Cu/W–N film stack is lower than of a sputtered Cu/TaSiN stack (Figure 5); *i.e.*, the r.m.s. roughness of the Cu/barrier interface is higher for the W–N films ($\sigma_{rms} \approx$ 0.8 nm).

During thermal stressing up to 550°C/1 h, no structural changes in the W–N–films were observed. Also a long-term anneal at 400°C/100 h left the amorphous barrier structure unchanged. First signs of barrier crystallization appeared at 600°C/1 h anneals, where weak reflections of α-W and β-W₂N arise. At further thermal stressing the crystallization proceeds. The intensity of the crystalline phases corresponds to the initial composition of the barriers, and also by depth profiling no compositional changes are observed. Since the crystallites of the W–N barriers grow preferentially within the barrier region, they leave layer stacking and interface morphology unchanged, what is favourable to the barrier performance.

As an issue for depth profiling, a strong tendency of the CVD-Cu layers to agglomerate occurred for anneals at 600°C for t_{an} > 4 h. This process complicates the interpretation of depth profiles due to the stepped surface shape for laterally macroscopic techniques like GDOES. However, after removal of the surface steps by Cu etching, clear GDOES depth profiles were obtained showing no increased Cu signal within the substrate, *i.e.* a possible onset of Cu trace diffusion remained below the GDOES detection limit of ~ 5 µg/g Cu in the substrate [13].

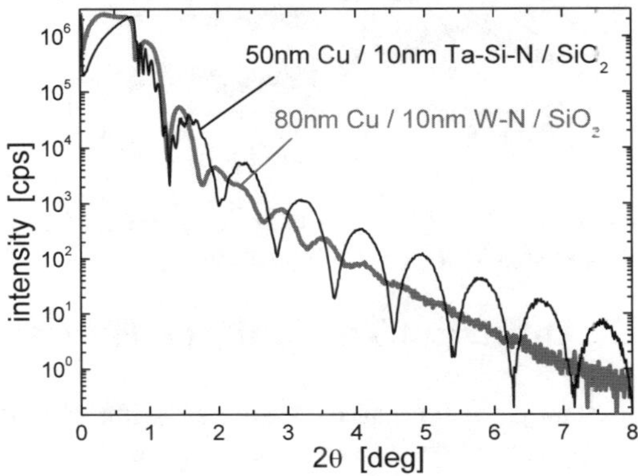

Figure 5. Comparison of XRR curves of a TaSiN1 and W–N film stack (20 at% N) with Cu cap layer. In the designation the nominal thicknesses are given.

Though the thermal stability of the binary W–N barriers is higher than that of a binary Ta–N barrier with 20 at% N content (*cf* [11]), it is below that of ternary Ta–Si–N barriers. Therefore, also films deposited with additional silane flow to obtain W–Si–N barriers were evaluated. To obtain effective barrier films by this process, a compromise between several deposition parameters had to be obtained. For example, the resistivity of the Si-containing films is significantly increased compared to binary W–N. Partially, this increase can be mitigated by lowering the N_2 flow. However, at too low N content in the barriers, their thermal stability is reduced. Furthermore, the films tend to form a bilayer structure with a W-poor lower sublayer, if the silane addition occurs simultaneously to the WF_6 inlet. Since the nucleation of W needs a relatively long incubation time, a more homogeneous barrier composition can be obtained by a retarded silane inlet. As an example, Figure 6 represents diffraction patterns obtained from a W–Si–N film deposited with a SiH_4/WF_6 ratio of 2 yielding a resistivity of 350 $\mu\Omega$•cm. The same crystallization products as in the case of binary W–N–films are observed, however at higher temperatures with first signs of crystallization not below 650°C/1 h anneals. A higher content of Si in the barriers leads to a steep increase in resistivity up to 600 $\mu\Omega$•cm. Therefore, further process optimization is promising in the region of low N_2 and silane flows, *e.g.* by varying the Ar flow during deposition, which has a strong impact on the film amorphization.

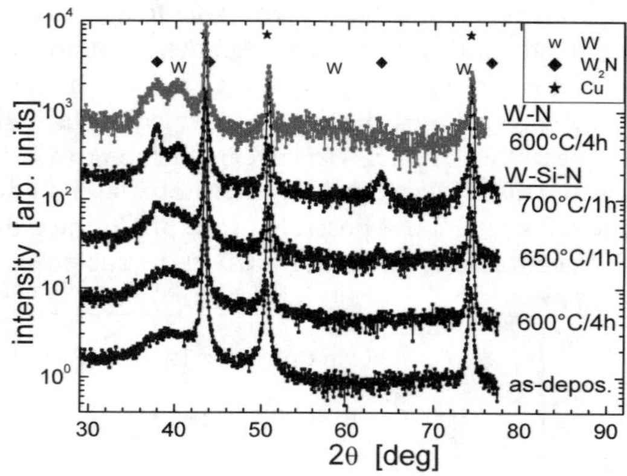

Figure 6. Diffraction patterns of W–Si–N barrier systems annealed at different temperatures in comparison with an annealed W–N system (upper curve).

Detection of Cu Diffusion Through the Barriers

Although structural changes of the barrier systems during annealing can obviously influence the barrier performance, a direct detection of Cu within and below the barriers is mandatory to prove the correlation between microstructure and barrier effect. The detection of the fast diffusing Cu in silicon and, in particular, in SiO_2 is a challenge for analytics. Several methods were applied to solve this problem:

(i)   In the case of a direct deposition of the barriers onto Si, Cu diffusion through the barrier leads to formation of Cu silicides, which for low N content barriers additionally triggers a reaction between the barrier and Si leading to Ta or W silicides. GDOS depth-profiling proves the Cu within the substrate and the silicides can be detected by XRD techniques. As an example, for 1 h anneals of the Cu/TaSiN0/Si system, 575°C has been determined as critical temperature for barrier degradation by this method [14].

(ii)   Electrical measurements on dedicated Schottky diode and MOS structures respond sensitively to Cu traces. In particular, the strong influence of applied electric fields on the Cu diffusion is taken into account. To determine concentrations of alkali ions in the substrate, which can be related to the barrier performance, shifts of the flatband voltage in capacitance-voltage curves after bias temperature stress (CV-BTS) were measured. For *in situ* detection of mobile ions during elevated temperatures, triangular voltage sweep (TVS) tests were performed. Additionally, for minority carrier lifetime determination capacitance–time (C–t) plots were implemented. These methods confirm the efficiency of the barriers in their amorphous state, and the correlation between observed structural changes like barrier crystallization and the onset of Cu diffusion [15].

(iii) Using AAS and secondary ion mass spectrometry (SIMS) very sensitive techniques are available, which allow Cu detection on a laterally macroscopic scale in the different layers. In case of AAS, also the quantification of the Cu content is straightforward to obtain (*cf.* Figure 7).

(iv)   For local analysis of Cu within the barrier and the SiO_2, electron energy loss spectroscopy (EELS) line scans at TEM cross sections were performed. The results of such EELS scans confirm the Cu trace detection from AAS measurements in the substrates of annealed samples. However, the unique spatial resolution of TEM techniques allows measurements in single grains, ultrathin sublayers and, for example, the proof that Cu diffuses into the SiO_2 at those local regions where it comes in contact with the dielectric due to barrier degradation.

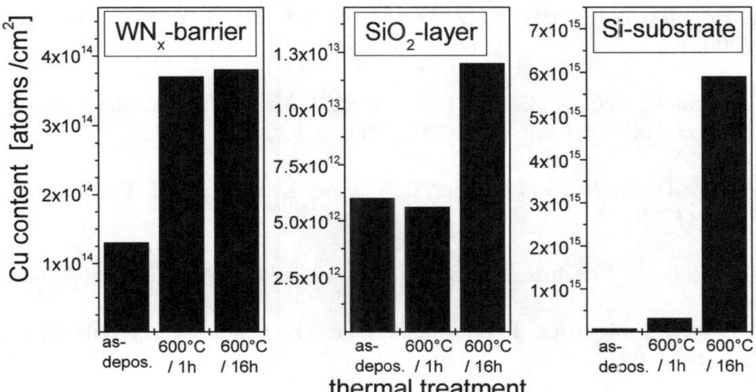

Figure 7. Cu content in barrier, dielectric, and Si substrate after deposition and after annealing for the binary W–N system as determined by AAS. Note that increased Cu concentrations are measured after barrier crystallization, and mark only small Cu traces not detectable with such methods like GDOES or wide-angle XRD.

Conclusions

Both Ta–Si–N and W–Si–N films have promising properties as diffusion barriers for copper metallization. For Ta–Si–N films with a silicon content of ~ 20 at%, as an optimised system regarding thermal stability and high conductivity a composition containing ~ 20 at% nitrogen was obtained. If the PVD techniques can be developed further to remain applicable for high aspect ratio structures, the PVD Ta–Si–N films are promising candidates for future technology nodes in semiconductor industry. Their integration behaviour for CVD and, in particular, ALD techniques has still to be investigated. The PE-CVD W–N barriers form amorphous barrier layers, which are expected to remain stable films for thicknesses down to 5 nm. Their thermal stability is slightly below that of PVD Ta–Si–N films, but can be improved by incorporation of silicon. Further optimisation of the ternary W–Si–N system is still ongoing.

Acknowledgements

This paper is based on three talks and two poster presentations at the EUROMAT 2003 conference in Lausanne. Financial support from BMBF (contr. No. 03N1067D) is gratefully acknowledged.

References

1.   International Technology Roadmap for Semiconductors (ITRS), Semiconductor International Association (2003), http://public.itrs.net/

2.   E. Zschech, Barrier and Nucleation Layers for Interconnects, in Metal Based Thin Films for Electronics (eds. K. Wetzig, C.M. Schneider), Wiley-VCH, Weinheim, p. 222, (2003).

3.   D. Edelstein, C. Uzoh, C. Cabral, P. deHaven, P. Buchwalter, A. Simon *et al.*, Proc. Int. Interconnect Technol. Conf. IITC 2001, Proc. p. 9 (2001).

4.   M. Stavrev, D. Fischer, F. Praessler, C. Wenzel, K. Drescher, J. Vacuum Sci. Technol. A 17, 993 (1999).

5.   F.A. Baiocchi, N. Lifschitz, T.T. Sheng, S.P. Murarka, J. Appl. Phys. 64, 6490 (1988).

6.   T. Riekkinen, J. Molarius, T. Laurila, A. Nurmela, I. Suni, J.K. Kivilahti, Microelectr. Eng. 64, 289 (2002).

7.   R. Hübner, M. Hecker, N. Mattern, V. Hoffmann, K. Wetzig, C. Wenger *et al.*, Thin Solid Films 437, 248 (2003).

8.   M. Traving, I. Zienert, E. Zschech, G. Schindler, M. Engelhardt, Appl. Surface Sci., submitted.

9. R. Ecke, S. Schulz, M. Hecker, N. Mattern, T. Gessner, Microelectr. Eng. 70, 364 (2003).

10. A.A. Istratov, E.R. Weber, J. Electrochem. Soc. 149, G21 (2002).

11. M. Hecker, D. Fischer, V. Hoffmann, H.-J. Engelmann, A. Voss, N. Mattern *et al.*, Thin Solid Films 414, 184 (2002).

12. R. Hübner, R. Reiche, M. Hecker, N. Mattern, V. Hoffmann, K. Wetzig *et al.*, Cryst. Res. Technol. 40, 135 (2005)

13. M. Hecker, R. Hübner, R. Ecke, S. Schulz, H.-J. Engelmann, H. Stegmann *et al.*, Microelectr. Eng. 64, 269 (2002).

14. R. Hübner, M. Hecker, N. Mattern, V. Hoffmann, K. Wetzig, C. Wenger *et al.*, Thin Solid Films 458, 237 (2004).

15. S. Zimmermann, R. Ecke, M. Rennau, S.E. Schulz, M. Hecker, A. Voss *et al.*, Proc. Adv. Metall. Conf. XIX, p. 397 (2004).

Synthesis and Characterization of Compounds Obtained by Crosslinking of Polymethylhydrosiloxane by Aromatic Rings

F. Sediri[a,b], F. Touati[b,c], N. Gharbi[b]

[a]Faculté des Sciences de Tunis, Tunis / Tunisia

[b]Laboratoire de Chimie de la matière Condensée, Tunis / Tunisia

[c]Unité des Matériaux, Tunis / Tunisia

Introduction

During the two last decades, sol-gel processes have received a great deal of attention. This technology offers broad ranges of possibilities leading to new products with high purity and homogeneity [1, 2].

The silicon alkoxides $Si(OR)_4$ have already been used with an outside source of carbon to produce silicon carbide by the carbothermal reduction of silica [3]. Trifunctional alkoxysilanes $R'Si(OR)_3$, in which the carbon source is internal, also appear to be convenient precursors to the preparation of the silicon carbide [4, 5]. The latter precursors lead to the formation of a silica network containing only terminal Si–C bonds. The pyrolysis of the obtained gels gives silicon oxycarbide glasses with various proportions of the SiC phase, depending on the nature of the starting reagents and essentially on the C/Si and C/O ratios [4].

The commercially available polymethylhydrosiloxane (PMHS) provides appropriate starting reagents for the synthesis of side-chain liquid crystalline polymers [6]. Compared with the trifunctional alkoxysilanes usually used, the PMHS precursor, having Si–H groups, readily allows changing the nature of the organic chain. Transparent and monolithic gels have been obtained from the chemical modification of PMHS precursor by some linear diamines $H_2N–(CH_2)_n–NH_2$ with $n = 3, 4$, and 6 or olefins via hydrosilylation reaction [7, 8].

In this paper we report the result of the PMHS chemical modification with mono- and bifunctional aromatic rings. The reaction leads to the formation of polymeric, transparent and colored networks that have been characterized by IR, ^{29}Si MAS-NMR, SEM, BET, X-ray diffraction, and thermal analysis. The presence of aromatic rings can apparently increase the pore size of the materials. In addition to that, this work is a part of a more extensive study on the synthesis of organic–inorganic materials, carried out by crosslinking of PMHS by 1,3,5-trihydroxybenzene and 2-hydroxybenzoic acid that will be reported in a future paper [9].

Experiment

Samples were prepared by mixing precursors (PMHS and 1,4-dihydroxyhenzene; hydroxybenzene) with stoichiometric proportions in tetrahydrofuran (THF). The hexachloroplatinic acid ($H_2PtCl_6.6H_2O$) was used as a catalyst (4×10^{-7} mol/g PMHS). The reaction occurred at room temperature in a closed vessel for 1 h. The gas bubbles formed in the reaction medium during synthesis may be due to the evolved hydrogen [7]. Transparent and monolithic gels were formed in several days, depending on the nature of the reagent (Table 1). The gels obtained were put in a drying oven to remove the solvent.

Table 1. Experimental conditions and the nature of obtained products

Reagents	Molar ratio: reagent/SiH	Gelification time	Nature of product
1,4-Dihydroxybenzene	0.5	3 d	$DB_{1,4}$: reddish gel
Hydroxybenzene	1	20 d	OB: yellowish gel

Infrared spectra were recorded on 550 Nicolet Magana Spectrometer as solids in KBr pellets.

^{29}Si MAS-NMR spectra were carried out on an MSL 400 Bruker Spectrometer. The thermal treatment was performed under an air atmosphere with a heating rate of 10°C min^{-1} using a thermal analyzer Netzsh STA 409 from 20°C up to 1500°C. X-ray powder diffraction data were collected on a CAD4 diffractometer using the CuK_α radiation (λ=1.54056 Å) and the graphite monochromator. The morphology of samples was determined by scanning electronic microscopy using a Cambridge Instruments Stereoscan 120.

Results and Discussion

Samples were obtained by coupling PMHS with 1,4-dihydroxybenzene ($DB_{1,4}$) and hydroxybenzene (OB), respectively, through hydrosilylation reactions. The bi-or monofunctional organic rings between siloxane chains ensured the reticulation and the formation of tridimensional network.

Infrared Spectroscopy
The IR spectra of the sample obtained by reaction of PMHS with trihydroxybenzene, 1,4-dihydroxybenzene is shown in the Figure 1. It shows the complete disappearance of the band at 2160 cm^{-1} (v_{Si-H}) and the broad bands corresponding to v_{OH} (3200–3500 cm^{-1}).

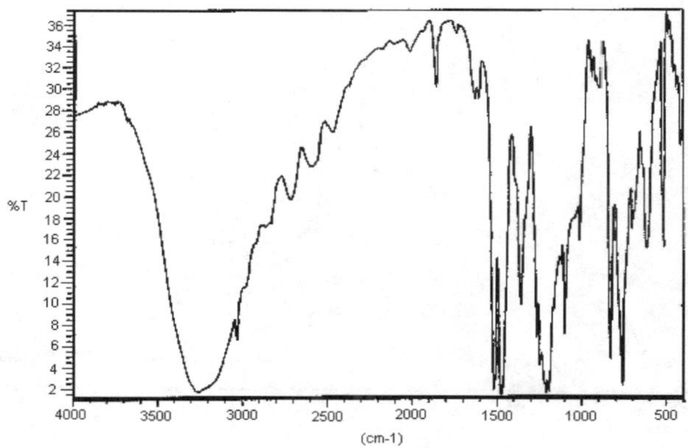

Figure 1. Infrared spectrum of DB$_{1,4}$

The presence of OH groups is attributed to the hydroxyl of aromatic chains un-reacted or linked at only side to Si–OH groups resulting from possible hydrolysis of the siloxane hydrogen [reaction (1)] [10], and the Si–O–C linkage [reaction (2)]. The IR exhibits sharp bands corresponding to organic (v_{C-H}, 2800–3020 cm^{-1}) and organometallic (v_{Si-CH3}, 1300 cm^{-1}; v_{Si-C}, 770 cm^{-1}), besides a broad band observed around 1150 cm^{-1} attributed to $v_{Si-O-Si}$ [4].

$$CH_3-Si(O)-H + H_2O \longrightarrow CH_3-Si(O)-OH + H_2 \tag{1}$$

$$CH_3-Si(O)-O-C- + H_2O \longrightarrow CH_3-Si(O)-OH + HO-C- \tag{2}$$

^{29}Si MAS NMR

^{29}Si MAS NMR is a convenient tool to display the different Si sites existing in the material. The ^{29}Si solid-state NMR spectrum of DB$_{1,4}$ (Figure 2) shows two main peaks at −56.4 and −65.4 ppm, and a small one at 10 ppm. The first is attributed to T$_2$ units [11], resulting from the substitution of Si-H by Si–O–C linkage [12].

The second one is assigned to T$_3$ units. T$_3$ units might result from partial hydrolysis of the silicon–hydrogen bond followed by intra- or intermolecular crosslinking of PMHS [11]. The small peak at 10 ppm is assigned to M units corresponding to (CH$_3$)$_3$Si–O– end groups.

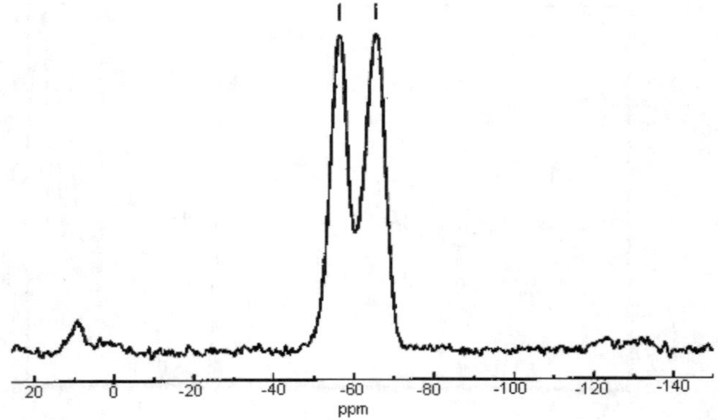

Figure 2. ^{29}Si NMR MAS spectrum of $DB_{1,4}$

Thermal Analysis

TGA and DTA curves are recorded under air atmosphere for the hybrid material $DB_{1,4}$ and OB. The curves of the hybrid material $DB_{1,4}$ (Figure 3) indicate that the combustion of organic groups at the surface linked only at one side is accompanied by a minor exothermic peak (DTA) that appears at 290°C. The broad exothermic peak (DTA) that occurs in the temperature range 450–460°C is attributed to the combustion of organic groups inside the network of hybrid material linked only at one side. The TGA curve shows a weight loss of 14.35% at 517°C, which correlates with an exothermic phenomenon assigned to the combustion of organic groups linked to the PMHS backbone.

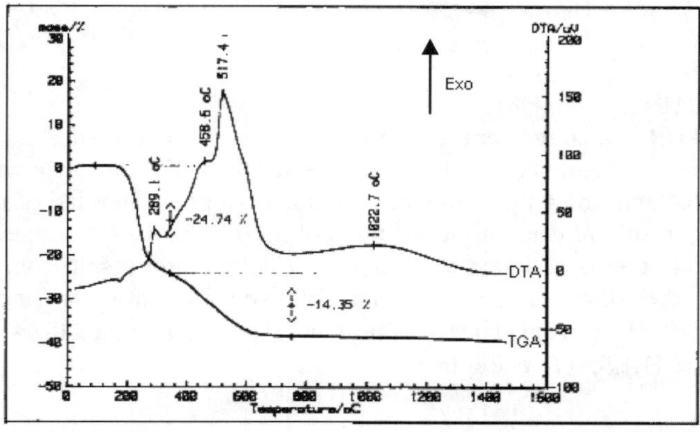

Figure 3. Thermal analysis of $DB_{1,4}$

Thermal analysis of all products shows clearly a retention of the siloxane backbone up to 600°C. For all materials no weight losses are observed between 600 and 1500°C. However, the DTA curves show an exothermic broad peak in the temperature range from 900 and 1200°C. This exothermic phenomenon corresponds to a crystallization process.

The X-ray diffraction patterns of $DB_{1,4}$ and OB after TGA-DTA treatment (Figure 4) show that we have the same crystallization phase. This result is in agreement with DTA and TGA curves. The peaks are sharpening, indicating the crystallization of the product. The observed peaks at d = 4.05, 2.84, and 2.49 Å are characteristic of the tetragonal SiO_2 [13].

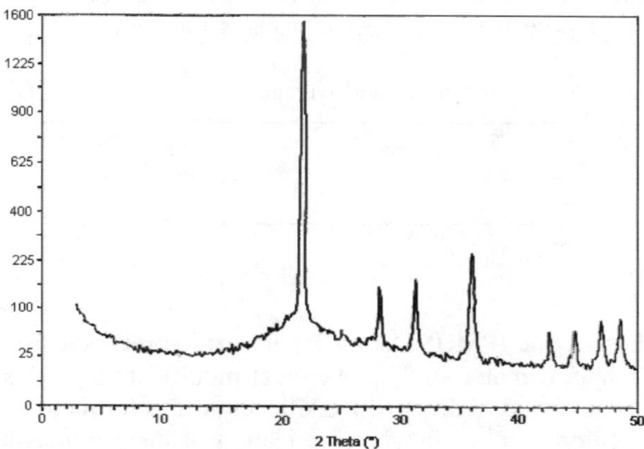

Figure 4. XRD patterns of $DB_{1,4}$ after thermal analysis

Scanning Electron Microscopy

Figure 5 represents an image from scanning electron microscopy of the hybrid material $DB_{1,4}$. The analysis of the image shows that the product presents a homogeneous phase constituted by blocks.

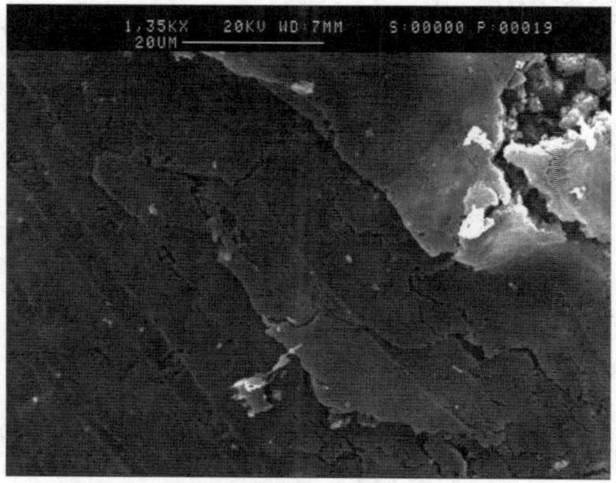

Figure 5. Scanning electron micrograph of $DB_{1,4}$

Brunauer–Emmett–Teller Surface Area

The Brunaer-Emmett-Teller (BET) surface area, the pore volume and the average pore size of materials OB and $DB_{1,4}$ are summarized in Table 2.

It seems that the number of OH groups linked to the organic precursors influences both the surface area, the pore volume, and the pore size. Indeed, the surface area tends to decrease as the number of hydroxyl groups is increased. However, it was found that the surface area and the pore volume values were of minor importance. On the other hand, the average pore size was found to be in the range 60–84 Å, reflecting a large pore size of the mesoporous frameworks.

Table 2. BET surface area, pore volume and average pore size of OB and $DB_{1,4}$

Samples	S_{BET} $(m^2.g^{-1})$	$V_{por.}$ $(cm^3. g^{-1})$	$d_{moy.}$ $(Å)$
OB	0.20	6.10^{-4}	83.7
$DB_{1,4}$	1.37	2.10^{-3}	59..3

Conclusion

Polymethylhydrosiloxane (PMHS) being an inorganic precursor, is used, as an inorganic–organic hybrid material. The chemical modification of this precursor by 1,4-dihydroxybenzene and hydroxybenzene leads to the formation of transparent, monolithic, and colored gels. The characterization of the products by IR, thermal analysis, and ^{29}Si MAS NMR showed that the modification of PMHS by the organic molecules is total and that the proportion of the T_2 sites and T_3 depends on the organic precursor. These inorganic–organic hybrids could find some applications as in the manufacture of electro-luminescent diodes and ion sensors [14]. The product heated to 1500°C is formed by a pure crystalline silicon dioxide SiO_2 phase.

Acknowledgement

We are grateful to Professor J. Livage, Dr. F. Babonneau, and J. Maquet, Laboratoire de Chimie de la Matière Condensée Université P. et M. Curie Paris VI, for MAS NMR and thermal analysis experiments.

References

1.   CJ. Brinker, J. Am. Ceram. Soc. 65, 4 (1982).

2.   S. Sakka, K. Kamiya, J. Non-Cryst. Solids 42, 393 (1980).

3.   KC. Chen, KJ. Thorne, A. Chemseddine, F. Babonneau, JD. Mackenzie, Mat. Res. Soc. Symp. Proc. 121, 571 (1988).

4.   F. Babonneau, K. Thorne, JD. Mackenzie, Chem. Mater. 1, 554 (1989).

5.   L. Bois, J. Maquet, F. Babonneau, H. Mutin, D. Balhoul, Chem. Mater. 6, 796 (1994).

6.   G. De Marignan, D. Teyssie, S. Boilleau, J. Maltete, C. Noel, Polymer 29, 1318 (1988).

7.   F. Sediri, F. Touati, N. Etteyeb, N. Gharbi, J. Sol-Gel Sci. Technol. 26, 425 (2003).

8.   F. Sediri, N. Etteyeb, N. Gharbi, J. Soc. Chim. Tunisie 4, 1479 (2002).

9.   F. Sediri, N. Gharbi, J. Sol-Gel Sci. Technol., (submitted).

10. L. Lestel, H. Cheradame, S. Boileau, Polymer 31, 1154 (1990).

11. M. Hetem, G. Rutten, B. Vermer, J. Rijks, L. Van De Ven, J. De Haan, C. Cramers, J. Chromatogr. 477, 3 (1989).

12. R. Kalfat, F. Babonneau, N. Gharbi, H. Zarrouk, J. Mater. Chem. 6, 1673 (1996).

13. ASTM Ficher No. 11-695.

14. R. Kalfat, N. Gharbi, S. Romdhane, J.-L Fave, A. Chaïeb, M. Majdoub, J. Sol-Gel Sci. Technol. 13, 439 (1998).

Revealing the Porous Structure of Low-k Materials Through Solvent Diffusion

D. Shamiryan, M. R. Baklanov, K. Maex

IMEC, Leuven / Belgium

Introduction

The performance of integrated circuits (IC) has been increasing for several decades. The most direct way to increase working speed of an integrated circuit is to pack faster and smaller transistors into an IC. For the last two decades device feature sizes have decreased from 1 µm down to 90 nm increasing the working frequency of microprocessors from 66 MHz to 4 GHz. However, not all IC components work faster as their sizes decrease. While continuous shrinking makes transistors work faster, it makes interconnections between transistors work slower.

A good figure of merit to characterize interconnects is RC, which is a unit of time. A signal propagating through the interconnection experiences resistance-capacitance (RC) delay. Shrinking the cross section of a wire increases its resistance and bringing wires closer together increases capacitance between wires. As a result, RC delay increases with device shrinkage and limits performance improvements resulting from device scaling.

The RC delay must be reduced as size reduction continues. It is predicted that soon RC delay will exceed transistor speed becoming a serious limitation to performance improvement. Since scaling down dimensions works against RC delay, the only way to bring down resistance and capacitance is to use other metals (with lower resistivity) and dielectrics (with lower dielectric constant) instead of conventional aluminium/SiO_2. The obvious candidate to replace Al is Cu (36% decrease in resistivity). Copper has been successfully integrated into IC manufacturing after considerable effort. Replacing SiO_2 as a dielectric has not been a straightforward process. In principle, any material with a dielectric constant k lower than 4.2 (so-called low-k dielectric) are of interest, but k-value is only one of many required properties.

One of the possible ways to reduce k-value of a material is reduction of its density. The density of a material can be decreased by increasing its free volume by rearranging the material structure or by the introduction of porosity. Porosity can be related in two different ways: constitutive and subtractive. Constitutive porosity refers to self-organization of a material. After manufacturing, such a material is porous without any additional treatment. Constitutive porosity is relatively low (usually less than 15%) and pore sizes are around 1 nm in diameter. According to IUPAC classification [1], pores with sizes less than 2 nm are called "micropores". Subtractive porosity, on the other hand, implies selective removal of part of the material. The removable part can be an artificially added ingredient (*e.g.* a thermally degradable substance called a "porogen", which is removed by an anneal leaving behind pores) or just part of the material that is removed by selective etching (*e.g.* Si–O bonds in SiOCH materials could be removed by HF). Subtractive porosity can be as high as 90% and pore sizes vary from 2 nm to tens of nm (pores with sizes larger than 2 nm are called "mesopores" according to IUPAC classification).

Although beneficial for *k*-value reduction, porosity creates a number of challenges when porous dielectrics are to be integrated in IC manufacturing. Among other issues (*e.g.* mechanical instability or low thermal conductivity), porosity allows unwanted diffusion through the film. It could be, for example, in diffusion of chemicals used in processing or diffusion of copper. Copper readily degrades the dielectric properties of the insulator, increasing leakage currents and decreasing the breakdown voltage. As a result, reliability of the devices significantly decreases making their lifetimes unacceptably short. Copper diffusivity drastically increases with porosity of the dielectric.

Copper diffusion must be stopped by a diffusion barrier between copper and a porous film. Due to technological reasons that will not be covered in this work, the diffusion barrier cannot be deposited on copper but must be deposited on a porous film instead. The barrier must be thin (nm scale) and fully dense (contain no pinholes). Covering a porous material with such a barrier is non-trivial. If the material is highly porous with large pores connected to each other, the barrier may be unacceptably thick in order to bridge all the exposed pores (see Figure 1). It should be noted that the barrier itself should not penetrate into the porous material, which is a possibility with some deposition techniques. In some approaches, porous surface is first sealed to prevent barrier penetration and then a diffusion barrier is deposited on the sealed surface. Deposition of a rigorous barrier tends to be easiest when pores are small and porosity is low.

From the above consideration we can conclude that not only porosity and pore size are important for characterization of a porous film, but also pore interconnectivity. There are plenty of techniques that are able to characterize total porosity and pore size distribution of a porous film: positron annihilation lifetime spectroscopy (PALS) [2, 3], small-angle neutron scattering (SANS) and small-angle X-ray scattering (SAXS) combined with specular X-ray reflectivity (XRR) [4], and ellipsometric porosimetry [5, 6]. However, characterization of pore interconnectivity is less obvious. EP is able to measure only interconnected pores since it is based on penetration of solvent into a porous film, and therefore, can provide an answer whether the pores interconnected or not. However, if pores are interconnected, the interconnection structure cannot be revealed. Another technique that is claimed to reveal interconnectivity is PALS where movement of a positronium through a porous film is studied. There is a concern, however, about ability of positronium to travel through the narrow channels since they may represent too high energetic barriers for such a quantum particle [7].

In this work, we propose a simple and effective method for studying interconnectivity of a porous film. The method is based on diffusion of solvent inside the porous film in the lateral direction and does not require any special equipment.

Figure 1. A schematic representation of a thin film deposited on a porous material with (a) separated mesopores connected by microchannels and (b) interconnected mesopores. As porosity increases, the mesopore connections make the deposition of a continuous film more difficult. It should be noted that the pore size is the same in both cases.

Proposed Experimental Technique

The studying of interconnectivity is based on lateral solvent diffusion in a porous film. The porous film is capped at the top surface by any film that would prevent solvent from penetration through the top surface (*e.g.* rather thick SiO_2, Si_3N_4 or SiC). It should be noted that the capping film should be transparent in order to allow observation of the solvent diffusion. Thus prepared, the samples were cleaved and introduced in liquid solvent. The solvent penetrates the porous film through the edges and diffuses laterally as schematically shown in Figure 2. The diffusion distance l is measured as a color variation in top-down view by optical microscope, a typical observation of such color variation is shown in Figure 3.

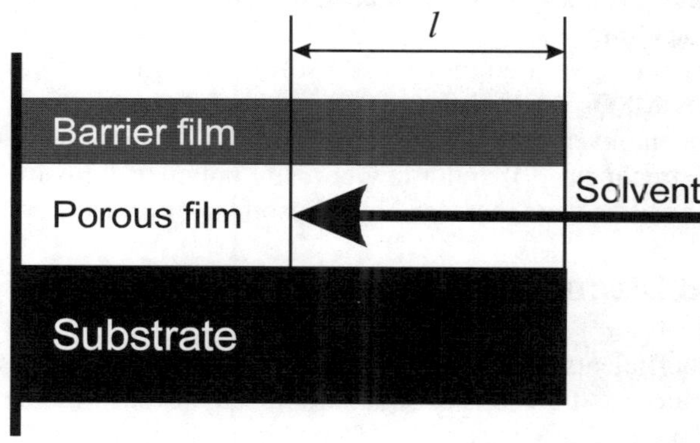

Figure 2. Schematic representation of solvent diffusion experiments. Diffusion distance l is measured as color change by optical microscope.

Figure 3. Example of color variation of a porous film due to solvent diffusion as observed by optical microscope.

The main measure of diffusion is a diffusion coefficient. It is known from diffusion theory [8] that the mean distance l travelled by a diffusion front is described (in a simplified form) by the following equation:

$$l = 2\sqrt{Dt} \qquad\qquad (1)$$

where D is a diffusion coefficient (in our case the diffusion coefficient of solvent in the porous film) and t is diffusion time. Therefore, measuring diffusion distance as a function of time it is possible to calculate a diffusion coefficient of the solvent in a particular porous film.

In our experiments, we used toluene as a solvent and four different low-k materials: two porous MSQs (methylsilsesquioxane, varied porosity from 15 to 40%), one SiOCH (silicon oxycarbide, initial porosity of 7%, increase up to 40% porosity achieved by HF treatment [9]) and a low porosity polymer. Porosity and pore size distribution of the films were measured by ellipsometric porosimetry.

Results and Discussion

The diffusion coefficients are calculated from the slopes of the curves representing penetration distances as functions of square root of the exposure time. Such curves for four studied materials (MSQs with 40% porosity, SiOCH with 7% porosity and the polymer) are shown in Figure 4. One can see that diffusion strongly depends on porosity.

Let us now see how diffusion changes when porosity of a film is gradually increased. For this observation we selected one of the MSQs with varied porosity. A few words should be said about the preparation of such material. Pristine as - deposited MSQ has constitutive porosity of about 15% consisting of micropores of about 1 nm in size. The porosity, however, can be increased by adding so-called porogen – thermally unstable organic molecules about 3 nm in size – during MSQ deposition. After thermal decomposition, the porogen leaves behind mesopores of about 3 nm size. Obviously, the higher the porogen content is, the higher the total porosity of the film will be. The results of characterization of such a film with varied porogen content by EP is shown in Figure 5, where total porosity is divided between microporous and mesoporous fractions. One can see that total porosity gradually increases as porogen content increases. It should be noted that the mean size of micro- and mesopores remains almost constant.

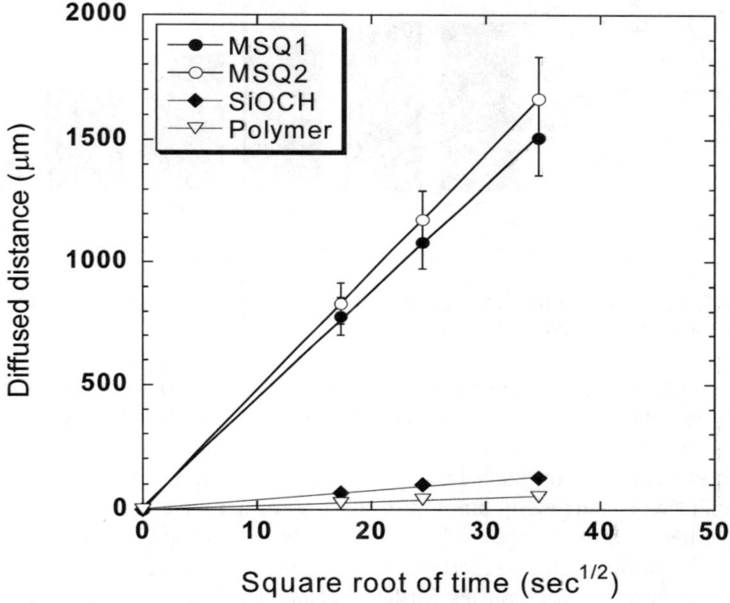

Figure 4. Diffused distances of toluene in different low-k materials (MSQs with 40% porosity, SiOCH with 7% porosity) as functions of square root of exposure time. Diffusion coefficients were calculated from the line slopes using equation (1).

The diffusion coefficient, however, shows a remarkably different behaviour. When the porogen load is low (less than 10%), the diffusion coefficient increases gradually. However, as the porogen load exceeds 10% an abrupt increase of the diffusion coefficient is observed, as shown in Figure 6.

Other materials with varied porosity show similar behavior to the one described above – diffusion drastically increases as porosity exceeds a certain threshold (see Figure 7). Steeper increase of the diffusion coeffcient in the case of SiOCH can be attributed to the fact that not only porosity but also mean pore size increases as this material is treated with HF, while both MSQs have constant pore size with only porosity increased.

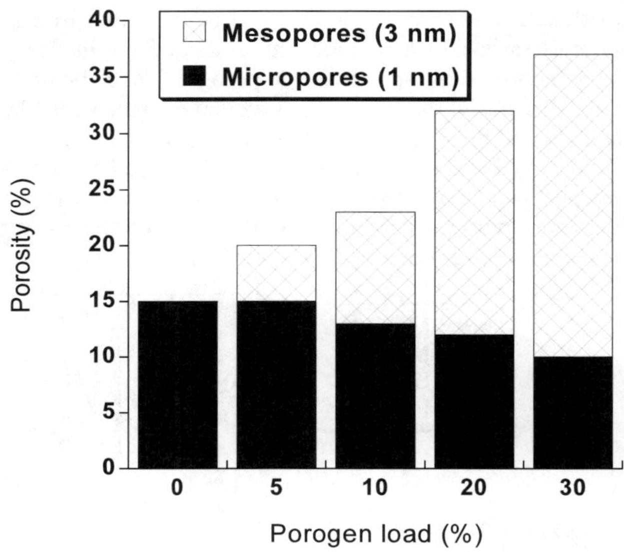

Figure 5. Porosity as a function of porogen load of MSQ. Relative volume of micro- and mesoporosity is indicated by different bar fill.

The following explanation can be proposed for the diffusion coefficient behaviour. When amount of mesopores is small they are connected to each other only by micropores and, therefore, diffusion is limited by movement of solvent through micropores resulting in rather low diffusion coefficient. Slight increase in diffusion coefficient with increase of amount of mesopores can be explained by the fact that distance between mesopores (where solvents travels through micropores) becomes smaller. This situation is described by the scheme represented by Figure 1a. As amount of mesopores increases further and passes a certain threshold, mesopores overlap creating rather large paths for diffusion. At this moment diffusion coefficient abruptly increases as micropores are not the limiting paths for diffusion anymore. The scheme represented by Figure 1b is realized.

One can conclude that low diffusion coefficients indicating the fact that mesopores are connected by micropores should predict easier diffusion barrier deposition (or, in other words, easier sealing of the porous surface). Indeed, such an effect was observed in experiments of sealing of the porous surface by plasma treatment [10]. The exposure of porous surface to plasma led in some cases to the conversion of the top porous surface to a dense sealing layer impermeable to solvents. It was found that a porous film can be sealed by plasma when the toluene diffusion coefficient does not exceed 100 $\mu m^2/s$.

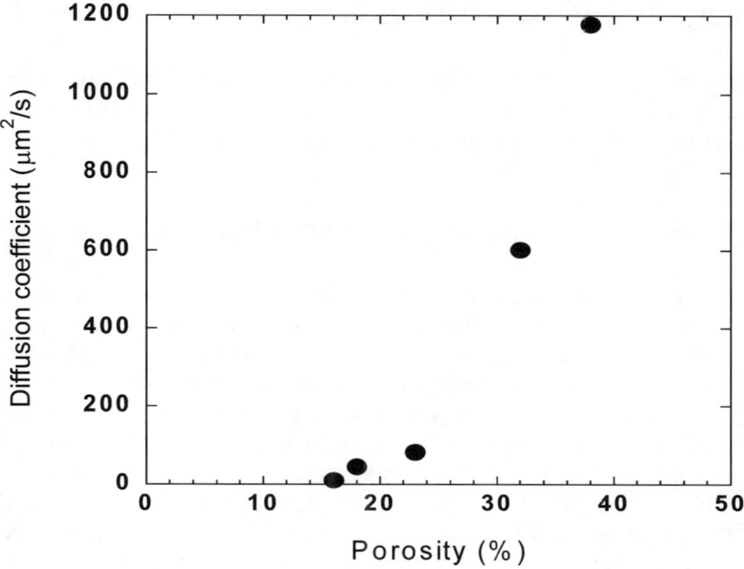

Figure 6. Diffusion coefficient of the porous MSQ with different porogen load as a function of total porosity.

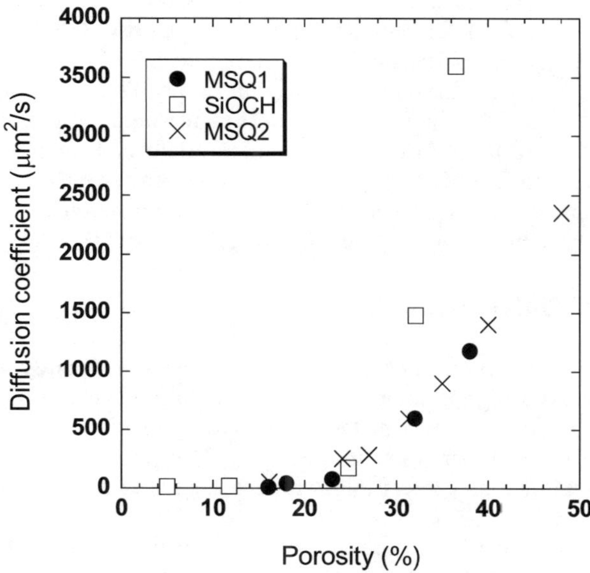

Figure 7. Diffusion coefficient as a function of porosity for all studied materials with varied porosity.

Other Applications of Diffusion

Besides revealing the pore interconnectivity structure, diffusion studies can be used for other characterizations of a porous film, for example, measuring quality of the sealing layer (or diffusion barrier) on top of porous surfaces, or studying affinity for water adsorption of a porous film.

Measuring the Barrier Integrity on a Porous Film

Recently, we reported a qualitative barrier integrity evaluation method [11] based on ellipsometric porosimetry. The method employs ellipsometric measurements of toluene adsorption in porous low-k films beneath the barrier. If the barrier is fully continuous, no adsorption is observed. It should be noted that although a barrier may appear fully dense in respect to toluene penetration it does not necessarily mean a barrier is impermeable to copper. If the barrier is highly porous, toluene is adsorbed inside the porous film that is registered by ellipsometer. Thus, the method is qualitative giving only an answer whether the barrier is "good" or "bad". There is, however, an intermediate case, when toluene fills the film beneath the barrier within a measurable time interval; in other words, there is a delay between exposure of the film to toluene and adsorption of toluene inside the porous film beneath the barrier. It was supposed that such a delay could be caused by diffusion of toluene beneath the barrier, when toluene enters the porous film through pinholes in the barrier and then fills the film by lateral diffusion [12]. Knowing the diffusion coefficient it is possible to calculate pinhole density and, therefore, to get a quantitative evaluation method for probing the barrier integrity. It was shown that the highest measurable pinhole density is limited by the ability of measuring short time intervals between toluene introduction and film filling. The lowest measurable pinhole density is limited by the throughput of the measurement tool since too much time is needed to fill the film through a limited number of pinholes. In that case, however, toluene diffusion around isolated pinholes creates rather large color spots that can be counted by a tool resolving light reflection differences. As a result, isolated pinholes become visible and can be counted.

Affinity for Water Adsorption

The presence of water molecules inside a porous film significantly increases the k-value as water has a high k (about 80) due to the high polarity of H_2O molecules. Although the majority of low-k materials are hydrophobic, they can contain a limited amount of water. Polar water molecules stick to polar centers inside a porous film. The amount of such centers might reflect the material's affinity for water adsorption.

Presence of polar centers inside a porous film results in different diffusion rates of polar and nonpolar solvents through the film. This effect is used in chromatography [13, 14] for separating mixtures of unknown components. Nonpolar solvents diffuse faster through porous film than polar solvents, as movement of nonpolar molecules does not affected by presence of polar centers in the porous medium. This observation can be applied to study porous films using diffusion of polar (*e.g.* ethanol) and nonpolar (*e.g.* toluene) solvents [15]. While chromatography uses diffusion of unknown solvents through known porous media to study solvents, we use diffusion of known solvents through unknown porous media (low-*k* films) to study the media.

The diffusion coefficient ratio $D_{toluene}/D_{ethanol}$ was selected as a measure of polar centers presence, which, in turn, would reflect the material's affinity for water adsorption. It was found that the lowest ratio of 1 occurred for the polymer (where concentration of polar centers were expected to be very low), while MSQs have the highest ratios. It was also found that increase of the porosity of SiOCH by HF treatment results in an increase of the affinity for water adsorption (higher ratio of $D_{toluene}/D_{ethanol}$) that was also confirmed by thermodesorption studies – more water was desorbed from SiOCH films modified by HF than from the pristine SiOCH film.

Conclusions

We propose a simple and effective method for studying the porous structure of a thin film. The method is based on the observation of lateral diffusion of solvents inside a porous film using an optical microscope. The diffusion coefficients of solvents calculated from such observations can be used for revealing the pore interconnection structure of the porous film in question.

Diffusion studies could also be used for characterization of other properties of a porous film. A known diffusion coefficient allows calculation of pinhole density of a barrier deposited on a porous film, when the film is filled with solvent that penetrates through the pinholes and then diffuses inside the film beneath the barrier.

The ratio of diffusion coefficients of polar and nonpolar solvents could be used as a measure for the film affinity for water adsorption. Polar centers inside a porous film could act as attraction sites for polar water molecules, which is detrimental for low-k properties of the film. The presence of polar centers causes polar solvents to diffuse slower than nonpolar ones.

References

1. J. Rouquerol, D. Avnir, C.W. Fairbridge, D.H. Everett, J. H. Haynes, N. Pernicone, J. D.F. Ramsay, K.S. W. Sing and K.K. Unger. Pure & Appl. Chem. 66, 1739 (1994)

2. D.W. Gidley, W.E. Frieze, T.L. Dull, J. Sun, A.F. Yee, C.V. Nguen, and D.Y. Yoon, Appl. Phys. Lett. 76, 1272 (2000)

3. M.P. Petkov, M.H. Weber, K.G. Lynn, K.P. Rodbell, and A. Cohen, Appl. Phys. Lett. 74, 2546 (1999)

4. W. Wu, W.E. Wallace, E. Lin, G. W. Lynn, C.J. Glinka, R.T. Ryan, and H. Ho, J. Appl. Phys. 87, 1193 (2000)

5.   F.N. Dultsev and M.R. Baklanov, Electrochem. Solid State Lett. 2, 192 (1999)

6.   M.R. Baklanov, K.P. Mogilnikov, V.G. Polovinkin, and F.N. Dultsev, J. Vac. Sci. Technol. B 18, 1385 (2000)

7.   K.P. Mogilnikov, M.R. Baklanov, D. Shamiryan, and M. Petkov. Japan J. Appl. Phys. 43, 247 (2004)

8.   P.W. Atkins. Physical Chemistry. 6th edn, Oxford University Press, p.753, (1998)

9.   D. Shamiryan, M. R. Baklanov, S. Vanhaelemeersch, and K. Maex, Electrochem. Solid State Lett. 4, F3 (2001)

10. T. Abell, D. Shamiryan, M. Patz and K. Maex, Proc. of Advanced Metallization Conference (AMC 2003), ed. by G.W. Ray, T. Smy, T. Ohta, and M. Tsujimura, 549 (2003)

11. D. Shamiryan, M.R. Baklanov, and K. Maex. J. Vac. Sci. Technol. B 21, 220 (2003)

12. D. Shamiryan, T.Abell, Q.T. Le and K. Maex, Microelctr. Eng. 70/2-4, 341 (2003)

13. L.R. Snyder and J.J. Kirkland. Introduction to Modern Liquid Chromatography. 2nd edn, Wiley, New York, 1979)

14. L.R. Snyder. Principles of Adsorption Chromatography, Dekker, New York, 1968

15. D. Shamiryan and K. Maex, Materials Research Society (MRS) Proc. (2003 MRS Spring Meeting, San Francisco, CA) 766, E9.11.1 (2003)

Carbon Nanotube Via Technologies for Future LSI Interconnects

M. Nihei[a, b], A. Kawabata[a, b], M. Horibe[a, b], D. Kondo[a, b], S. Sato[a, b], Y. Awano[a, b]

[a] Fujitsu Limited / Japan

[b] Nanotechnology Research Center, Fujitsu Laboratories Ltd., Atsugi / Japan

Introduction

Carbon nanotubes (CNTs) [1] are attractive as nanosize structural elements from which devices can be constructed by bottom-up fabrication. A carbon nanotube is a macromolecule of carbon and is made by rolling a sheet of graphite into a cylindrical shape. Carbon nanotubes exhibit not only a unique nanoscale structure but also interesting physical properties, such as the ability to be either metallic or semiconductive, depending on the twist or chirality of the tube [2]. They offer unique electrical properties such as current densities exceeding 10^9 A/cm^2 [3], thermal conductivity as high as that of diamond [4], and ballistic transport along the tube [5]. They have been reported to be potentially useful as channels in field-effect transistors (FETs) [6, 7] and as wiring materials [8–10]. Even with the current limitations on device dimension, LSI circuits have problems that originate from stress and electromigration that occurs in Cu interconnects, particularly in vias and their surroundings. One potentially effective solution to these problems is to use a CNT, which has a large migration tolerance, as a via. It is difficult, however, to obtain sufficient current for LSI interconnects when using only one CNT. We have therefore explored the feasibility of using bundles of metallic CNTs as a wiring material for future LSI interconnects (Figure 1).

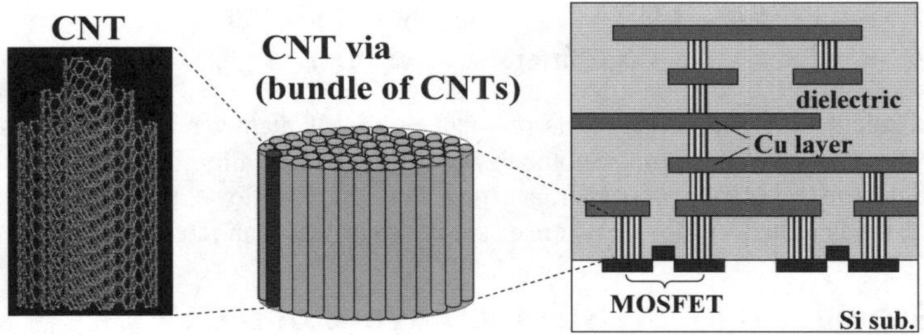

Figure 1. Structure of future LSI interconnects consisting of CNT vias.

Low-Resistance of CNT-Bundles in Scaled-down Vias

The advantage of CNT bundles we would like to discuss in this chapter is their low resistance, which may be the solution to the problem of high resistance in scaled-down vias. As shown in Figure 2, we have estimated the resistance of a 0.1-μm-diameter via tightly packed (filling rate of 91%) with CNTs of various diameters [11]. In this estimation we assumed that the CNT via had the parallel quantum resistance R = h/4e2 = 6.45 kΩ (conductance $G = 2G_0 = 4e^2/h$, which reaches the maximum conductance limit for ballistic transport in two channels of a single-wall carbon nanotube (SWNT) [5]), that current flows through each wall of multiwall carbon nanotubes (MWNTs), and that there is no dependence of ballistic transport on CNT length. This figure shows, for example, that if a bundle of single-wall carbon nanotubes with an outer diameter of less than 3 nm or a bundle of six-wall CNTs with an outer diameter of less than 7 nm were packed tightly in a via 0.1 μm in diameter, the via resistance would be comparable to or less than that of a Cu via. Moreover, the conventional Cu vias need a barrier layer, which increases its resistance. Since CNT vias do not need barrier layers, there is a clear advantage to using CNT vias instead of conventional Cu vias.

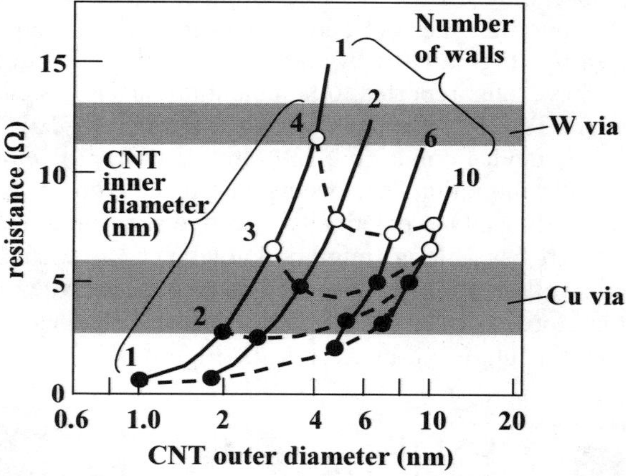

Figure 2. Calculated dependence of via resistance on CNT diameter [11]. The resistance of a 0.1-μm-diameter via filled with nanotubes was estimated assuming that CNT vias have the parallel quantum resistance of nanotubes and that current flows through each wall of MWNTs. Filled circles (•) show resistances similar to or less than those of Cu vias.

CNT Growth on Metal-Silicide Layers of MOSFET

As shown in Figure 3, we have grown vertical MWNTs directly on a Ni-silicide layer, which can be used as electrodes for metal-oxide-semiconductor field-effect transistors (MOSFETs) [12]. Using a Ni-silicide layer as a catalyst, we grew nanotubes with diameters smaller than can be grown using a Ni-film catalyst. We think that the composition of Ni-silicide composition plays an important role in controlling the diameter of the nanotubes. In this chapter, we report the diameter-controlled growth of vertically aligned CNTs on catalytic metal-silicide substrates.

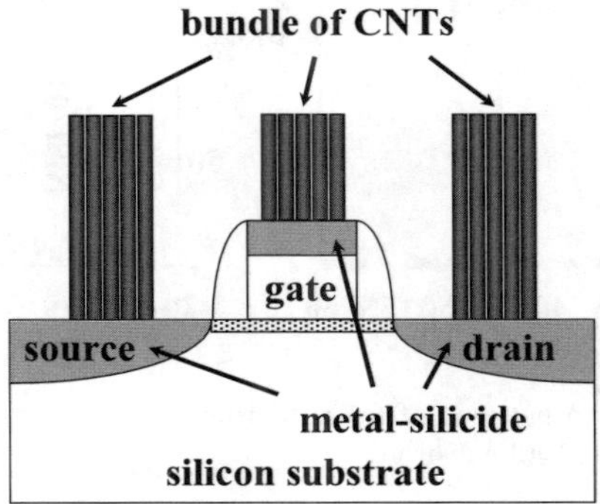

Figure 3. Schematic of the MOSFET with CNT vias.

Although the growth of CNTs on $CoSi_2$ has been reported [13], there are very few reports on the use of a metal-silicide layer as a catalyst for growing vertically aligned CNTs. This is probably because CNT growth is generally thought to be suppressed by silicidation of the catalyst metal [14]. Our experiment using the plasma-enhanced chemical vapor deposition (p-CVD) method, however, confirmed that CNTs could be grown on Ni-monosilicide (NiSi) substrates without using any catalysts other than a metal-silicide layer. It is particularly noteworthy that the diameter of CNTs grown on Ni-monosilicide substrates can be made smaller than that of CNTs grown on Ni film substrates.

Our p-CVD apparatus has a 2.45-GHz 2-kW microwave power source, a substrate holder that can hold wafers of up to eight inches diameter, and a substrate heating mechanism. A mixture of CH_4 and H_2 was used as the reactive gas source. After the vacuum chamber was pumped down to 3×10^{-4} Pa, the substrate temperature was set to 400°C. The actual substrate temperature during CNT growth, however, was not measured. The gas mixture was introduced at a pressure of 266 Pa, and the CH_4 and H_2 flow rates were, respectively, 40 and 60 sccm. Vertically aligned CNTs can be grown by establishing an electric field between the substrate and the grounded chamber during growth [15], and we did that by applying −400 V to the substrate.

Figure 4 shows the X-ray diffraction (XRD) patterns obtained from a Ni film on a silicon substrate and a Ni-monosilicide layer produced by heating a Ni film on a silicon substrate to 700°C for 30 s [12]. Since this rapid thermal annealing (RTA) resulted in the appearance of several diffraction peaks related to the Ni-monosilicide phase, it was clear that a uniform Ni-monosilicide phase was formed by the solid-state reaction at 700°C and that the unreacted Ni film had been removed.

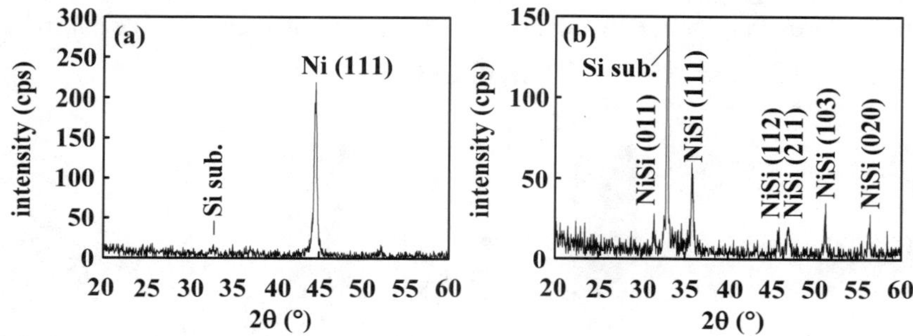

Figure 4. X-ray diffraction patterns of (a) a Ni film on a Si (100) substrate and (b) a Ni-monosilicide layer on a Si (100) substrate.

Figure 5. SEM images of CNTs grown for 30 min on (a) the Ni film and (b) the Ni-monosilicide layer. (c) Dependence of the outer diameter of CNTs on growth time [12].

Figures 5a and 5b show SEM images of CNTs grown for 30 min on the Ni film and the Ni-monosilicide layer. The almost vertically aligned CNTs in both samples were densely aligned. We estimated the diameter of the CNTs on the Ni film to be about 100 nm and the diameter of the CNTs on the Ni-monosilicide layer to be about 50 nm. The dependence of the CNT diameter on growth time is shown in Figure 5c for both the Ni film and the Ni-monosilicide layer. We found that for both samples the CNT diameter saturated within the first 2 min. Our results suggest that the Ni catalyst determines the CNT diameter at the initial growth stage. We think that CNT diameter is smaller on the Ni-monosilicide layer because the Ni-monosilicide layer is decomposed to Ni and Si, and a smaller amount of Ni is supplied as a catalyst. We have found that CNT diameter can be controlled by forming various Ni-silicide phases such as Ni_2Si, NiSi, and $NiSi_2$.

CNT Growth on Cu Wiring

CNT growth technology cannot be used in Si LSIs unless the growth temperature can be kept below 450°C, a limit established by the thermal tolerance of low-k dielectrics. Growing ideal MWNTs by conventional thermal CVD, however, typically requires a growth temperature higher than 800°C [2]. The growth temperature can be lowered to 600°C by using hot-filament chemical vapor deposition (HF-CVD) and to 450°C by using thermal CVD. Figure 6a shows a schematic cross-section of our CNT via test sample. Figure 6b is a SEM image of CNT vias formed by using the same p-CVD method used to grow the CNTs on the Ni-silicide layer. The substrate temperature was initially set to 400°C, but was not measured during the CNT growth. Figures 6c and 6d show SEM images of CNT vias formed by the HF-CVD methods with a growth temperature of 600°C and by the thermal CVD method with a growth temperature of 450°C [16]. We have succeeded in growing vertically aligned MWNT-bundles in via holes about 2 μm in diameter with a catalytic metal at the bottom.

Figure 6. (a) Schematic cross section of CNT vias. SEM images of the top of a vertically aligned MWNT-bundle in a via hole on a 100-nm-thick Cu layer. These bundles were grown (b) on a 100-nm Ni catalyst layer by p-CVD [12], (c) on a 30-nm Ni catalyst layer and 50-nm Ti contact layer by HF-CVD at 600°C [11], and (d) on a 2.5-nm Co catalyst layer, a 2.5-nm Ti contact layer, and 5-nm Ta barrier layer by thermal CVD at 450°C [16].

Before the HF-CVD growth, the substrate was heated to 510°C and cleaned for 10 min in H_2 gas at 1 kPa. A mixture of C_2H_2, Ar, and H_2 was then introduced as the gas source. The pressure of the CVD chamber was set to 1 kPa, the hot filament 6 mm above the substrate was turned on, and MWNT growth occurred. The stage temperature was set to 510°C, but the heat radiated from the filament (which had a temperature above 2000°C) increased the substrate temperature to 600°C during the 10-min MWNT growth period.

Before the thermal CVD growth, the substrate was heated to 450°C, and cleaned for 10 min in H_2 gas at 1 kPa. A mixture of C_2H_2 and Ar was introduced, the pressure of the CVD chamber was set to 1 kPa, and MWNT growth occurred for 40 min.

Each part of Figure 7 shows a transmission electron microscopy (TEM) image of a MWNT in a hole. The MWNT grown by p-CVD, with an outer diameter of about 60 nm, is a multiwall bamboo-like structure containing many defects on the nanotube walls. On the other hand, we could grow MWNTs of better quality, with an outer diameter of about 10 nm and with well-graphitized graphen sheets, by using HF-CVD [11] and thermal CVD [16]. The electrical properties of these higher-quality MWNTs were suitable for wiring materials.

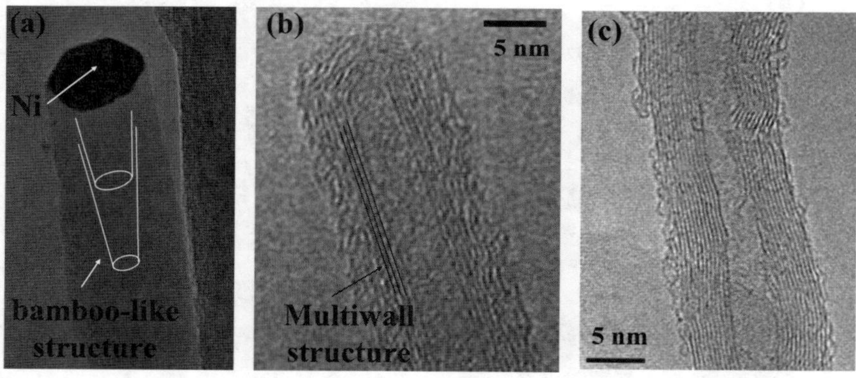

Figure 7. TEM images of an MWNT in a via hole. These MWNTs were grown (a) on a 100-nm Ni catalyst layer by p-CVD, (b) on a 30-nm Ni catalyst layer and 50-nm Ti contact layer by HF-CVD at a growth temperature of 600°C [11], and (c) on a 2.5-nm Co catalyst layer, a 2.5-nm Ti contact layer, and 5-nm Ta barrier layer by thermal CVD at a growth temperature of 450°C [16].

To use CNT vias in Cu interconnects, we have to grow the MWNTs on a layer of Cu. When the catalyst layer is very thin, like the 2.5-nm Co layer, the metal diffusion between the catalyst and Cu makes it hard to grow MWNTs directly on Cu. The Ta barrier layer suppressed this metal diffusion and enabled us to grow MWNTs on a Cu layer.

Low-Resistance Ohmic Contacts Between CNTs and Ti Contact Layers

Forming low-resistance ohmic contacts between CNTs and metal electrodes is important for exploiting the excellent features of CNTs. This is generally done by covering the ends of the CNTs with a Ti electrode and then forming titanium carbide (TiC) ohmic contacts by annealing [17]. Fabricating the CNT and the ohmic contact to the bottom electrode at the same time, However, is preferable for vertically embedded CNT structures like CNT-bundles in via holes. We have therefore grown the MWNTs and formed their end-bonded ohmic contacts to Ti electrodes at the same time by using HF-CVD and a double-layer (Ni or Co on Ti) electrode; that is, by using a catalyst layer on a Ti electrode.

In a preliminary experiment comparing 10-nm-Ni/100-nm-Ti double-layer electrodes and 100-nm-Ni electrodes without Ti [11], we used HF-CVD at 600°C to grow MWNTs that bridged 5-μm spaces between metal electrodes as shown in Figure 8a. Two-terminal current-voltage measurement was performed for one nanotube bridging Ni/Ti electrodes, three nanotubes bridging Ni/Ti electrodes, and one nanotube bridging Ni electrodes. Figure 8b shows a SEM image of the sample with Ni/Ti electrodes after MWNT growth. The SEM image shows that MWNTs grew on the Ni/Ti electrodes and that one of the MWNTs bridged the gap between the electrodes.

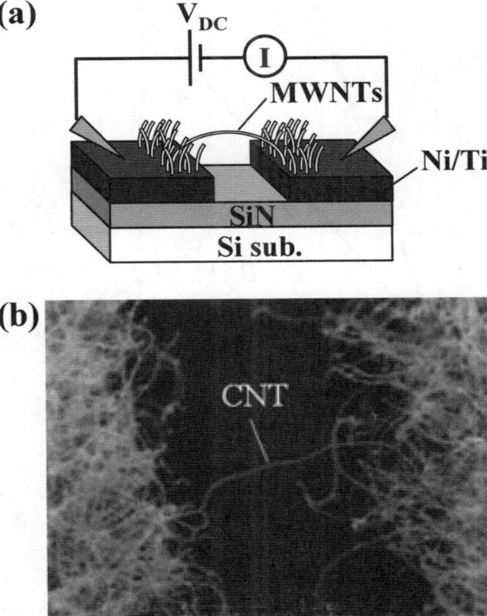

Figure 8. (a) Schematic and (b) SEM image of a MWNT bridging a 5-μm gap between Ni/Ti electrodes [11].

The results of the two-terminal current-voltage measurements are shown in Figure 9 [11]. The sample with 100-nm Ni electrodes without Ti, with a one-nanotube bridge, exhibited a resistance of 15–32 MΩ. This large resistance was due to the large contact resistance at the MWNT/Ni-electrode interface. On the other hand, the resistance of the sample with 10-nm-Ni/100-nm-Ti electrodes with a one-nanotube bridge was only 134 kΩ. To verify this result, we measured the resistance of Ni/Ti electrodes with three-nanotube bridges. We obtained a resistance of 54 kΩ, which is about 1/3 of the resistance we measured in the sample with a one-nanotube bridge. Consequently, we would like to emphasize that the contact resistance of a MWNT with Ni/Ti electrodes is two orders of magnitude smaller than that of one with Ni electrodes.

Figure 10a is a cross-sectional SEM image of a nanotubes/(Ni/Ti)-electrode interface. We observed Ni nanoclusters with a diameter of approximately 10 nm on the surface of Ti electrodes by using TEM and energy dispersive X-ray spectroscopy (EDX) analysis [11]. MWNTs with Ni nanoclusters inside the ends of the nanotubes grew vertically on the Ti electrodes. These results suggest that low-resistance ohmic contacts can be fabricated by forming TiC at the ends of MWNTs while the MWNTs are being grown.

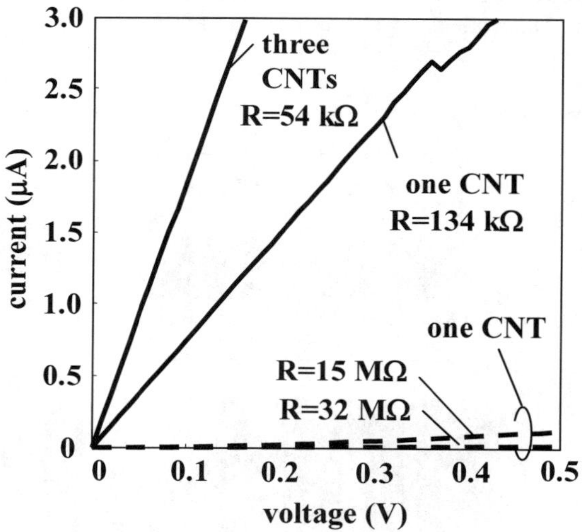

Figure 9. Two-terminal current-voltage characteristics of MWNTs bridging 5-μm gaps between metal electrodes. Solid lines: results measured when the electrodes were Ni/Ti electrodes. Dashed lines: results measured when the electrodes were Ni electrodes without Ti [11].

On the basis of these results, we propose a model for simultaneously forming MWNTs and end-bonded low-resistance ohmic contacts. It is shown in Figure 10b. As reported previously, low-resistance ohmic contacts could be made by forming TiC, which is made by annealing at temperatures above 800°C [17]. Our results of the electrical measurement suggest that low-resistance ohmic contact was achieved by forming TiC at the MWNT ends during the growth at a temperature of only 600°C. The results of an experiment testing this model confirmed the presence of TiC at the MWNT ends by XRD analyses (data not shown here). Turning now to the MWNT's tips, that is, to the other ends of MWNTs, we could not observe the contacts between the MWNT's tip and the other electrode distinctly. We think that there is a possibility that MWNTs grown on two electrodes cross each other, or the MWNT's tip contacts the other Ti electrode directly. Our experimental data on two-terminal current-voltage measurements could include a larger contact resistance of the MWNT's tips than that of the MWNT's ends because there could be an electrical barrier between crossing MWNTs, or between a MWNT's tip and a Ti electrode.

(a)   (b)

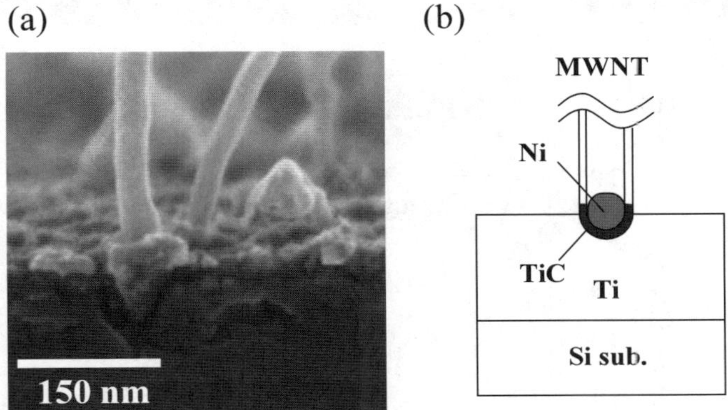

Figure 10. (a) Cross-sectional SEM image of MWNTs/(Ni/Ti)-electrode interface. (b) Model of simultaneous formation of MWNTs and end-bonded low-resistance TiC ohmic contacts.

Electrical Properties of CNT Vias

We grew bundles of MWNTs in via holes to preliminarily demonstrate CNT vias [10]. As shown in Figure 11a, each test sample consisted of a 100-nm Ti contact layer for TiC contact formation, a 10-nm Ni catalyst layer, and a 350-nm SiO_2 dielectric layer on a Si substrate. Via holes were patterned using conventional g-line lithography and anisotropic dry-etching with fluorine-based gases. After the lift-off process using photo-resist, we obtained via holes that had a Ni catalyst layer exposed at the bottom. On the substrate we grew the bundles of MWNTs by HF-CVD at 600°C. We used a 100-nm Ti contact layer for the upper electrodes. We then made two-terminal I–V measurements on CNT bundles end-bonded to the upper and lower Ti electrodes. The total resistance of a CNT via consisting of about 1000 tubes is shown in Figure 11b, which also shows the total resistance of one CNT and three CNTs bridging the gap between electrodes (shown in the previous chapter). The total resistance of the CNT via was only 100 Ω, which is three orders of magnitude less than that of one CNT. This indicates that the current flows in parallel through about 1000 tubes.

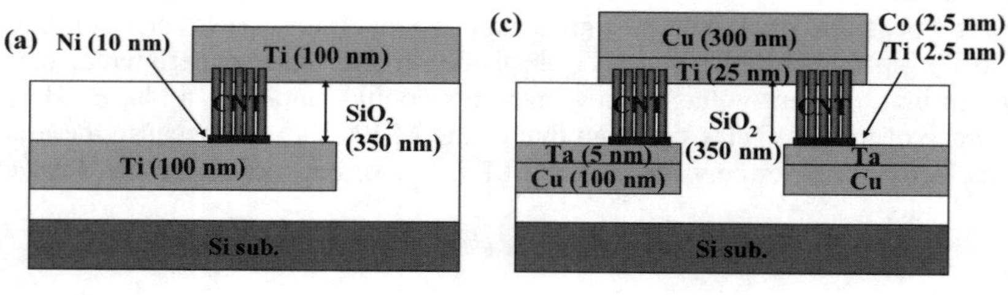

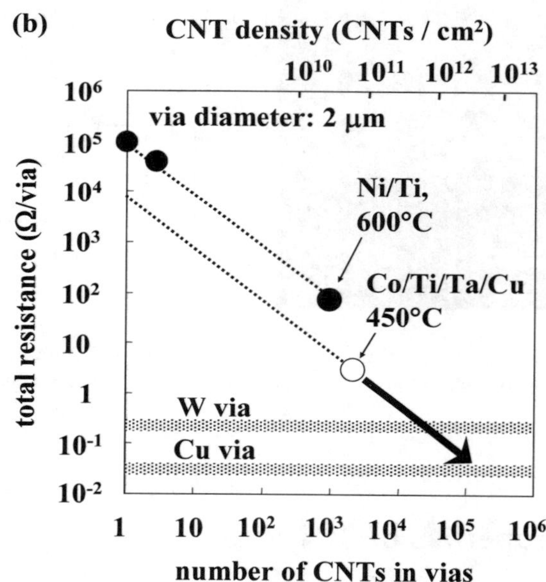

Figure 11. (a) Schematic of CNT via structure using Ni/Ti layers. (b) Dependence of total resistance of CNT vias on number of CNTs [18]. (c) Schematic of CNT-via-chain structure using Co/Ti/Ta/Cu layers.

In the latest improvement of our CNT vias on a Cu wiring [18], we have decreased the via resistance by using a low-temperature CVD method with a Co catalyst, a TiC ohmic contact, and a Ta barrier layer and fabricated 1000 via chains with the structure shown in Figure 11c. The resistance of a 2-μm-diameter via consisting of about 1000 tubes is shown in Figure 11b. The lowest resistance we obtained was about 5 Ω, which is one order of magnitude greater than the resistance of a 2-μm-diameter with tungsten and two orders of magnitude greater than that of one filled with Cu. So it's important to produce smaller-diameter nanotubes that pack more densely. We are therefore going to try using a catalytic nano-particle technique [19] to increase the current roughly 10^{10}/cm^2 packing density of the nanotubes to 10^{12} /cm^2.

Conclusion

We have developed CNT vias consisting of about 1000 tubes by using a low-temperature CVD method with metal catalysts (Ni or Co), TiC ohmic contacts, and Ta barrier layers on Cu wiring. The total resistance of the CNT via was three orders of magnitude lower than that of one CNT, indicating that the current flows in parallel through about 1000 tubes. We believe this is the first trial demonstration of CNT vias for future LSI interconnects.

Acknowledgements

We thank Dr. N. Yokoyama, who is Director of the Nanotechnology Research Center of Fujitsu Laboratories Ltd., and Professor M. Niwano of Tohoku University for their support and useful suggestions. Our discussions with Drs. M. Ohfuti, N. Shimizu, and T. Nakamura are also appreciated. This work was completed under the management of JFCC as part of the METI R&D program (Advanced Nanocarbon Application Project) supported by NEDO.

References

1.   S. Iijima, Nature 354, 56 (1991)

2.   R. Saito, M.S. Dresselhaus and G. Dresselhaus, Physical Properties of Carbon Nanotubes (Imperial College Press, London, 1998)

3.   Z. Yao, C. L. Kane and C. Dekker, Phys. Rev. Lett. 84, 2941 (2000)

4.   J. Hone, M. Whitney and A. Zettle, Synth. Met. 103, 2498 (1999)

5.   J. Kong, E. Yenilmez, T.W. Tombler, W. Kim and H. Dai, Phys. Rev. Lett. 87, 106801 (2001)

6.   S.J. Tans, A.R.M. Verschueren and C. Dekker, Nature 393, 49 (1998)

7.   R. Martel, T. Schmidt, H.R. Shea, T. Hertel and Ph. Avouris, Appl. Phys. Lett. 73, 2447 (1998)

8.   W. Hoenlein, International Microprocesses & Nanotechnology Conference 2001, Proceedings, p. 76 (2001)

9.   J. Li, Q. Ye, A. Cassell, J. Koehne, H. T. Hg, J. Han and M. Meyyappan, International Interconnect Technology Conference, IITC 2003, Proceedings, p. 271 (2003)

10. M. Nihei, M. Horibe, A. Kawabata and Y. Awano, International Interconnect Technology Conference, IITC 2004, Proceedings, p. 251 (2004)

11. M. Nihei, A. Kawabata and Y. Awano, Japan J. Appl. Phys. 43, 1856 (2004).

12. M. Nihei, A. Kawabata and Y. Awano, Japan J. Appl. Phys. 42, L721 (2003)

13. J.M. Mao, L.F. Sun, L.X. Qian, Z.W. Pan, B. H. Chang, W.Y. Zhou *et al.*, Appl. Phys. Lett. 72, 3297 (1998)

14. T. de los Arcos, F. Vonau, M.G. Garnier, V. Thommen, H.-G. Boyen, P. Oelhafen *et al.*, Appl. Phys. Lett. 80, 2383 (2002)

15. H. Murakami, M. Hirakawa, C. Tanaka and H. Yamakawa, Appl. Phys. Lett. 76, 1776 (2000)

16. A. Kawabata, M. Horibe, D. Kondo, M. Nihei and Y. Awano, International Conference on the Science and Application of Nanotubes, NT04, Proceedings, p. 31 (2004)

17. Y. Zhang, T. Ichihashi, E. Landree, F. Nihey and S. Iijima, Science 285, 1719 (1999)

18. Y. Awano, M. Nihei, S. Sato, A. Kawabata, D. Kondo, M. Ohfuti *et al.*, International Conference on Solid State Devices and Materials, SSDM2004, Proceedings, p. 40 (2004)

19. S. Sato, A. Kawabata, M. Nihei and Y. Awano, Chem. Phys. Lett. 382, 361 (2003)

Nickel Nanowires Obtained by Template Synthesis

I. Z. Rahman, K. M. Razeeb, M. A. Rahman

University of Limerick, Materials and Surface Science Institute (MSSI),
Limerick / Ireland

Introduction

High aspect ratio magnetic nanowires and other nanostructures in particular, compacted in a matrix need to be properly investigated for possible applications in the field of sensors, high-density storage media, electronics, separation chemistry and biomedical engineering. Metallic nanowire structures of Ni, Au, *etc.* have potential wide-ranging applications as interconnects where their electronic conductivity behaviour can be exploited. In addition, these nanostructures are interesting to study the fundamental magnetic or electronic phenomena in small dimensions. For example, Nielsch *et al.* [1] found that bulk magnetic properties of nickel nanowires arrays show a strong magnetic anisotropy along the columnar axes, having a coercive field of 1200 Oe and nearly 100% squareness. Template synthesis has recently proved to be an elegant chemical approach for the fabrication of nanoscale materials and are proven to be a cheaper alternative to sophisticated methods like lithography, sputtering or molecular beam epitaxy. In this paper we focus on fabrication of nickel nanowires (grown through the pores of nanochannel alumina or NCA templates) by electrodeposition and discuss potential problems for their use as interconnects.

Fabrication of Nanostructured Materials

In nanotechnology one key issue is the development of simple fabrication techniques for mass production of identical nanostructures. The ideal synthesis technique should also provide control and ease over particle or crystallite size, distribution of particle size and interparticle spacing, efficiency and cost effectiveness. Synthesis and processing of metallic and ceramic nanostructured materials are usually related to particular applications and have been reviewed by several authors [2–4]. In the microelectronics industry, lithography is a proven technique in the fabrication of nanostructures due to its efficiency and ease of controlling the process parameters above the optimal dimensions of the nanostructure. With new generation e-beam/ X-ray lithography, features as small as 3–4 nm can be produced. This "top-down" approach to fabricate nanostructure is better from the perspective of simplicity as compared to the "bottom up" processes like scanning tunnelling microscopy (STM) that realises nanostructures by manipulating individual atoms.

This also requires an ultra-high vacuum, which makes the processing expensive and time consuming. With deceasing feature size, present lithographic techniques grow exponentially more expensive, and may never reach the dimensions required for nanotechnology. This conventional approach is also wasteful because many production steps involve depositing unstructured layers and then patterning them by removing most of the deposited films. There is also a challenge in wiring the components together in nanoscale electronic devices. Conventional lithography techniques cannot produce metal contacts much smaller than 100 nm. These are an order of magnitude larger than components made from nanowires and nanotubes. Based on these limitations, it has been suggested that future device integration may be based on alternative approaches that rely on the assembly of nanometre- scale colloidal particles such as isotropic and anisotropic metallic particles [5, 6], carbon nanotubes [7], and nanopatterning [8] and can involve laser or electrochemical techniques as complementary tools. The competing methods for fabricating nanowires are lithography [9], deposition into tracks of high-energy particles [10], deposition at step edge [11], synthesis by self-assembling technique and STM [12]. On the other hand, an alternative approach that exploits a naturally occurring process, namely the self-assembly inside porous media, may be used as a vehicle for fabrication.

Template Synthesis of Nanowires

Patterning materials using template the synthesis technique is very cheap and efficient. Templating is not an old technique for magnetic media in microscale though interest on this technique is growing rapidly due to the ease of fabrication of materials on the nanoscale range with efficiency and cost effectiveness. This method can be much more economic if the template itself can be produced using a cheaper technique rather than lithography. One possible route towards cheaper template mass production are the chemical synthesis techniques. Until recently, it was not feasible to fabricate templates with homogeneity in nano- and microscale on a large surface area. Several groups [13–16] came out with some innovative solutions that have popularised the chemical synthesis technique and thus opened up new paths towards homogeneous template synthesis en route to fabrication of nanostructured materials.

Fabrication Process of Templates

The template synthesis technique requires a template with pores or discontinuity where the desired material could be deposited. If these templates have pore structure with diameter in the nanometre range, the deposited material inside the pores with the same diameter in nanoscale and depending on the length of the pores the ultimate nanostructure could be formed. Structures like pillars, fibrils, dots and wires could be fabricated using different templates and pore structures. In the synthesis of the nanostructured media using templates, physical or chemical synthesis techniques can be adopted. The physical synthesis includes techniques like sputtering or vacuum evaporation, while the chemical synthesis comprises electrochemical deposition [17], electroless deposition, sol gel deposition, chemical vapour deposition [18], polymerisation [19], *etc.* The template synthesis again can be sub-divided into two categories:

- Pattern transfer on the resist or polymer using a template by the nanoimprint technique and [20–25].
- Synthesis of desired material within the pores or on the wall of the pores of a porous template [18, 26, 27].

The synthesis and processing of magnetic and ceramic nanowires showed that all these techniques have inherent limitations [2]. Most of them suffer from the inhomogeneity in the matrices that are fabricated. It is difficult and very laborious to place such nanowires in precisely controlled locations, as would be needed in micro-superconducting quantum interference devices (SQUID) for magnetic measurements. These also limit the materials that can be used for fabrication. For example, it is not possible to fabricate non-conductive (insulator) nanostructured materials (nanowires) using electrodeposition technique but can be synthesised using sputtering or lithography techniques. The same analogy can be applied in the synthesis of magnetic nanostructures and also on nanowires. In case of the fabricating magnetic nanowires, the ideal synthesis technique should provide control over particle or crystal size, distribution of particle sizes and interparticle spacing. For arrays of magnetic nanowires, the material requires patterning to go down to a size that is comparable to that of a single domain of the magnetic material of interest.

A wide variety of templates, like track etch membrane [18, 26], porous alumina (NCA), nanochannel glass (NCG) [28] and also self-assembled polymers may be used. Nanochannel alumina and nanochannel glass, may be used to fabricate both magnetic nanonetworks [20, 29, 30] and nanowires [31]. Some authors reported deposition of a magnetic layer on NCA with enhanced coercivity suitable for information storage although no work on using this type of template for pattern transfer purposes has been reported [32]. In the fabrication of these templates different techniques are adopted from lithography to self-organising processes. Two popular patterning techniques are:

- Lithography and
- Chemical synthesis/self-organised techniques

Lithography Technique for Patterning Templates

The lithography technique currently involves the formation of a pattern and then transfer of that pattern to form an electronic, optical or magnetic structure. In this technique, by shining light through a stencil or "mask" of the pattern and passing this light through a series of lenses, the size of the image is reduced. Then this image is projected onto a substrate (*e.g.* silicon) covered by a photosensitive resist. Chemicals are then used to wash away the exposed areas of the resist, leaving behind the pattern of the mask on the substrate. There are several types of lithography available namely, optical lithography, electron beam lithography (EBL), ion beam and X-ray beam lithography and proximal probe lithography. Electron beam lithography and X-ray lithography are capable of patterning features as small as 10 nm but the process is time consuming and expensive. There are several drawbacks of e-beam lithography [33], it suffers from low throughput because of slow processing speed. X-ray lithography can be a quicker alternative to EBL; it is able to print nanofeatures in one flood of exposure by X-rays. The shorter wavelength also gives better resolution in patterning. The proximal probe is another addition to nanolithography. A proximal probe tip, like atomic force microscope or scanning tunnelling microscope tips can be used to define patterns in the resist [33].

Despite the popularity in industry and potential for high resolution, there are drawbacks associated with lithography techniques in the fabrication of nanostructures and nanowires. The length scale of the optical wavelength used has limited feature spacing in light masks. Thywissen and Prentiss [34] have overcome this problem in principle by exposing a silicon substrate to a beam of "metastable" argon atoms. These metastable atoms exist in a naturally excited state and release their energy when they strike the substrate. A layer of hydrocarbons that resides on the substrate functions as the resist by adhering more strongly to the substrate when energised. To create a pattern in the atom beam, the researchers used a mask made of laser light. By modifying the frequency rather than the intensity of the light, the resolution of the mask can be extended beyond the optical diffraction limit. Atom lithography has demonstrated feature sizes smaller than the diffraction limit of light. The spacing between the features can also be much smaller, even when using light to do all the patterning. But this potential will require more laser power and a more finely collimated atomic beam.

Chemical synthesis/Self-organisation Techniques for Patterning

In the fabrication of templates, self-organisation techniques usually behave in the way that the final template structure is self-organised and form nanostructured pores or voids due to the physical or chemical treatment during the fabrication. Applying a high electric field on the film of diblock copolymers to build nanopores [13] and anodisation to fabricate pore structures on pure metals and semiconductors are popular techniques among nanotechnology researchers. Chemical synthesis techniques have also been used to fabricate NCA templates.

Nano Channel Alumina (NCA)

The controlled anodisation of aluminium has been an industrial process since the 1920s [35], but the main emphasis was anodisation at above pH 7 to yield thick corrosion-resistant passivated oxide "barrier type films." When Al is anodised in acidic solutions, pores form that have only an approximately hexagonal order [36], which is idealised as truly hexagonal as shown in Figure 1. This structure has been called "alumite" [37]. Since the pores were relatively uniform, their use as hosts for magnetic nanowires of Fe or Co received early attention by many groups worldwide [37–40]. The mechanism for pore formation and its hexagonal ordering (a mesoscopic phenomenon) remains unknown. Early interests by the hard disk industry in alumite faded in the early 1990s due to the heterogeneity of pore structures and inter pore distances. Since then, a dramatic breakthrough was achieved when Masuda and co-worker [16] showed that prolonged anodization, followed by stripping of the thick oxide, and re-anodization produced ordered nanopores with perfect hexagonal domains. This result was confirmed by others [41–44].

The process for the preparation of NCA is well documented in several publications. Depending on the processing steps and parameters, either alumina with ordered or disordered pores could be formed. All anodization processes adopted so far start with electropolishing pure Al substrate. In this regard, electropolishing is an essential step for forming ordered hexagonal patterns [43]. Other patterns are also formed during electropolishing [45,46]. Using single-crystal Al (110) samples, stripes (1–7 nm high, with repeat distances of 42±2 nm at 20 V dc and 62±2 nm at 40 V dc) were found to extend uniformly throughout a 1 cm^2 sample [47], which was a truly macroscopic ordering.

Anodization in Acid: Disordered Pore

The recipe for forming porous structures in Al by anodising in acid is simple. Using pure Al (or even the Al–1%Mg alloy used for commercial soda cans, but not commercial Al plate [48]), acids such as H_2SO_4, $H_2C_2O_4$, H_3PO_4, or H_2CrO_4 (but not HCl) and prolonged anodization with a dc power supply (10-60 V), creates a porous structure. The structure shown in Figure 2 only crudely resembles the structure depicted in Figure 1. The pore growth rate is 0.1 µm min^{-1}. Amorphous and anhydrous Al_2O_3 forms, but the pore at the bottom of Al_2O_3 may be partially crystalline. Within 10 s of anodization, the pore bottoms become highly resistive (10 MΩ cm^{-2}). Thereafter, the current increases about five-fold and remains steady over time, as pores form, until the Al is fully exhausted. The pore spacing D_s is in-

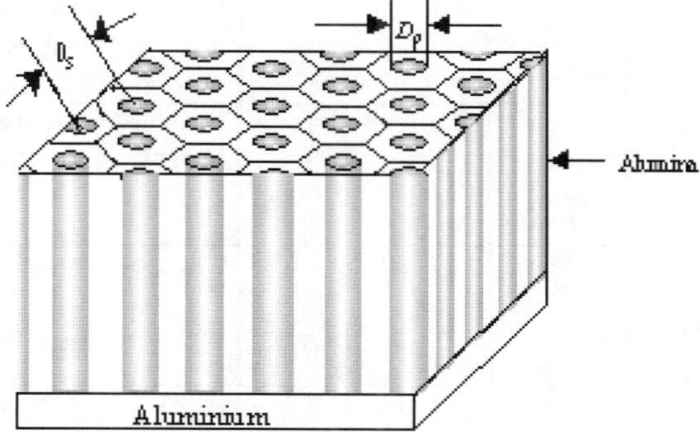

Figure 1. Idealised hexagonal array [32] of porous Al_2O_3 film formed by DC anodization of Al in acid (sulphuric, oxalic, or phosphoric, below pH 4).

dependent of the electrolyte and pH, and is about 2.8 nm per volt. The pore diameter D_p varies somewhat with acid concentration and temperature: the D_s/D_p ratios are found to be 1.74 to 2.15 for H_3PO_4, 3.33 for H_2CrO_4, 4.88 for H_2SO_4, and 3.01 for $H_2C_2O_4$. O. Jessensky and his group [43, 44] concluded for the process described by Masuda [16] that anodization voltage varied between 30 to 60 V and was able to produce a pore growth rate of 1–2 µm per hour. This is equivalent to an etching time between 2 to 4 days. At this rate, a layer thickness between 100 to 200 µm and aspect ratios (ratios of pore diameter to depth) of more than 1000 can be achieved.

Anodization in Acid: Ordered Pore

In the so-called one-step anodization process, aluminum is anodised only once in an acidic electrolyte whereas in 1995 Masuda and his coworkers displayed a two-step anodization technique to fabricate ordered pore structure. The original recipe by Masuda [49]: after degreasing an Al plate (99.99%) in acetone, a mirror surface was achieved by electropolishing in a solution of perchloric acid and ethanol. Table 1 summarises a few cell characteristics for anodization as well as electropolishing of aluminium adopted by several groups. The anodising of Al was carried out at a constant voltage of 40 V in a 0.3 M oxalic acid solution at 17°C for 10 h. Then the anodic distorted oxide layer was removed in a mixture of phosphoric acid (6 wt%) and chromic acid (1.8 wt%) at 60°C for 14 h that formed a textured pattern of concaves on the surface of the Al substrate. This textured Al specimen was anodised again for 5 min under identical conditions to those of the first anodization. Each concave of the surface resulted in the ordered formation of the holes, that is, holes were produced at the bottom part of each convex due to its geometrical effect.

Table 1. Different electropolishing techniques and different bath characteristics for anodization [49–52].

Electropolishing	First anodization				Removing oxide	Second anodization
Parameters	Bath	Current Density/ Voltage	Temp (°C)	Time (h)	Parameters	Parameters
0.24 M Na_2CO_3 at 85 °C for 1 min and neutralised in 1:1 HNO_3: H_2O for 20 s	0.23 M $H_2C_2O_4$	11–20 V	24-25	1	None	None
H_2CrO_4: H_3PO_4=2:8 At 200 mA cm^{-2} for 8 min	0.4 mol dm^{-3} $H_2C_2O_4$	18 mA cm^{-2} (46–47 V)	25	-	None	None
Perchloric acid and Ethanol	0.3 M $H_2C_2O_4$	40 V	17	10	H_3PO_4 (6 wt %) H_2CrO_4 (1.8 wt%) at 60°C for 14 h	Same as 1st one but only for 5 min
5% NaOH at 60°C for 20 s and rinsed in 35% HNO_3	85% H_3PO_4	2 mA cm^{-2}	20±2	-	2 M H_3PO_4	0.4 M H_3PO_4

In order to study the cellular growth of highly ordered porous anodic films on aluminum L. Zhang *et. al.* [41] varied the second anodization time from 10 min to 10 h. Jessensky *et. al.* [43] thoroughly investigated the growth kinetics and the influence of experimental conditions. They followed the track of Masuda *et. al.* [49] for sample preparation but varied the anodization voltage 30 to 60 V and pretreatment parameters of the aluminum foil. The morphology of the top of the alumina showed disordered pores is shown Figure 2 whereas, the bottom layer showed densely packed nearly perfectly ordered hexagonal structures. The AFM image analysis of the bottom surface of the alumina showed a roughness less than 30 nm over the image area as shown in Figure 3.

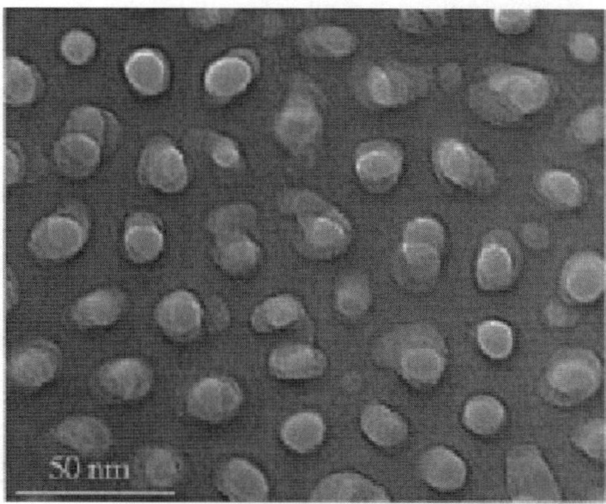

Figure 2. TEM top view of disordered alumite film formed in 15% H_2SO_4 [47].

Electrodeposition of Nickel Nanowires

Nickel is generally deposited from sulphate, sulphate-chloride, or sulfamate electrolytes with or without additives. For magnetic applications, nickel based thin film and multilayers have been deposited to investigate the phenomenon of giant magnetoresistance (GMR) [54]. Among all the electrodeposition process parameters, pH and the electrolyte temperature had the most influence on the crystallographic structure of the film. Also other parameters like the cathode current density or additives in the electrolyte may change the deposition characteristics. Several types of electrolytes have been reported so far to fabricate thin magnetic films or multilayers of nickel (see Table 2). A consideration of the effect of deposition conditions on the morphology of nickel deposits has long been studied from theoretical as well as experimental standpoint [55–57].

To fabricate microstructures by electroplating, a conductive plating base or seed layer and a means to pattern the electrodeposits are needed. Since the localised electrodeposition rate is proportional to the localised current density, a uniform current density over the entire seed layer is needed to obtain an electrodeposit of uniform thickness. For deposition of nickel nanowires, aluminium acting as the seed layer is sputter-deposited on one side of the NCA template. The wall of the alumina pores will work as the plating mask. Alumina, being a non-conductive material (resistivity in the range of $<10^{14}$ Ω-cm), does not allow any electrodeposition. On the other hand, aluminium is exposed through the pores of alumina and allows the electrodeposition to occur through the pores of NCA on sputtered aluminium. Usually the constant current (galvanostatic) or constant voltage (potentiostatic) deposition has been applied for fabricating nanowires, though deposition by alternating current or voltage is not unusual. It has been reported that using alternating current various metals have been deposited inside the pores of NCA [37, 51, 58, 59]. For example, using sulphate electrolyte Co could be electrodeposited [39] at V_{rms} = 20, 200 Hz. Homogeneous nucleation at high current density can be favoured by applying a short initial pulse (40–50 V for <1 s), promoting homogeneous growth of nanowires. Filling of the nanometric pores can also be initiated by sonication [49].

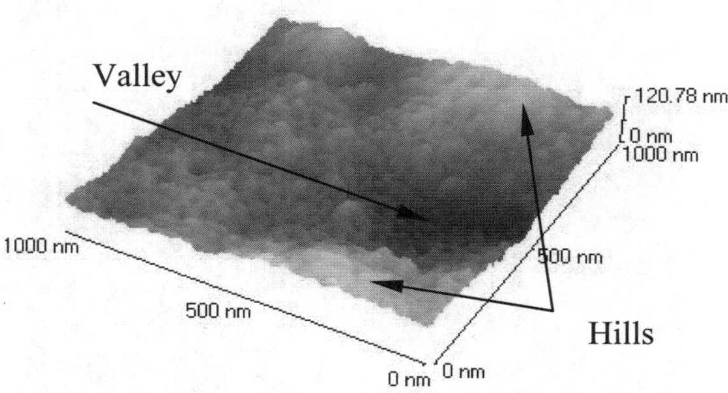

Figure 3. Atomic force microscopic topography of NCA: 3-D image of NCA 20 showing surface convolution: hills and valley [53].

In another deposition technique [39], nickel was directly plated onto nearly insulating barrier oxide at the bottom of the pore from a Watts-type electrolyte and by current pulses. Alternating current is needed either for rectifying the nature of the Al metal/oxide junction or for diffusion barriers inside the deep nanopores. These barriers or the metal oxide junction are formed during the two-step anodization process adopted for the fabrication of homogeneous nanochannel alumina as described before. The rectifying properties of the barrier layer allow the pores to be filled uniformly by ac or pulsed electrodeposition. Experimental investigations have shown that pores fill on-uniformly when direct current for the deposition is used [53]. For dc deposition, the applied potentials are smaller than for pulsed or ac deposition. The potential of the electrons, which drops across the barrier layer during the deposition, depends on the thickness of the barrier layer. Therefore, a higher proportion of the applied deposition potential drops across the barrier layer in the case of dc deposition. Thus thickness fluctuations of the barrier layer influence the deposition rate in each pore more strongly. In the case of commercial NCA templates, direct current or direct voltage deposition techniques (galvanostatic or potentiostatic) can be used [39, 62–63], as there is no thick oxide layer (barrier layer) forms between the sputtered aluminium and the pores.

Detailed work carried out by Rahman's group [53, 63–65) in fabricating Ni nanowires using three different nominal pore diameters of 200 nm, 100 nm and 20 nm commercial NCA templates (Anodisc 13 by Whatman plc). The electrolyte used throughout this study is a mixture of $NiSO_4.6H_2O$ and H_3BO_3 in deionised water, which has a resistivity of 18 MΩ. The electrodeposition is conducted using galvanostatic deposition process. The ohmic contact is made using a sputtered layer of aluminium on the back side of NCA templates that acted as a conductive substrate for electrodeposition. The pores in the template are found to be arranged in a hexagonal array having a thickness of 60 µm. Pore densities as high as 10^8 to 10^{12} cm^{-2} are observed. There are two sides of these NCAs namely, the "filtration side" and the "reverse side" as shown in Figure 4. Table 3 shows the comparison of pore sizes measured by different experimental techniques (AFM and SEM.)

Table 2. Different bath characteristics for nickel deposition

Bath ID	Name of the baths	Component salts	Concentration (g l^{-1})	Temperature (°C)	pH	Cathode current density (A dm^{-2})	Deposits property
N1 [10]	Watts bath	$NiSO_4.6H_2O$ $NiCl_2.6H_2O$ H_3BO_3	240–340 30–60 30–40	45–65	1.5–4.5	2.5–10	Moderate stress, presented strain and flaws
N2 [11]	Semi-bright	$NiSO_4.7H_2O$ $NiCl_2.6H_2O$ H_3BO_3	300 10 45	50	3.5–4.5	0.5–1	Good coverage but larger flaws
N3 [12]	Chloride	$NiCl_2.6H_2O$ H_3BO_3	300 38	50–70	2.0	2.5–10	Stressed film
N4 [11]	Strike	$NiSO_4.6H_2O$ $NiCl_2.6H_2O$ H_3BO_3	100 30 30	50	4.5–5.0	0.5–1	Lowest coverage of the exposed area and less flaws
N5 [17]	Sulphate bath	$NiSO_4.6H_2O$ Na_2SO_4 H_3BO_3	40 150 Varied	40–60	2.0–3.5	2	Strong <220> orientation

The average roughness of the surface of these NCA templates turned out to be quite high with a range of 16.52 nm on the filtration side and 70.68 nm on the contact side. Hills and valleys are also observed (see Figure 3) on the contact side of the NCAs. On the reverse side of the 200 nm nominal pore size NCA, the pores have an average diameter ranging from 250 to 300 nm, while on the filtration side, the values ranged from 190 to 215 nm.

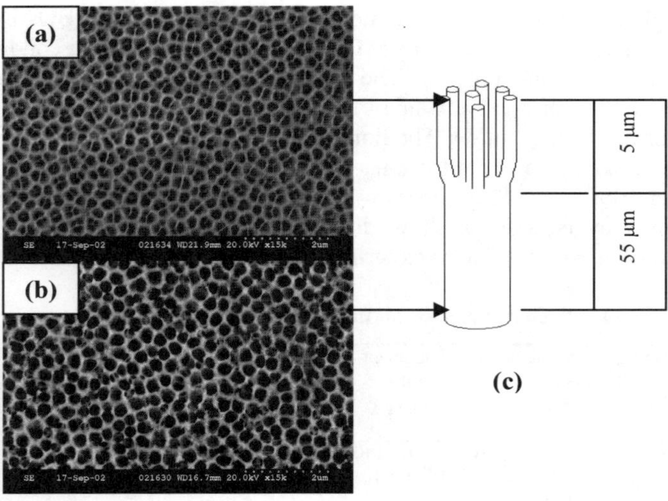

Figure 4. Pore and wire growth characteristics of NCA (200 nm pore diameter): (a) Filtration side of NCA; (b) Reverse side of NCA; (c) Schematic representation of branching of a single pore [60].

Table 3. Pore size data using SEM and AFM topographic images *(F= filtration side and R= reverse side).

Template	*Sides	SEM (nm)	Standard deviation (nm)	AFM selective line analysis (nm)	Standard deviation (nm)	AFM peak valley analysis (nm)	Standard deviation (nm)
NCA 200	F	209.4	39	213	37.7	220	55
	R	308.51	76.6	304	66.9	290	60
NCA 100	F	111.9	28.6	107	6.12	104	38
	R	241.8	81	238.5	20.9	175	39.6
NCA 20	F	29.8	12	28.3	5.3	40.5	13.5
	R	201.9	85.7	165.5	76.5	144	68.7

In Figure 5a the SEM image of the cross section shows Ni nanowires electrodeposited inside the pores of NCA200 template. Nickel nanowire arrays inside the alumina could be freed by chemical treatment: the electrodeposited alumina template was put in 10% NaOH solution for 30 min to etch away the aluminium backing plate. Then the template was rinsed with deionised water for 10 min and placed in 0.4 M H_3PO_4 solutions for 12 h. The alumina matrix was completely dissolved with nickel nanowires dispersed in the solution. The wires were then cleaned in deionised water using ultrasonication. A drop of that water was placed on the top of the Si wafer (sputtered with Al) and dried with a nitrogen gun. The wires are then observed under the polarisation optical microscope and the SEM (shown in Figure 5). In Figure 6, a single nanowire and its compositional analysis by EDX technique have been depicted. The measured average diameter of nickel nanowires is found to be 323.3 nm for wires deposited inside NCA200. The pore diameter of NCA200 that is measured by other technique showed a mean diameter of 308.5 nm (see Table 3 on the reverse side of NCA200).

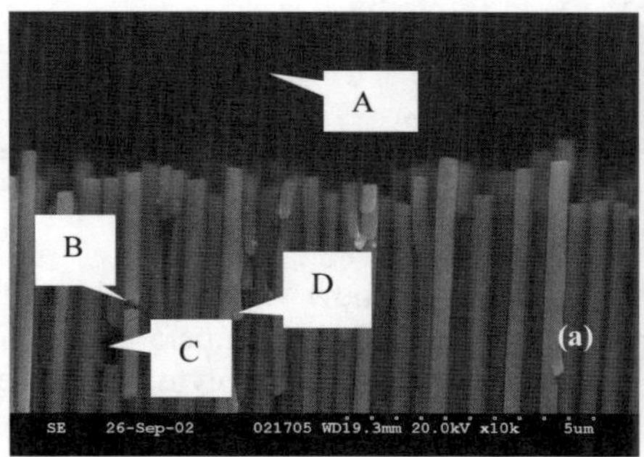

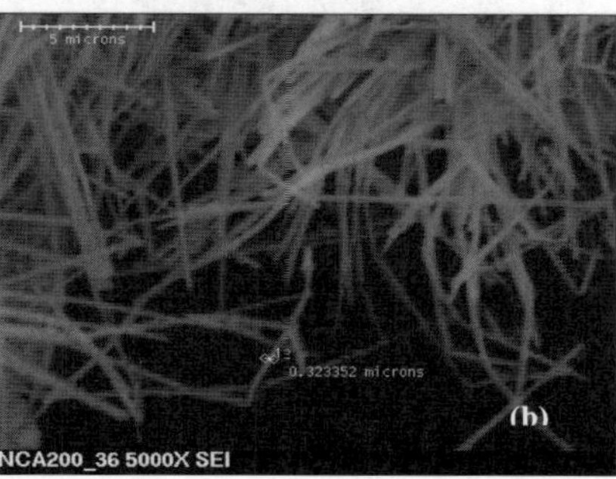

Figure 5. SEM images of (a) cross section of NCA200: electrodeposition for 2 h showing growth of nickel nanowires from aluminium backing plate where, voids and gaps (denoted by B and C) in the grown wires and some branching in the wires (D) and (b) SEM image of dispersed nickel nanowires on Si wafer [63].

The EDX analysis from Figure 6b confirmed that the wires were pure nickel, though the amount of oxide is not revealed from the analysis. The peaks of aluminium and Si were from the Si substrate on which aluminium was sputtered beforehand to minimise the charging effect while the images were analysed using SEM. Figure 7a shows the TEM image of a cross section of NCA200 with nanowires and Figure 7b shows the heterogeneity in the growth of wire diameter and length. Further TEM analysis on single nanowires also confirmed that there was a difference in diameters at the two ends of the NCA templates (filtration side and reverse side). For wires grown using NCA200, on the reverse side the wire has an average diameter of 311.11 nm, whereas, on the filtration side of the same wire, the average diameter is found to be 142.85 nm.

The XRD analyses revealed that the nickel nanowires were polycrystalline with preferred orientation of (111) plane. It was also observed that changing the pH of the electrolyte did not affect this orientation. During deposition from a bath at 40°C, (220) orientation dominated and increasing the electrolyte temperature from 40°C to 60°C resulted in the development of a (200) plane. The development of preferred planes in nickel deposits is a result of adsorbed hydrogen, which may be enhanced by the electrolyte temperature while the pH of the solution remains constant [65].

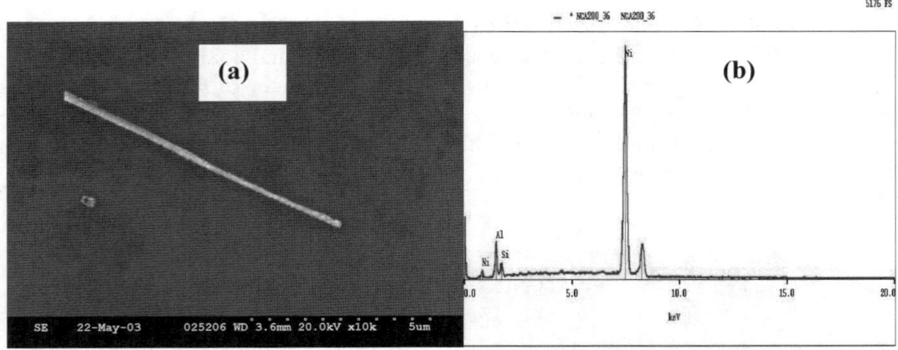

Figure 6. (a) SEM image of single nanowire dispersed on Si wafer and (b) EDX analysis on the dispersed wires on Al-sputtered Si wafer [54].

The XRD patterns of nickel nanowires in Figure 8 show varying peak width ranging from 0.2703° up to 0.4271°, depending on the pH and temperature of the electrolytes. It is important to point out that the pH of the electrolyte has a profound effect on the crystallite size at the nanometric scale as can be observed from Figure 9. The crystallite size decreased with the pH of the electrolyte and increases for both 30 and 60 min deposition time at 40°C, which is in accordance with the investigation by Ebrahimi *et. al.* [67]. Above of pH 2.99, the average crystallite size increased again. Deposition from a bath temperature of 60°C showed different trends where the average crystallite size decreased after a sudden increment at pH 2.99 and 3.48 for 30 min and 60 min deposition times respectively. However, deposition for longer time (120 min) showed different trends both for 40°C and 60°C electrolyte temperatures. Average crystallite size decreased when the pH was increased to 3.7 at 60°C. For 40°C bath temperature, the crystallite size decreased almost linearly with the increment of pH value at 2.5.

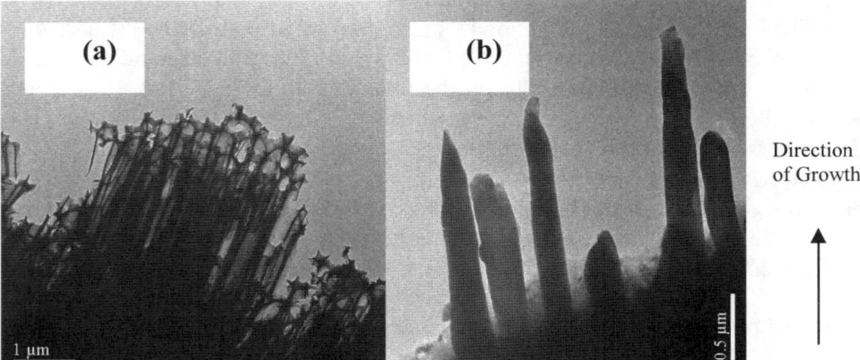

Figure 7. (a) TEM image of the cross section of nickel nanowires inside NCA200 and (b) Image of nickel nanowires showing the heterogeneity in wire diameters [53].

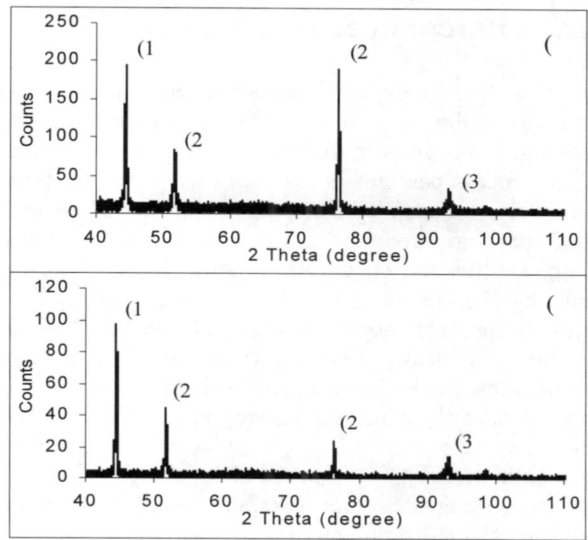

Figure 8. XRD of samples deposited at process parameters (a) Bath temperature = 40°C, time = 30 min and pH 3.48 and (b) Bath temperature = 60°C, time = 30 min and pH 3.48 [65].

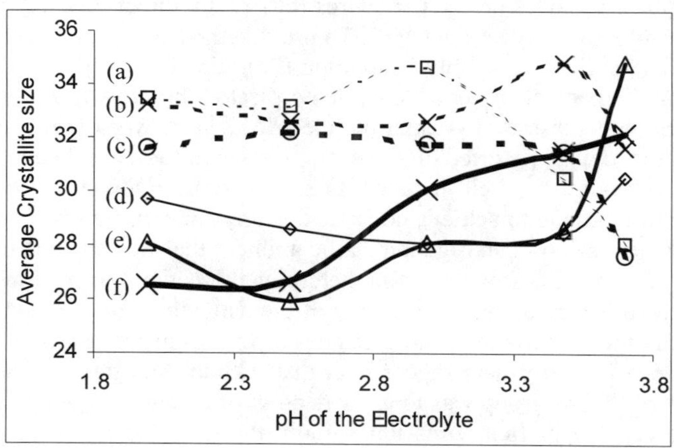

Figure 9. Variation of average crystallite size as a function of pH, time and temperature of the electrolyte deposited at 60°C: (a) 30 min, (b) 60 min and (c) 120 min and at 40°C: (d) 30 min, (e) 60 min and (f) 120 min [53].

Comments on Ni Nanowires Grown By Template Synthesis and Interconnect Technology

Pores of commercial NCA templates have a heterogeneous shape due to the branching of the pores on the filtration side. Nickel nanowires grown through these pores showed homogeneous growth, though in a small number of pores the wires showed voids. The process parameters, *e.g.* bath temperature, influences the preferred orientation of electrodeposited nanowires regardless of pH of the electrolyte. The pH parameter plays an important role in controlling the formation of secondary phases. A quick reflection on this study is that as a fabrication process for nanowires, templating is a simple and cost-effective technique. The alumina template itself makes the process more cost effective due to their easy fabrication steps, which are done chemically. Except for the usefulness of the commercially available NCA templates, the major problems associated are the branching of the pores, distribution of pore sizes and the surface roughness. These along with the electrodeposition process parameters (*e.g.* pH and temperature) define the uniformity in nanowire length. Nielsch [1] reported that ordered porous alumina arrays with shapely defined pore diameter and interpore-distance can be obtained the by two-step electrochemical anodization process [41, 43, 49].

A number of other techniques have been reported to grow nanowires. Nevertheless, due to the small dimensions, manipulation of nanowires into position and providing electrical contacts pose a difficult challenge. The most popular technique for providing electrical contacts to an *ex situ* grown nanowire is by dispersing the nanowire on a substrate followed by electron beam lithography and metallization to fabricate gold electrodes on the nanowires [68] or, by direct-focused ion beam-induced deposition of the electrodes [69]. To provide the contacts, other techniques are: electric field-assisted assembly in solution [70], direct growth of the nanowire on the substrate, where Zhang *et al.* [71] demonstrated the growth of aligned carbon nanotubes on a substrate by using the electric field between two biased electrodes. Calleja *et al.* [72] reported on conducting AFM to form a gold nanowire between two electrodes by field-induced mass transport. Oon and Thong [73] reported a new technique to achieve controlled single nanowire growth on pointed objects by field emission in an organometallic ambient that was adapted to achieve localised single nanowire growth at a predetermined location (such as metal electrodes) and initial electrical characterization of the nanowires for growing tungsten nanowires and other materials. Initial estimates of the nanowire resistivity indicate values about one to two orders higher than that of bulk tungsten. An alternative idea put forward by this group was to grow nanowires on sharp-pointed nanostructures that are capable of field emission without the need for anode contact, and thereby may provide a method for interconnection to such structures.

Acknowledgements

The Higher Education Authority, Ireland under the PRTLI scheme supported this work.

References

1. K. Nielsch, R.B. Wehrspohn, J. Barthel, J. Kirschner, S.F. Fischer, H. Kronmüller *et al.* J. Magn. Magn. Mater, 249, p. 234 (2002)

2. D.L. Leslie-Pelecky, *Chem. Mater.*, 8, p. 1770 (1996)

3. R.W. Siegel in Mechanical Properties and Deformation Behaviour of Materials Having Ultra-Fine Microstructures, M. Nastasi *et al.* (eds.), Kluwer Academic, Dordrecht, The Netherlands, 1993

4. R.P. Andres, R.S. Averback, W.L. Brown, L.E. Brus, W.A. Goddard III, A. Kaldor *et al.*, Mater. Res., 4, p.704 (1989)

5. V.P. Roychowdhury, D.B. Janes and S. Dandyopadhyay, Proc. IEEE, 85, p.574 (1997)

6. R. Service, Science 286, p. 2442 (1999)

7. T. Hertel, R. Martel and P. Avouris, J. Phys. Chem. B 102, p.910, (1998)

8. Marchi, D. Tonneau, H. Dallaporta, R. Pierrisnard, V. Bouchiat, V.I. Safarov *et al.*, Microelectronic Eng., 50, p. 59-65, (2000)

9. J.F. Smith, S. Schultz, D.R. Fredkin, D.P. Kern, S.A. Pishton, H. Schmid *et al.*, J. Appl. Phys., 69, p.5262, (1991)

10. A. Fert and L. Piraux, J. Magn. Magn. Mater., 200, p. 338-358, (1999)

11. L. Lin, D.Y. Petrovykh, A. Kirakosian, H. Rauscher, F.J. Himpsel, P.A. Dowben, Appl. Phys. Lett., 78/6, p. 829-831, (2001)

12. B. Rodell, V. Korenivski, J. Costa and K.V. Rao, Nanostructure. Mater., 7/1-2, p.229-235 (1996).

13. T. Thurn-Albrecht, J. Schotter, G.A. Kästle, N. Emley, T. Shibauchi, L. Krusin-Elbaum *et al.,* Science, 290, p. 2126-2129, (2000)

14. M. Vázquez, Physica B, 299, p.302–313, (2001)

15. F. Favier, E.C. Walter, M.P. Zach, T. Benter, and R.M. Penner, Science, 293/ 5538, p.2227, (2001)

16. H. Masuda and K. Fukuda, Science, 268, p. 1466-1468, (1995)

17. F. Ronkel, J. W. Schultzeand and R. Arens-Fischer, Thin Solid Films, 276, p. 40-43, (1996)

18. J.C. Hulteen and C.R. Martin, Template Synthesis of Nanoparticles in Nanoporous Membranes, Nanoparticles and Nanostructured films: Preparation, Characterization and Application, (ed.) J. H. Fendler, Wiley-VCH, p. 235-262, (1998)

19. R.V. Parthasarathi and C.R. Martin, *Nature*, 369, p. 298-301, (1994)

20. S. Y. Chou, *Proc. IEEE*, 85, p. 652-671, (1997)

21. R.M.H. New, R.F.W. Pease and R. L. White, *J. Vac. Sci. Technol. B*, 12, p. 3196, (1994)

22. S.Y. Chou, M.S. Wei, P.R. Krauss and P.B. Fischer, *J. Appl. Phys.*, 76, 6673-6675, (1994)

23. S.Y. Chou, P.R. Krauss and L. Kong, *J. Appl. Phys.*, 79, 6101-6106, (1996)

24. S. Y. Chou, P. R. Krauss and P. J. Renstrom, J. Appl. Phys. Lett., 67, p. 3113-3116, (1995)

25. L. Kong, L. Zhuang and S.Y. Chou, Proc. of the IEEE International Magnetics Conference, New Orleans, LA, 1997.

26. C.R. Martin, *Science*, 266, p. 1961-1966, (1994)

27. J.L. Weston, A. Butera, D. Otte, and J.A. Barnard, *J. Magn. Magn. Mater.*, 193, p. 515-518, (1999)

28. C.R. Eddy, Jr., R.J. Tonucci and D.H. Pearson, *Appl. Phys. Lett.*, 68/10, 1397-1399, (1996)

29. A. Butera, J.L. Weston and J.A. Barnard, *J. Appl. Phys.*, 81, p. 7432-7436, (1997)

30. J.L. Weston, A. Butera, V.R. Intury, J.D. Jarratt, T.J. Klemmer, and J. A. Barnard, *IEEE T Magn.*, 33 /5, Part 2, p. 3628-3630, (1997)

31. W. Schwarzacher, O.I. Kasyutich, P.R. Evans, M.G. Darbyshire, G. Yi, V.M. Fedosyuk *et al., J. Magn. Magn. Mater.*, 199, p. 185-190, (1999)

32. J.A. Barnard, H. Fujiwara, V.R. Inturi, J.D. Jarratt and J.L. Weston, *Appl. Phys. Lett.*, 69 /18, p. 2758-2760, (1996)

33. E.A. Dobisz, F.A. Buot and C.R.K. Marrian, Nanofabrication and Nanoelectronics, Nanomaterials: Synthesis, Properties and Application, A.S. Edelstein and R.C. Cammarata (eds.), Institute of Physics Publishing, Bristol, p. 497-540, (1996)

34. http://nanotechweb.org/articles/news/1/10/2/1

35. M. M. Lohrengel, Mater. Sci. Engg., R11, p. 243–294, (1993)

36. F. Keller, M.S. Hunter, and D.L. Robinson, J. Electrochem. Soc., 100, p. 411–419, (1953)

37. N. Tsuya, T. Tokushima, M. Shiraki, Y. Wakui, Y. Saito, H. Nakamura *et al.*, IEEE Trans. Magn., 22/5, p. 1140–1144 (1986)

38. G. E. Thompson and G. C. Wood, Treatise on Materials Science and Tech., (ed.) J.C. Scully, New York: Academic Press, ch. 5, 205–329, (1983)

39. G. Zangari and D. N. Lambeth, IEEE Trans. Magn., 33/5, p. 3010-3012, (1997)

40. H. Daimon, O. Kitakami, and H. Fujiwara, J. Magn. Soc. Japan. 13/S1(1989), p. 795–800.

41. L. Zhang, H. S. Cho, F. Li, R. M. Metzger, and W. D. Doyle, J. Matrls. Sci Lett., 17/4 (1998), p. 291–294.

42. F. Li, L. Zhang, and R.M. Metzger, Chem. Mater., 10/9(1998), p. 2473– 2480.

43. O. Jessensky, F. Müller, and U. Gösele, J. Electrochem. Soc., 145/11, (1998), p. 3735–3740.

44. O. Jessensky, F. Müller, and U. Gösele, Appl. Phys. Lett., 72/10 (1998), p. 1173–1175.

45. R.E. Ricker, A.E. Miller, D.-F. Yue, G. Banerjee, and S. Bandyopadhay, *J.* Electron. Mater., 25 (1996), p. 1585–1592.

46. V.V. Yuzhakov, H.-C. Chang, and A.E. Miller, Phys. Rev., B56 (1997), p. 12608–12624.

47. V.V. Konovalov, G. Zangari, and R.M. Metzger, Chem. Mater., 11/8 (1999), p. 1949–1951.

48. X. Bao, F. Li, and R.M. Metzger, J. Appl. Phys., 79 (8), 1996, 4869–4871.

49. R.M. Metzger, V.V. Konovalov, M. Sun, T. Xu, G. Zangari, B. Xu *et al.*, *IEEE Trans. Magn.*, 36/1 (2000), p. 30-35.

50. H. Masuda and M. Satoh, Japan. J. Appl. Phys., 35 (1996), p. L126-L129.

51. D. AlMawlawi, N. Coombs and M. Moskovits, J. Appl. Phys., 70/8 (1991), p. 4421.

52. Mozalev, S. Magaino and H. Imai, Electrochimica Acta, 46 (2001), p. 2825-2834.

53. N. Stein, M. Rommelfangen, V. Hody, L. Johann and J. M. Lecuire, Electrochimica Acta, 47 (2002), p. 1811-1817.

54. K.M. Razeeb, PhD thesis, 2003, "Electrodeposition of Magnetic Nanowires in NCA Templates", University of Limerick, Limerick, Ireland.

55. W. Schwarzacher, K. Attenborough, A. Michel, G. Nabiyouni and J.P. Meier, J. Magn. Magn. Mater., 165 (1997), p.23-29.

56. K.N. Reddy and S.R. Rajagopalan, J. Electro. Chem., 6 (1963), p.153-158.

57. Epelboin, M. Joussellin and R. Wiart, J. Electroanal. Chem., 119 (1981), p. 61-71.

58. E. Vallés, R. Pollina and E. Gómez, J. Appl. Electrochem., 23 (1993), p. 508-515.

59. H. Daimon, O. Kitakami, and H. Fujiwara, J. Magn. Soc. Japan, 13/ S1 (1989), p. 795–800.

60. C.R. Martin, Acc. Chem. Res., 28 (1995), p. 61–68.

61. K. Nielsch, F. Müller, A.P. Li and U. Gösele, Adv. Mater., 12/8 (2000), p. 582-586.

62. H. Zhu, S. Yang, G. Ni, D. Yu and Y. Du, Scripta Mater., 44 (2001), p. 2291-2295.

63. J.M. García, A. Asenjo, J. Velázquez, D. García, M. Vázquez, P. Aranda *et al.*, J. Appl. Phys., 85 (8), 1999, p. 5480-5482.

64. I.Z. Rahman, K.M. Razeeb, and M. Serantoni, 2005, to be published in Journal of Materials Processing Technology (In press), Available online since 4 June 2004.

65. K.M. Razeeb, I.Z. Rahman and M.A. Rahman. "Growth of Ni-nanowires by Electrodeposition Technique" Journal of Metastable and Nanostructured Materials, 17 (2003), p. 1-8.

66. Z. Rahman, K.M. Razeeb, M.A. Rahman and Md. Kamruzzaman, Journal of Magnetism and Magnetic Material. 262/1 (2003), p. 166 –169. N. Reddy, Journal of Elec. Chem., 6(2) (1963), p.141-152.

68. F. Ebrahimi, G.R. Borune, M.S. Kelly and T.E. Matthews, Nanostructured Materials, Vol. 11(3) (1999), p. 343-350.

69. Huang Y, Duan X, Wei Q and Lieber C M 2001 Science, 291, p. 630

70. Lin J, Bird JP, Rotkina L and Bennett B A Appl. Phys. Lett. 82 (2003), p. 802.

71. P.A. Smith, C.D. Nordquist, T.N. Jackson and T.S. Mayer, Appl. Phys. Lett. 77 (2000), p. 1399.

72. Y. Zhang, A. Chang, J. Cao, Q. Wang, K. Woong, Y. Li *et al.*, Appl. Phys. Lett. 79 (2001), p. 3155.

73. M. Calleja, M. Tello, J. Anguita, F. Garcia and Garcia R 2001 Appl. Phys. Lett. 79, p. 2471

74. C. H. Oon and J. T. L. Thong, Nanotechnology, 15 (2004), p. 687-691

Part IV

Materials for Assembly/Packaging

The Importance of Polymers in Wafer-Level Packaging

M. Töpper

Fraunhofer IZM, Berlin / Germany

Introduction

Polymeric coatings such as polyurethanes, acrylic, epoxies and silicones have been used for over 40 years to protect printed wiring boards (PWB) from moisture, handling and environmental influences [1]. Special semiconductor grade polymers have been developed for chip passivation layers. Especially for epoxy resins multiple distillation procedures were introduced to remove the sodium and chloride ions, which are by-products in the standard epoxy synthesis. Polyimide became the standard passivation layer for memory chips and other devices with the need of surface protection for the handling and testing procedure. Photosensitive resins have been developed to reduce processing cost. Dry-etching requires a masking process with either a hard mask, which is a physically deposited and structured metal layer or a thick photoresist coating. Due to cost savings programs the in-depth characterization of materials that was prevalent in the 1960s through the 1990s has slowed down considerably. In addition, due to mergers, spin-offs and low economic margins the continuity of polymeric products is not always given. In contrast to that, polymeric materials became more important with the introduction of new packaging concepts for ICs. The package is by definition the protection for bare dice. It has to protect the IC for environmental influence, support the IC performance (operating speed, power, signal integrity, *etc.*), handle the thermal management and has to compensate all kind of stress. Therefore packaging technologies determine the size, weight, ease of use, durability, reliability, performance and cost of electronic products.

World-wide the trend in SCP has been from the dual in-line package (DIP), the plastic quad flat package (PQFP) and the ball grid array (BGA) to a package which is only marginally larger than the chip itself, namely the chip size package (CSP) [2]. The highest level of electrical performance will be attaching a bare die onto the printed circuit board (PCB) called direct chip attach (DCA), which has been used practically long before CSP. But FC on a PCB is not reliable without underfiller due to the high CTE mismatch of Si and laminates. The main issues for DCA are assembly, standardization of size or footprint, reduced protection of the die, testing and rework. For the assembly of electronic components, moving to the array arrangement of solder balls (ball grid array packages, BGAs) was a revolution. CSP evolved as the result, combining FC-technology with SMT and BGA. Wafer level packaging (WLP) is the optimum synergy of wafer processing and CSP. If the dice on the wafer have a peripheral pad layout a redistribution process becomes necessary to reroute the peripheral pads to the area array pads (Figure 1).

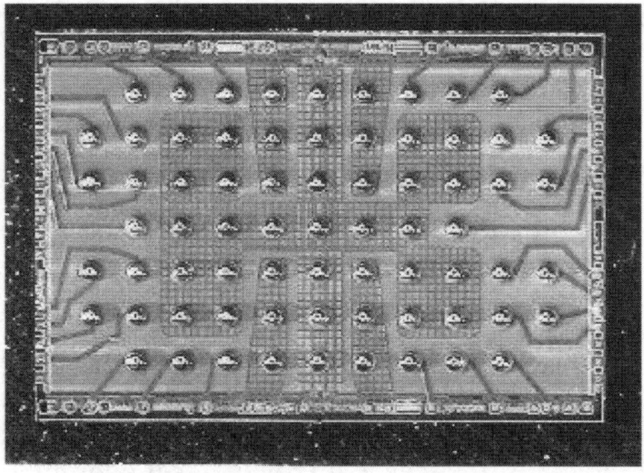

Figure 1. Wafer level package: The peripheral pads are redistributed into an area array of solder balls

A redistribution layer is a combination of polymer and metal layers. The importance of polymers for WLP will be discussed in more detail.

Redistribution for WLP

Overview of Thin Film Polymers

Thin film polymers have proven to be an integral material basis for many different types of advanced electronic applications. First used as IC stress buffer layers, then established for MCMs, they are now used in various new packages, especially in the field of WLP. The requirements for the selection of a given polymer are quite broad: high decomposition or glass temperature for the high temperature processes in packaging like solder reflow, high adhesion, high mechanical and chemical strength, excellent electrical properties, low water up-take, photosensitivity and high yield manufacturability. Only thermosets are therefore the polymer class for packaging applications. An important process difficulty is due to the fact that these high-end thermosets are nearly insoluble in organic solvents. Therefore, pre-polymers are manufactured, which have a molecular weight in the hundred thousands and are dissolved in an organic solvent. These solutions are commonly called pre-cursors and are ready for the spin-on process. The final polymerization is done on the wafer by thermal curing.

The performance of polymers plays a major role in the build-up structure of WLP because it is one of the key layers that act as a buffer between IC and PWB. Low-k materials are preferred because a high capacitance reduces the computing speed between integrated circuits. In addition, the selection of the optimal polymer for a given application depends not only on its physical and chemical properties and processability, but also on its intrinsic interfacial characteristics. Table 1 gives a selection of common types of photosensitive spin-on dielectric materials.

Polyimide (PI), benzocylobutene (BCB) and polybenzoxazole (PBO) are common re-passivation materials. However, BCB has gained certain dominance in the market for these applications. BCB and polyimides (with only a few exceptions) are negative acting materials requiring organic developer, while PBO is positive acting. All materials require a bake to remove solvents but no rehydration step. In some cases a post exposure bake is necessary to enhance the photoinitiated polymerization. All materials are exposed by the broadband spectrum and achieve a sidewall angle of approximately 40–60°. Resolution is usually not an important requirement because only vias of 20 microns or larger in diameter have to be opened over larger I/O pads.

Process Technology for Pad Redistribution Layers (RDL)

Several different redistribution processes have been developed but main process steps are similar to each other. Differences exist mainly in the material selection. As an example, the redistribution technology of Fraunhofer-IZM/TU Berlin will be described in more detail (Figure 2).

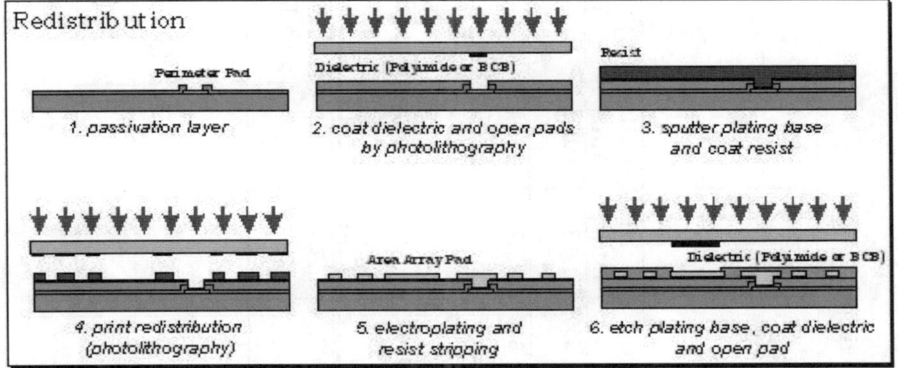

Figure 2. Process flow of an RDL process (Fraunhofer IZM)

First a dielectric layer is deposited on the wafer to enhance the passivation layer of the die. Pinholes in an inorganic passivation would give shorts in the rewiring metallization. The polymer layer under the rewiring metallization also acts as a stress buffer layer for the bumping and assembly processes.

Fraunhofer IZM/TU Berlin uses photo-BCB. Compared to other polymers BCB has a low dielectric constant and dielectric loss, minimal moisture uptake during and after processing, very good planarization and a low curing temperature. The rewiring metallization consists of electroplated copper traces to achieve a low electrical resistivity. A sputtered layer of Ti:W–Cu (200/300 nm) serves as a diffusion barrier to Al and as a plating base. A positive acting photoresist is used to create the plating mask. After metal deposition the plating base is removed by a combination of wet and dry etching. The copper process is shown in Figure 3.

Table 1. Overview of thin film polymers and their characteristics

	Trade-name	Photosensitiv-ity	Developer	Base chemistry	Curing T (for 1–2 h)	Diel. const	Loss factor	T_g/decomposition T	CTE	Tensile strength	Elongation to break	Residual stress	Youngs modulus	Water uptake
					[°C]	[1 kHz–1MHz]	[1 kHz–1MHz]	[°C]	[ppm/K]	[MPa]	[%]	[MPa]	[GPa]	[%]
Arch	Probimide 7000	negative	or-ganic	PI	> 350	3.3	0.007	> 350	27	170	73	30	2.9	1.3
	348	negative	or-ganic	PI	> 350	3.2	0.004	> 350	23	123	8		3.2	1.8
Asahi	Pimel G7621	negative	or-ganic	PI	> 350	3.3	0.003	355	40 – 50	150	30	40 - 50		0.8
Dow Chemical	Cyclotene 4000	negative	or-ganic	BCB	210 - 250	2.65 2.55 (1 GHz)	0.0008 0.002 (1 GHz)	> 350	45	87	8	28	2.9	< 0.2
Dow Corning	Photoneece PWDC 1000	positive	aque-ous	PI	320	2.9		290	36	130	40	28	3	
	WL 5150	negative	or-ganic	Sili-cone	250	3.2	0.0070		236	6	37	2.6	0.16	
HDM	PI 2730	negative	or-ganic	PI	> 350	2.9	0.003	> 350	16	170	?	18	4.7	> 1.0
	HD 4000	negative	or-ganic	PI	> 350	3.2	0.006	350	35	200	45	35 - 37	3.5	

Table 1. *Continued*

Property	WLP 1200	HD 8000	Cardo VPA	Intervia 8000	PBO	Photoneece BG 2400	Photoneece UR 5480	Photoneece PW 1000	Photoneece PN 1000
Manufacturer	JSR		Nippon Steel	Rohm and Haas	Sumitomo	Toray			
Photosensitivity	negative	positive	negative	negative	positive	negative	negative	positive	positive
Developer	aqueous	aqueous	aqueous	aqueous	aqueous	organic	organic	aqueous	aqueous
Base chemistry	nano-filled Novolac	PI	modified acrylic resin	modified resist	PBO	PI	PI	PI	PI
Curing T (for 1 – 2hrs)	170	>350	200	175	320	>350	>350	>350	180 – 200
Diel. const	3.8	3.4	3.4	2.9 (1 GHz)	2.9 - 3.5	3.2	3.2	2.9	2.9
Loss factor	0.036	0.0097	0.03	0.026 (1 GHz)			0.002		0.002
T_g/decompostion T	> 210	300	180	150 – 200	380	255	> 350	290	280
CTE	51	47	80	58	55	25	16	36	36
Tensile strength	75	122	7.5		100	180	150	130	110
Elongation to break	7.5	11	11.5		18	40	40	40	10
Residual stress		29	14 – 30						
Youngs modulus	1.6	2.5	2.5	4.5	2.5	3.9	4.2	3	3.2
Water uptake			1.6	<0.38	0.3				1.5

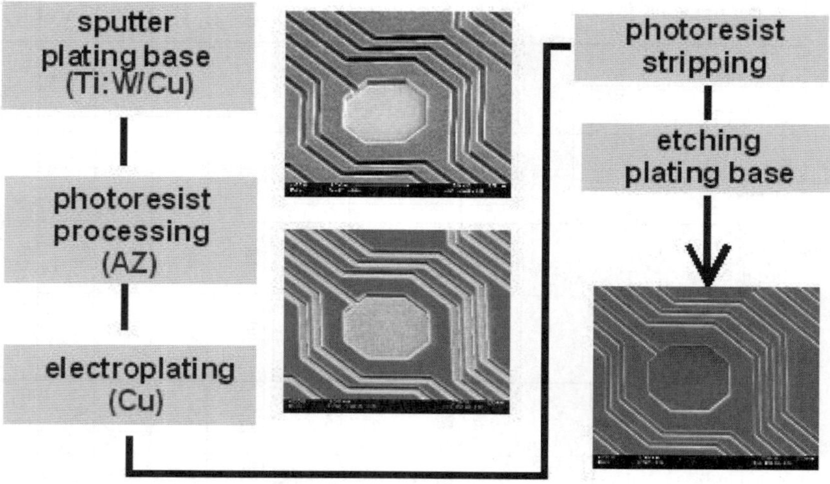

Figure 3. Thin film copper process at Fraunhofer IZM / TUB

A second photo-BCB layer is deposited to protect the copper and to serve as a solder mask. BCB can be deposited directly over the copper metallization without any additional diffusion barriers. Electroplated Ni/Au is used for the final metallization. Solder balls (PbSn or lead-free) are deposited by solder printing directly on the redistributed wafers. Then the solder paste is reflowed in a convection oven under nitrogen atmosphere and the flux residues are removed in a solvent adapted to the used solder paste (Figure 4).

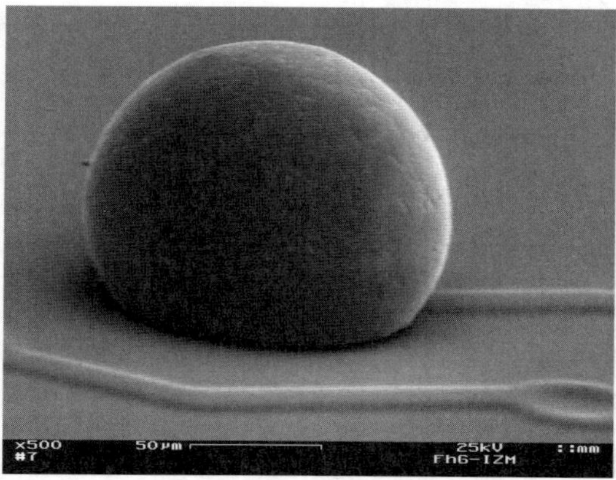

Figure 4. Solder-printed PbSn on redistributed wafer (photo-BCB/Cu)

Mean value of the solder ball diameter can be adapted to the assembly and board requirements between ~ 100 and 250 µm depending on the ball pitch. Shear testing is the method of choice for a first quality check. Values for these ball diameters should be higher than 130 cN per bump. Dicing the wafer with a standard wafer saw completes the CSP-WL build-up. The reliability of the redistribution layer was evaluated for consumer, medical, automotive and space applications.

RDL Design

There are two major different designs for redistribution: The bump and the UBM are separated to the chip surface by an additional polymer layer or the UBM is deposited directly on top of the inorganic chip passivation (Figure 5a and b).

The company FCI (former FCT) came up with the classifications of bump on polymer (BOP) (Figure 5a) and bump on nitride (BON) (Figure 5b) for the different built-up structures. It is still under discussion whether the type of polymer under the bump plays a critical role for the reliability. The BOP type is preferred if underfiller is used. Different geometric values are possible for the design. The opening in the first polymeric passivation (a in Figure 5a) should be similar but not the same size than the peripheral pad passivation. The UBM and the ball size (b and c in Figure 5a) can be adjusted to the pitch of the area array. The maximization of the ball size is limited by the pitch. Today minimum ball pitches of 500 µm and even 400 µm are in production.

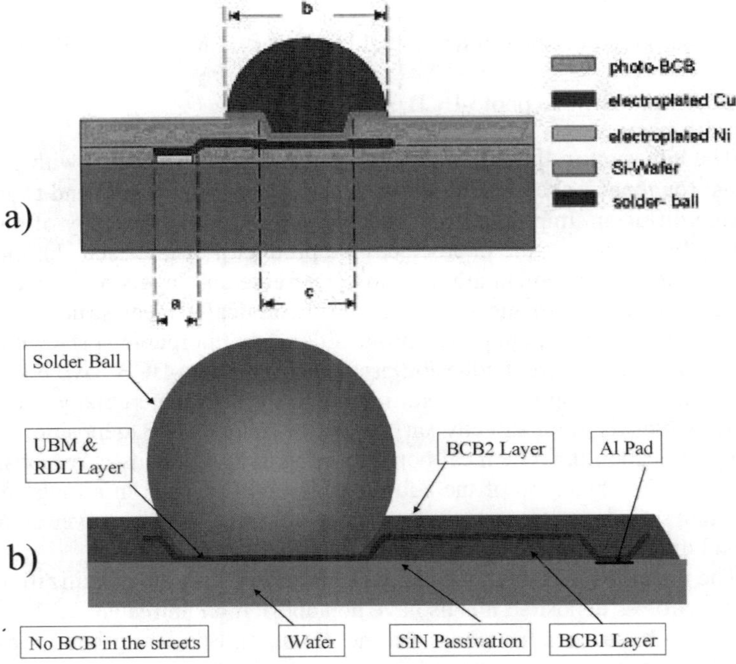

Figure 5. Two different RDL layouts: Polymer between chip and bump (a) (Fraunhofer IZM) and UBM directly on top of the inorganic passivation (b) (courtesy of FCI)

Layer Adhesion

The reliability of a multi-layer thin film structure strongly depends on the adhesion between the different layers [3]. In Figure 6 the different interfaces in a photo-BCB/TiW/Cu/ technology-based WLP are shown.

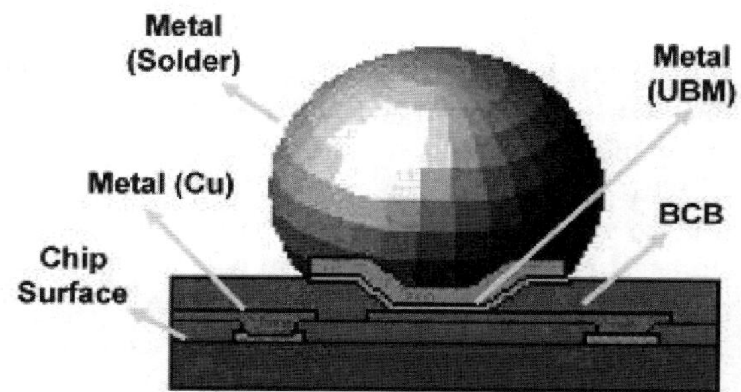

Figure 6. Structure of the WLP for the adhesion study

1. Photo-BCB to the inorganic chip passivation (silicon-oxide, nitride *etc.*) and to metal (chip pad (Al) or redistribution metallization (Cu))

2. Metal (*i.e.* redistribution metallization or UBM) to Photo-BCB

3. Photo-BCB to photo-BCB

The adhesion in these thin film structures can be described with three mechanisms: roughness, chemical bonding using adhesion promoters and chemical interlocking/diffusion. Important for a reliable package is the integrity of the interfaces during the lifetime of the microelectronic product. For interface (1) organo-silane-based adhesion promoters are used to create essential layers to couple the organic dielectric to the inorganic surfaces. The theoretically ideal structure would be a monomolecular layer, coupling on one side to the inorganic surface and the other to the polymer. Different adhesion promoters for photo-BCB have been evaluated. High adhesion strengths on several inorganic surfaces were obtained using a vinyl-silane, which is spun directly on the wafer before BCB deposition. There is a strong indication that chemical bonding through Si–O bonds is responsible for the adhesion. The thickness of the adhesion promoter layer is in a range of 0.5–5nm. This adhesion promoter layer improves the adhesion to values over 60 MPa on Al, Cu and different inorganic chip passivations.

The metal to BCB interface (2) depends strongly on the metallization technique [4]. Electroless deposited metals have no adhesion on untreated BCB films with its very smooth surfaces (roughness in the Å range). Sputtering is used as a reliable metallization process for the redistribution of the WLP because the metal atoms have a penetration depth of around 100 nm. This guarantees the strong adhesion of over 80 MPa between the sputtered metal on the photo-BCB. Paik *et al.* [5] described the stable interface between Cu and BCB. In addition, there is nearly no influence of the descum process (RIE) using a fluorine gas/oxygen mixture, which is necessary to achieve a 100% electrical yield in via holes. Only pure O_2-plasma reduces the adhesion due to an oxidization of the BCB surface [6-7].

For the photo-BCB to photo-BCB interface (3) high adhesion is obtained by performing a partial cure of the underlying BCB layers followed by a final full cure of the whole stack. A correlation between the adhesion strength and the degree of cure was found. The roughness and the surface chemistry of the Photo-BCB layer modified by RIE had no significant effect on adhesion. The mechanisms of chemical interlocking and inter-diffusion are the driving forces for the adhesion between BCB layers. Therefore, the degree of cure is the key to high adhesion.

In conclusion, the surface chemistry and physics should be carefully analyzed for the reliable built-up of thin film structure. Special emphasise has to be made to all modification processes like sputtering, plasma, *etc.*

Redistribution with Resilient Interconnect Elements

Fraunhofer IZM and TUB, together with Motorola, started a program in 1996 to develop new concepts for highly reliable wafer level packages (fab integrated package, FIP-CSP) [8]. A modified Mitsubishi package which could be manufactured totally on the wafer-level showed promising results of finite element method simulations. The technology consists of a stacked solder ball array with a stress compensation layer in between (Figure 7).

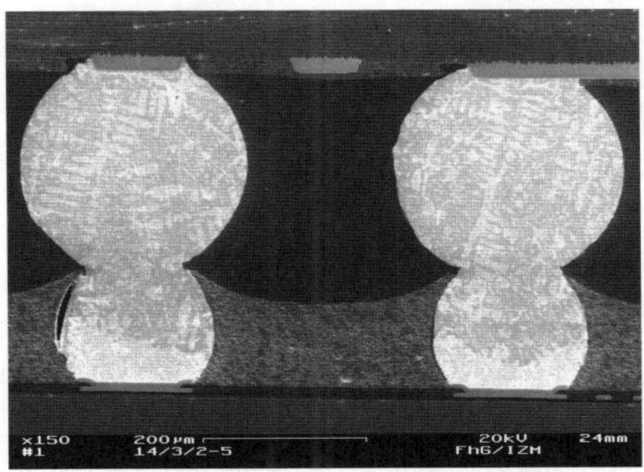

Figure 7. Cross section of solder balls using preformed eutectic PbSn balls (cross section was made after thermal cycling)

As a CSP, the FIP-CSP eliminates underfill operation during flip-chip bonding using high throuput SMT assembly lines. The technological structure of this FIP-CSP is a pad redistributed die with a solder ball array. A stress compensation layer (SCL) embeds the solder balls before second solder balls are stencil printed or placed on top of embedded balls.

An important aspect of this package was the choice of the polymeric-based SCL. Figure 8 shows the accumulated equivalent creep strains in the outermost free solder balls with respect to the SCL material properties. The highlighted numbers give an impression of the estimated mean cycles to failure. Lifetime results have been evaluated using the modified Manson–Coffin concept. The thermal cycle used simulates a 1 h air to air cycle process between 125°C and −55°C with dwell times of 10 min and ramp times of 20 min.

The 3-D-FE results of investigated variations concerning Youngs's modulus and CTE of the SCL lead to the following conclusion: an SCL material with a Youngs's modulus of 3 to 4 GPa and a CTE of 20 to 35 ppm/K should be favoured for use in the package development. The highest accumulated equivalent creep strains are in outermost free balls of this model. The polymer selection has been proven as one of the major influences in this package. Polymers with high Young's modulus reduced the board-level reliability dramatically.

Infineon is using a silicone bump under the UBM for stress compensation in their WLP called ELASTec [9]. This "Elastic Bump on Silicon Technology" or ELASTec is based on resilient interconnect elements on the wafer. These interconnect elements consist of printed silicone bumps with redistribution traces routed from the I/O pads onto the top of the silicon bumps (Figure 9).

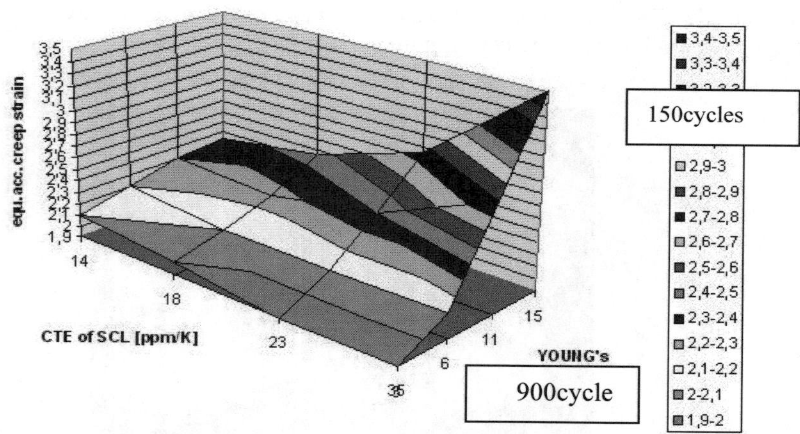

Figure 8. Behaviour of the accumulated equivalent creep strains in outermost free solder balls and mean cycles to failure depending on SCL materials properties — 3-D FE-simulation (number of mean cycles to failure for AATC –55/+125°C). RDL: each layer 20 μm thick.

The traces on the silicone bumps form a spiral pattern. The redistribution traces on top of the resilient bumps have a gold finish as contact surface for test and burn-in and as solder pad in second level assembly. The traces of the redistribution are electroplated where the geometry of the traces is defined by photolithography on a sputtered seed layer. The bump height of the ELASTec package is 170 μm to realize packages of smaller height down to 500 μm. This is a factor of 2 in height reduction compared to conventional interposer-based packages. Compared to conventional solder bumped WLP with ball diameters larger than 300 μm the ELASTec package reaches a smaller package form factor at high reliability. The design of the ELASTec package has various advantages: the bumps are flexible in the z-direction for reliable test contact and in the x,y-direction for compensation of thermal mismatch to the printed circuit board after board assembly.

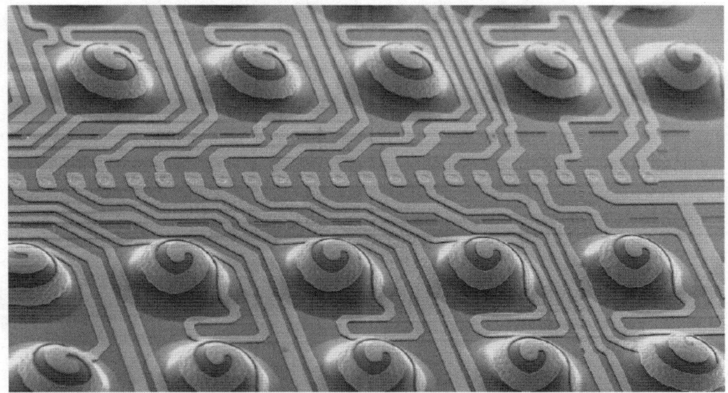

Figure 9. The ELASTec WLP package consists of metal traces routed on top of resilient silicon bumps (courtesy of Infineon Technologies)

Wafer Level Technologies with Higher Integration

3-D Integration

Stacking of chips for 3-D integration can be done using wiring bonding, flip chip or vertical integration by wafer stacking or chip-to-wafer stacking. Figure 10 shows a stacked FC-BGA using RDL by BCB/Cu with a flip chip mounted microcontroller on a silicon chip with redistributed IC pads. The interconnection from the interposer to the board is done using wire bonding.

In this approach a base chip on wafer-level is used as an active substrate for FC-bonding of a second die. The electrical and mechanical interconnection is done using eutectic solder balls, which are deposited by electroplating. The base chip is redistributed to an area array of UBMs. A low electrical resistivity of the redistribution is achieved by electroplating copper. The dielectric isolation is achieved using the low-κ photo-BCB.

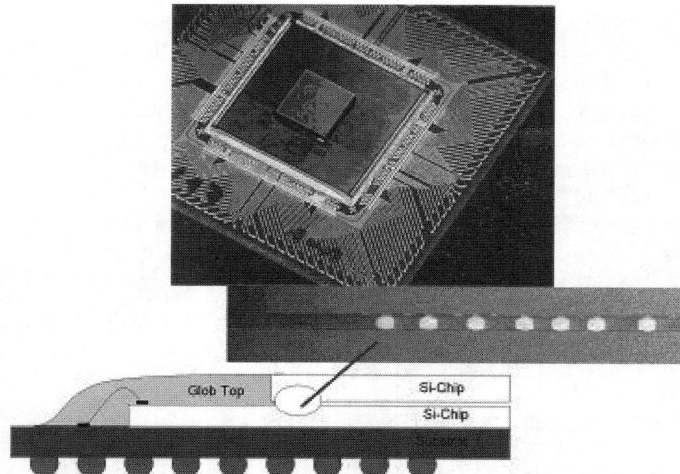

Figure 10. Chip-on-chip integration using FC-bonding (courtesy of Fraunhofer IZM and Infineon)

Integration of Passives

Resistors, capacitors and inductors are generally referred to as passive components. Compared to the developments in integrated circuit technology, passive components at the circuit board level have made only marginal advances in decreasing size [10]. Therefore, they are a major hurdle for further miniaturization of electronic products.

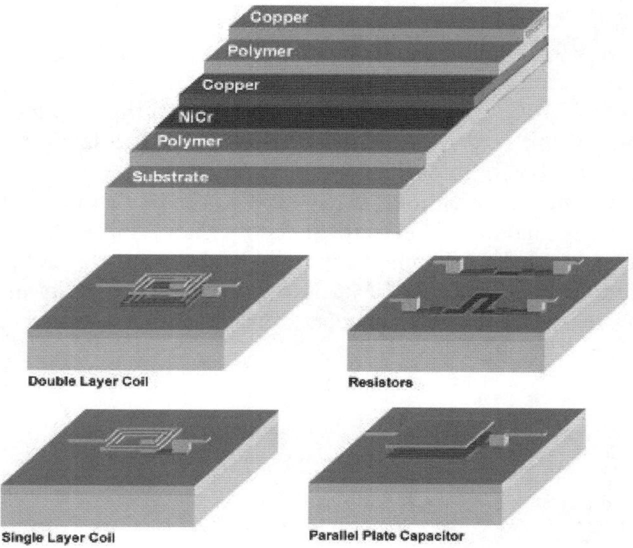

Figure 11. Thin film builtd-up for integrated passive components (Fraunhofer IZM) [11]

Three basic material classes are needed for the realization of integrated passive elements: conductors, resistors and dielectrics [10]. These can be made of metals, polymers or ceramics. The main difference between polymer and ceramic technologies is the maximum process temperature, which can be up to 300°C for polymers, but can reach 700°C and above for the firing of ceramics. Metals like Cu, Au or Al or metal-filled polymer thick films with a resistivity of less than 0.1 Ω/square are used for the conductors to avoid high parasitic resistance. Alloys like NiCr, CrSi, or TaN, cermets (ceramic-metal composites) or carbon-filled polymers are the materials of choice for resistors having values of 100 to 10,000 Ω/square. Polymers with a dielectric constant κ of 2–5, amorphous metal oxides (κ=9–50) or crystallographic ordered mixed oxides with κ>1000 are the central building blocks for capacitors.

An integrated passive filter based on thin film technology is given as an example. The build-up process is similar to the redistribution for WLP, therefore the same production line can be used. Copper/BCB technology in conjunction with NiCr sputtering is the core process steps. In Figure 11 the build-up process for inductors, resistors and capacitors are shown.

The low capacitance values of 0.2–2.5 pF are due to the BCB usage with its low dielectric constant of 2.6. Sputtered high-κ materials have to be used for the nF-range. A combination of these passive elements can be used for the realization of filter components. An example for Bluetooth band (2.4 GHz) is given in Figure 12.

One filter consists of three inductors of 3.9 nH and two capacitors of 1.8 nF. Microstrip lines with 50 Ω impedance are used for the interconnect of the single elements resulting in two low-pass filters of second order and one single inductor within an area of 1.3 mm × 2.6 mm. PbSn or lead-free solder balls are used for the board assembly.

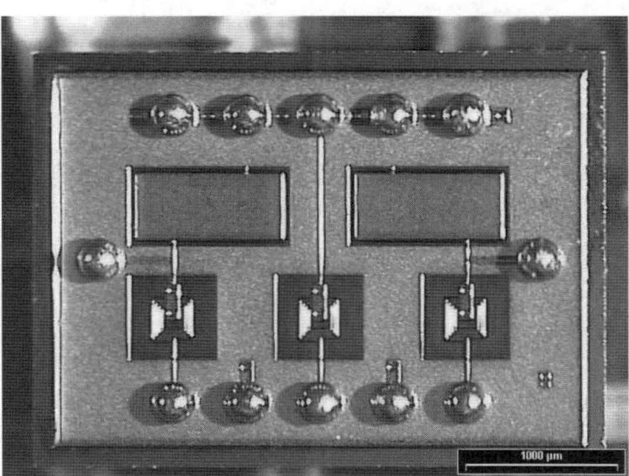

Figure 12. An FC mountable integrated filter for Bluetooth band

Outlook: System in Package (SiP)

The definition of SiP has not been well established across the literature. Basically, SiP requires breaking up the boundaries between system design, semiconductor front end and back end in order to have more flexibility for overall product design in order to enhance the product performance and reduce cost. A paradigm shift in product design and manufacturing is necessary because Moore's law is now more and more restricted by packaging and substrate technology than by nanotechnology. A close cooperation between semiconductor, interconnection, integrated passives, substrate and system designers will be necessary [12]. Stacked memory dice should not fall into the class of SiP, a minimum combination of logic and memory is required [13]. The increasing role of packaging for microelectronic systems is schematically shown in Figure 13.

Even though the SMT and area array technology were major developments in the electronic industry, the gap between semiconductor technology and packaging has been increasing. System in package will be the packaging wave for the next ten years to keep Moore's law alive.

The ever-decreasing size of electronic packages has brought the technology from simple plastic housing to material and equipment challenges soon touching the nanoworld. Simulation of electrical, thermal and thermo-mechanical nature will gain higher importance to identify the interaction of packaging materials. The concept of SiP will not only be the key to further miniaturization but also to higher reliability due to less interfaces and interconnects. New materials based on polymers will play a major role in up-coming new technologies.

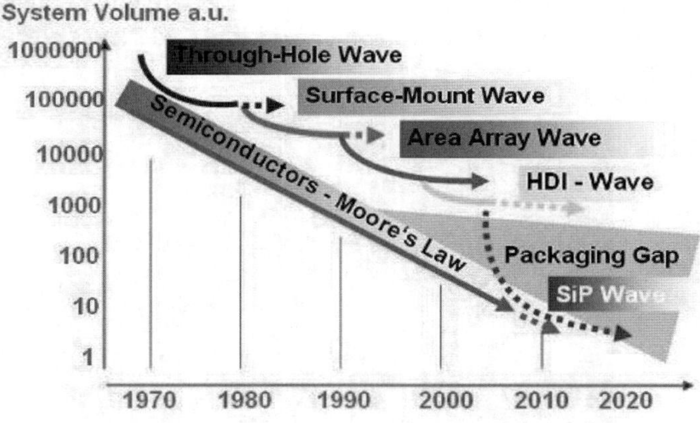

Figure 13. The increasing role of packaging for microelectronic systems

References

1.   J. Licari "Coating Materials For Electronic Applications" Noyes Publications, 2003

2.   M. Töpper, M. Schaldach, S. Fehlberg, C. Karduck, C. Meinherz, K. Heinricht *et al.*, "Chip Size Package - The Option of Choice for Miniaturized Medical Devices" Proceedings IMAPS Conference San Diego, CA, 1998

3.   M. Töpper, A. Achen, H. Reichl, "Interfacial Adhesion Analysis of BCB / TiW / Cu / PbSn Technology in Wafer Level Packaging" Proceedings ECTC, 2003

4.   M. Töpper, Th. Stolle, H. Reichl, "Low Cost Electroless Copper Metallization of BCB for High-Density Wiring Systems", Proceedings of the 5th Internat. Sympos. on Advanced Packaging Materials, Braselton, USA, March 1999

5.   K.W. Paik, R.J. Saia, J.J. Chera, "Studies on the Surface Modification of BCB Film", Proceedings MRS, Boston, MA November 1990

6.   F. Krause, K. Halser, K. Scherpinski, M. Töpper, "Surface Modification due to Technological Treatment Evaluated by SPM and XPS Techniques", Proceedings MicroMat 2000, Berlin, April 2000

7.   P. Chinoy, "Reactve Ion Etching of Benzocyclobutene Polymer Films", IEEE Trans. Comp. Packaging Manuf. Technol., Part C, 20, 3, pp 199-206 (1997)

8.   M. Töpper, J. Auersperg, V. Glaw, K. Kaskoun, E. Prack, B. Keser *et al.*, "Fab Integrated Packaging (FIP) A New Concept for High Reliable Wafer-Level Chip Size Packaging" Proceedings IEEE, ECTC, May 2000

9.   T. Meyer, H. Hedler, "A New Approach to Wafer-Level Packaging Employs Spin-On and Printable Silicons", Chip Scale Review, pp 65-71, July 2004

10.   R. Ulrich, L. Schaper "Integrated Passive Component Technology" Wiley Interscience/IEEE Press, 2003

11.   11. ThK. Zoschke, J. Wolf, M. Töpper, O.Ehrmann, Th. Fritzsch, K. Scherpinski, F.-J. Schmückle, H.Reichl "Thin Film Integration of Passives - Single Components, Filters, Integrated Passive Devices" Proceedings of the 54th ECTC Conference, Las Vegas, USA, pp 294-301, June 2004

12.   van Roosmalen, "There is More Than Moore", 5th Int. Con. on Mech. Sim. and Exp. in Microelectronics and MST, EuroSim 2004

13.   B. Levine, "System-in-Package: Growing Markets, Ongoing Uncertainty", Semiconductor International pp. 47 -61, March 2004

Electrically Conductive Adhesives as Solder Alternative: A Feasible Challenge

G. Luyckx, G. Dreezen

ICI Belgium N.V., Emerson & Cuming, Westerlo / Belgium

Introduction: Electrically Conductive Adhesives

Electrically conductive adhesives are composite materials in which an organic matrix is filled with an extremely large volume, typically 70 to 80% by weight, of an (electrically conductive) inorganic filler. The organic matrix primarily determines the mechanical properties of the end product, while the final electrical properties depend on the type of filler used. Intrinsic conductive polymers have as yet no practical significance as an alternative for traditional solder and remain out of consideration here. Most polymers have dielectric properties and are therefore used frequently as insulators. Two fundamental types of polymers are distinguished: thermoplast and thermoset materials. Thermoplast polymers consist of long chains that are physically entangled. Heating to temperatures far above the glass transition temperature (T_g) allows these chains to move independently of each other so that the polymer becomes liquefied. Cooling to below the T_g "freezes" the polymer structure and the chains have insufficient mobility to move. These polymers can become liquid again through heating up to above their T_g and can consequently be reworked like solder. But in general, they do not have sufficient suitable properties to be able to act as an alternative for solder.

Thermoset materials consist of polymer resins: non-polymerised monomers with a low molecular weight or partially polymerised oligomers. These resins are fluids or low-melting solid substances. The hardener, eventually in the presence of a catalyst (accelerator), can initiate chemical bonds with the polymer resins so that a chemically crosslinked network with a high molecular weight is formed; this is the curing of the polymer. This polymer network will retain its form when heated above the T_g. Adding heat will allow the chains between the cross-links to move, resulting in a "softening" of the polymer, but the material will be unable to flow and consequently cannot be easily reworked. A thermoset material therefore consists of a resin and a hardener, possibly with a catalyst. Different additives can be added for varying purposes. Depending on the system's reactivity, thermoset adhesives are available as one or two components.

The electrical conductivity of an adhesive depends to a large degree on the type of filler, the filler's production process itself and the filler content that is used. Silver is mostly used as a conductive filler. It can be easily shaped into the desired form such as powder or flake and the silver oxide is also a good conductor; oxidation of the silver does not significantly influence the conductivity of the material. Silver flakes are generally used in electrically conductive adhesives. Gold is much more expensive than silver. Non-precious metals like nickel show a fast increase of the resistance due to the formation of non-conducting oxides. In certain cases non-precious metals with a precious covering (*e.g.* silver-plated copper) are used. Under the influence of damp heat the resistance increases however much faster in comparison to solid precious metal due to the porosity of the precious covering. The content of a conductive filler must be sufficient to make the adhesive conductive. On the other hand, when the concentration is too high, processing will become very difficult and the mechanical strength after cure will be very poor. There is always a critical volume from which the material becomes conductive. According to this percolation theory, the resistance drops very quickly to a constant value once this critical volume is reached. The critical volume for silver-filled adhesives generally lies between 65 and 75% by weight [1].

Electrically conductive adhesives can be classified into two types: isotropic and anisotropic conductive adhesives. Isotropic conductive adhesives (ICA) produce approximately equal electrical conductivity in all three dimensions (*xyz*) while anisotropic conductive adhesives (ACA) are only conductive in the Z-direction. The electrical contact between the adhesive and the contact surfaces is created through the metallic contact of the filler particles with the polymer in between, meaning that the effective contact surface area is only a part of the total adhesive surface area. The polymer has an insulating effect and this explains why the electrical conductivity of the adhesive is always lower than the conductivity of the inorganic filler. The polymer further determines the cohesion in the adhesive, the adhesion through the correct interaction with the contact surface areas and the protection of the electrical contacts against humidity. As a result, the adhesive has two functions working alongside each other: providing electrical contact and mechanical strength. The isotropic conductive adhesives are mainly seen as the alternative for traditional solder because the electrical conduction runs in three dimensions for both. Most isotropic conductive adhesives contain Ag as a filler and are epoxy-based because of their many-sided favourable properties and the ability to modify them. Epoxy resins offer the benefit that they can be cured at lower temperatures (< 200°C).

Background

Traditionally, electrically conductive adhesives have been used in hybrid applications typically using noble metallisations such as gold and silver palladium on both substrates and components. Benefits of electrically conductive adhesives over traditional solder are the lower processing temperatures, fine pitch capability and improved thermo-mechanical performance [2,3]. An adhesive joint is less sensitive to thermal fatigue and shows no brittleness upon long-term exposure to high temperature. Until recently, components with non-noble metallisations such as Sn/Pb and Sn have traditionally been assembled using solder. Sn/Pb eutectic solder has been the dominant interconnecting material in electronic assemblies. The drive towards lead-free assembly has increased strongly and legislative measures have been proposed to ban or limit the use of lead in solders. The disadvantage of most of these lead-free solders is the higher melting point thus requiring higher reflow temperatures up to 250–260°C. This may damage temperature-sensitive components and limit the lifespan of the product. This is an additional driver for using low-temperature curing adhesives. Although electrically conductive adhesives are already commonly used in specific applications, there have been two major obstacles preventing conductive adhesives from becoming a general replacement for metal solders in the electronics industry: the unstable contact resistance on common electronic metallisations such as copper and Sn/Pb under elevated temperature and humidity and the inferior impact resistance (drop test performance). Much progress has been made in identifying the causes for these limitations. Recently, the development of adhesives with stable contact resistance in combination with non-noble metals including OSP copper, Sn/Pb alloys and even 100% Sn has renewed interest in electrically conductive adhesives in applications that were traditionally reserved for solders.

The Mechanism of the Unstable Contact Resistance

The contact resistance instability of electrically conductive adhesives in combination with non-noble metallisations is due to the occurrence of electrochemical corrosion at the interface between the adhesive and the metallisation. It has been shown during 85°C/85%RH temperature/humidity testing that the electrically conductive adhesive displays a stable bulk resistivity but that the contact resistance at the metal adhesive interface increases [4,5]. Galvanic corrosion at the interface of the silver in the conductive adhesive and the non-noble metal rather than simple oxidation of the non-noble metal is the key mechanism for performance instability over time in past conductive adhesives. Electrochemical corrosion occurs in the presence of two metals with dissimilar reduction potential and water (Figure 1). When silver, with a high reduction potential, is in contact with traditional electronic metals such as copper, tin and lead or other low reduction potential metals, the addition of small amounts of water completes a galvanic cell and the following reactions may occur.

- The metal with the higher electrochemical potential (Ag/0.80V) acts as the cathode at which following reaction can take place:
  $$2H_2O + O_2 + 4e^- \rightarrow 4\,OH^-$$
- The metal with the lower electrochemical potential acts as the anode:
  $$M - ne^- \rightarrow M^{n+}$$

Normal potentials for different metals are shown in Table 1. For metals such as Ag, Au and Pt with similar or electrochemical potentials > 0.40 V (potential of the cathode reaction), corrosion is not taking place and the assembly is providing theoretically a stable contact resistance. For metals such as Cu, Sn, Pb and Ni with a potential below 0.40 V, electrochemical corrosion will occur and the contact resistance will increase after exposure to humidity due to the formation of a non-conductive oxide at the interface. An example of the contact resistance increase during 85°C/85% RH testing will be discussed in Figure 8.

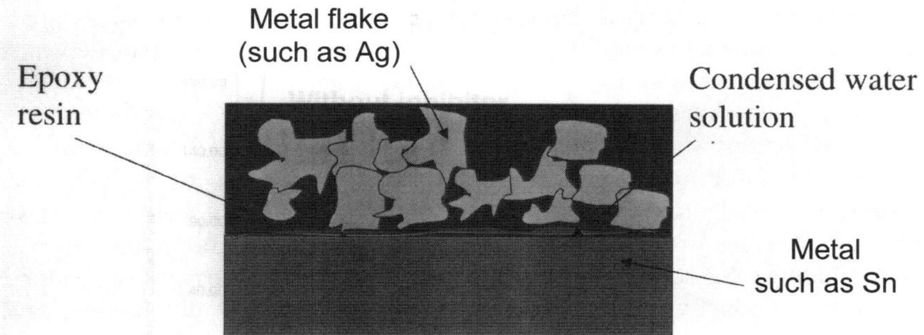

Figure 1. Formation of an electrochemical cell at the interface between an electrically conductive adhesive and the non-noble metallisation.

Based on the above fundamental understanding, Emerson & Cuming has been developing new and unique materials using organic corrosion inhibitors which exhibit exceptional contact resistance with non-noble metallisations [6,7]. Organic corrosion inhibitors act as a barrier layer between the metal and the environment by absorbing onto the metal surfaces. Some chelating compounds are especially effective in preventing corrosion. Most organic corrosion inhibitors, however, can react with epoxy under certain conditions. This is not desirable because the corrosion inhibitor will then be consumed by the epoxy and loses its effectiveness. Many chelating corrosion inhibitors were tested and effective inhibitors were identified. The discovery of effective corrosion inhibitors was the first breakthrough of this program.

The first metallisations to be evaluated during the development of these adhesives were Cu and Sn/Pb. Complete Sn metallisations were tested in a later stage as the drive towards lead-free also resulted in the elimination of lead from the component terminations. Until very recently the compatibility of the new electrically conductive adhesives with components having 100% Sn terminations was not fully studied.

Table 1. Electrochemical potential of some electrode reactions.

Electrode reaction	Normal potential (V)
$Au - 3e^- = Au^{3+}$	1.50
$Pt - 2e^- = Pt^{2+}$	1.20
$Ag - 1e^- = Ag^+$	0.80
$H_2O + O_2 + 4e^- = 4OH^-$	0.40
$Cu - 1e^- = Cu^+$	0.52
$Cu - 2e^- = Cu^{2+}$	0.34
$Pb - 2e^- = Pb^{2+}$	$^-0.13$
$Sn - 2e^- = Sn^{2+}$	$^-0.14$
$Ni - 2e^- = Ni^{2+}$	$^-0.25$

Adhesives and Components

Components from different suppliers have been tested and are listed in Table 2.

Table 2. Overview of different components tested.

Component	Type	Size	Metallisation
A	0603	MLCC	100% Sn
B	0603	MLCC	100% Sn
C	0603	MLCC	100% Sn
D	0603	MLCC	100% Sn
E	0805	MLCC	100% Sn
F	0805	Resistor	100% Sn
G	0805	Resistor	100% Sn
W	0805	Resistor	80/20 Sn/Pb
X	0805	Resistor	80/20 Sn/Pb
Y	0805	Resistor	100% Sn
Z	0805	Resistor	100% Sn

A select group of adhesives was chosen for comparison. Adhesive A2 has been commercially available for a number of years and is widely used in hybrid applications with noble metallisations. Adhesive C uses specific corrosion inhibitors. Adhesives E and F are two adhesives containing both corrosion inhibitors and low-melting alloys. Eutectic Sn/Pb solder was used as a reference allowing a comparison between these adhesives and solder paste.

Contact Resistance Test Devices

A simple test device was constructed for ease of repeatability. The device consisted of a daisy-chain silver palladium (AgPd) pattern on a ceramic substrate or a gold pattern (Cu/Ni/Au) on an organic FR-4 substrate. The conductive pads were separated from one another by a 1.27 mm suitable for placing 0805 or 0603 components. Ten such separations were included in each daisy chain or loop. The test boards were completed by printing the adhesives onto the metallised pads using a stencil with a thickness of 100 μm and by placing the components. The test boards were then cured for 1 h at 150°C.

Initial Contact Resistance With Different Components

The contact resistance of adhesive C has been evaluated with 0805 components from four different suppliers W, X, Y and Z. Components W and X contain a 80/20 Sn/Pb metallisation, components Y and Z contain a 100% Sn metallisation. The contact resistance measured after cure for the different components is shown in Figure 3. Components W and X show a very typical low single joint contact resistance of about 25 mΩ. However, for components Y and Z a much higher resistance has been measured; 780 and 120 mΩ, respectively. The origin of this high initial resistance is suspected to be due to oxidation of the Sn metallisation before attaching the component to the test board.

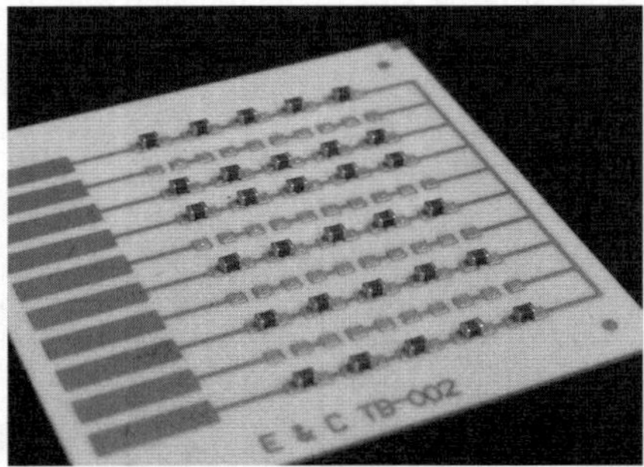

Figure 2. Example of a AgPd daisy chain test pattern on a ceramic substrate with Sn terminated null-ohm resistors.

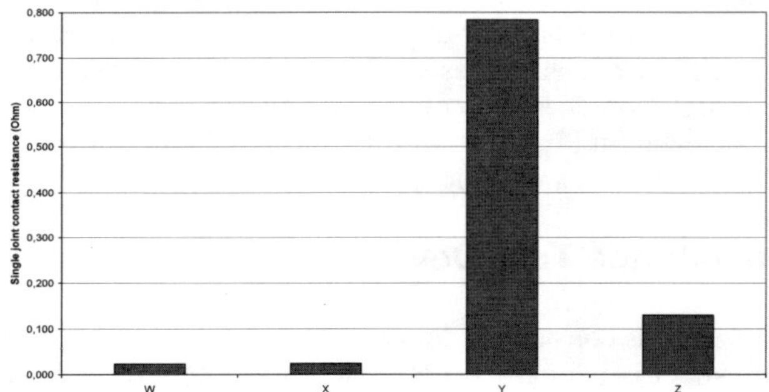

Figure 3. Initial contact resistance of adhesive C with different component types.

Contact Resistance Stability Under Thermoshock Testing

Further work has been performed on the contact resistance stability after thermoshock testing between −40°C and +150°C. Figure 4 shows the resistance per joint after 850 thermoshocks for the four evaluated component types attached to the test board using adhesive C. The high initial contact resistance with components Y (780 mΩ) and Z (120 mΩ) decreases significantly after 125 thermoshocks. Probably breakdown of the thin insulating layer on the component metallisation takes place due to mechanical stress caused by the differences in thermal expansion coefficient of the various materials. Components W and X also display a minor decrease in contact resistance after 125 thermoshocks due to post cure of the adhesive. The more detailed plot of Figure 4 shows that for all component types a slight increase in contact resistance up to an average value of 30 to 60 mΩ occurs between 125 and 850 thermoshocks. Although a very stable joint resistance has been established in 85°C/85%RH temperature/humidity testing using specific anti-corrosion agents in the newly developed conductive adhesives some potential issues have been defined in a typical automotive reliability test such as −40°C/+150°C thermoshock, especially for the components with the highest Sn levels.

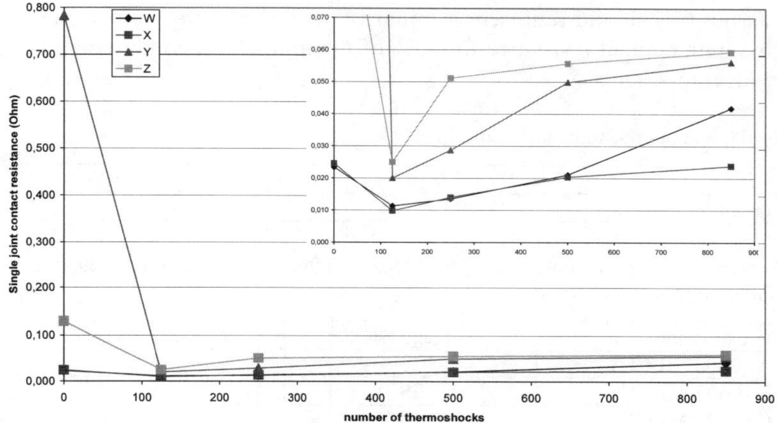

Figure 4. Full scale and detailed plot of the contact resistance stability of adhesive C with different component types during thermoshock testing from −40°C to +150°C

Improved Contact Resistance and Contact Resistance Stability

The initial contact resistance data obtained with three different adhesives and eutectic solder in combination with components Y are shown in Table 3. All adhesives show an initial contact resistance that is higher than the initial contact resistance obtained with eutectic solder. Adhesive C shows unacceptably high contact resistance data indicating that this material may not be fully compatible with 100% Sn metallisations. Adhesive E, which contains low-melting alloys in order to reduce the contact resistance, shows a better result. It should be emphasized that the reported data are an average of the total resistance measured for one daisy-chain loop.

Adhesive E has then been tested in combination with seven different types of components in order to get a better understanding of its performance with 100% Sn terminations. The contact resistance figures have been measured for each adhesive joint individually. The average single-joint contact resistance (SJCR) data and the standard deviation on 40 connections are presented in Table 4. For components C, D, E and G the single joint contact resistance is similar to solder. For three component types (A, B and F) a higher average single joint contact resistance and significantly larger standard deviations were measured. Among these poor-performing components, type A has been identified as the most critical one for achieving a low contact resistance. There appears to be no direct relation between good or bad performance and the component size or type. Figure 5 shows the individual data measured for adhesive E with components from type A; 15% of all contacts have a contact resistance above 100 mΩ. For adhesive E in combination with components from type C all contacts are below 100 mΩ (see Figure 5). It should be noted that some values above 50 mΩ could be assigned to inaccurate manual placement of the components but this should represent a minority of the measured joints. Figure 5 shows also a high contact resistance for 85% of the joints for a combination of adhesive C with components A.

Table 3. Initial contact resistanc after cure in mΩ.

	A2	C	E	Solder
Component Y	50	990	37	15

Table 4. Average single joint contact resistance (SJCR) after cure in mΩ for adhesive E with different components.

Component	Average SJCR	Standard deviation
A (0603 MLCC)	50	64
B (0603 MLCC)	29	94
C (0603 MLCC)	15	15
D (0603 MLCC)	19	17
E (0805 MLCC)	11	11
F (0805 resistor)	49	57
G (0805 resistor)	13	14

A scanning electron microscopy investigation of a connection with higher contact resistance did not show clear artifacts within the adhesive joint (see Figure 6).

Literature reports potential issues when soldering 100% Sn-terminated components [8]. Two potential issues could be defined to explain the high contact resistance with some of the components. Both ideas are related to the formation of an oxide layer at the component surface. Firstly, growth of Ni_3Sn_4 or $NiSn_3$ intermetallics within the component termination can occur. Oxidation will take place if these intermetallics grow until the surface of the component. It has been shown that intermetallic growth in Pb-free metallisations is more pronounced as compared to Pb-containing metallisations. Secondly, the formation of SnO and SnO_2 is inevitable. Fresh components typically have an oxide layer of 3 to 6 nm, whereas the oxide layer thickness can increase until 10 to 20 nm after aging. The test results shown above indicate that the presence of a too-thick oxide layer will result in a high initial single joint contact resistance. It is clear that the formation of oxides is more severe with 100% Pb-free components and this explains why similar issues were not observed earlier when evaluating Pb-containing components [9,10]. The occurrence of a thicker oxide layer may also depend on the components production process and the storage conditions. This may explain why higher resistance figures are only observed for a limited number of components and suppliers.

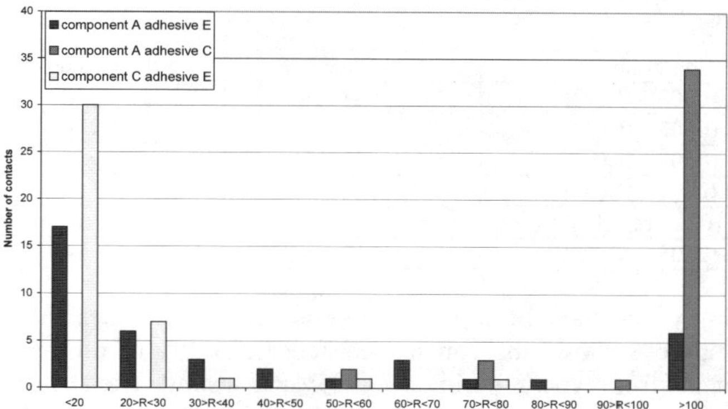

Figure 5. Number of contacts within a certain range (mΩ) for adhesive E with components A and C and adhesive C with components A.

Adhesive F containing oxide-removing additives has been evaluated with the same series of components. All contact resistance data are listed in Table 5. Figure 7 shows the SJCR distribution for adhesives E and F in combination with components A. The contact resistance has been reduced significantly for all components showing that the conductive adhesive F is able to remove the thicker oxide layer even on the most critical parts; components from type A, B and F. Using adhesive F, 60% of all contacts with components A have a contact resistance below 20 mΩ. These data show clear evidence that adhesives E and F can provide a low initial contact resistance approaching the same level as for eutectic solder. Furthermore, adhesive F is also compatible with the most critical 100% Sn components.

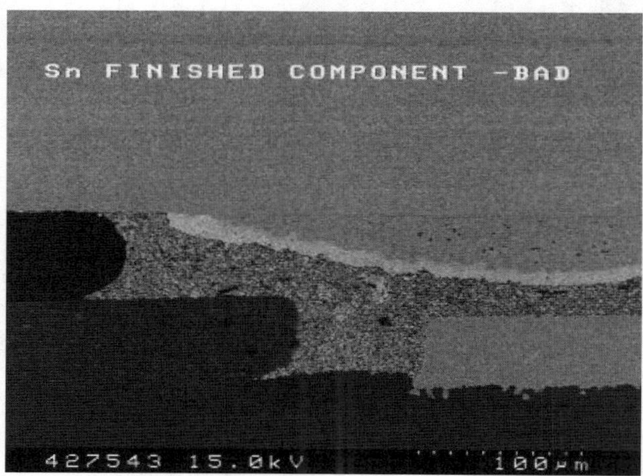

Figure 6. Scanning electron microscopy picture of a connection with high-contact resistance.

Table 5. Average single-joint contact resistance after cure in mΩ for adhesive F with different components.

Component	Average SJCR	Standard deviation
A (0603 MLCC)	21	12
B (0603 MLCC)	14	8
C (0603 MLCC)	11	7
D (0603 MLCC)	14	10
F (0805 resistor)	18	12
G (0805 resistor)	17	14

All reliability tests have been performed on ceramic substrates with AgPd metallisations. Figure 8 shows the contact resistance stability of different adhesives with components from type Y during 85°C/85%RH testing. The standard adhesive A2 shows a significant increase in single joint contact resistance within the first 200 h of the test. Adhesive E shows a low initial contact resistance which remains stable up to 1000 h. The SJCR data for adhesive E and eutectic solder are comparable. These results show that the anti-corrosion agents used provide a stable assembly for components with 100% Sn terminations.

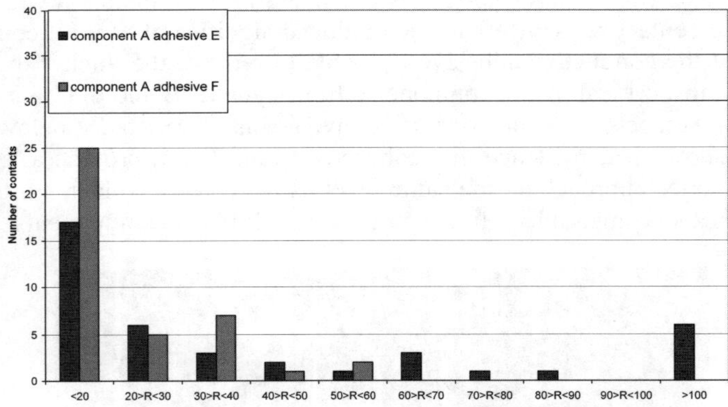

Figure 7. Number of contacts within a certain range (mOhm) for adhesives E and F with components A.

The contact resistance stability of the different adhesives in combination with components Y after 850 thermoshocks from −40°C up to +150°C is shown in Figure 9. Adhesive A2 shows a strong increase in single joint contact resistance during thermo shock test. Probably moisture accumulates at the interface and initiates corrosion of the Sn component metallisation. Adhesive E provides a low and stable contact resistance up to 850 thermoshocks. For solder the single joint resistance increases from 15 to 30 mΩ during the test. The use of conductive adhesives in combination with non-noble Sn metallisations can clearly be favourable upon solder when severe thermoshock reliability is required.

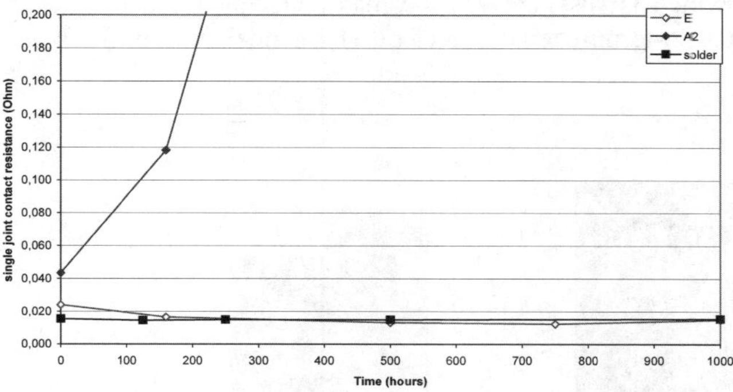

Figure 8. Single joint contact resistance (Ω) with component Y during 85°C/85%RH temperature/humidity reliability testing for 1000 h.

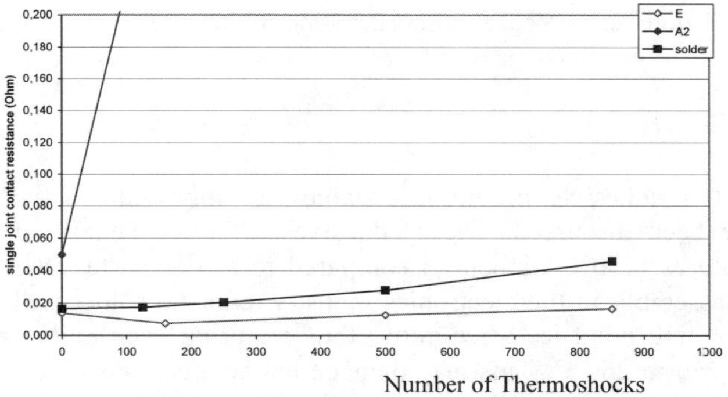

Figure 9. Single joint contact resistance (Ω) with component Y during −40°C/+150°C thermo shock testing.

Adhesion Stability after Thermoshock Testing

One of the other main requirements for the new conductive adhesives is adhesion and adhesion stability. Thermoshock testing has been defined as the most critical reliability test for components having non-noble metallisations. The performance of adhesive E has been evaluated against standard eutectic solder using component type Y. Figure 10 shows the adhesion data before and after 850x thermoshock testing between −40°C up to +150°C. It is clear that adhesive E displays very good adhesion retention. Whereas the adhesion using eutectic solder drops about 25% from 4.2 to 3.2 kg, adhesive E only shows a decrease of 10% 3.8 kg. The improved performance of adhesive E might be explained by: (a) the soft low-melting alloy stopping the formation of microcracks at the interface and (b) the low-melting alloy can form a metallic layer at the interface preventing diffusion of Sn into the conductive adhesive which is suspected as the major driving force behind adhesion drop in contact with Sn-terminated components (Kirkendale voiding).

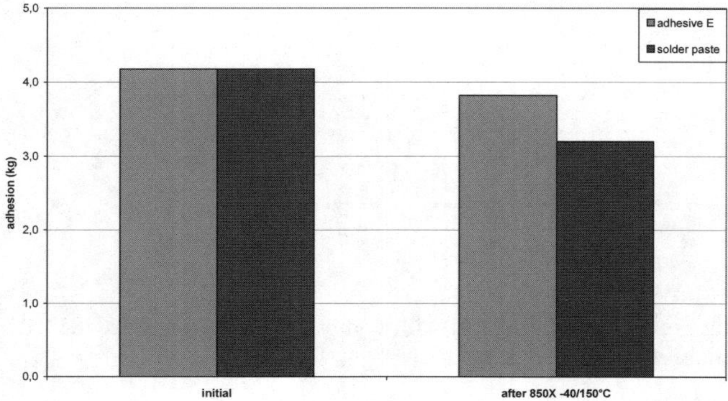

Figure 10. Adhesion before and after thermoshock testing of adhesive E as compared to standard solder paste.

Conclusions

Electrically conductive adhesives providing a stable assembly with non-noble Sn metallisations have been discussed. These adhesives offer the benefit of lower processing temperatures (150°C or below) compared to solder paste. Due to the drive for lead-free assembly in the electronics industry, eutectic solder will be replaced by higher melting point lead-free alloys thus requiring even higher reflow temperatures up to 250 to 260°C. This may damage temperature-sensitive components and/or temperature-sensitive substrates and limit the lifespan of the product. Also the terminations of surface-mount components have to become lead-free. Standard and new developed electrically conductive adhesives have been evaluated with different components from different suppliers all having 100% Sn metallisations. For some of these components oxidation of the component terminations can result in an increased initial contact resistance. The new electrically conductive adhesives are able to provide a low and stable single joint contact resistance with the most critical components under 85°C/85%RH temperature/humidity testing. The SJCR data for these adhesives are comparable with traditional eutectic solder joints. Under severe thermoshock test conditions in the temperature range from −40°C up to +150°C, the contact resistance stability of these adhesives even outperforms solder. The excellent performance of these adhesives is based on patented technology. The anti-corrosion agent is preventing electrochemical corrosion at the interface with the non-noble metal and the low-melting alloy improves the contact resistance and mechanical strength from the formation of an intermetallic layer at the interface.

References

1. D. Kloserman, L. Li, J. E. Morris, IEEE Trans. Comp. Packag. Manufact. Technol. A 1998, 21, 23

2. K. Gilleo : Surface Mount Technol. 1995, 9.1, p. 39 – 45

3. K. Gilleo.: Soldering Surface Mount Technol., 1995, 19, p. 2 – 17

4. D. Lu, C.P. Wong, Q. Tong : IEE Trans. Elec. Pack. Man., July 1999, 2-3, p. 223

5. Q. Tong, D. Markley, G. Fredrickson, R. Kuder and D. Lu: 1999 Electronic Components & Technology Conference, 1999, p. 347 – 352

6. G. Dreezen, G. Luyckx, IMAPS 2001 Proc. International Symposium on Microelectronics, 2001, p. 19

7. G. Dreezen, E. Deckx., G. Luyckx., IMAPS 2003, Proc. European Microelectronics Packaging & Interconncetion Symposium, 2002, p. 164

8. M. Arra., D. Shangguan, D. Xie, J. Sundelin, T. Lepisto, E. Ristolainen., IPC Soldertec International Conference on Lead Free Electronics Proc., p. 422, 2003

9. G. Fredrickson, C.-M. Cheng, Eletronic Components and Technology Conference, 2001

10. A. Grusd, Circuits Assembly, August 1999, p. 32

The Role of Au/Sn Solder in Packaging

H. Oppermann

Fraunhofer Institute for Reliability and Microintegration, Berlin / Germany

Introduction

Packaging is commonly known as a process to protect a single integrated circuit from the environment and to connect the electrical terminals to the printed circuit board. With increasing demands for miniaturisation and higher performance, technologies for system in package (SiP) and hybrid system integration become more important. To address the interaction of package, integrated passive components and active ICs co-design methods have to be implemented into the design flow. In order to achieve the desired system functionality different semiconductor technologies have to be combined, *e.g.* digital, analogue, RF, MEMS, and optoelectronics. This adds new challenges as higher level design and test capability, but also for hybrid system integration. Flip chip is a key technology for higher hybrid integration as it provides smallest footprint area and also, due to the short interconnection, lower parasitic for higher frequency applications.

Flip Chip Interconnection

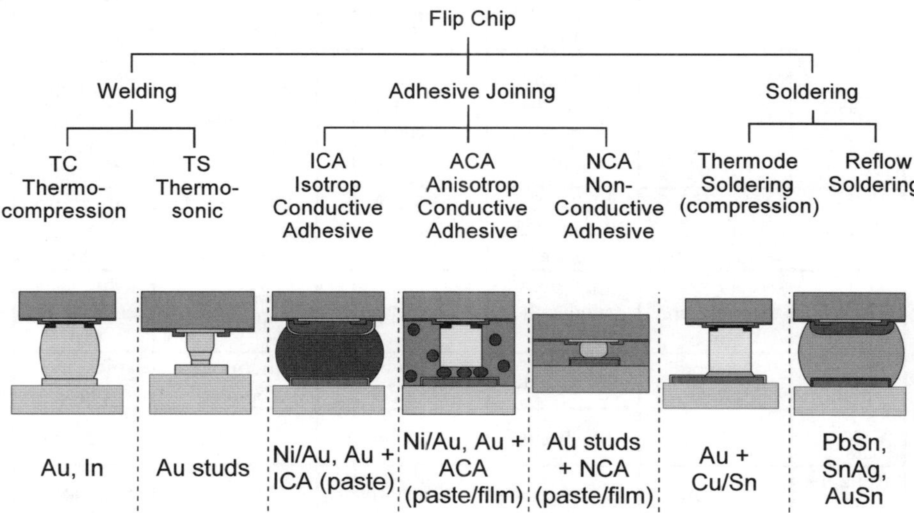

Figure 1. Different methods to form a flip chip interconnect.

There are three different methods to form the flip chip interconnect: welding, adhesive joining and soldering. Metals with high ductility such as gold are suitable for welding or fusion bonding. For thermocompression bonding of gold a force is applied to the chip at an elevated temperature of approximately 300°C. With thermosonic bonding ultrasonic power is applied in addition, which allows to reduce temperature or bonding force. A thermode bonder has to be used for both processes.

Adhesive joining methods can be distinguished into three types: isotrop conductive adhesive ICA, anisotrop conductive adhesive ACA, and non-conductive adhesive NCA. ICA pastes have a high filler particle loading, usually silver flakes in an epoxy matrix, and the material is patterned by stencil printing, dispensing or a dipping and transfer process. Electroless Ni/Au, electroplated Au or mechanical Au stud bumps have to be formed on the aluminum electrodes of the chip to prevent the joint from resistance degradation by oxidation. The flip chip assembly can be done by simple pick and place. The adhesive is cured in a second step. In contrast, ACA has a lower loading of conductive particles in the polymer matrix and the material is applied as film or paste over the whole chip area. A large bump area is advantageous to assure that each electrode is connected by a clamped conductive particle. A thermode bonder has to be used to apply force and temperature to clamp the particles and to cure the ACA matrix. With NCA no particles are used. Instead that the mechanical stud bumps are deformed during the bonding process, they are planarized, pressed and fixed by a thermode bonder while curing the polymer.

Table 1. Characteristic properties of various flip chip joints.

	Metal Bump	Isotropic Conductive Adhesive	Anisotropic Conductive Adhesive	Non-Conductive Adhesive	Solder Bump
Basic materials	Au	highly filled epoxy/ thermoplastic paste	filled epoxy/ thermoplastic paste / b-stage film	unfilled epoxy paste / b-stage film	SnPb37, SnPb90, SnAg, SnCu, AuSn20,
Bonding	250 – 350°C 1 min – 5 s ~50 g/bump	80 – 175°C 90 min – 15 s no load	150 – 200°C 1 min – 5 s 10 – 30 kg/cm²	150 – 200°C 1 min – 5 s 10 – 30 kg/cm²	230 – 310°C 1 min – 5 s no load
Process limits	high temperature, force	fine pitch, underfill needed	coplanarity	organic substrates	higher temperature, flux
Equipment	thermode bonder	pick & place	thermode bonder	thermode bonder	pick and place
Resistance	1...2 mΩ	20...30 mΩ	10...15 mΩ	5...10 mΩ	1...2 mΩ
Reliability	Avionic	Consumer	Consumer	Consumer	Automotive

Two methods of flip chip soldering can be distinguished: reflow and thermode soldering. With thermode soldering, either the height of the solder joint or the force is controlled. If a compressive force is applied only a thin layer of solder must be used to avoid shorts between neighboring bumps and substrate planarity should be excellent. Reflow soldering uses pick and place equipment for chip placement and the solder reflow can be done for all components in one step. The substrate pads usually offer flux or solder paste where the solder bumps are placed in.

The different bonding methods will result in corresponding joint structures. Table 1 summarizes the characteristic process parameters and the field of application. Isotropic conductive adhesive and solder bumps do not require a bonding force during thermal treatment and therefore simple pick and place equipment can be used. Adhesive joints are in general of interest for low temperature processing, *e.g.* for low-cost temperature sensitive substrates. Metal and solder joints have better electrical conductance and are used for higher reliability demands.

Most common solder alloys are based on eutectic SnPb37 having their melting temperature at 183°C. High-lead solder alloys like PbSn5 have a higher melting point around 300°C and are very ductile, which means that the stress is released by plastic deformation. But with the EC directive on Waste from Electrical and Electronic Equipment (WEEE) lead in solder alloys will be banned and a lot of investigations took place for finding an equivalent replacement by lead-free solders.

None of the solder alloys that have been investigated satisfy all the requirements. Some eutectic solder alloys have a lower melting point than SnPb37: SnBi58 (138°C), SnIn52 (118°C) and BiIn33 (109°C). But indium is rather expensive and the availability of bismuth is limited. Another candidate below 200°C is eutectic SnZn9, which melts at 198°C. But zinc tends to rapidly oxid and corrosion problems are expected with zinc alloys. Nonetheless, this alloy type might play a larger role for low-cost consumer applications. With increasing melting temperature the eutectic alloys SnAg3.5 (221°C), SnCu0.7 (227°C) and SnAg3.8Cu0.7 (217°C) come into focus for replacement of eutectic SnPb37. They are recommended for automotive and telecommunication application. At higher operating temperatures SnSb5 (240°C) and AuSn20 (280°C) are of interest for replacement, but, on the other side, they are much harder than high-lead solders. No replacement is found that provides high melting point and similar mechanical properties, therefore the concepts for high lead solder applications must be revised.

Au/Sn Solder

Au/Sn solder with an eutectic composition of 80 wt% (70 at%) gold and 20 wt% (30 at%) tin is preferred over other compositions and has a melting point at 280°C. Bumps with such gold-rich composition are commonly used for flip chip assembly of optoelectronic and RF devices because this allows a fluxless assembly, which is required to avoid contamination at optical interfaces. Also, the solder alloy is compatible to thick gold pads and conductor lines as used with III/V semiconductors and RF substrates.

The phase equilibrium diagram of Au–Sn in Figure 2 shows that there is another tin-rich eutectic of 90 wt% tin and 10 wt% gold, which has an eutectic transformation at 217°C. The η-phase (AuSn$_4$) is known to be brittle [1], if it reacts with nickel to (Au,Ni)Sn$_4$ the adhesion becomes weak [2].

The brittleness of the η-phase is evident as shown in the cross sections of LED assemblies in Figure 3a and b. The reflow temperature must be above 280°C to achieve the eutectic composition of AuSn20 (Figure 3d) with fabulous properties. Otherwise, below that temperature, tin-rich phases will occur (Figure 3c).

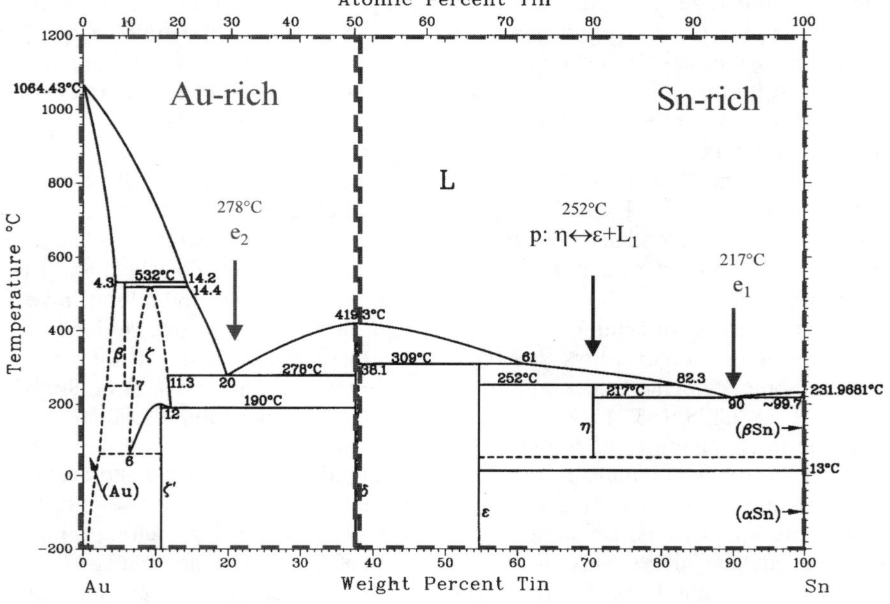

Figure 2. Phase equilibrium diagram of binary Au–Sn system [3]. The most important reactions of the gold-rich and tin-rich subsystem are indicated.

Au/Sn Bumping and Solder Formation

Au/Sn bumping can be performed either by evaporation techniques or by electroplating. For evaporation a lift-off process is used that limits the total deposition height to ~10 μm and the wafer must be cooled to keep the temperature low for the resist. It is much more cost effective to deposit Au/Sn by electroplating [4,5]. In our case TiW is sputtered on the entire wafer to serve as adhesion and diffusion barrier layer and Au to protect it from oxidation and to carry the current during plating. A thick resist is spun on the wafer and patterned by photolithography. Gold and tin are subsequently plated into the resist opening. The resist is stripped and the plating base removed by wet etching.

The bumps consist of gold and a tin layer. Intermetallic compounds as AuSn$_4$ (η),AuSn$_2$ (ε), and AuSn (δ) are even formed at room temperature (Figure 4). The bumps are reflowed above 280°C on wafer level to form the eutectic solder (Figure 5a). The transformation can be investigated by calorimetric measurement

(Figure 5b). The tin-rich eutectic ($\eta+Sn{\rightarrow}L_1$) forms first at 217°C, the peritectic reaction ($\eta+L_1{\rightarrow}\varepsilon$) takes place at 251°C and the gold-rich eutectic transformation ($\delta+\varepsilon{\rightarrow}L_2$) at 281°C. Obviously, an exothermic reaction occurs just before the second eutectic transformation and is contributed to the mixing of two liquids of different composition: a tin-rich L_1 and a gold-rich L_2.

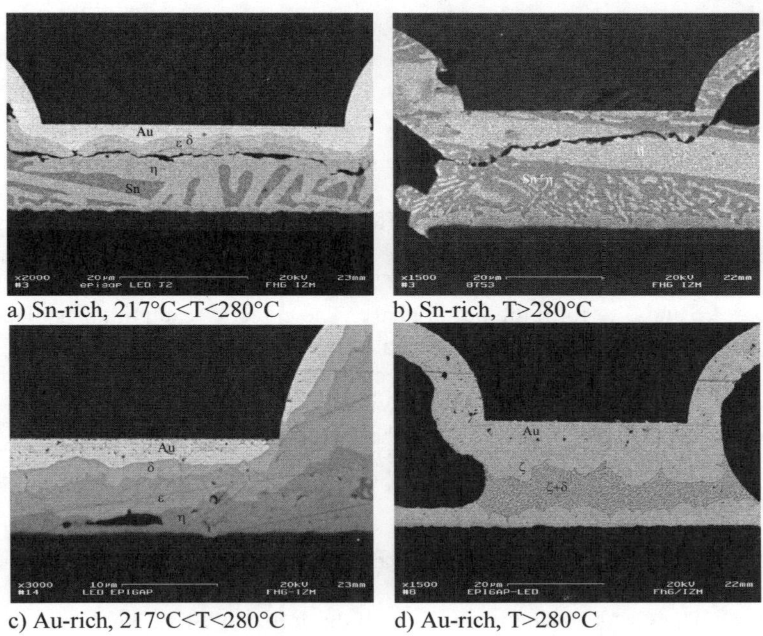

Figure 3. Au/Sn microstructure of LED solder joints with different composition and assembly temperature. Cracks are visible in the brittle η-phase.

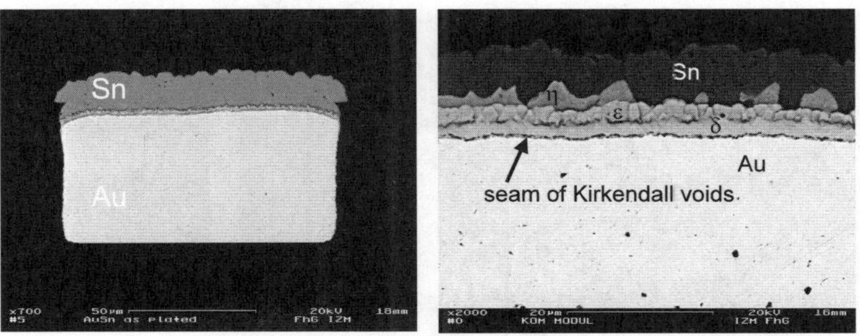

Figure 4. Au/Sn as plated, IMC formed after storage at room temperature.

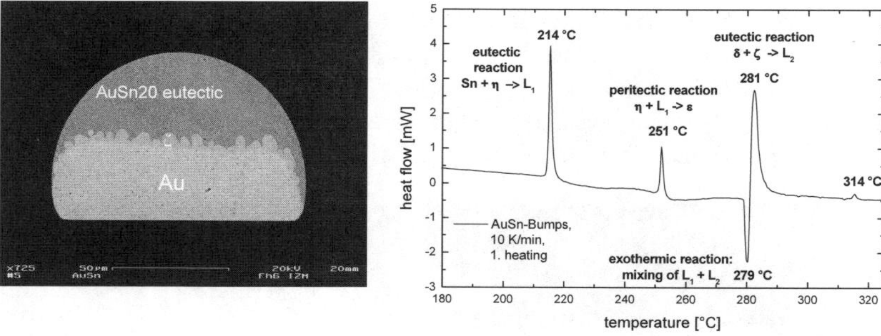

Figure 5. Transformation of Au/Sn bumps during reflow.

It was previously shown how the different Au–Sn phases occur and disappear by stopping the transformation sequence during bump reflow [6,7]. ε-, δ- and ζ-phase are present at 270°C, and at 278°C the eutectic transformation starts by the reaction of δ and ζ to form the liquid phase. The eutectic reaction takes place until all δ-phase has been transformed completely. With further heating the ζ-phase grows, which has been studied at 290°C and has been compared with growth rates below the eutectic temperature [6]. Obviously, the growth mechanism does not change and the diffusion through the ζ-phase determines the growth rate.

Large Bumps

Chiaro Networks Ltd. has developed a GaAs device which is the integral part of an optical switch. The chip has a size of 31×12 mm² and 2608 inputs/outputs [I/Os] (Figure 6). The GaAs device works as a "light deflector", moving a beam of light in free space. Every 128 waveguides (WG) forms a deflector. The GaAs device comprises 18 such deflectors. Putting an array of 64 fibers a few mm away behind and a fiber in front of a deflector, light can be directed through the deflector to any output fiber wanted. As 18 such deflectors are in one GaAs device it is an 18 × 64 switch.

Figure 6. 18×64 optical switch. Right: cross section of flip chip assembled GaAs chip. A small via connects the bump with the waveguide (courtesy of Chiaro Networks Ltd.).

The principle of operation of each deflector is the same as a phased-array-antenna: A WG grating is formed in the GaAs. Light is injected from one fiber to a group of 128 WG called a deflector, which forms a grating. The deflector output goes into free space, where a grating-far-field-diffraction-pattern is formed. The maximal intensity of the diffraction pattern can be directed left and right in the plane of the deflector by changing the phase of the light coming out of each single WG. The phase of the light coming out of each WG is changed by applying a voltage to the WG, thereby changing the effective index of refraction.

The GaAs chip is flip chip soldered to an Al_2O_3 ceramic manufactured by Kyocera. The routing of the conductors requires 25 signal layers. The pads of the ceramic are finished with Ni/Au. A SEM micrograph of a cross section through an assembled Au/Sn bump is illustrated in Figure 6. The WG structure on the GaAs device is clearly visible on top. The bump and the discrete WG is connected by a small via through an organic layer on the GaAs device. Surface planarity and shrinkage tolerance on the ceramic in the flip chip attachment area arising from the very complex manufacturing must be compensated by the bump solder volume. The solder joint shown measures 125 μm in diameter and 50 μm in height.

Large bumps with 33 μm gold and rather thick layer of 17 μm tin (Figure 7, left) are used to address these tolerances. After reflow, the total bump height is increased to 70 μm and consists of approximately 5 μm Au, 6 μm ζ-phase and a 59 μm eutectic dome. With even larger bump diameter of 300 μm the eutectic domes increase to 72 μm and the total bump height to 83 μm (Figure 9).

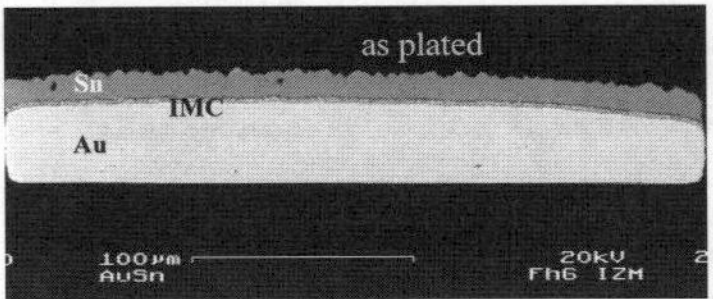

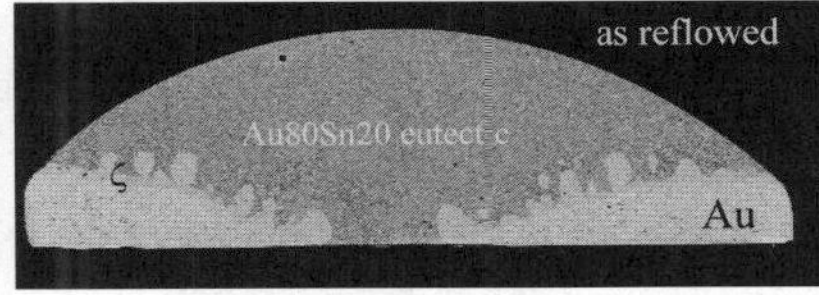

Figure 7. Large Au/Sn bump, as plated (left) and after reflow (right).

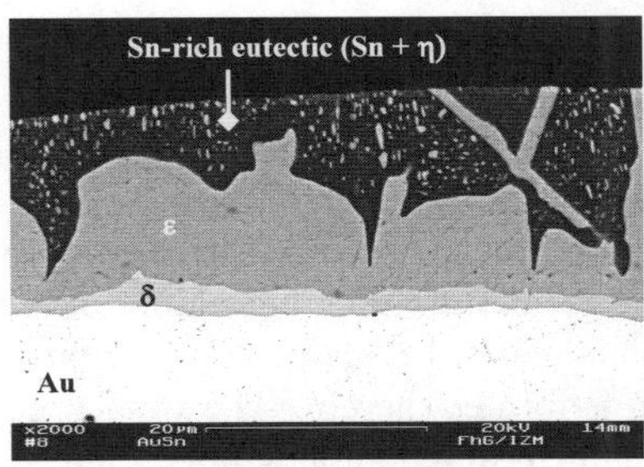

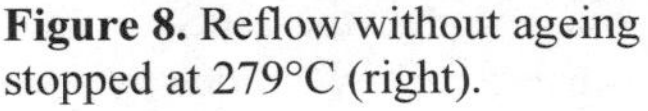
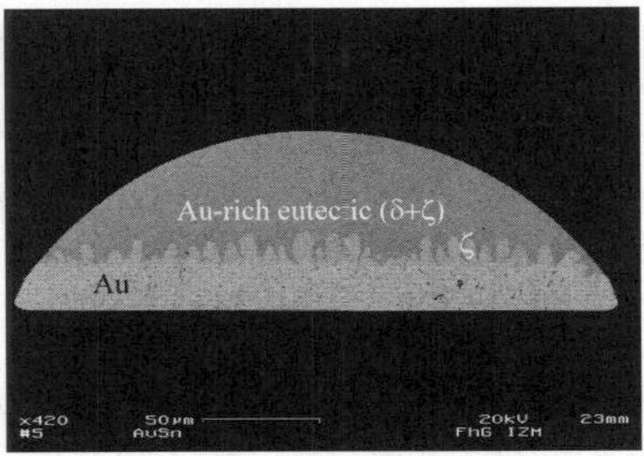

Figure 8. Reflow without ageing stopped at 279°C (right).

Figure 9. Large reflowed Au/Sn bump after ageing.

Depending on the substrate pad diameter the bump collapses during reflow soldering and compensates for coplanarity tolerances. With a thicker tin layer of 10 μm and above a non-planar reaction front and incomplete eutectic transformations was observed [8] (Figure 7, right). At 217°C the first eutectic reaction starts. For further transformation the gold must diffuse through the intermetallic layers δ, ε and η. If the tin layer is thicker, then the diffusion time is longer to complete the reaction, and consequently there is a tin-rich melt still present before the second gold-rich eutectic reaction starts (Figure 8). The non-uniform reaction is due to the mixing of two liquids of different chemical composition, which results also in an exothermic heat, as can be seen by the calorimetric measurements of Figure 5, right.

If the bumps are aged, the exothermic heat does not show up and regular bumps are formed with the reflow (Figure 9). The aim of the ageing is the transformation of the whole tin into intermetallics in order to avoid the formation of tin-rich eutectic during heating. In Figure 10 no tin is observed. Successful soldering could be demonstrated even when the tin, η and ε were completely transformed to δ-phase, which was present only on top of gold sockets. The progress in the transformation of intermetallics depends on temperature and time and on the initial tin layer thickness; an example is given in Figure 11.

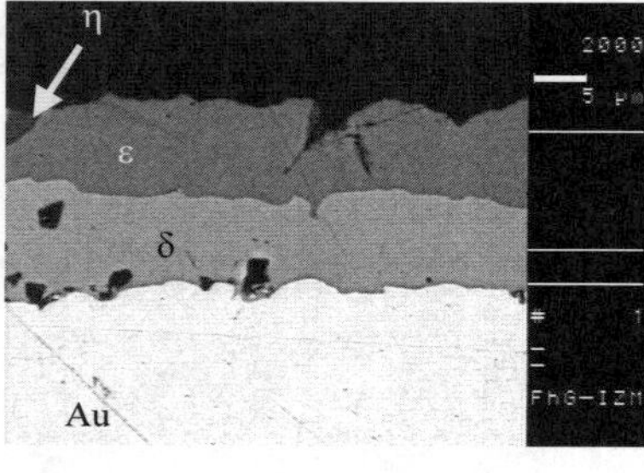

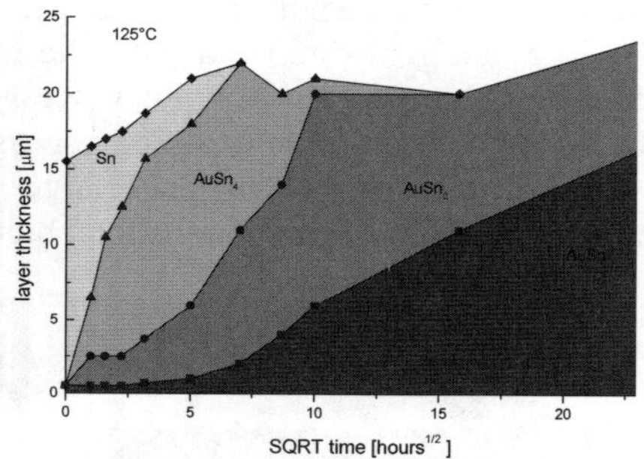

Figure 10. Aged Au/Sn bump at 200°C for 4 h.

Figure 11. Layer thickness of intermetallic phases and tin at 125°C.

Small Bumps

Usually a dome of eutectic AuSn solder is formed during bump reflow on top and a layer of ζ-phase separates it from the gold socket (see Figure 5, left). But with a smaller bump diameter of 50 µm or less the liquid solder wets the sidewall during reflow and the height planarity from bump to bump varies depending on volume fraction which spreads down the socket. This results in a larger interface of solder to the gold socket and a larger amount of ζ is formed thereby. The smaller eutectic dome and the bump height variation lead to opens and poor assembly yield.

In order to increase the yield and to avoid these open joints, the AuSn solder bumps are assembled without prior bump reflow. During reflow soldering, the bumps are reflowed and the substrate pads are wetted at the same time. But the soldering process with non-reflowed bumps can only be applied if the pad geometry is adapted: substrate passivation must be lower than substrate pad metal or the substrate pad must be larger than the bump diameter due to placement tolerances.

Under Bump Metallization and Pad Metal

The under bump metallization (UBM) and the substrate pad metallization serve as adhesion layer, diffusion barrier, and solderable layer. The solderable metal is dissolved into the liquid solder and intermetallic compounds (IMC) are grow during soldering and solidification. If the solderable layer is consumed completely, then the layer underneath directly contacts the solder. As a result, dewetting of solder could occur or the device could be destroyed (Figure 12, right). Therefore the knowledge of phase equilibrium, IMC formation and dissolution rates are important to design the UBM and pad metal layers. Of course, the minimal layer thickness will depend on the bonding method and their specific time-temperature profile. Phase equilibria of Au–Sn–Cu [9] and Au–Sn–X (X=Ni, Pt, Pd) [10] and growth rates of IMC have been published for Ni, Pt, Pd, Cr and Cu [10].

It should be mentioned that the AuSn solder shows a different wetting behavior on chromium or titan in contrast to tin-rich solders like SnPb, SnAg or SnCu due to the phase formation of Au–Cr and Au–Ti phases (Figure 12, left).

Optoelectronic Components

Planar lightguide circuits (PLC) and optical silicon benches are usually offered with thin evaporated AuSn layers for the high-precision mounting of laser and monitor diodes or photodetectors. High-precision flip chip thermode bonders with a post-bond accuracy of 1 µm or less are normally used to meet the chip-to-fiber or chip-to-waveguide tolerances of single mode applications [11]. During the bonding process the temperature has to be applied very fast to keep the soldering time short. This is required to avoid excessive reaction of the solder with the barrier layers as can be seen in Figure 12, right. The AuSn, solder is very suitable as it allows a fluxless processing, and creep or relaxation is extremely low compared to tin-rich solders or indium. Due to the high melting point of AuSn further soldering steps below this temperature are possible without loss in positioning accuracy.

High-power laser bars having an optical output power of 50 W and more are soldered to thermal heat spreaders by the same method [12]. Further decrease of solder thickness below 3 µm is targeted to reduce evaporation cost and thermal resistance. The subassembly can further be soldered to a micro channel cooler with a tin-rich solder at lower temperature.

High-brightness LEDs are mainly used for illuminations and today they dissipate about 2 W over an area of 1 mm^2. They are often die-bonded to heat sink materials with evaporated or electroplated AuSn layers [13].

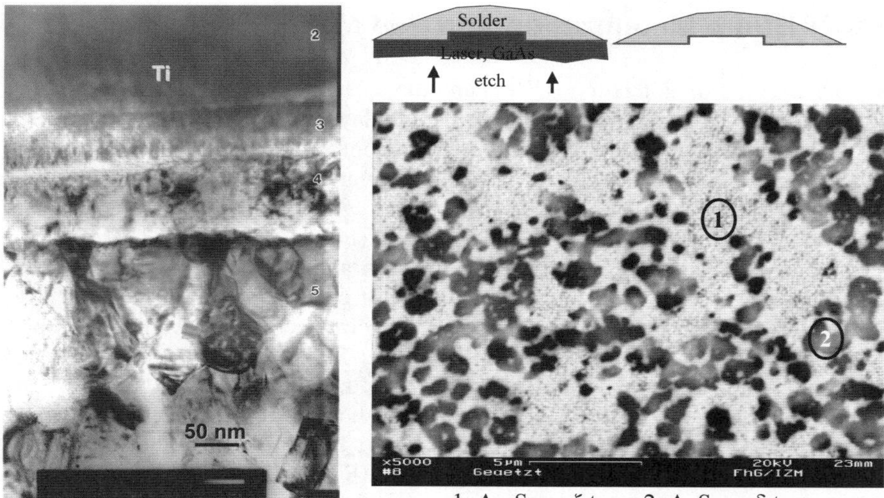

Figure 12. Left: TEM picture of AuSn solder attached to titan: Ti (2), Ti66 Au32 Sn2 (3), Ti60 Au36 Sn4 (4) and Ti6 Au88 Sn6 (5) (all in at%). Right: Laser etched away down to the solder, barrier layer (Ti/Pt/Au) has reacted completely with the solder.

Figure 13. Wafer level assembly of high-brightness LEDs.

Figure 14. Receiver module with photodetector assembled with 3 µm AuSn on silicon bench.

Flip Chip Self-alignment

In order to couple the laser or the photodetector diode to a waveguide or an optical fiber they are usually actively operated while the fiber or the microlens are moved to find the optimum position and finally they have to be fixed. For single-mode applications alignment tolerances of approximately 1 µm are required to achieve a considerable coupling efficiency. As this active alignment is still causing considerable costs in optoelectronics packaging, viable concepts for passive alignment are required. A concept for achieving high-precision positioning is using the self-alignment of liquid solder based on surface tensions [14].

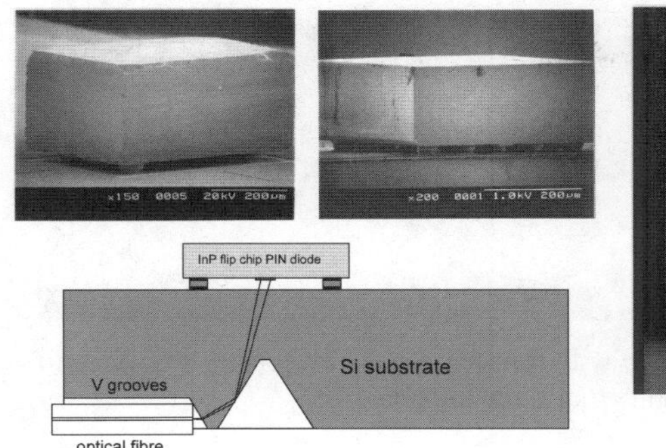

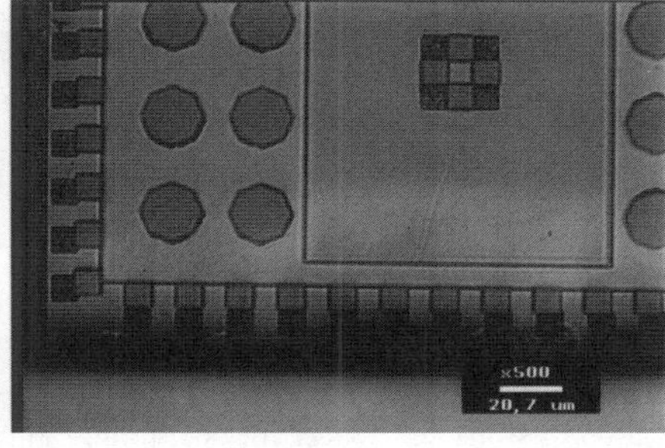

Figure 15. Photodetector and laser modules and principal arrangement.

Figure 16. Self-alignment test chip assembled on glass substrate with Vernier pattern.

AuSn solder has an excellent wetting behavior and no liquid flux is required that could disturb the self-alignment and contaminate the optical interfaces. The chips are flipped onto the substrate and the assembly is then heated. Alignment takes place immediately when the solder melts. The solder spreads over the whole substrate pad and the chip moves to a position where the surface tension reaches the minimum. The chip is fixed when the solder solidifies during cooling. Typical modules achieved by this method are presented in Figure 15. An accuracy of better than 2 µm can be obtained [6] by using glass substrates and Vernier pattern (Figure 16). Compared to soft solders the AuSn solder does not creep significantly, which gives a stable final positioning.

In order to reduce the impact of the accuracy required for bump deposition and substrate pad definition micro mechanical stops can be introduced. The chip and substrate are misaligned to a certain extent, the chip starts to move driven by surface tension in order to minimize the surface area of the solder joint. The final position of the laser with regard to the PLC is solely determined by micromechanical stops that had been etched in the laser diodes as well as in the PLC (Figure 17).

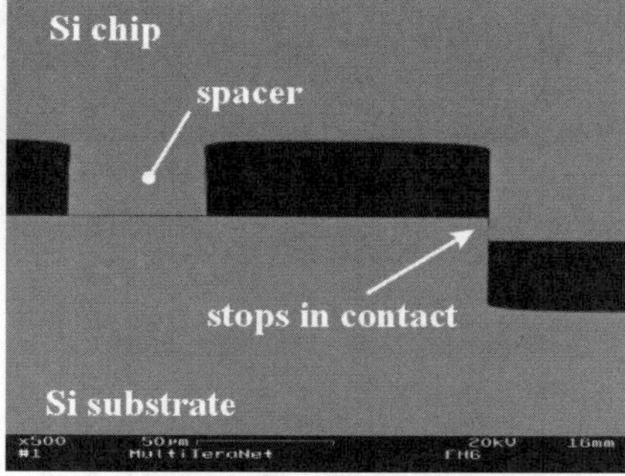

Figure 17. Left: Cross section of aligned chip with micro mechanical stops. Right: Magnification of stops and spacer for positioning in two of three dimensions.

Hermetic Sealing

Hermetical sealing of glass or silicon lids to housings or to MEMS have been fabricated with AuSn solder. With the fluxless process no residuals are enclosed and even vacuum sealing has been demonstrated for absolute pressure sensors [15]. The AuSn solder can be applied by electroplating, evaporation or using AuSn preforms and on the opposite side Au or Ni/Au metallization has been used (Figure 18). For wafer-to-wafer solder bonding without any compression load the process is limited to the wafer bow. In this case only single parts or wafer tiles can be assembled to the substrate wafer before a common reflow in vacuum or inert atmosphere is applied.

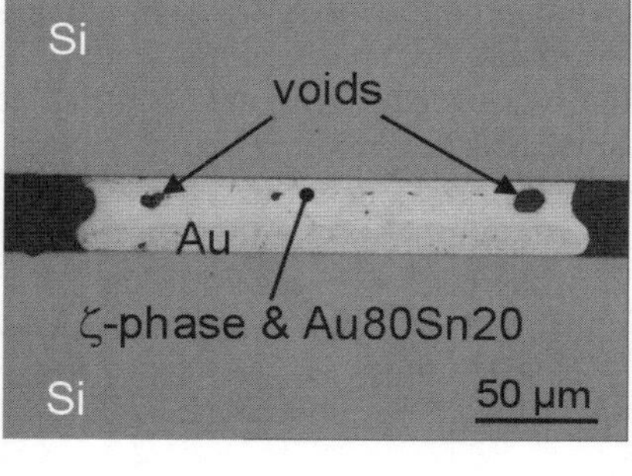

Figure 18. Left: X-ray picture of hermetical sealing ring. Right: Cross-sectional view on sealing ring with AuSn solder.

References

1. Zama, S. *et al*, "Flip Chip Interconnect Systems Using Wire Stud Bumps and Lead Free Solder," Proc 50th Electronic Components and Technology Conf, Las Vegas, May 2000.

2. A. Zribi, *et al*. "Solder metallization interdiffusion in microelectronic interconnects," in Proc. 49th IEEE Electron. Comp. Technol. Conf., 1999, p. 451–457.

3. H. Okamoto, T.B. Massalski, "The Au-Sn(Gold-Tin) System" in Phase diagram of binary gold alloys, p. 278-289, ASM International, Metals Park, Ohio, 1987

4. Engelmann, G., Ehrmann O., Simon, J., Reichl, H., "Development of a Fine Pitch Bumping Process," Proc 1st International Conference on Micro, Electro, Opto, Mechanic Systems and Components, Berlin, Germany, 1990.

5. Dietrich, L., Engelmann, G., Ehrmann, O., Reichl, H., "Gold and Gold-Tin Wafer Bumping by Electrochemical Deposition for Flip Chip and TAB," 3rd European Conference on Electronic Packaging Technology (EuPac'98), Nuremberg, Germany, June 15-17, 1998.

6. Kallmayer, C., Oppermann, H.H., Kloeser, J., Zakel, E., Reichl, H.: "Experimental Results on the Self-Alignment Process using Au/Sn Metallurgy and on the Growth of the ζ-Phase During the Reflow". Proceedings of the International Flip Chip, Ball Grid Array, TAB and Advanced Packaging Symposium (ITAP '95), February 1995, San Jose, California, p. 225-236.

7. Zakel, E., Reichl, H.: "Flip Chip Assembly Using Gold, Gold-Tin, and Nickel-Gold Metallurgy", in *Flip Chip Technologies*, ed. John Lau, Chap. 15, McGraw-Hill, 1995.

8. M. Hutter, H. Oppermann, G. Engelmann, J. Wolf, O. Ehrmann, R. Aschenbrenner et al., Calculation of Shape and Experimental Creation of AuSn Solder Bumps for Flip Chip Applications. 52nd ECTC Electronic Components and Technology Conference, 2002, p 282-8.

9. Zakel, E.; Reichl, H.: "Au-Sn bonding metallurgy of TAB contacts and its influence on the kirkendall effect in the ternary Cu-Au-Sn", IEEE Transactions on Components, Hybrids and Manufacturing Technology, 16, 3, p. 323-332 (1993).

10. Anhöck, S., Oppermann, H., Kallmayer, C., Aschenbrenner, R., Thomas, L., Reichl, H.: "Investigations of AuSn Alloys on Different End-Metallizations for High Temperature Applications". 22nd IEEE/CPMT International Electronics Manufacturing Technology Symposium (IEMT Europe 98), 1998, p. 156-65.

11. G. Elger, J. Voigt, H. Oppermann: "Application of Flip-Chip Bonders in AuSn Solder Processes to Achieve High After Bonding Accuracy for Optoelectronic Modules". LEOS 2001. Lasers and Electro-Optics Society, 2001. The 14th Annual Meeting of the IEEE Conference Proceedings, Volume: 2, November 11-15, 2001, San Diego, USA, p. 437-8.

12. A. Bärwolff, J.W. Tomm, R. Müller, S. Weiß, M. Hutter, H. Oppermann, H. Reichl: "Spectroscopic Measurement of Mounting-Induced Strain in Optoelectronic Devices". IEEE Transaction on Advanced Packaging, 23, 2, p. 170-5 (2000).

13. G. Elger, R. Jordan, M. v. Suchodoletz, H. Oppermann: Development of an Low Cost Wafer Level Flip Chip Assembly Process for High Brightness LEDs Using the AuSn Metallurgy. 35th International Symposium on Microelectronics, September 4-6, 2002, Denver, Colorado, p. 199-204.

14. H. Oppermann, E. Zakel, G. Engelmann, H. Reichl: "Investigation of Self-Alignment During Flip-Chip Assembly Using Eutectic Gold-Tin Metallurgy". 4th Int. Conf. and Exhibition on Micro Electro, Opto and Mechanical Systems and Components, MST '94.

15. B. Rogge, D. Moser, H. Oppermann, O. Paul, H. Baltes: "Solder-Bonded Micro-machined Capacitive Pressure Sensors". Proceedings of SPIE's Symposium and Continuing Education on Micromachining and Microfabrication, Santa Clara, September 1998, 3514.

Packaging Materials: Organic–Inorganic Hybrids for Millimetre-Wave Optoelectronics

N. Rapún

University of Surrey, Chemistry Division, Surrey / UK

Introduction

The developments in optical communications have been considered as one of the most impressive events of the second millennium [1]. The advantages in microelectronics are growing into our life styles, thus computers require more capacity and faster processing and additionally, smaller, faster, more functional and more reliable electronics products are required. Nowadays, improvements in microelectronics have been produced in a very short period of time. However, owing to the velocity of the electrons there will be a limit and therefore, photons will be the transport medium of the future. Optical interconnects on-chip are more complicated because this technology must be compatible with the semiconductor process, and therefore hybrid technologies have been explored where the integration of microoptical elements, like waveguides, on semiconductor, glass or quartz wafers can be realised by UV lithography [2].

These hybrid technologies include two main approaches: the planar lightwave circuit approach, which involves silica-on-silicon technology, and the possibility fabricating optical interconnects on a fully-processed wafer forming multi-chip-modules [3].

Recent studies focused on the encapsulation of multichip modules by organic–inorganic hybrids for frequencies up to 100 GHz. Modern high-frequency optoelectronic systems make use of a wide range of optoelectronic, microwave and millimetre wave devices. For frequencies up to 100 GHz there are many applications, such as high bit rate telecommunication circuits, radio-over-fibre systems and antenna signal distribution.

Packaging as a Key Area

Packaging of these microwave and millimetre-wave circuits requires great care in order to avoid the possibility of parasitic components compromising circuit performance [4]. The addition of optoelectronic devices to the package, with the additional complexity of ensuring good optical coupling, adds further constraints to the circuit and package design.

Furthermore, when sizes are reduced the probability of contamination or damage increases and the packaging step acquires great importance.

At the same time, the production of intelligent devices involves the development of the isolation regions and the electrical contacts of the device. Thus, a demand emerges on the fabrication techniques, which requires thin film deposition of insulators [5].

Materials for Packaging

The advances emerging in the field of optoelectronics have induced the rapid development of optical interconnects and have also generated a need for optically transparent materials for packaging. The plastic encapsulation of the chip is critical to the reliability of the package. The most remarkable properties this material should embody, apart from high optical quality and transparency [6,7], include low thermal expansion, sufficient rigidity to protect the encapsulated structures, heat resistance, ease of processing [8] and control of refractive index and thickness.

Different materials have been studied for this application such as silica and polymers. However, these materials show a few drawbacks.

Polymers embody potential properties for optoelectronic applications such as inexpensive materials, easy processing technology, acceptable loss performance [9], low dielectric constant [10], transparency, adhesive properties, light weight and possible molecular tailoring through controlled synthesis [11,12]. However, these materials present some drawbacks such as low thermal stability and increased loss which make them unsuitable for telecommunication applications [9].

In contrast, *silica glass* shows potential for optical communications [13] due to its low optical losses, low dielectric constant [14], and ability to tune light scattering and its low refractivity. Furthermore, sol-gel processing is a promising deposition technique. The major drawback is the shrinkage, which may be circumvented by the use of new composite organic–inorganic materials based on crosslinked polymers.

Thus, *organic–inorganic materials* appear as suitable materials providing the best combination of properties for optoelectronic applications. In recent years, organically modified silica (ormosils) materials, a type of organic–inorganic hybrids, have begun to receive attention for integrated optical applications. Ormosils offer some advantages, such as the possibility of required thickness, about microns, in a single step. They can be made photosensitive for direct UV writing of waveguides [15] and confer compatibility with both organic and inorganic dopants for active device applications. There is much interest in the field of optoelectronics to apply these materials as passive or active layers in optoelectronic devices [16, 17].

Organic–Inorganic Hybrids (properties and types) Overview

Organic–inorganic hybrids emerge as a mixture of two dissimilar phases. Organic materials can offer structural flexibility, convenient processing, tuneable electronic properties, photoconductivity, efficient luminescence, and the potential for semiconducting and even metallic behaviour. On the other hand, inorganic compounds provide the potential for high carrier mobilities, band gap tunability, a range of magnetic and dielectric properties and thermal and mechanical stability.

Their final properties are not just the sum of the individual phases as the interface also plays an important role and new phenomena can arise as a result of the interface between organic and inorganic components [18]. From the chemical point of view, the potential is based on controlling the interface. The interfacial forces between both components are influenced by the nature and relative content of the organic and inorganic components as well as the experimental conditions used for preparing the hybrids [19].

Depending on what kind of interaction is employed between organic content and inorganic components these hybrids can be divided into two main classes, which are illustrated in Figure 1.

1.   Type I: only weak interactions give cohesion to the whole structure (hydrogen, van der Waals or ionic bonds).

2.   Type II: the two phases are kept together by covalent bonds (covalent or iono-covalent bonds).

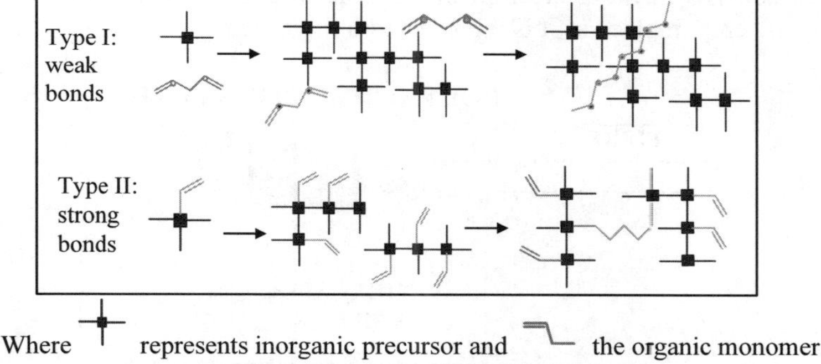

Figure 1. Types of hybrids according to the nature of the interactions [20].

The precursors utilised to prepare type-II hybrids present two different functionalities: alkoxide groups (M-OR), which undergo hydrolysis-condensation reactions and M-C bonds in which their stability depends on the nature of the metal. They are stable through the hydrolysis when the metal is silicon, tin, mercury, lead or phosphorus, but not when transition metals are involved. The modified ceramics obtained from their hydrolysis and condensation are usually called ormosils and ormocers (organically modified ceramics).

Sol-Gel Process: Method of Synthesis Hybrids

Sol-gel process allows the preparation of organic–inorganic hybrids at temperatures which ensure the survival of the organic component [21]. This preparation of organic–inorganic hybrids has attracted enormous attention due to the fact that their properties can be tuned through the functionality or segment size of each component. Thus, this process is widely used with applications in high technology fields and protective coatings. This method offers advantages over other ones in terms of compositional tailoring, functionality and cost.

The sol-gel process is a method for preparing metal oxo polymers from metallo-organic precursors (salts or alkoxides) *via* continuous hydrolysis and condensation reaction steps which are governed by pH, solvent, water-to-alkoxy ratio, concentration of reactants, catalyst, temperature and pressure [22,23].

The chemical steps involved in the sol-gel process consist of simultaneous hydrolysis and condensation of a metal alkoxide in a solution to result in the formation of a three-dimensional network: the gel. The most important reaction of this type involves the catalytic hydrolysis of tetraethoxysilane [TEOS].

Optoelectronic Multichip Modules 1–100 GHz

The development and design of Optoelectronic Multichip Modules between 1 and 100 GHz was produced on glass substrates. It allows the fabrication in the same substrate of both optical and electrical waveguides. The function of the encapsulant material consists in the protection of the chip and the waveguides as well as in holding the chip in place as it is represented in Figure 2.

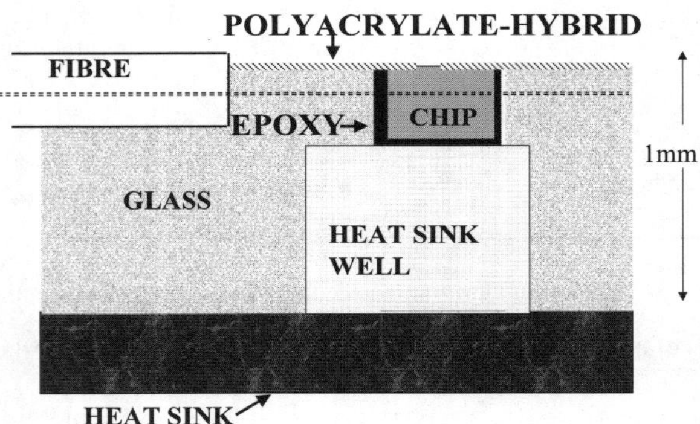

Figure 2. Scheme of millimetre-wave optoelectronic multichip modules.

The packaging of the above chip consists of two main materials:

1. The material that will cover the materials on the surface of the multichip as well as the chip, which is marked with diagonal lines (▨▨▨▨) in Figure 2.

2. The material which will hold the chip on place and furthermore will glue to the glass and waveguide, which in Figure 2 is marked on black.

Both materials, types of organic–inorganic hybrids, are summarised in Figure 3.

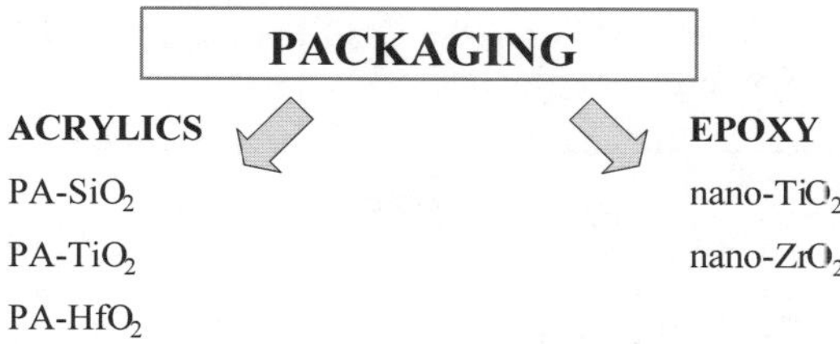

Figure 3. Scheme of materials studied for the packaging devices, where PA means poly-acrylates.

Polyacrylate-Silica Hybrids Synthesis

The synthesis of polyacrylate-silica hybrids provides excellent advantages over other materials. On one side, polyacrylate offers excellent optical properties, and additionally, its mechanical and thermal properties are improved by hybridisation with silica. Consequently, the synthesis of silica-acrylic hybrids *via* sol-gel synthesis and photopolymerisation considered for the protecting material offers improved compatibility with the fibre and the chip substrate and furthermore, properties may be tuned by varying the ratio of organic to inorganic components. The range of monomers available offers the potential for refractive index control in the hybrid by varying the composition. These properties are based on the network structure of inorganic matrix linking to organic domains. Besides, UV curing provides a rapid processing route and photopolymerisable liquid encapsulants offer advantages over the traditional transfer moulding process, due to the lower thermal stress involved.

Preparation techniques are shown in Figure 4. The synthesis of polyacrylate-silica hybrids carried out under basic conditions leads to opaque white films. This may be due to the fact that under basic conditions condensation is faster than hydrolysis and therefore dense, colloidal particles were formed and as a result no strong absorption was achieved between organic and inorganic phases. In contrast, changing to synthesis in an acid medium leads to transparent films as the relative rates are reversed and highly ramified structures were formed.

Raman as a Useful Technique for Studying the Acrylate Curing Process

Raman spectroscopy may be used to follow the UV-curing process. Thus, Raman spectroscopy has been used to follow the real-time curing of acrylates by monitoring the disappearance of the signal at 1635 cm⁻¹, due to the C=C bond of the acrylate. Representative Raman spectra in the region from 1500 cm⁻¹ to 1800 cm⁻¹ are displayed in Figure 5 and it is possible to notice the disappearance of the signal at 1635 cm⁻¹ with time due to butyl acrylate polymerisation. After 20 min, this signal was no longer appreciable, which indicates that the polymerisation has finished; a similar result has been found following this polymerisation with Fourier transform infrared spectroscopy. At the same time the C=O bond slightly shifts to higher frequencies because it is not longer conjugated.

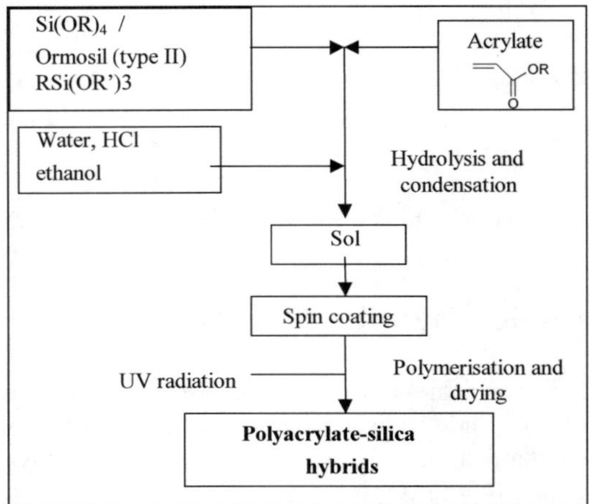

Figure 4. Preparation of processing for polyacrylate-silica coatings.

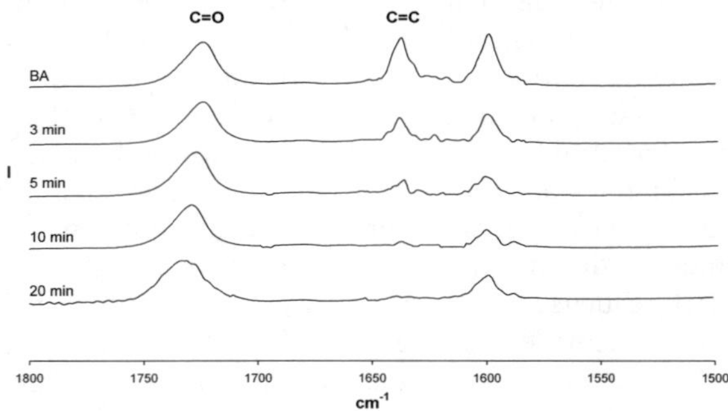

Figure 5. Raman spectrum region from 1500 cm^{-1} to 1800 cm^{-1} for butyl acrylate (BA) polymerisation.

Furthermore, confocal Raman microscopy, a technique used to optimise UV formulation and curing conditions [24, 25], has been employed in order to study the cross-section behaviour and the structure of the films. It combines the chemical information of vibrational spectroscopy with the spatial resolution of confocal microscopy [26].

The spectra in these regions show satisfactory homogeneity within the coating throughout its thickness, as can be concluded from the negligible deviations between ratios of the peak areas for corresponding chemical bond deformations: C–H bending of the CH_3 (1380 cm^{-1}) and CH_2 (1470 cm^{-1}) and C=O stretching (1730 cm^{-1}). Depth profile spectra of poly(butyl acrylate) [PBA] are shown in

Figure 6 for different depths of the film. The upper line shows a superficial depth profile and it gets deeper through the film as the lines get lower.

The absence of an obvious peak corresponding to C=C stretching in the spectra of the coating suggests that UV curing of the acrylate matrix was successful throughout the coating thickness.

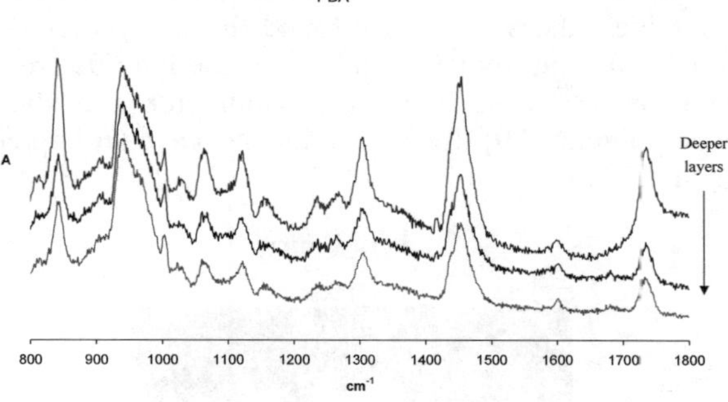

Figure 6. Poly(butyl acrylate) Micro-Raman depth profile spectra for deeper layers as the lines get lower.

Polyacrylate-Silica Hybrids Structure: Miscible Phases

A combination of solid-state ^{13}C NMR and ^{29}Si NMR spectroscopy has been used as a very attractive technique to characterise the polyacrylate-silica hybrids. ^{13}C NMR spectroscopy is used to study the organic domains and ^{29}Si NMR spectroscopy the inorganic domains. The interaction of both phases by weak bonds as hydrogen bonds give cohesion to these structures.

The incorporation of polymer chains does not significantly affect the condensation reaction of silanol groups during the sol-gel process. Q^3 is the predominant signal, the silica network contains Si–OH groups, which facilitates the interaction with the polymer by the carbonyl groups [27].

The structure is studied by scanning electron microscopy (SEM) and by mapping the elements it is possible to appreciate that they are well distributed throughout the film. Transparent hybrid films indicate that the polymer chains were distributed in the silica network with high uniformity. Optical transparency is used as the initial criterion for the formation of a homogeneous phase composition of the organic–inorganic constituents [28].

The polyacrylate-silica films show a good miscibility as in the case of poly(2-hydroxy-ethylacrylate)-silica [PHEA–SiO$_2$] hybrid shown in Figure 7.

Polyacrylate-Metal Oxide Hybrids Controlling Refractive Indices

Ellipsometry has been used as a technique for measuring the refractive index and thickness of the films.

The refractive index and thickness of hybrid materials can be controlled with the variation of composition, for instance the thickness of the coatings increases with the amount of TEOS.

Furthermore, the refractive index and the thickness of hybrid systems can be tuned by modifying the acrylate backbone in the hybrid. Different hybrid structures are displayed in Table 1, and different values have been obtained that are displayed in Table 2. Refractive indices at 633 nm found for the synthesised hybrids are in the range 1.43 to 1.70. The refractive indices of the hybrids are lower than their corresponding polymers because the refractive index for pure silica is smaller than that of the acrylic polymer [29]. Refractive indices have an imprecision of ± 0.01 and thicknesses of ± 2 nm.

PHEA-SiO$_2$

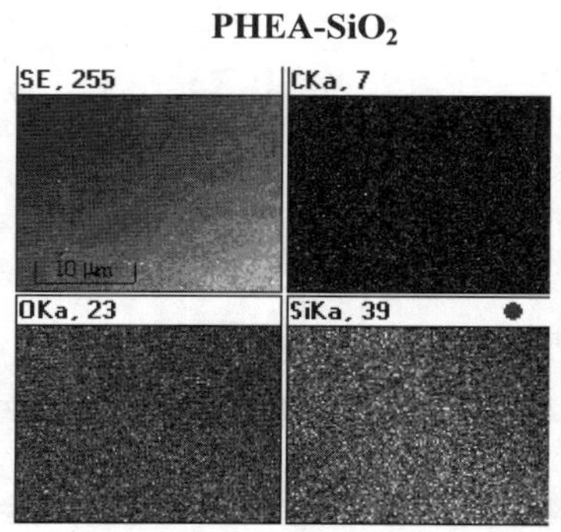

Figure 7. PHEA-SiO$_2$ SEM pictures. SE shows the mapping region.

Organic–Inorganic Hybrids: A Method for Minimising Polymer Microwave Losses

Silica presents low optical losses, hence polymer microwave loss values decrease when the polymer is hybridised with silica. In addition, it can be observed that the types of interactions in the hybrids determine microwave loss indices. When both types of hybrids are compared, type-II hybrids show lower loss indices.

Organic–Inorganic Hybrids: A Method for Decreasing Inorganic Shrinkage

A reduction of the inorganic shrinkage of films such as silica occurs when organic groups are integrated. Shrinkage reduction responds to the presence of bulky organic groups filling the pores. The use of organically modified alkoxysilanes precursors, type-II hybrids, allows for even greater shrinkage reduction levels, allowing for shrinkages as low as <1% [30].

Table 1. Structure of polyacrylates.

Poly(methyl acrylate) **PMA** 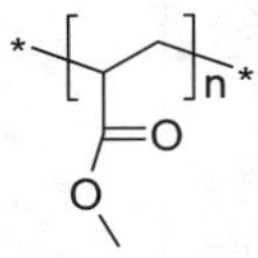	Poly(*tert*-butyl acrylate) **PtBA** 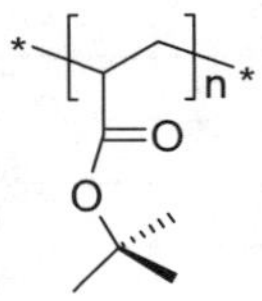	Poly(isooctyl acrylate) **PIA** 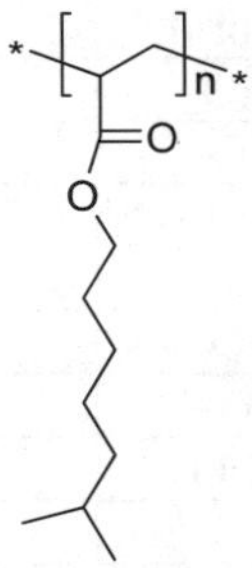	Poly(lauryl acry-late) **PLA** 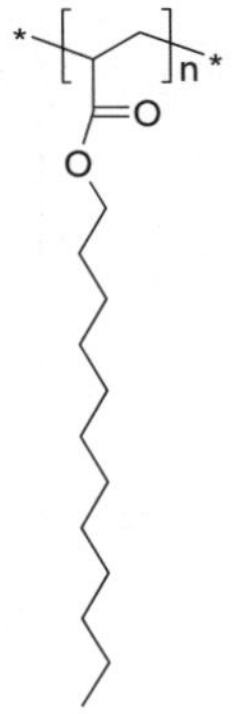
Poly(ethyl acrylate) **PEA** 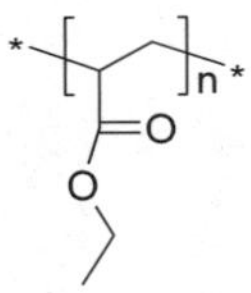	Poly(butyl acry-late) **PBA** 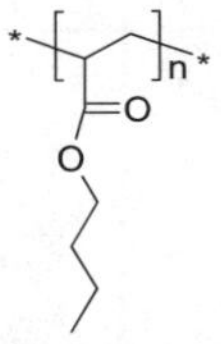		
Poly(2-hydroxy ethyl acrylate) **PHEA** 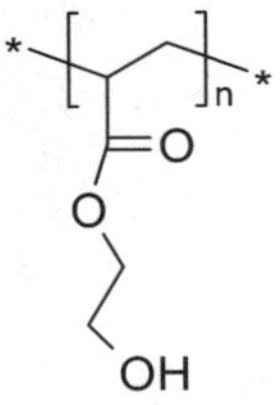	Poly(ethylene glycol phenyl ether acrylate) **PEGPA** 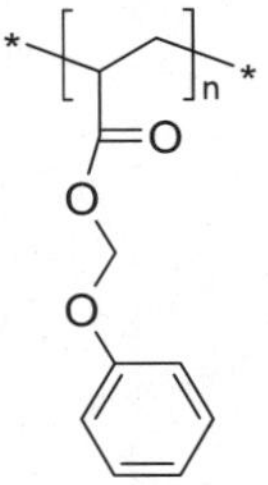	Poly(dodecafluoro heptyl acrylate) **PFA** 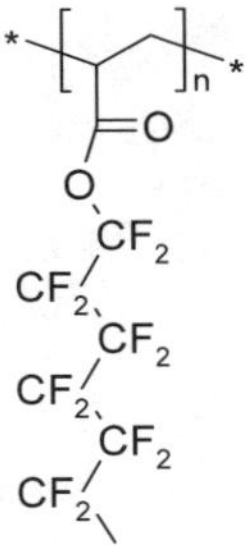	Poly(3-meth acry-loxy propyltrimeth-oxysilane) **PMEMO** 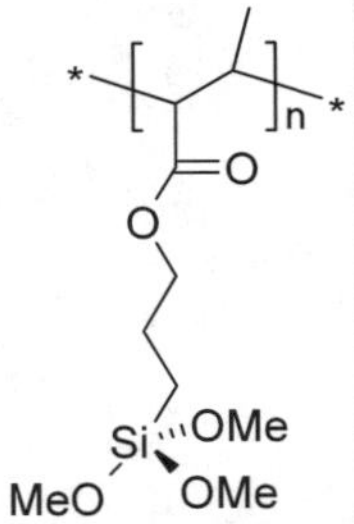

Polyacrylate-Silica Hybrids Thermostability

Thermostability is an important parameter for these materials to determinate the temperature range of application. Polyacrylate-silica hybrids push the T_g (glass transition temperature) to higher temperatures compared with the pure polyacrylates. Thus, the temperature application of the hybrids is broadened. T_g of the hybrid material is associated with co-operative motion of long chain segments, which may be hindered by the inorganic oxide network formed from the hydrolysis of

TEOS by decreasing the free volume [31,32]. Figure 8 shows a shift in the T_g for poly(tert–butyl acrylate) (PtBA) when the hybrid is formed.

Table 2. Refractive index and thickness of PA-SiO$_2$ and silica.

1Polyacrylate-1 silica hybrid	Refractive index	Thickness (nm)
PMA-SiO$_2$	1.44	290
PEA-SiO$_2$	1.43	320
PBA-SiO$_2$	1.44	350
PtBA-SiO$_2$	1.44	380
PLA-SiO$_2$	1.46	250
PIA-SiO$_2$	1.50	120
PHEA-SiO$_2$	1.47	760
PFA-SiO$_2$	1.46	380
PEGPA-SiO$_2$	1.70	429

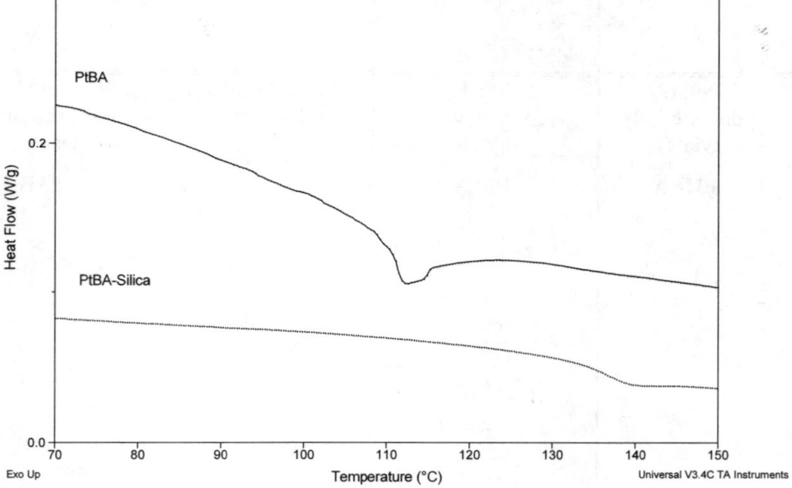

Figure 8. DSC scan for PtBA and PtBA–SiO$_2$ hybrids.

The glass transition temperature range of the composites depends on the composition, in the degree of mixing and the morphology [33]. The effect of adding organically modified alkoxysilanes is to increase the T_g value: there is just one transition which indicates the good miscibility of the copolymer and also confining effect within the SiO$_2$ network and as a result the T_g shifts to higher temperatures.

Polyacrylate-Inorganic Oxide Hybrids: An Alternative to Silica

One way of expanding these materials is by using alternative precursors to silica. Organic–inorganic hybrids based on titania and hafnia have been studied as materials with potential applications in optoelectronics. These metal oxides possess high refractive indices and may be used to tailor this parameter, for instance titania films have been prepared with a range of refractive indices between 1.95 and 2.1 [34]. Hafnium also belongs to group 4 and its oxide appears to be a promising material for these applications [35].

Polyacrylate-titania/hafnia hybrids were prepared by sol-gel process (1 h) followed by UV polymerisation to yield transparent hybrid films.

This attractive method of synthesis is based on the dual role of the acrylates. They act as complexing agents allowing the formation of M-O-C coordinative bonds and consequently leading to transparent films as well as acting as monomers in the hybrid network tailoring refractive indices by varying the acrylate monomer (in a range from 1.5 to 1.75).

Epoxy Filled with Nanoparticles

When evaluating the other material that will glue the chip to the glass, it is not only necessary to consider its bonding abilities, but also its optical compatibility with the bonding substrates. Epoxy resins offer an ability to bond well to various materials including metals, glass, ceramics, wood and other polymers and have been used as optoelectronic adhesives. Additionally, they have the versatility to be modified by varying the properties such as thermal stability, refractive index and flame retardance, *etc.* The refractive index depends especially on the presence of heteroatoms (Ti, Zr) and, in order to broaden the range of refractive index of the epoxy, nanoparticles of zirconia and titania have been incorporated into the resin.

When the nanoparticles are introduced, two phases were obtained and transparency was not achieved. For nanoparticles without surface pre-treatment, by decreasing the particle size the specific surface area increases and the probability of agglomeration increases [36]. Actually, the strong attractions between them make their dispersion in the polymer a difficult process to achieve. To enhance the interactions between filler and elastomer, some coupling agents were added, Figure 9. The surface modification of the nanoparticles improves the dispersibility of the filler in the organic medium.

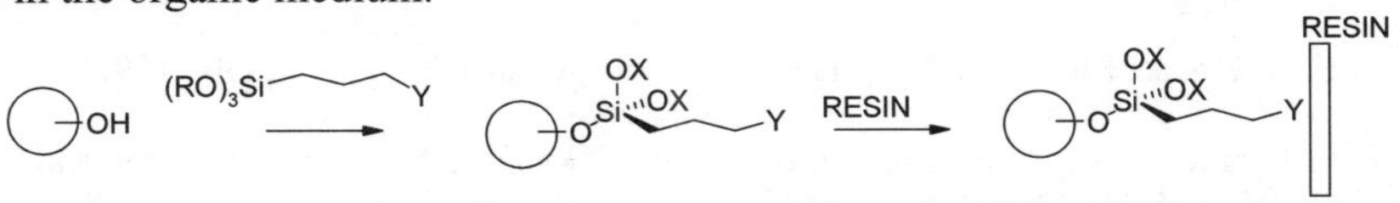

Figure 9. Treatment of nanoparticles by a silica coupling agent [37]. Where Y = organic group, X = R, H, Si or the mineral surface.

Conclusion

A new kind of material as organic–inorganic hybrids with tuneable properties depending on the composition and synthesis conditions, have been used to encapsulate multichip devices.

These organic–inorganic materials offer promising properties for millimetre optoelectronic packaging such as the control refractive indices and thicknesses, thermostability, good miscibility, low shrinkage, optical transparency and low microwave losses.

Type-II hybrids appear to be more suitable materials for optoelectronic application due to the low shrinkage, low thermal expansion coefficients, low microwave losses and an increase in the T_g value, which emerges as a single value indicating good miscibility. These effects may be attributed to the covalent bonds between the organic and inorganic domains.

Acknowledgment

The author thanks Prof. John N. Hay for his guidance and the UK Engineering & Physical Sciences Research Council for funding and Agilent for providing a CASE studentship.

References

1.   G. Masini, L. Colace and G. Assanto, Mater. Sci. Eng., B 89, 2-9 (2002).

2.   P. Dannberg, L. Erdmann, A. Krehl, C. Wächter and A. Bräuer, Mater. Sci. Semicon. Proc., 3, 437-441 (2000).

3.   K.K. Baikerikar and A.B. Scranton, Polymer, 42, 431-441 (2001).

4.   G.R. Atkins, R.M. Krolikowska and A. Samoc, J. Non-Cryst. Solids, 265, 210-220 (2000).

5.   J. Singh, Semiconductor Devices, MacGraw-Hill, London (1994).

6.   Y.-Y. Yu, C.-Y. Chen and W.-C. Chen, Polymer, 44, 593-601 (2003).

7.   C. Chan, S. Peng, I. Chu, S. Ni, Polymer, 42, 4189-4196 (2001).

8.   C. A. Kovac, Electronic Materials Handbook, ASM International, OH (1998).

9.   F. Tooley, N. Suyal, F. Bresson, A. Fritze, J. Gourlay, A. Walker *et al.*, Optical materials, 17, 235-241 (2001).

10.   G. Maier, Prog. Polym. Sci. 26, 3-65 (2001).

11.   S. Rimdusit and H. Ishida, Polymer, 41, 7941-7949 (2000).

12. W.-F. Su, Y.-C. Fu and W.-P. Pan, Thermochim. Acta, 392-393, 385-389 (2002).

13. E.M. Yeatman, K. Pita and M. M Ahmad, A. Vannucci and A. Fiorello, J. Sol-Gel Sci and Tech., 13, 517-528 (1998).

14. H. Fan, H. R. Bentley, K. R. Kathan, P. Clem, P. Clem, Y. Lu *et al.*, J. Non-Cryst. Solids, 285, 79-83 (2001).

15. P. Etienne, P. Coudray, Y. Moreau and J. Porque, J. Sol-Gel Sci. Tech., 13, 523-527 (1998).

16. G. R. Atkins, R. M. Krolikowska and A. Samoc, J. Non-Cryst. Solids, 265, 210-220 (2000).

17. F. Gan, J. Sol-Gel Sci. Tech., 13, 1998, 559-563.

18. D. B. Mitzi, Chem. Mater., 13, 2001, 3283-3298.

19. K. Kim, K. Adachi and Y. Chujo, Polymer, 43, 2002, 1171-1175.

20. P. Judeinstein and C. Sanchez, J. Mater. Chem., 6, 1996, 511-525.

21. P. Calvert, Nature, 353, 501-502 (1991).

22. C.J. Brinker, G.W. Sherer, Sol-Gel Science, The Physics and Chemistry of Sol-Gel Processing, Academic Press, London (1990).

23. B.E. Yoldas, J. Mater. Sci., 14, 1843-1849 (1979).

24. W. Schrof, E. Beck, G. Etzrodt, H. Hintze-Bruning, U. Meisenburg, R. Shchwalm and J. Warming, Prog. Org. Coat., 43,1-9 (2001).

25. M.E. Nichols, C. M. Seubert, W.H. Weber and J.L. Geerlock, Prog. Org. Coat., 43, 226-232 (2001).

26. W. Schrof, W. Beck, R. Königer, W. Reich and R. Schwalm, Prog. Org. Coat., 35, 197-204 (1999).

27. T. Çaykara and O. Göven, Polym. Degrad. Stabil., 62, 267-270 (1998).

28. C. J. T. Landry, B. K. Coltrain, J. A. Wesson, N. Zumbuladis and J. L. Lippert, Polymer, 33, 1496-1506 (1992).

29. Y.-Y. Yu and W.-C. Chen, Mater. Chem. Phys., 82, 388-395 (2003).

30. I. Cunningham, Hay, J.N.; Rapun, N.; Weiss, B., Proceedings of Nanocomposites 2004, San Francisco, 1-3 September (2004).

31. Y. Wang, T.C. Chang, Y.S. Hong and H.B. Chen, Thermochim. Acta, 397, 219-226 (2003).

32. C.K. Chan and I. M. Chu, Polymer, 42, 6089-6093 (2001).

33. J. Habsuda, G.P. Simon, Y.B. Cheng, D. G. Hewitt, D.A. Lewis and H. Toh, Polymer, 43, 4123-4136 (2002).

Wafer-Level Three-Dimensional Hyper-Integration Technology Using Dielectric Adhesive Wafer Bonding

J.-Q. Lu, T. S. Cale, R. J. Gutmann

Center for Integrated Electronics, Rensselaer Polytechnic Institute,
New York, NY / USA

Introduction

Although scaling of planar CMOS ICs, combined with copper/low-k interconnect technology, continues to drive productivity enhancements in integrated electronics, scaling beyond the 45-nm technology node is projected to be increasingly expensive and difficult [1]. Increasing demands for ever higher functionality in integrated systems will be met by introducing new materials and processing protocols into planar CMOS-based technology. To help meet these challenges, various three-dimensional (3D) technology platforms have been investigated, including hybrid die-to-wafer [2,3] and monolithic wafer-to-wafer approaches [4–11]. Wafer-level monolithic 3D integration is predicted to have a performance advantage from reduced interconnect parasitics [12]. It also enables high-density multifunctional integration or hyper-integration [4–6], and offers the potential for the highest volumetric density of integrated electronics and optoelectronics with a high density of high-performance, vertical interconnections.

This paper briefly compares alternative technologies for 3D ICs in order to provide a perspective of the 3D technologies, then focuses on a specific 3D hyper-integration platform using dielectric adhesive wafer bonding and Cu damascene inter-wafer interconnects. Results of collaborative research with industrial partners are summarized, demonstrating that the bonding and thinning processes do not significantly affect the electrical and mechanical properties of the most advanced IC technologies. After a discussion of some issues with such 3D integration, this paper concludes with a comparison of wafer-level 3D integration with other high-density integration technologies, namely system-on-a-chip (SoC) and system-in-a-package (SiP).

3D IC Technologies

Three major 3D IC approaches being pursued are: (1) die-to-die, (2) hybrid die-to-wafer, and (3) monolithic wafer-to-wafer 3D integrations, as shown in Figure 1. Die-to-die 3D integration or SiP is currently used to increase functionality and to reduce form factor [13–14]. The other two major approaches have the potential for higher interconnect density, higher performance capability and lower cost for high-

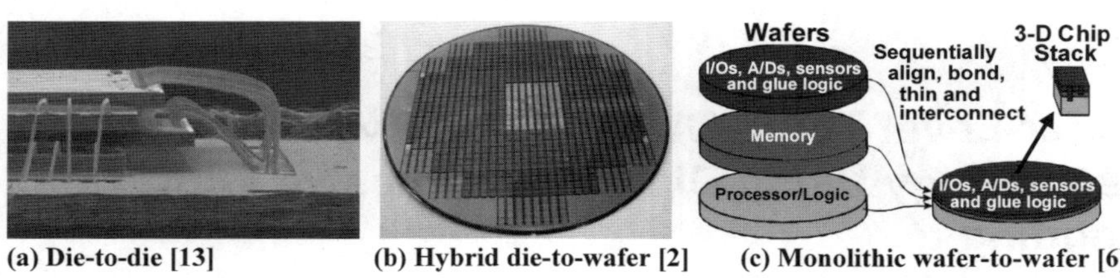

Figure 1. Three major 3D IC approaches [2, 6, 13].

throughput production [14] due to lower parasitics (both RC delay and inductance) with shorter inter-chip connections. Both approaches also allow the integration of sub-blocks within circuits for added design flexibility and optimization capability due to denser inter-chip interconnectivity, while maintaining the separation of incompatible materials and processes on different wafers. However, there are significant differences between these two approaches.

The die-to-wafer approach offers the use of known-good-die (KGD) [2-3], and the assembly flexibility allows high yield for products in low-to-medium quantities. This approach has a higher cost compared to monolithic wafer-to-wafer approaches for large production volumes, since alignment and bonding is done by "pick-and-place" assembly process; *i.e.*, in die-by-die fashion. While die yield issues need to be considered further, smaller die and wafer-specific processing indicate that yield may not be a limiting factor with a robust monolithic wafer-level 3D process. Moreover, the monolithic wafer-to-wafer approaches have many performance advantages, such as a very high density of low parasitic inter-chip interconnects for high bandwidth and increased noise immunity. Most important, lower high-volume interconnect cost should be possible with monolithic wafer-level processes.

Wafer-level 3D IC technologies are being actively pursued [2–12] because of these advantages. Process technologies to achieve 3D integration of ULSI CMOS ICs include high-performance thin film transistors (TFTs) fabricated using lateral crystallization [8] or wafer bonding using vias to internally connect stacked wafers. Wafer bonding technologies using copper [9, 15], silicon dioxide [7], or dielectric adhesive glues [4-6] for bonding are being developed and provide micron-sized vias for inter-wafer interconnection.

The 3D IC approaches based on wafer bonding and inter-wafer interconnects generally involve four major processing steps: 1) wafer-to-wafer alignment, 2) wafer-to-wafer bonding, 3) wafer thinning, and 4) inter-wafer interconnection. All processes must be compatible with IC back-end-of-the-line (BEoL) fabrication constraints. The most differentiating process steps are the wafer bonding approach and the inter-wafer interconnect (via) processing. Three major approaches to wafer bonding are under investigation and are described briefly below (Figure 2):

1. Metal-to-metal bonding, where metal bonds also serve as inter-wafer interconnects; copper [9, 15], solder [2] or micro-bump [11] with or without underfill have been utilized. Issues with surface oxidation, interface contamination, and particularly the surface non-planarity may cause micro-voids,

thereby impacting the inter-wafer interconnect yield and reliability. Using solder or other alloy for the metal-to-metal bonding may alleviate the stringent bonding surface requirement; however, compatibility with subsequent metallization and packaging process steps, as well as next-wafer level 3D processing, would likely be compromised or limited. This approach has been demonstrated for two or more wafers in the 3D "wafer-stack" [15].

2. Direct oxide bonding at low temperature, which requires damage-free atomic scale surface planarization with proper surface activation for bonding [3, 7]. The sensitivity to particles and surface warp makes this approach more amenable to bonding of wafers with only lower levels of IC interconnect structures.

3. Dielectric adhesive bonding using curable polymers, polyimides, or other adhesives [4, 10, 16, 17], where the dielectric adhesive can accommodate wafer non-planarity, sub-micron-sized particles and some processing-induced stress. Depending on the adhesive properties, 3D ICs with more than two wafers are possible.

In terms of inter-wafer via processing, there are two approaches: via first or via last, depending on whether the inter-wafer vias are fabricated before or after bonding, respectively. To date, both direct oxide bonding and dielectric adhesive bonding use a via-last process, while metal-to-metal bonding uses a via-first process.

While not yet demonstrated as production-ready, no technology show-stoppers have been delineated to date. However, various manufacturing and design issues need to be addressed, as discussed below.

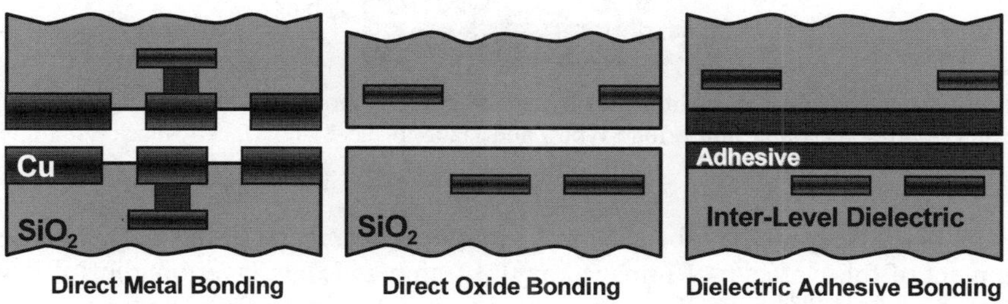

Figure 2. Three major bonding approaches for wafer-level 3D ICs. Common issues include BEoL IC process compatibility and bonding integrity.

3D Technology Platform Using Dielectric Adhesive Wafer Bonding and Copper Damascene Inter-Wafer Interconnects

The Rensselaer approach to a monolithic wafer-level 3D technology platform is depicted in Figure 3 [4]. Two fully processed wafers are aligned face-to-face (or interconnect structure-to-interconnect structure) to tolerances within 1 micron and bonded using a dielectric adhesive under conditions compatible with CMOS processing. The top wafer in the two-wafer stack is thinned to ~1 μm by a three-step baseline thinning process, *i.e.*, backside grinding, chemical-mechanical polishing (CMP) and selective wet-etching to an etch stop, *e.g.*, an ion-implanted layer, a graded SiGe epitaxial layer, or a buried oxide (BOX) layer with silicon-on-insulator (SOI) wafers. Subsequently, bridge-type and/or plug-type inter-wafer interconnects are formed using a Cu damascene patterning process, developed jointly by Rensselaer and the University at Albany [6, 18]. Thus, a long distance interconnect that might run a centimeter across a conventional 2D chip may be replaced by a 2–10 μm vertical inter-wafer interconnect (via) between chips. Repeating this process flow, additional wafers can be aligned, bonded, thinned and inter-wafer interconnected without changing the processing approach; most proposed applications require two- or three-wafer stacks. Besides the advantages described in the previous section, this technology platform does not require handling wafers as thinned silicon is not transferred, as in some of the other technology platforms.

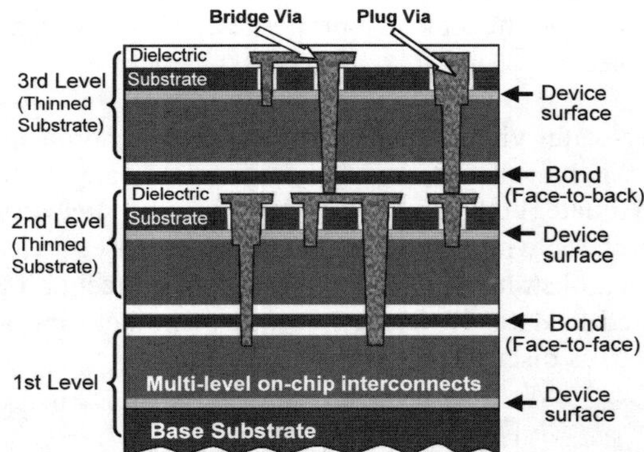

Figure 3. Schematic of a monolithic 3D hyper-integration, showing bonding interface, vertical inter-chip vias (plug- and bridge-type), and "face-to-face" and "face-to-back" bonding [4].

Key processing metrics for such a 3D platform have been defined: (1) precise alignment of full wafers (≤ 1 μm accuracy); (2) thin adhesive-layer bonding at low temperature (≤ 400°C); (3) precision thinning and leveling of top wafer (~1 μm thick); and (4) inter-wafer connection by high-aspect-ratio (> 5:1) vias. As mentioned previously, these processes must be BEoL-compatible.

An inter-wafer interconnect via-chain structure has been designed and fabricated to demonstrate the feasibility of our approach to 3D processing (Figure 4). The starting wafers have a 2-μm thermal oxide. Conventional Cu damascene BEoL processes are used for Cu metallization on the bottom and top wafer (M1 and M2). The wafers with Cu patterns are aligned with an alignment accuracy of ~1 μm using an EVG SmartView aligner, then bonded in an EVG bonder using a baseline bonding process with benzocyclobutene (BCB) as the bonding adhesive, bonding pressure of ~ 3 bar, and bonding temperature of 250°C. The silicon substrate on the top wafer is thinned and completely removed using the baseline three-step process, with the final etch stopped at the oxide layer. Inter-wafer vias and bridge metal (M3) are formed using an inter-wafer damascene process, *i.e.*, deep via etching and clean, chemical vapor deposition (CVD) of TaN liner, Cu CVD fill, and CMP. Additional fabrication process information is described elsewhere [4, 18].

Results obtained with the inter-wafer via-chains are summarized in Figure 4, including a focused ion beam (FIB) cross section after the processing. Although the via-chain contact resistance of ~5 $\mu\Omega$-cm^2 is much larger than anticipated, the linear relationship between the chain resistance and chain length indicates that continuous and uniform 3D via-chains are demonstrated for nominal inter-wafer via sizes of 2, 3, 4, and 8 μm.

Thermal, Mechanical and Electrical Evaluation of Wafer Bonding and Thinning

With the feasibility of our approach to wafer-level 3D integration processing demonstrated, it is important to evaluate the integrity of wafer bonding and thinning processes as well as compatibility with IC industry requirements. Results of thermal, mechanical, and electrical robustness and compatibility with BEoL processes and subsequent inter-wafer interconnection and packaging are summarized in this section.

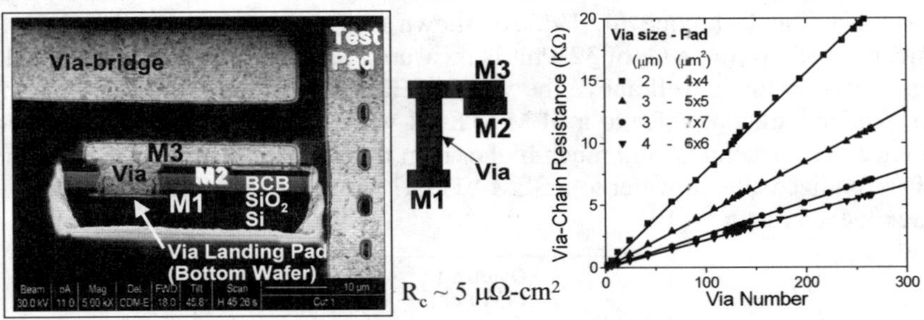

Figure 4. A FIB cross section of the via-chain structure (left) and via-chain resistance *vs.* via-chain length (via number) for nominal via size of 2, 3, 4 μm (right) [18].

Based on screening tests of various dielectric adhesives, BCB was selected as our baseline bonding adhesive. BCB is a low-k polymer, which has been used in various semiconductor applications and is compatible with most CMOS processes. We have routinely obtained void-free bonding of coefficient of thermal expansion (CTE) matched glass substrates with both blanket and patterned 200-mm diameter silicon wafers. For example, Figure 5 illustrates a two-level copper interconnect test structure (provided by SEMATECH) after bonding to a CTE-matched glass wafer using 2.6-μm thick BCB and removal of the bulk silicon using the three-step thinning process [19]. Although the surface profile shows a step of ~850 nm across the aluminum bond pads, void-free bonding is obtained and maintained after the thinning process. The interconnect structures are completely damage-free with this process, and even maintain structural integrity after grinding and polishing through the bulk silicon.

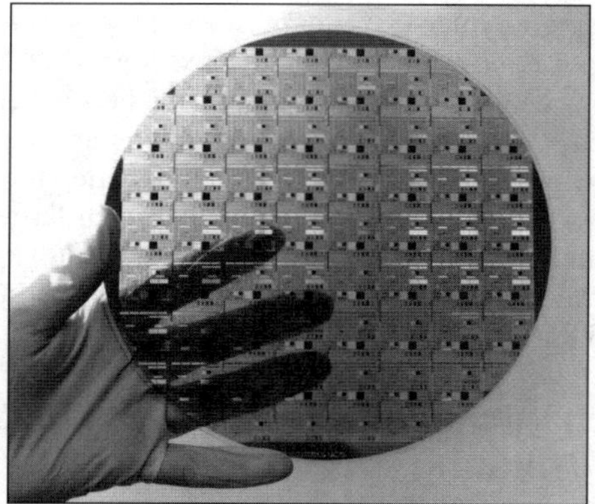

Figure 5. Photo image of a semi-transparent Cu/oxide interconnect wafer bonded to a CTE-matched glass after removal of Si substrate on SEMATECH wafer [19].

Besides the defect-free bonding interface, it is critical to obtain high wafer bonding strength, or critical adhesion energy, G_c, which is determined by a four-point bending technique [4, 19]. As shown in Figure 6, for our baseline BCB thickness of 2.6 μm, a G_c of 32 J/m^2 is measured for BCB bonding of oxidized silicon wafers. This is well above the value of 22 J/m^2 obtained for BCB bonding of an oxidized silicon wafer to a SEMATECH wafer with two-level Cu/oxide interconnect test structures, and much higher than the value of 6 J/m^2 for BCB bonding of an oxidized silicon wafer to a SEMATECH wafer with Cu/JSR porous low-k interconnect structures [19].

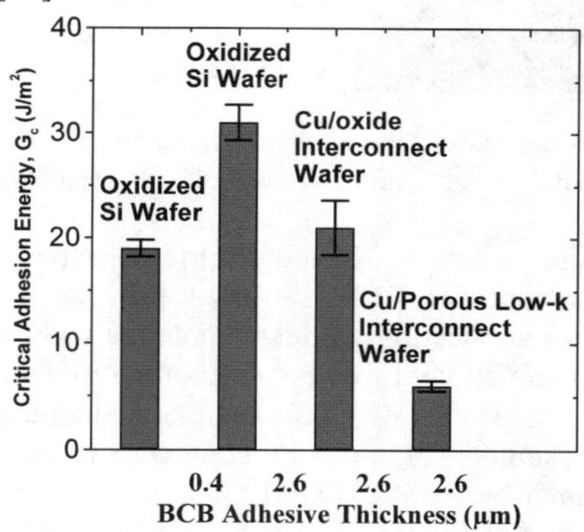

Figure 6. Critical adhesion energy, G_c, measured by a four-point bending technique, for different wafers bonded to an oxidized Si wafer using BCB [4, 19].

Direct observations indicate that failure occurs within the interconnect test structure with either oxide or JSR porous low-k dielectrics, and not at either bonding interface. Even when as thin as 0.4 μm BCB, a measured G_c of 19 J/m^2 is sufficiently high, *i.e.*, much higher than that of Cu/porous low-k interconnect structures, indicating that a range of BCB thicknesses can be implemented.

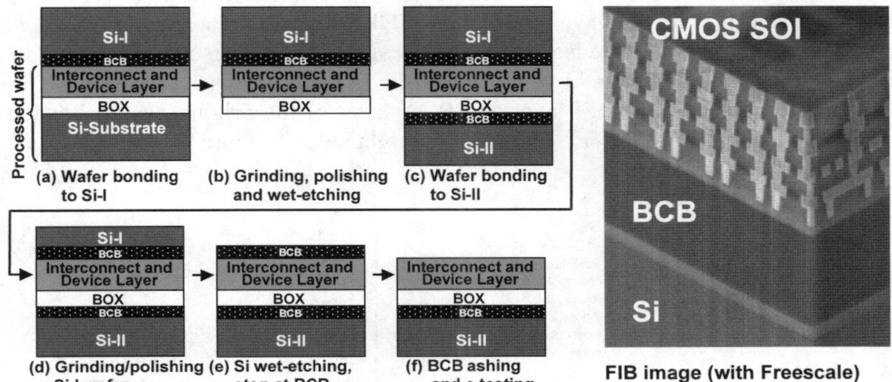

Figure 7. Process flow of a typical double bonding/thinning procedure (left), permitting bonding and thinning impacts on electrical properties of processed wafers without inter-wafer interconnect processing [19]. Right image is a FIB cross section of CMOS SOI wafer after such a procedure [5].

This high G_c for wafers bonded using BCB is maintained after standard IC package reliability tests performed at Freescale, such as die-level autoclave and liquid-to-liquid thermal shock (LLTS) tests, and sawing of BCB-bonded wafers [20]. The autoclave tests were conducted at conditions of 100% humidity, 2 atm, and 120°C for 48 h or 144 h. The LLTS tests were conducted between −50 and 125°C for 1000 cycles.

In addition, the thermal cycling tests conducted in nitrogen ambient indicate that the wafer bonding using BCB is stable up to at least 400°C [17].

While BCB as a dielectric bonding adhesive provides a defect-free bonding interface and sufficient bond strength with high temperature stability and packaging reliability as well as insensitivity to surface conditions (particles, roughness, and planarity), it is also critical to evaluate bonding and thinning impacts on the electrical performance of processed wafers. Figure 7 depicts a procedure of double bonding/thinning and BCB ashing for such an evaluation, with electrical properties of wafers measured before and after this procedure.

Joint work with SEMATECH focused on copper back-end interconnect structures. Both electromigration tests as well as performance tests (via chain resistance and surface leakage current with comb-like structures) were conducted, with performance tests depicted in Figure 8. While some degradation in via-chain resistance and surface leakage was observed after the double bonding/thinning procedure, particularly with the Cu/ultra-low-k structures, the final results were considered mostly within specification. There was no attempt to reduce process-induced stress with the Cu/ultra-low-k structures; we believe the small degradations observed can be reduced with further process refinement. Electromigration data was similarly promising [20].

Similar electrical tests have been done with Freescale state-of-the-art 130-nm technology CMOS SOI test structures with four-level copper/low-k (organosilicate glass) interconnects and aluminum bond pads [5]. Results indicate very little change in ring-oscillator delay, n-FET and p-FET threshold voltage and n-FET and p-FET sub-threshold voltage leakage. Figure 9 shows typical results of (a) ring oscillator delay and (b) threshold voltages of n-FET. An FIB cross section of the interconnect structure, active Si layer and BOX layer of the original wafer, bonded with BCB to a prime Si wafer is shown in the right side of Figure 7.

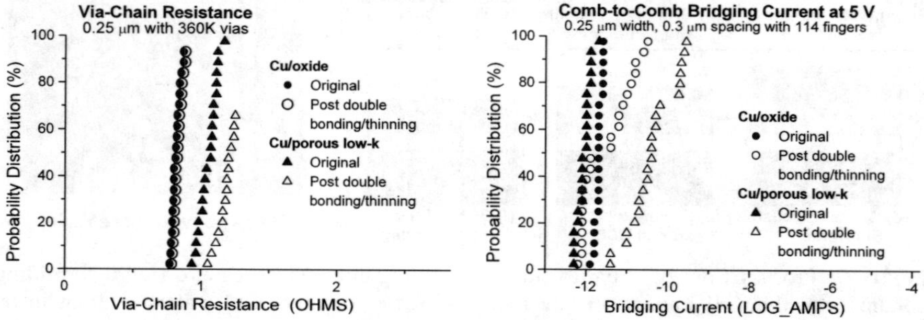

- **Slight changes with Cu/oxide interconnects**
- **Larger changes with Cu/porous low-k interconnects, but within accepted limits**

Figure 8. Electrical tests of SEMATECH copper interconnect wafers before and after double bonding/thinning processing [19].

Key Issues with Wafer-Level 3D Technologies

Wafer-level 3D IC technology is a significant addition in IC processing, although the complexity of subsequent packaging might be reduced. While the individual process steps of wafer-to-wafer alignment, wafer bonding, wafer thinning and inter-wafer interconnection are within IC process capabilities at this time, all such processes must be introduced simultaneously. For commercial markets, this would be either at a given technology node or a specific large-volume market, *e.g.*, a large-size memory stack integrated with a processor, or smart imagers with pixel-by-pixel processing. While 3D design tools that properly account for parasitic coupling and signal integrity issues are needed, their development is considered within current capabilities.

One of technical concerns with 3D integration is how to align wafers with adequate alignment accuracy across the full wafer. The 3D approach we have focused on in this chapter generally does not require sub-micron wafer alignment accuracy because the wafers are stacked after BEoL processing. With this approach, a relatively small number of inter-wafer vias are needed to substitute for the global long interconnects that exist on conventional 2D chips; the via landing pads on top of the BEoL-processed "bottom wafer" can be made large (micron scale); vias can be precisely aligned to the "top thinned wafer" using a conventional stepper lithography, therefore allowing some wafer-to-wafer alignment tolerance. In contrast, wafer-level 3D approaches in which front-end-of-the-line (FEoL) processing (and perhaps with first-level metal BEoL processing) is completed, require very high inter-wafer interconnectivity, and therefore very high wafer alignment accuracy. High alignment accuracy is also essential for applications where very large numbers of inter-wafer interconnects are needed.

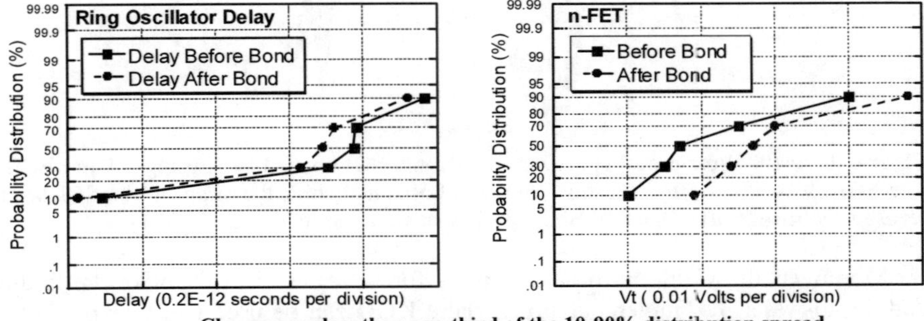

- **Changes are less than one-third of the 10-90% distribution spread**
- **No degradation after double bonding/thinning**

Figure 9. Electrical tests of Freescale CMOS SOI Cu/low-k wafer before and after double bonding/thinning processing [5].

For 3D integrated systems with any wafer bonding approach, the issue with thermal-mechanical stress must be considered because the active thin silicon device layers bonded on top of a 3D stack (*e.g.*, the third-level active wafer or higher as shown in Figure 3) are far from the rigid silicon substrate on the bottom of the stack. This is particularly true for sub-100-nm devices with Cu/low-k interconnects, which are usually more sensitive to the local stress than that of larger devices with Cu/oxide interconnects. Some stress was observed in a nanoscale SOI layer that was BCB-bonded to another silicon wafer, forming a layer structure of SiO_2-TEOS/SOI/BOX/BCB/Si-substrate, as shown in Figure 10. So far, the strain measured in the active SOI (70 nm or 140 nm) is small; therefore, shifts in the silicon electronic properties that cause changes in transistor performance are not anticipated [21].

Another key issue is thermal management, which is already limiting high-performance processor design. However, by proper design partitioning (*e.g.*, processor and part of memory cache at lower level and large-size memory at upper levels), the thermal constraints can be relaxed, particularly with extra copper inter-wafer interconnects for thermal flow (as well as electrical grounding) [22].

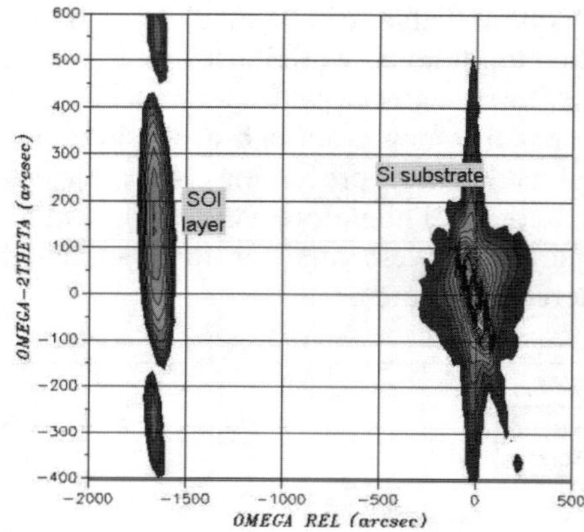

Figure 10. Si (004) reciprocal space map for 70 nm SOI bonded to Si using a 1-μm BCB. The SOI tilt relative to the substrate causes a horizontal shift; the strain in the SOI layer leads to a vertical shift of the SOI Bragg peak relative to the substrate peak [21].

Moreover, die yield can be a key issue for wafer-level 3D. Unlike die-to-die and die-to-wafer 3D implementations where KGD can be used (fully or partially, respectively), wafer-level 3D implies a yield hit. However, there exist yield enhancements inherent to the 3D integration scheme that arise from reductions of both die size and processing/material complexity, allowing the use of mature, optimized and high-yield processes/materials specific to each wafer in the stack. 3D integration can also eliminate high-risk technologies as required for interconnects on 2D chips since small die size and vertical vias reduce global wire lengths, hence RC delay and power consumption. Moreover, 3D integration allows the use of more flexible architectures to further enhance the yield, such as, partitioning of functional circuits/technologies on separate layers, increased redundancy design, and 3D routing.

A key limitation is time-to-market of new IC designs with die scaling within a technology node. Wafer-level 3D requires equal die area or wasted area in some wafers in a stack, which is contrary to IC design practice. Furthermore, wafers from different manufacturers would have to be carefully controlled for wafer alignment and bonding. Integrated device manufacturers and foundries will benefit from wide-spread acceptance of wafer-level 3D technology platforms.

Comparison with SiP and SoC

SoCs and SiPs are idealizations that emphasize high functional density at the die (or chip) level, or in a package containing multiple dies (or chips) and passive components. SiPs can either be in planar 2D implementations (*e.g.*, multichip modules) or in 3D stacks (either bare die or chip-scale packaged die) with peripheral wire-bonded contacts. This is distinct from die-to-die 3D with micron-scale areal inter-die interconnects.

SoCs are similar to wafer-level 3D ICs in that monolithic wafer-level processing is used throughout (prior to single-die packaging), but dissimilar to wafer-level 3D in that signal integrity and power delivery issues can be limiting in leading-edge high-performance designs. SiPs are similar to wafer-level 3D ICs in that heterogeneous integration is readily achieved. But a key difference is that SoCs and SiPs have on-going technology platforms with well-defined technology roadmaps. Wafer-level 3D is in a relatively embryonic stage, with many research and development activities, but no known manufacturing commitment to date.

While wafer-level 3D is not currently in manufacturing, the anticipated high-performance, high-component density and ease of heterogeneous technology integration capabilities have created much research interest. Wafer-level 3D has the potential for low manufacturing cost, as the monolithic processing is extended to another level of integration. However, more than a robust technology platform will be needed to launch this technology (see a summary of the comparison in Table 1). The design community will need to fully utilize the enhanced interconnectivity potential, not only map 2D designs into 3D. Moreover, technology scaling to lower feature sizes and design enhancements must be coordinated so that die size in the various wafers in the stack are changed synchronously.

Table 1. Comparison of wafer-level 3D integration with SoCs and SiPs.

	SoCs	SiPs	3D ICs
Performance (speed, frequency, power)	M	W	B
Signal process packing density	M	W	B
Heterogeneous integration	W	B	M
Manufacturing ready	M	B	W
Manufacturing cost in low-medium quantities	B	M	W
Manufacturing cost in high quantities	M	W	B

Summary and Conclusions

A viable wafer-level 3D hyper-integration technology platform with dielectric adhesive bonding and copper vias has been described. An inter-wafer via-chain structure has been fabricated, demonstrating the feasibility of this 3D technology, with a baseline process flow of one-micron wafer-to-wafer alignment, BCB wafer bonding, three-step wafer thinning and copper damascene patterned inter-wafer interconnection. Evaluations indicate the thermal, mechanical, and electrical robustness of the baseline wafer bonding and thinning processes as well as compatibility with BEoL processes and packaging. The key advantages of BCB wafer bonding include the ability to accommodate wafer-level non-planarity (*e.g.*, surface topography, wafer bow) and particulates at the bonding interfaces, high bond strength, and relatively low temperature bonding process as well as high temperature stability after bonding.

Wafer-level 3D hyper-integration offers potentially significant benefits in performance and functionality (with less material/processing constraints) over 2D ICs (including SoCs), and performance, interconnectivity, and cost benefits over die-to-wafer or die level solutions (including SiPs). Issues related to wafer alignment accuracy, thermal-mechanical stresses, thermal management, yield and common die size need to be more fully considered. However, the wafer-level 3D integration, as a future information technology, has clear advantages in performance, heterogeneous integration, and in low-cost high-volume manufacturing compared to SoCs and SiPs.

Acknowledgements

This work was supported in part by DARPA, MARCO, and NYSTAR through the Interconnect Focus Center (IFC).

References

1.   International Technology Roadmap for Semiconductors (ITRS), 2003 Edition (Semiconductor Industry Association, 2003, http://public.itrs.net/).

2.   A. Klumpp, P. Ramm, R. Wieland, and R. Merkel, in Presentation CD of International Workshop of 3D System Integration, Fraunhofer-Institute, Munich, Germany, 2003.

3.   Ziptronix, Inc., http://www.ziptronix.com/.

4.   J.-Q. Lu, Y. Kwon, J.J. McMahon, A. Jindal, B. Altemus, D. Cheng *et al.,* in Proceedings of 20th International VLSI Multilevel Interconnection Conference, T. Wade (ed.), pp. 227-236, IMIC (2003).

5.   R.J. Gutmann, J.-Q. Lu, S. Pozder, Y. Kwon, A. Jindal, M. Celik *et al.,* in Advanced Metallization Conference 2003 (AMC 2003), G.W. Ray, T. Smy, T. Ohta and M. Tsujimura (eds.), pp. 19-26, MRS (2004).

6.   J.-Q. Lu, Y. Kwon, A. Jindal, K.-W. Lee, J. McMahon, G. Rajagopalan *et al.*, in Proceedings of 19th International VLSI Multilevel Interconnection Conference, T. Wade (ed.), pp. 445-454, IMIC (2002).

7.   K. Guarini, A. Topol, M. Ieong, R. Yu, L. Shi, D. Sing *et al.*, in International Symposium on Thin Film Materials, Processes, and Reliability, G.S. Mathad, T.S. Cale, D. Collins, M. Engelhardt, F. Leverd, and H.S. Rathore (eds.), PV2003-13, pp. 390-404, ECS (2003).

8.   S.J. Souri and K.C. Saraswat, in 1999 IEEE International Interconnect Technology Conference (IITC), pp. 24-26, IEEE (1999).

9.   A. Rahman, A. Fan, J. Chung, and R. Reif, in 1999 IEEE International Interconnect Technology Conference (IITC), pp. 233-235, IEEE (1999).

10.   P. Ramm, D. Bonfert, R. Ecke, F. Iberl, A. Klumpp, S. Riedel *et al.*, in Advanced Metallization Conference 2001 (AMC 2001), A.J. McKerrow, Y. Shacham-Diamand, S. Zaima, and T. Ohba (eds.), pp. 159-165, MRS (2002).

11. K.W. Lee, T. Nakamura, T. One, Y. Yamada, T. Mizukusa, H. Hasimoto *et al.*, in International Electron Device Meeting (IEDM) 2000, pp. 165-168, IEEE (2000).

12. J.A. Davis, R. Venkatesan, A. Kaloyeros, M. Beylansky, S.J. Souri, K. Banerjee *et al.*, Proc. IEEE, 89, 3, 305-324 (2001).

13. K. David, in SEMI - Strategic Business Conference (SBC) 2003, Orlando, FL, April 2003.

14. S. List, C. Webb, S. Kim, in Advanced Metallization Conference 2002 (AMC 2002), B.M. Melnick, T.S. Cale, S. Zaima, and T. Ohta (eds.), pp. 29-36, MRS (2003).

15. Tezzaron Semiconductor: http://www.tezzaron.com/.

16. J. Burns, L. McIlrath, C. Keast, C. Lewis, A. Loomis, K. Warner *et al.*, in 2001 IEEE International Solid-State Circuits Conference (ISSCC 2001), p. 268, IEEE (2001).

17. Y. Kwon, J. Seok, J.-Q. Lu, T.S. Cale and R.J. Gutmann, in Journal of The Electrochemical Society, 2004 (in press).

18. J.-Q. Lu, A. Jindal, Y. Kwon, J.J. McMahon, K.-W. Lee, R.P. Kraft *et al.*, in International Symposium on Thin Film Materials, Processes, and Reliability, G.S. Mathad, T.S. Cale, D. Collins, M. Engelhardt, F. Leverd, and H.S. Rathore (eds.), PV2003-13, pp. 381-389, ECS (2003).

19. J.-Q. Lu, A. Jindal, Y. Kwon, J.J. McMahon, M. Rasco, R. Augur *et al.*, in 2003 IEEE International Interconnect Technology Conference (IITC), pp. 74-76, IEEE (2003).

20. S. Pozder, J.-Q. Lu, Y. Kwon, S. Zollner, J. Yu, J.J. McMahon *et al.*, in 2004 IEEE International Interconnect Technology Conference (IITC04), pp. 102-104, IEEE (2004).

21. Private conversation with Michelle Rasco at SEMATECH.

22. J. Kalyanasundharam, R.B. Iverson, E.T. Thompson, V. Prasad, T.S. Cale, R.J. Gutmann *et al.*, in Advanced Metallization Conference 2002 (AMC 2002), B.M. Melnick, T.S. Cale, S. Zaima, and T. Ohta (eds.), pp. 59-65, MRS (2003).

Part V

Advanced Materials Characterization

Challenges to Advanced Materials Characterization for ULSI Applications

A. C. Diebold

International SEMATECH, Austin, TX / USA

Introduction

The semiconductor industry has moved into the realm of nanoelectronics [1]. The shrinking of device and structure sizes has also required new materials and is expected to soon require new device designs. All of these factors have driven the need for both evolutionary and revolutionary advances in materials characterization. In this chapter, we discuss the requirements for new materials characterization capability and some of the advances in this area.

The semiconductor industry has always relied on materials characterization to enable both research and manufacturing. Materials characterization methods provided microscopy as well as elemental and chemical analysis. Recently, traditional microscopy methods such as scanning electron microscopy (SEM) and transmission electron microscopy (TEM) have been augmented by scanned probe microscopies such as atomic force microscopy. Chemical and elemental analysis has been done on liquids, bulk solids, and solid surfaces. A variety of chemical information has proven critical to the semiconductor industry from trace contaminant detection and quantification to quantification of the elemental content. Often this information must be determined in the form of a depth profile. Electrical properties such as carrier concentration are routinely determined. In this book, advances in a number of materials characterization methods are discussed in detail. This chapter is an overview of the present and future materials characterization requirements for the semiconductor industry and some of the advances in the science and technology of the characterization methods themselves.

The continuous advance of the semiconductor industry toward smaller devices and greater clock speeds is forcing the introduction of new materials, structures, and devices. The most advanced transistors have gate lengths below 50 nm and use thin films less than 1.5-nm thick. Interfacial layers either dominate or greatly influence a majority of the physical and electrical properties of transistors and on-chip interconnect. Over the past several years, the rapid reduction in feature size has driven the use of higher dielectric constant materials in the transistor and lower dielectric constant materials in interconnect. Soon, the heavily dope poly-silicon used for gate electrodes will be replaced by metal gates, and silicon oxynitride will be replaced by the so-called high-k dielectrics. Copper metal lines have replaced aluminum metal lines in logic chips. Copper interconnect structures are fabricated processes using Damascene (also known as inlaid metal). The semiconductor industry roadmap for the future (the International Technology Roadmap for Semiconductors) indicates a continued need for new materials and structures [2]. These topics will be covered in the first section.

New materials and reduced device sizes will not solve all of the problems associated with future transistors. Already, local stress is being used to improve carrier mobility and thus transistor performance. The ITRS also projects the need for non-classical complementary metal oxide semiconductor (CMOS). A comparison of CMOS, non-classical CMOS, and novel transistors is shown in Figure 1. Fabricating transistors on ultra-thin silicon layers on top of a buried oxide layer (silicon on insulator wafers – SOI) is one part of the solution. Multi-gate and surround gate transistor designs such as the FINFET are considered likely candidates for manufacturing. These topics will be covered in the second section.

Materials characterization has also steadily advanced. Microscopy has seen the introduction of new electron lens technology that reduces lens aberration [3]. Use of lower energies has improved the depth resolution of ion-beam-sputtering-based depth profiling. Surface analysis methods have greatly improved their sensitivity and repeatability. Optical and X-ray methods have proven their ability to measure thin films and new materials. In the final section, we will review recent advances in materials characterization methods and project some future needs.

Materials Characterization Needs for New Materials

Over the past several years, the semiconductor industry has introduced more new materials and processes into manufacturing than it did over the past 20 years combined. Logic applications have switched from aluminum-silicon dioxide metallization materials and processes to copper – lower κ-Damascene (inlaid) processes. Chemical mechanical planarization would have been considered inappropriate fifteen years ago, and now it is standard. Transistor and capacitor dielectric materials have switched to silicon oxynitride in manufacturing and high-κ with metal gate in R&D. Lithographic exposure systems have moved from 248-nm to 193-nm light sources in manufacturing, and immersion lens systems are expected to extend 193-nm photolithography into the technology range previously covered by 157-nm exposure. New photoresist and anti-reflective coatings have been introduced to meet this need. In addition, reflective photomasks and photoresists sensitive to 11 nm are being developed for EUV photolithography [4]. In this section, the materials characterization requirements that are driven by new materials are discussed.

Photolithography

The introduction of new photoresist materials for dry and immersion 193-nm exposure systems require development of new optical measurement and modeling capability. Reflection and ellipsometry measurements have been extended to the UV beyond 157 nm. Laboratory-grade systems reach from 140 nm in VUV to 1700 nm in the near IR. Optical models in visible and IR are well known, especially for traditional materials. The introduction of low- and high-κ materials has required considerably more sophisticated modeling. These materials absorb light in UV. The situation in photolithography is even more complicated. Knowledge of the optical properties of new photoresist materials in the UV is a critical part of engineering the entire process.

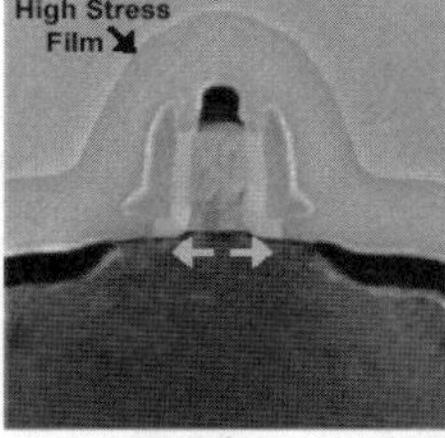

90-nm Node Transistor - 2004

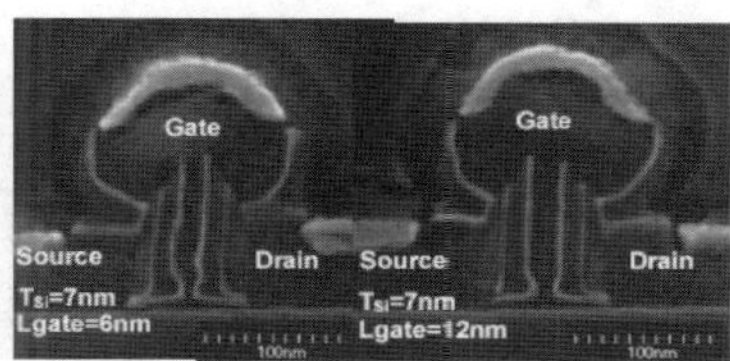

Sub-10-nm Beyond CMOS Transistor on thin SOI

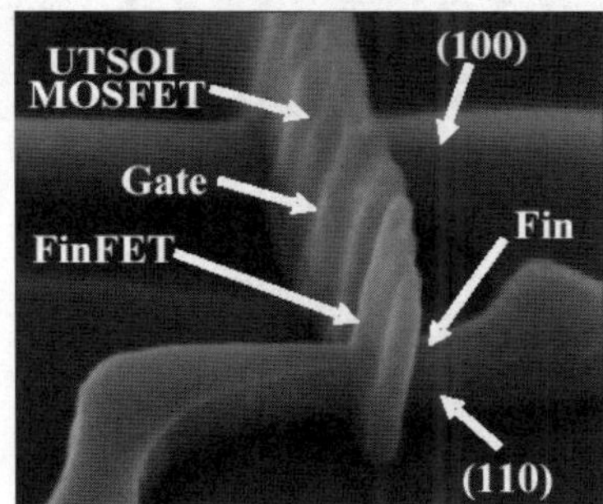

nMOS MOSFET
pMOS FINFET

Figure 1. Comparison of CMOS, non-classical CMOS, and novel transistors. Top right figure courtesy Mark Bohr (Intel) [9]. Other figures courtesy Bruce Doris (IBM) [18]. Figures © IEEE and used with permission.

EUV photolithography technology is seen as a key enabler for future photolithography processes. The reflective mask technology is considerably different from traditional transmission masks used for visible and UV exposure [4]. Multilayers of Mo and Si provide a mirror to reflect the 11 nm EUV radiation. X-ray diffraction and reflectivity are used to insure the periodicity and quality of these mirrors. A TEM cross section of a typical (R&D) EUV mask is shown in Figure 2. Reflectometers must now function at the 11-nm wavelength so that the total reflectivity of these mirrors can be measured [4]. The atomic level smoothness requirements are driving applications improvements in flatness measurements. Atomic level resolution in all three dimensions is possible using atomic force microscopy. Atomic level resolution normal to the surface is also measured at larger surface wavelengths using interferometry and other methods. In Figure 3, we show the relationship between spatial frequency ranges and the three flatness measurement methods: AFM, surface profiling, and interferometry.

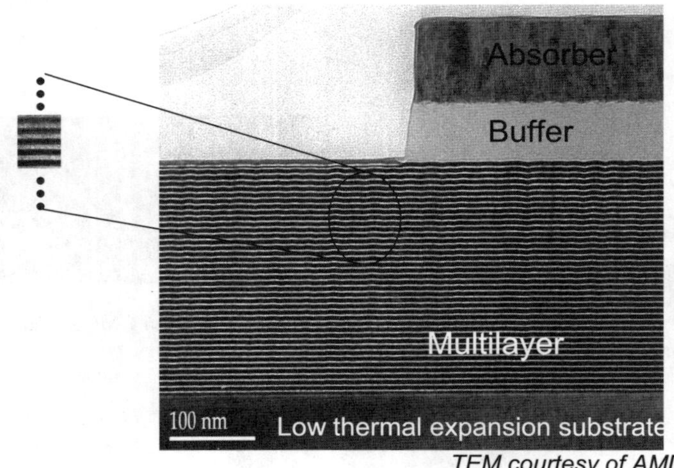

Figure 2. Cross-sectional view of multi-layer EUV masks
Figure courtesy Phil Seidel (International SEMATECH) [4.]

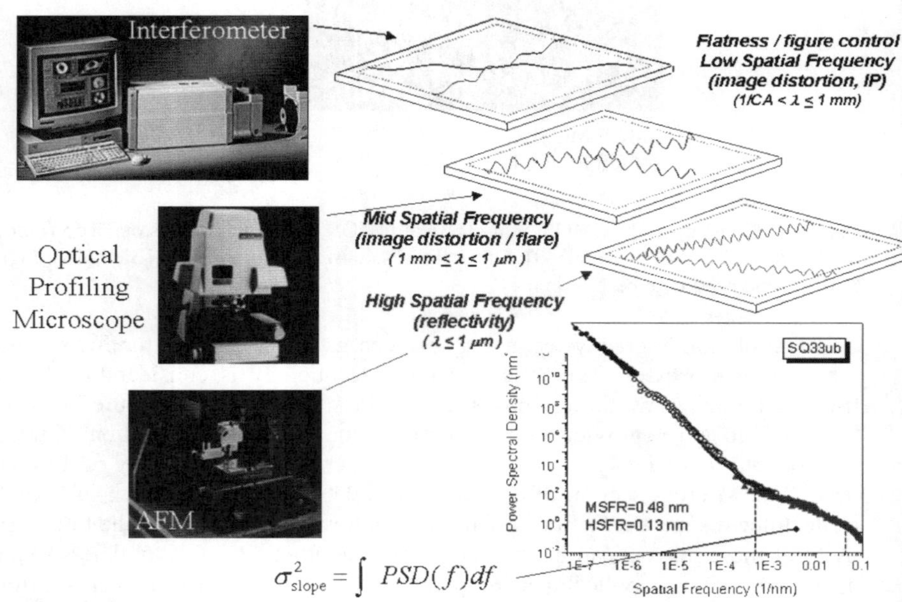

$$\sigma^2_{slope} = \int PSD(f)\,df$$

Figure 3. Measurement of EUV mask flatness. Figure courtesy Phil Seidel (International SEMATECH) [4].

Front-End Processes

New materials and structures have been critical enablers in the continued scaling of transistor and capacitor technology. In Figure 4, the materials characterization needs for transistors are depicted. In order to continue reducing the effective oxide thickness (EOT) below 2 nm, silicon oxynitride is now used as the gate dielectric material [2]. Further reduction of the EOT will require the use of metal gate electrodes and higher κ-gate dielectrics. The interface between the gate dielectric and transistor channel as well as the interface between the gate dielectric and the gate electrode play a key role in the capacitance of the transistor gate stack, and thus the electrical performance of the transistor. The need for atomic level characterization of these interfaces and the thin films in the gate stack is a key motivator for improving the spatial resolution of TEM and the optical models used for ellipsometry. *Ideally, the industry needs atomic maps of the interfaces and thin films in the transistor gate stack.* Advances in TEM capability are discussed below in the section on advances in materials characterization methods.

The use of metals gates adds some complexity to the materials characterization and in-line metrology of the transistor gate stack. Although it is often forgotten, the thickness of thin metal films less than ~15 nm can be measured by ellipsometry [5]. Thicker films and multi-film stacks require the use of X-ray reflectivity (XRR). Applications development for XRR includes the need to understand the limitations of this method and improved models that enable multiplayer measurement in the presence of interfacial roughness [6]. New substrate materials are discussed below.

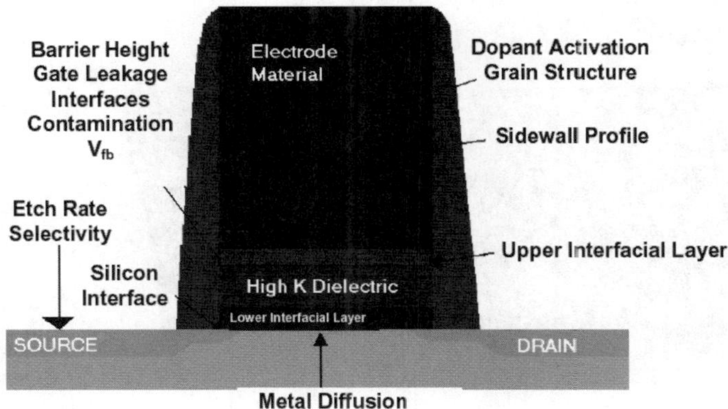

Figure 4. Materials characterization requirements for transistors: ability to understand interface structure – atom by atom, ability to observe interdiffusion — Imaging and spectroscopy, ability to determine elemental composition – ELS and EDS, ability to understand chemical state — ELS. Figure courtesy Howard Huff.

On-Chip Interconnect

Ideally, the semiconductor industry would like to know the microscopic properties of patterned line and via structures. In Figure 5 we illustrate the materials characterization requirements for interconnect. The grain orientation and point by point film thickness of barrier layers on trench and via/contact sidewalls is just one example of an interconnect materials characterization need that is becoming more difficult as feature sizes shrink. Previously, the diameter of contact/via structures was large enough that one could know when a cross-sectional view was through the middle of the via cylinder. *As films become thinner, the need for 3D atomic mapping can also be applied to interconnect structures.*

A great variety of measurements are required for complete characterization of low-k materials [7]. Optical and X-ray methods have proven capable of routine measurement of thickness and porosity. Pore size distribution can be measured by small angle X-ray scattering. A number of methods have been used to determine the elastic properties of low-κ materials such as acoustic methods and nano-indentation [7]. Other properties such as the coefficient of thermal expansion remain difficult to measure.

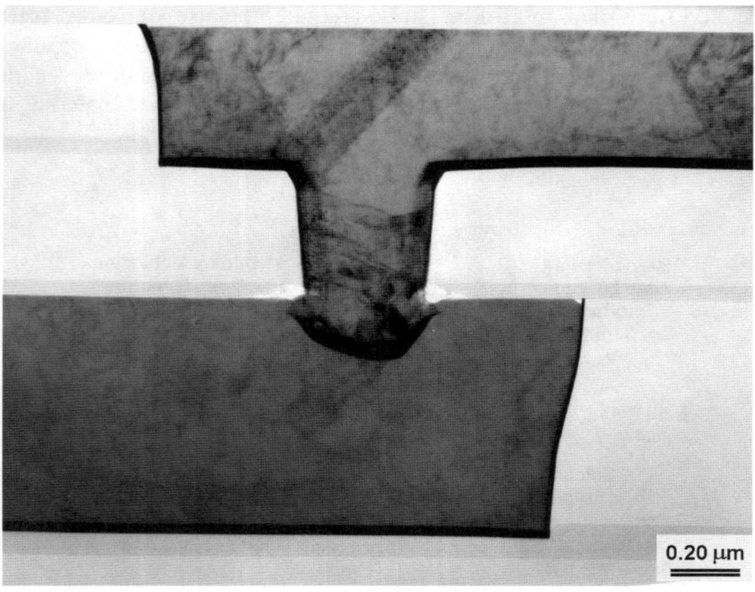

Figure 5. Materials characterization requirements for interconnect. (Characterization of barrier layers, vias, and copper metallization is becoming increasingly more difficult as feature sizes shrink. Cross sectioning through a via often results in a section that does not represent the middle of the via. Figure courtesy Brendan Foran (International SEMATECH).

Materials Characterization Needs for New Structures and Substrates

Transistors and Capacitors

Shrinking transistor gate lengths allow both increases of device density and transistor switching speed. It is useful to discuss why channel engineering to improve mobility has become a critical need for the semiconductor industry. The transistor delay (switching speed) τ is proportional to the power supply voltage V_{dd} and inversely proportional to the saturation drive current I_{dsat} as follows (C_{load} is the capacitive load) [8]

$$\tau = C_{load} \, V_{dd} \, / \, I_{dsat} \qquad .$$

For long channel devices, I_{dsat} is inversely proportional to gate length and the EOT under inversion conditions. It can also be increased by improving the carrier mobility. For very short channel devices, the gate length does not appear in the approximate expression for drive current I_d . An approximate expressions for I dsat are [8]

$$I_{dsat} = (W/L) \, \mu_{eff} \, C_{ox} \, [0.5 \, V_{Dsat}^2] \text{ long channel transistors } \quad V_{Dsat} = (V_G - V_T)$$

$$I_d \sim W \, \mu_{eff} \, C_{ox} \, (V_G - V_T) \, v_{sat} \quad \text{very short channel devices}$$

Here, μ_{eff} is the carrier mobility, W is the transistor width, C_{ox} is the capacitance of the gate dielectric, and V_G and V_T are the gate and threshold voltages, respectively. Thus, the industry has focused on channel engineering as a critical part of maintaining I_{dsat} values ~ 1 mA/μm. The saturation carrier velocity is v_{sat}.

At least three means of increasing carrier mobility are already in manufacturing at the 90 nm technology node. Transistor gate lengths in logic devices are 45 nm at the 90-nm node. These methods use tensile stress to increase electron mobility and compressive stress to increase hole mobility. One approach [9] uses thicker silicon nitride layers above the NMOS and replaces the source and drain in the PMOS with silicon germanium as shown in Figure 6. Another approach uses the stress induced by the shallow trench isolation to stress PMOS channels [10]. The amount of stress can be altered by changing the distance between the STI and the edge of the gate electrode. Critical modeling work has shown how compressive stress along the <110> direction of silicon increases hole mobility [11, 12]. Measurement of localized stress is very difficult. One of the issues is that the area of interest is the transistor channel which is buried under the transistor gate. For NMOS, the silicon nitride layer is above the transistor, and stress can be measured in that layer. The difficulty with this approach is correlating layer stress with that in the channel. Recent work has demonstrated that the principle of no increase in stress with layer thickness holds true for the silicon nitride layer. However, layer thickness changes do increase stress in the channel region [13]. The relationship between the stress in the measurable layer (silicon nitride) and that in the channel must be model based. Stress in the PMOS channel is even more difficult to measure locally. The process-based methods of increasing carrier mobility use uni-axial stress (or mostly uni-axial) stress.

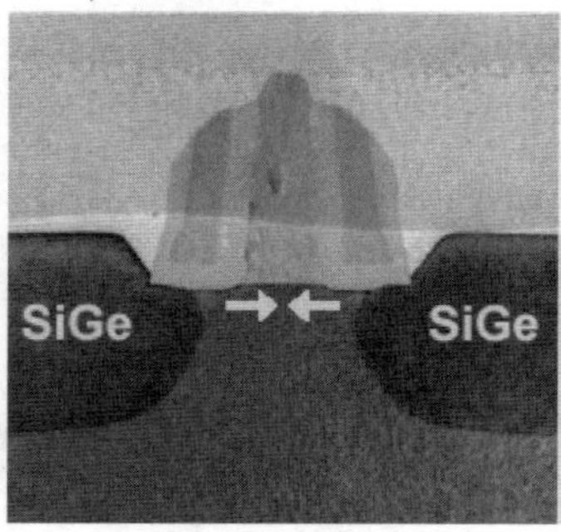

PMO
Compressive strain
increased hole mobility

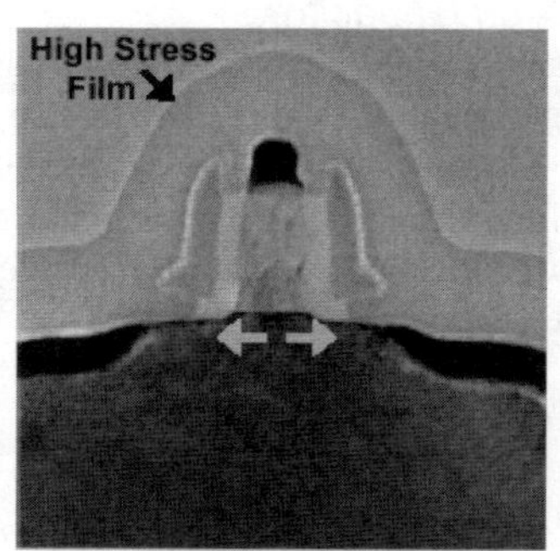

45-nm NMOS
Tensile stress SiN layer
increased electron mobility

Figure 6. Compressively stressed PMOS and tensile stressed NMOS Transistors. Process induced stress is already used in volume manufacturing in the 90-nm node. Figures courtesy Mark Bohr of Intel. Figures © IEEE and used with permission [9].

Future approaches to stress-induced improvement of carrier mobility include the use of metal gates [14] and strained silicon substrates. Strained silicon substrates are grown on top of silicon germanium layers where the lattice mismatch provides the stress. Strained silicon can either be left on top of the SiGe or transferred to a wafer with a surface oxide layer producing the so-called sSOI. It is important to note that these layers are bi-axially strained, while the channel engineering provide nearly uniaxial stress along one crystallographic direction. It is also important to note that the process-based approach allows one to use compressive stress for P channel devices and tensile stress for n-channel devices, while the substrate approach presume that both n- and p-channel devices can improve mobility using the same type of stress. The complete description of the stress imparted by the silicon nitride layers includes uniaxial tensile stress along the channel and compression out of the plane of the transistor [11].

The complete explanation for how the stress increases carrier mobility seems to controversial. Fischetti *et al.* [15] have investigated the standard explanation of bulk electron mobility increases coming mainly from the removal of degeneracy. This mechanism accounts for only part of the increase in channel mobility for the high carrier concentrations found in the inversion layer at the channel surface. This group believes that the increased mobility may be due to reduced interfacial roughness resulting in less carrier scattering. Lundstrom and Ren have proposed that the current in short channel devices is limited by the amount of thermal carriers injected from the source into the channel [16]. In short, metrology must turn to the modeling community for a usable relationship between the stress in a film that one can measure and the stress in the channel region.

In the future, maintaining I_{dsat} will require additional improvements in transistor technology [17]. Functioning sub-10-nm CMOS transistors have been fabricated by Doris and coworkers on SOI substrates [18, 19]. Characterization of the ultra-thin silicon layers used in these devices is extremely challenging. Most notable of these is the need for either multiple gates on the same transistor or wrapping the transistor gate around the sides of the channel. The FINFET is the most widely discussed of these designs, and it is shown in Figure 7. The gate dielectric is on the sidewalls of these devices. Other challenges to the metrology community include measurement of doping in new structures such as FINFETs.

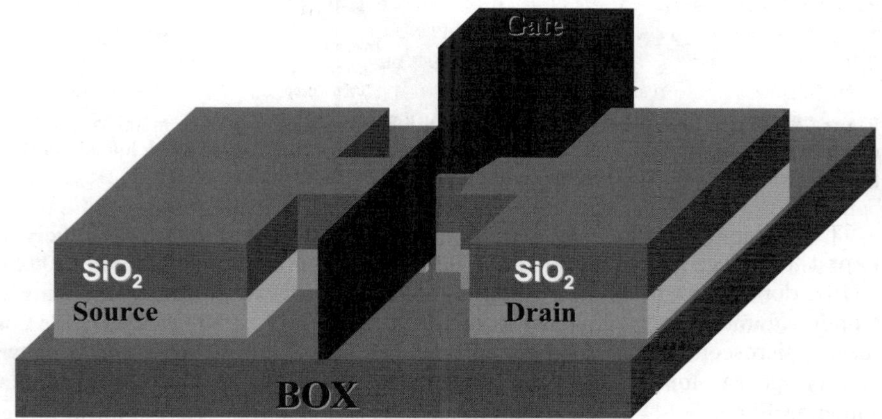

Figure 7. FINFET transistor. Figure courtesy T.J. King, UC Berkeley.

The scaling of capacitor structures used in memory ICs also challenges materials characterization. In Figure 8, we show both trench capacitor and stacked capacitor cross sections. Higher dielectric constant materials are also needed for capacitors.

Advances in Materials Characterization Methods

Transmission Electron Microscopy (HR-TEM and ADF-STEM)
The invention of aberration correctors for electron lenses has resulted in spectacular advances in the spatial resolution of both HR-TEM and ADF-STEM [3, 20–25]. Along with the ability to place a very finely focused beam on a single column of atoms, the energy resolution of electron energy loss spectroscopy has been greatly improved by energy filtering the electron source. Both the Wien filter and the Omega filter have been used as energy filters [25, 26]. Along with advances in the microscope hardware, simulations have played an important role in understanding the imaging and the best approach for analysis [27–30].

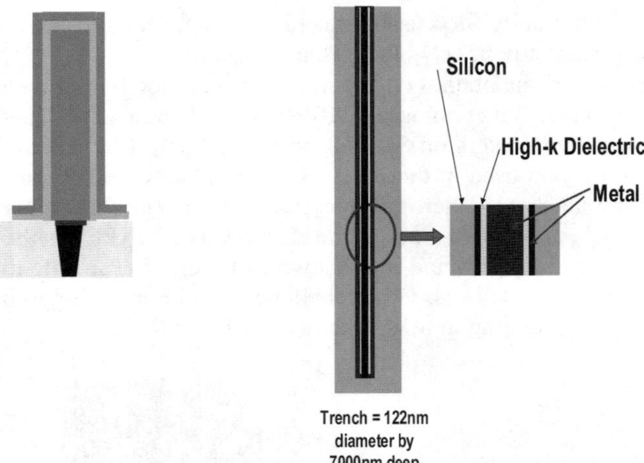

Figure 8. Trench and stacked capacitors TEM characterization of thin metal layers is becoming more challenging as scaling continues. Figure courtesy FEP technical working group of the International Technology Roadmap for Semiconductors.

The advantages of STEM for microanalysis include the ability to intuitively interpret the image and the localization of the image to an atomic column [20–25]. STEM done with an annular dark field detector provides Z contrast imaging of atomic columns in crystalline materials [3, 21–23]. The first STEM aberration corrected microscope resolved the silicon dumbbell structure, which consists of two closely spaced atomic columns when imaging down the Si <110> zone axis as shown in Figure 8 [3]. Aberration corrected STEM was soon used to locate which atomic column a La atoms was in $CaTiO_3$ [20, 21]. The latest advance brings hope for 3D atomic resolution [22, 23]. Recently, Pennycook showed the utility of aberration corrected STEM in the analysis of atomic diffusion of Hf atoms in high-k gate stack materials [31]. The depth of focus of aberration corrected lens enables 3D location of single atoms. In Figures 9 and 10, this ability is illustrated. HRTEM has also profited by use of the aberration correction lens [24]. Again, images resolving the silicon dumbbell structure were used to illustrate resolution improvement.

As aberration correctors reduced the beam diameter to below 0.1 nm, the microscopy community used multi-slice simulations to understand the physics of STEM imaging using reduced beam size [27–30]. One example of how the simulations can assist image and understanding of localization of ELS spectra is the work of Dwyer and Etheridge for 60-nm-thick silicon [29]. Their simulations show match experimental observation that the Si dumbbell structure of Si (110) is not resolvable for beam diameters of 0.2 and 0.12 nm, while it is resolvable for a beam diameter of 0.07 nm [29]. They also find that a beam diameter of 0.2 nm is not able to resolve the atomic columns of Si (100), while beam diameters of 0.14 nm and 0.07 nm do resolve the atomic columns. These simulations found an oscillatory nature of the beam intensity on adjacent atomic columns vs sample depth [29]. This implies that ELS spectra can contain contributions from adjacent atomic columns. The oscillatory nature of the signal on the atomic column originally impacted can result in unusual effects. For example, a dopant atom at 8 nm down the atomic column may not produce any signal when the 0.007 nm beam diameter is used even though image resolution is improved [29]. Muller *et al.* have shown that after traveling 30 nm down one atomic column the beam can jump to adjacent atomic columns when imaging the Si dumbbell structure [28]. This simulation includes the effect of lattice vibrations, and was done using an aberration corrected beam.

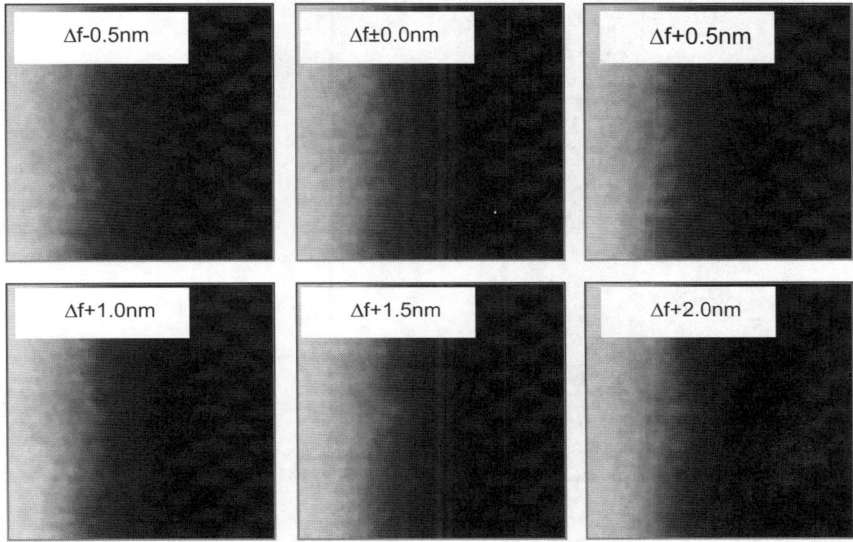

Figure 9. 3D imaging by Aberration corrected STEM.
The ability to focus at different depths enables observation of Hf atom diffusion into thin oxide interface layers between the silicon substrate and the HfO_2 high-k layer. Here, six images from a through focus series of 21 images, are used to show a single Hf atom going in and out of focus (see the red circle). The focus step size was 0.5 nm per frame. Figure courtesy Steve Pennycook (Oak Ridge National Laboratory), work done for the US Department of Energy and not subject to copyright [31].

They further showed that the depth of a dopant atom in one column of the dumbbell structure can influence which of the two atomic columns has the higher intensity in an image [28]. In addition, their simulation showed that beam spreading is considerably larger in amorphous Si. *This information is critical to understanding ELS spectroscopy in ADF-STEM. At some sample depths and imaging conditions, an atom resident in the atomic column may not contribute to the ELS signal [28]. It also shows that the ELS signal may come from an atom in an adjacent atomic column.* Simulations of other materials also exist as well as simulations of the effect of atomic disorder on the columns [30]. Another important and under-utilized capability is the characterization of strain at the interface by STEM [32].

Figure 10. 3D View of HfO_2 – silicon interface. A graphical image of the interface was created using the through focus series of STEM images such as those shown in Figure 9.
The yellow color is the Hf oxide high k and the dark gold areas represent the silicon atoms. The colored atoms between these layers are Hf atoms in the silicon dioxide interface layer. Figure courtesy Steve Pennycook (Oak Ridge National Laboratory), work done for the US Department of Energy and not subject to copyright [31].

The ultimate goal of 3D atomic resolution mapping of atomic locations and physical properties remains elusive. As discussed above, aberration corrected STEM has proven capable of atomic location in interfacial structures. Other methods are also working toward this goal. An important, new method know as local electrode atom probe [LEAP] provides 3D atomic maps of conductive samples. LEAP has already mapped the location of boron atoms in silicon [6].

Conclusion

The semiconductor industry has already entered the world of nanotechnology. The impact of nanodimensions on materials and device properties has driven the introduction of new materials such as low-k inter-level dielectrics. This change in dimension and properties has also resulted in the need to develop new materials characterization capabilities. In addition, it is important to note that materials characterization will continue to play a critical role in the development of new nano-electronic technology and in manufacturing process control.

References

1.   G.D. Hutchinson, The First Nanochips, Scientific American, April, 2004, p 48 – 55.

2.   Semiconductor Industry Association (SIA), International Roadmap for Semiconductors 2003 and subsequent editions, Austin, TX: International SEMATECH, 2003. (This is available for printing and viewing from the Internet, with the following URL: http://public.itrs.net.)

3.   P.E. Batson, N. Dellby and O.L. Krivanek, Sub-ångstrom resolution using aberration corrected electron optics, Nature 418, 617 (2002).

4.   P. Seidel, EUV Mask Blank Fabrication and Metrology, In: Characterization and Metrology for ULSI Technology 2003, edited by D.G. Seiler, A.C. Diebold, T.J. Shaffner, R. Mcdonald, S. Zollner, R.P. Khosla *et al.*, AIP conference Proceedings 683, (AIP, New York, 2003), p 371 –380.

5.   H.G. Tompkins and W.A. McGahan, Thin Metal Films In: Spectroscopic Ellipsometry and Reflectometry, Wiley, New York, (1999), pp 181 – 187.

6.   A.C. Diebold, P.Y. Hung, J. Price, B. Foran, H. Celio, and T. Kelly, Interface Characterization in the Semiconductor Industry, The 204th Electrochemical Society's International Symposium on Interfaces in electronic materials (H1), 2003.

7.   M. Kiene, M. Morgan, J-H Zhao, C. Hu, T. Cho, and P.S. Ho, Characterization of Low-Dielectric Constant Materials, In: Handbook of Silicon Semiconductor Metrology, edited by A.C. Diebold, Dekker, New York, (2001), pp 245 – 278.

8.   B.G. Streetman and S. Banerjee, Solid State Electronic Devices, 5th edn, Prentice Hall, Upper Saddle River, (2000), pp 452 – 461.

9.   T. Ghani, M. Armstrong, C. Auth, M. Bost, P. Charvat, G. Glass *et al.*, A 90-nm High Volume Manufacturing Logic Technology Featuring Novel 45-nm Gate Length Strained Silicon CMOS Transistors, IEDM Tech. Digest 2003, pp 978 – 980.

10.   V. Chan, R. Rengarajan, N. Rovedo, W. Jin, T. Hook, P. Nguyen *et al.*, High Speed 45 nm Gate Length CMOSFETs incorporated into a 90 nm Bulk Technology Using Strain Engineering, IEDM Tech. Digest (2003), pp 77 – 80.

11.   M.D. Giles, M. Armstrong, C. Auth, S.M. Cea, T. Ghani, T. Hoffmann *et al.*, Understanding Stress Enhanced Performance in Intel 90nm CMOS Technology, 2004 Symposium on VLSI Technology Technical Digest of Papers, pp 118 - 119.

12. R.A. Bianchi, G. Bouche, and O. Roux-dit-Buisson, Accurate Modeling of Trench Isolation Induced Mechanical Stress Effects on MOSFET Electrical Performance, IEDM 2002 Tech. Digest, pp 117-120.

13. R. Khamankar, H. Bu, C. Bowen, S. Chakravarthi, P. R. Chidambaram, M. Bevan *et al.*, An Enhanced 90nm High Performance Technology with Strong Performance Improvements from Stress and Mobility Increase through Simple Process Changes, pp 162-163.

14. Q. Xiang, J-S Goo, J. Pan, B. Yu, S. Ahmed, J. Zhang, and M.-R Lin, Strained Silicon NMOS with Nickel-Silicide Metal Gate, 2003 Symposium on VLSI Technology Digest of Technical Papers, pp 101-102.

15. M.V. Fischetti, F. Gamiz, and W. Hansch, On the Enhanced Electron Mobility in Strained-Silicon Inversion Layers, J. Appl. Phys. 92, 7320 (2002).

16. M. Lundstrom and Z. Ren, Essential Physics of Carrier Transport in Nanoscale MOSFETs, IEEE Trans. on Electron. Devices 49, (2002), pp 133 – 141.

17. P.M. Zeitzoff, J.A. Hutchby and H.R. Huff, MOSFET and Front-End Process Integration: Scaling Trends, Challenges, and Potential Solutions Through The End of The Roadmap, Int. J. High-Speed Electron. Syst., 12, 267-293 (2002).

18. B. Doris, M. Ieong, T. Kanarsky, Y. Zhang, R.A. Roy, O. Dokumaci *et al.*, Extreme Scaling with Ultra-Thin Si Channel MOSFETs, IEDM Techn. Digest 2002, pp 267-270.

19. B. Doris, M. Ieong, H. Zhu, Y. Zhang, M. Steen, W. Natzle *et al.*, Device Design Considerations for Ultra-Thin SOI MOSFETs, IEDM Techn. Digest 2003, pp 631-634.

20. L.J. Allen, S.D. Findlay, A.R. Lupini, M.P. Oxley, and S.J. Pennycook, Atomic-Resolution Electron Energy Loss Spectroscopy Imaging in Aberration Corrected Scanning Transmission Electron Microscopy, Phys. Rev. Lett. 91, (2003), 105503 1-4.

21. M.Varela, S.D. Findlay, A.R. Lupini, H.M. Christen, A.Y. Borisevich, N. Dellby *et al.*, Spectroscopic Imaging of Single AtomsWithin a Bulk Solid, Phys. Rev. Letters 92, (2004) 95502 1-4.

22. S. Wang, A.Y. Borisevich, S.N. Rashkeev, M.V. Glazoff, K. Sohlberg, S.J. Pennycook *et al.*, Dopants adsorbed as single atoms prevent degradation of catalysts, Nature Mat. 3, (2004), p143 –146.

23. S.J. Pennycook, A.R. Lupini, A. Borisevich, Y. Peng and N. Shibata, 3D Atomic Resolution Imaging through Aberration-Corrected STEM, Microsc. and Microanal. (2004), p 1172 CD.

24. C.-L. Jia, M. Lentzen, and K. Urban, High-Resolution Transmission Electron Microscopy Using Negative Spherical Aberration, Microsc. Microanal. 10, 174–184, (2004).

25. M. Mukai, T. Kaneyama, T. Tomita, K. Tsuno, M. Terauchi, K. Tsuda *et al.,* Performance of the MIRAI-21 Analytical Electron Microscope, Microsc. Microanal. 10, CD858-859, 2004.

26. L.F. Allard, D.A. Blom, M.A. O'Keefe, C. Kiely, D. Ackland, M. Wantanabe *et al.,* First Results from the Aberration-Corrected JEOL 220FS-AC STEM/TEM, Microsc. and Microanal. 10, (2004), pp 110 111.

27. Earl J. Kirkland, R.F. Loane and John Silcox, Simulation of Annular Dark Field STEM Images using a Modified Multisclice Method, Ultramicroscopy 23 (1987) 77—96.

28. P.M. Voyles, D.A. Muller, and E.J. Kirkland, Depth-Dependent Imaging of Individual Dopant Atoms in Silicon, Microsc. Microanal. 10, 291-300 (2004).

29. C. Dwyer and J. Etheridge, Scattering of Atomic Scale Electron Probes in Silicon, Ultramicroscopy 96, (2003), p 343 – 360.

30. T. Plamann and M.J. Hytch, Tests on the Validity of the atomic column approximation for STEM Probe Propagation, Ultramicroscopy 78, (1999), p 153 – 161.

31. K. van Benthem, M. Kim, S.J. Do, J.T. Luck, A.R. Lupini, and S.J. Pennycook, Three Dimensional Imaging of Individual Hafnium Atoms at a $Si/SiO_2/HfO_2$ Dielectric Interface, submitted for publication.

32. Z. Yu, D.A. Muller, and J. Silcox, Study of Strain Fields at a-Si˚ c-Si Interface, J. Appl. Phys. 95, (2004), pp 3362-3371.

Advanced Material Characterization by TOFSIMS in Microelectronic

T. Conard, W. Vandervorst

IMEC, Material and Component Analysis Department, Leuven / Belgium

Introduction

With the ever-demanding shrinking of the dimensions of electronic devices, and according to the ITRS roadmap [1], the number of materials involved in IC manufacturing will only increase in the near future (including high- and low-k dielectrics, copper, polymers, *etc.*). Also, as development cycles become shorter and shorter, this presents a heavy challenge for characterisation techniques that have to be able to provide, in a quantitative manner, a full characterisation of the different layers and materials involved in device processing.

Unfortunately, no single technique is capable of providing a full quantitative 3-D characterisation of all materials. Among the techniques that are recently gaining in interest, time of flight-secondary ion mass spectrometry (TOF-SIMS) is a significant one due to the wide range of information that it can provide both on organic and inorganic systems. It can indeed not only be used in a static (surface sensitive) mode but also, combined with a second ion source, in a profiling mode. However, in general, only semi-quantitative information can be retrieved from a TOF-SIMS experiment due to the large matrix effect present in SIMS experiments and/or lack of references.

In this review, we will present a number of applications of TOFSIMS in advanced materials characterisation in microelectronic research, both in static and in dynamic mode.

Experimental

Secondary ion mass spectrometry (SIMS) is the mass spectrometry of ionised particles that are emitted following the bombardment of a surface by an ion beam. In its time of flight version, the primary beam is pulsed for a very short time ($\sim$ns). The emitted ions are accelerated to a given potential and then allowed to drift in a field-free space. The flight time of the secondary ions in this space to the detector can then be converted to their masses using the classical kinetic energy relation.

Commonly used ions as primary beams are Ar^+ and Ga^+ but more recently, heavier ion (clusters) such as C_{60}^+, Au_n^+ or Bi_n^+ have been investigated and have shown significant improvement is signal quality, mostly at large masses compared to the Ga^+ ions.

In the dynamic mode, a second primary ion beam is used in order to erode the sample, and analysis is performed in the formed crater in order to obtain depth information of the analysed sample. This second ion beam is typically from noble gas (Ar, Xe), O or Cs, similarly to the usual dynamic SIMS study.

In this review, all results presented were acquired from a TOF-SIMS IV from IONTOF with a primary analysis Ga 15 keV source and Xe 350 eV–1 keV as a sputtering source for the dynamic data.

Metal Contamination

The measurement of surface metal contamination during device fabrication is increasingly important with the new era IC technology. It is traditionally measured on unpatterned wafers by total X-ray reflection fluorescence spectroscopy (TXRF). The large area needed for TXRF analysis, however, renders this technique useless when a specific area of a wafer has to be analysed. In addition, in a number of cases, the sensitivity of conventional TXRF tools does not reach the necessary levels and sensitivity enhancement that has to be obtained through vapour phase decomposition (VPD) over the whole wafer, giving only an average contamination level at wafer level. On the other hand, using TOFSIMS most of the metallic elements can be detected and, with an adequate procedure, also quantified on the Si surface. It can be of particular advantage when, for instance, Al contamination has to be analysed on Si. Indeed in this case, due to the high Si intensity measured by TXRF, the detection limit for Al is very poor, while the good mass resolution of the TOFSIMS spectra allows easy identification of Al intensities. Even in the analysis of Fe, which is often considered as a test case, the mass resolution of present TOFSIMS instruments are good enough to separate Fe^+ ions from Si_2^+ ions.

In order to assess the use of TOFSIMS as an analytical tool, it has been compared to other established techniques by several groups [2–5]. These works all showed the importance of reference sample preparation in order to compare the results with TXRF. Indeed, when using standard methods of chemical analysis used for the validation of the vapor phase decomposition-atomic adsorption technique (VPD-AA), large inhomogeneities in the surface concentration prevents the comparison of the TOFSIMS results with the TXRF results [5]. When careful experiments are performed, very reliable correlation with TXRF can be obtained and relative sensitivity factors (RSF) can be determined. An exhaustive list of RSF has been published by Douglas et al. [6].

In the implementation of Cu interconnects, due to the difficulty of dry etching this material, a damascene metallization scheme is used in combination with chemical mechanical polishing (CMP). This potentially induces Cu contamination of the insulating layers. However, due to the small dimension, TXRF cannot be applied to such measurements. In this case, TOFSIMS is a very useful alternative due to its high lateral resolution possibilities [7]. From a comparison of XPS/TXRF and TOFSIMS data on Cu-contaminated wafers, we determined a RSF of 8.6E13, very close to the value published by Douglas and Chen [6]. Figure 1 shows typical images that can be obtained on an array of Cu and SiO_2 lines from which full quantification can be obtained by normalizing the ^{30}Si and ^{63}Cu images, clearly showing a large dependence of the Cu contamination coverage at the center of the SiO_2 lines as a function of the line spacing (Figure 2).

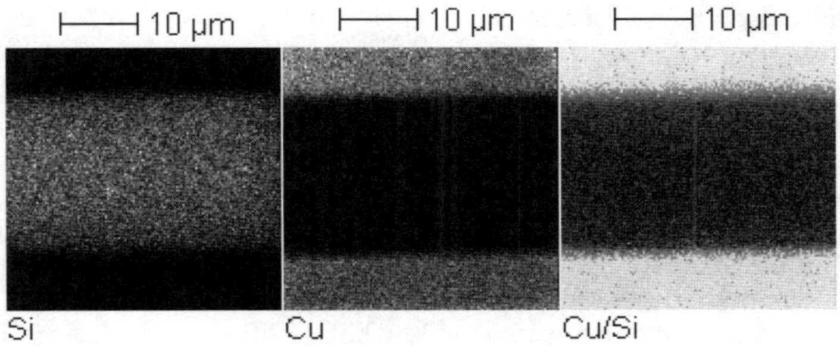

Figure 1. Si and Cu intensities measured by TOFSIMS after a CMP process. The right image presents the Cu/Si intensities, used for Cu coverage measurements.

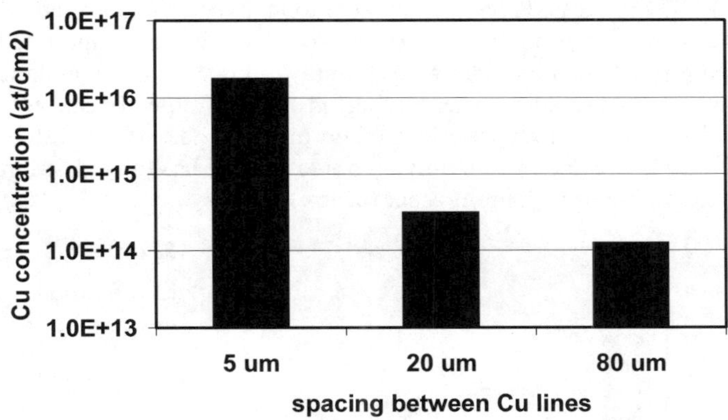

Figure 2. Cu contamination concentration dependence on the line spacing in a SiO_2/Cu line pattern after CMP.

Cleaning of Silicon and Drying Technology

The cleaning of the Si surface is one of the major challenges for future technologies. It is not only the metal contamination that is of concern, but also organic contamination should be at a very low level, in the order of 1E13 atm/cm^2. On the one hand, the primary use of TOFSIMS has historically been the analysis of organic materials and on the other hand, most traditional techniques such as XPS, FTIR, *etc.* have too high detection limits to be useful for such analyses. It is thus natural to consider TOFSIMS as a good candidate for the analysis of organic contamination. This approach has already been studied by several groups [8–10].

This method is used, for instance, in the analysis of the presence of organic contamination or residue after post-CMP cleaning. Figure 3 shows the intensities measured for some specific ion clusters typical of a complexing agent used in a cleaning procedure normalised to the intensity measured on a reference sample subjected to a standard clean. It is readily observed that for all ion clusters considered, the optimised clean (clean B) produces a surface as clean as a reference clean on Si wafers. When a non-optimised clean is used (clean A), a difference in the contamination level is observed depending on the drying method used.

The evaluation of cleaning/drying procedure also has to be performed on patterned wafers. For this purpose, the analysis of metal contamination with TOFSIMS in the imaging mode is particularly useful in order to identify possible problems in the drying uniformity as a function of the patterned structure. It is indeed possible to use patterned wafers with a TEOS-SiO$_2$ pattern that are first immersed in KCl solution in order to contaminate the surface and then to apply a drying method. An inhomogeneous drying or residue accumulation can thus be observed through the observation of K or Cl intensities. However, as topography also influences the ion intensities collected by TOFSIMS, the patterns are first removed by a VPD process, leaving a flat area to analyse. We analysed two types of sample, one with a hydrophobic pattern between the SiO$_2$ area and one with a hydrophilic pattern between the SiO$_2$ area. Figure 4 shows the original pattern as well as the K intensities measured after drying on the two types of surface. It clearly shows that the residues are mostly located on the SiO$_2$ part of the wafers but that a much better dry is obtained when only an oxide surface has to be dried than when a mixture of oxide and bare Si are present (centre).

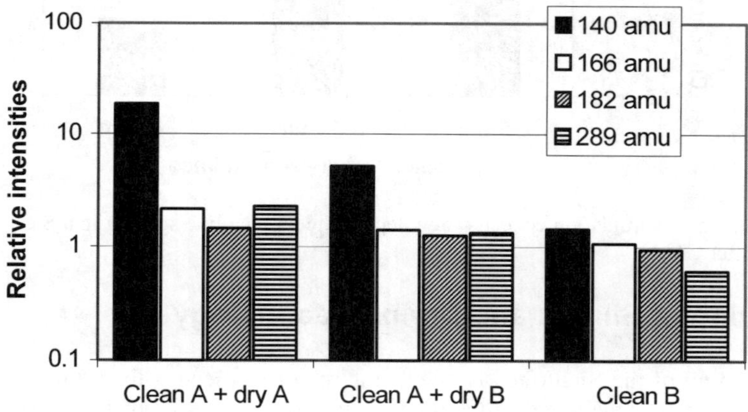

Figure 3. Relative intensity of specific complexing agent intensities compared with reference cleans for different cleaning and drying methods.

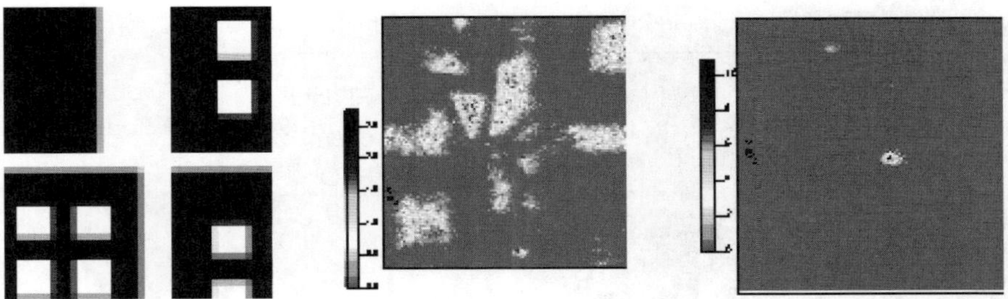

Figure 4. K contamination on a patterned wafer after drying. The patterned is presented at the left side, with TEOS being the white area. The two TOFSIMS intensity maps are for a hydrophobic pattern between the TEOS (centre) or a hydrophilic one (right).

High-k Analysis

One of the major issues in the route to higher performances in transistor operation is to decrease the equivalent oxide thickness (EOT) of the gate oxide. For that purpose, materials with higher dielectric constant (high-k) are integrated into the IC processing. This poses a number of challenges such as the possibility to grow ultra-thin layers with high quality, reducing contamination levels in the grown layers, and so on, that TOFSIMS can analyse (see, for instance, references [11–14]).

Static Mode: Layer Integrity

One of the main techniques used for growing high-k layers is atomic layer CVD (ALCVD) in which alternating self-terminating reactions are made at a surface using for example $HfCl_4$ and H_2O as precursors for the growth of HfO_2 high-k layers. One of the main problems with this technique is that the chemical state of the starting surface plays a major role in the nucleation of the layer and thus smooth pinhole-free, ultra-thin layers are difficult to obtain. Low energy ion spectroscopy (LEIS) is a useful technique in order to study the layer closure in ALD due to its ultimate surface sensitivity but suffers from both a slow throughput and a lack of sensitivity. We introduced the use of TOFSIMS as an alternative in the study of the growth of ZrO_2 on HF-last Si and on hydrophilic Si surfaces [11].

This work has been extended to other systems such as HfO_2. Figure 5 shows the change in the relative Si intensity measured on a series of HfO_2 layers grown on different starting surfaces. It is seen that, for a given number of Hf atoms present at the surface, three clearly different sets of growth quality are observed. On HF-last surfaces (OH-free), very poor growth is observed with the substrate intensities only decreasing very slowly with increasing Hf coverage, indicating significant islanding of the HfO_2 layer at the surface of Si. On the other hand, the presence of a chemical oxide strongly favours the growth and hence a very fast decay of the Si intensities is observed, indicating a much smoother growth than on Hf-last Si. When a dry oxide is present as the starting layer, some intermediate condition is observed.

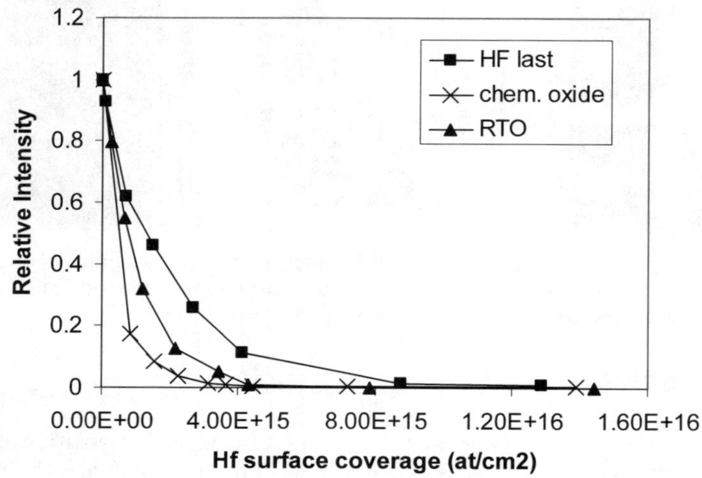

Figure 5. Relative Si intensity from the growth of ALCVD HfO_2 on different starting surfaces

Profiling Mode: Contamination and Thickness-Diffusion

TOFSIMS profiling in high-k research is principally used for two purposes: for the study of contamination levels within the layer and in the determination of layer modifications during processing.

The chlorine impurities resulting from the ALCVD $HfO_2/HfCl_4$-based process are believed to be detrimental for electrical performance of devices. It is thus important to optimise process conditions regarding the impurity content. Figure 6 presents the Cl profile measured on three different HfO_2 layers. It shows that compared to standard processing conditions, it is possible to significantly reduce the amount of Cl in the films as well as at the HfO_2/Si interface by increasing the length of the H_2O pulse during processing or even more by increasing the process temperature.

During the full processing, gate oxide layers have to sustain some thermal budget. It is thus important to understand the modifications occurring in the high-k stack upon annealing. Figure 7 shows the profiles measured on an AlZrO layer grown on thin rapid thermal oxidation (RTO) layer after different annealing under nitrogen. One can easily observe that no significant modification occurs in the layer up to an annealing temperature of 700°C but that from 800°C, significant diffusion of the Si into the AlZrO layer is observed.

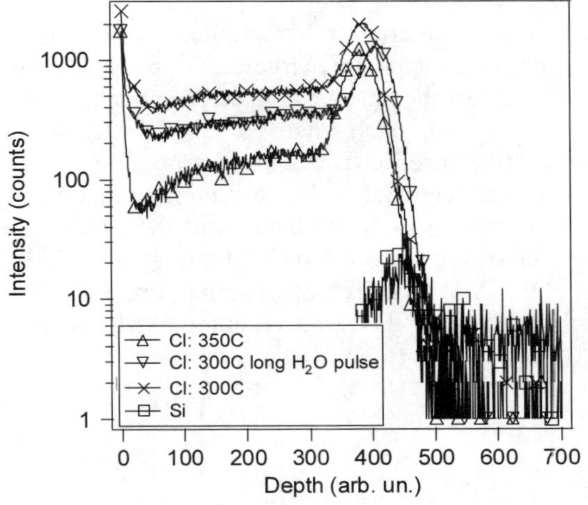

Figure 6. Cl profile from ~10 nm ALCVD HfO$_2$ layer processed under different conditions.

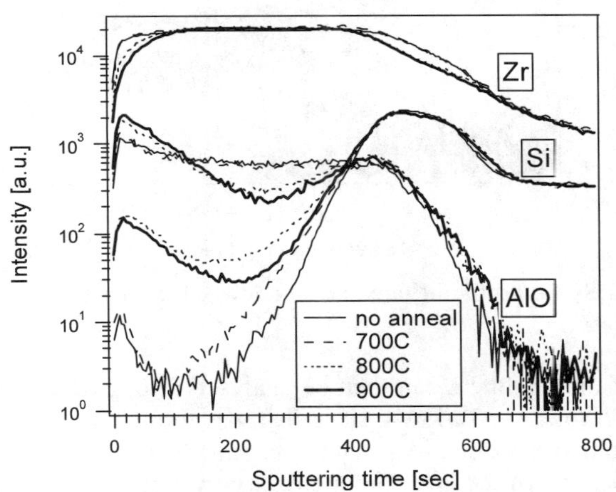

Figure 7. TOFSIMS depth profile of an AlZrO layer annealed under N$_2$ at different temperatures.

Back-End-of-Line Application

A number of problems in back-end-of-line (BEoL) integration can also be studied using TOFSIMS (see, for instance, [15–18]). Here we will present three examples of the use of TOFSIMS in BEoL studies.

Surface Plasma Modification

The development of low-k materials often implies the use of porous materials. However, metal penetration into the pore structure is an issue. In order to avoid this problem, several methods are being investigated in order to seal the pores at the surface of such films. One approach involves using a plasma treatment to seal. Figure 8 shows the intensity profile from a porous organic low-k deposited on Si that received different treatments before being submitted to a N_2/H_2 plasma. While surface modification is important in order to seal the porous dielectric, ideally modification of the bulk of the dielectric itself should be minimal. It can easily be observed from the TOFSIMS data that for treatment one modifies the bulk of the dielectric the least thus favouring its use as a method of film sealing.

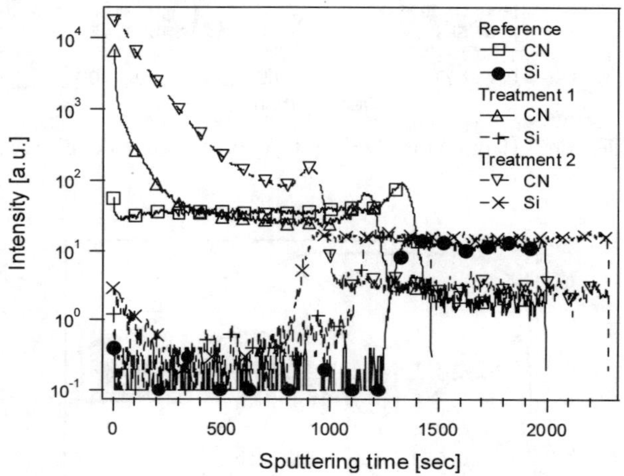

Figure 8. CN and Si ion profile of three organic low-k layer on Si treated with a N_2/H_2 plasma.

The effect of the treatment can then be analysed through the full stack as presented in Figure 9 for a Cu/Ta/low-k stack that was formed on a low-k with different treatments. In this case, one sees that the treatment containing a plasma induces too much modification of the layer producing a very bad interface, most likely due to substantial roughness, while the treatment without a plasma is closer to the reference sample, with a sharper interface.

Copper Electroplating

Electroplating is used for copper metallisation. To achieve the so-called bottom-up copper filling in small features, organic and inorganic additives are widely used in a plating bath. Consequently, various types of impurities are trapped in the narrow trench during plating. These impurities originate mainly from the additives and intermediate cathodic products. A small impurity concentration can influence the electrical property of electroplated copper significantly.

The influence of the line width on the impurity content has been seldom investigated. Using TOFSIMS depth profiling in arrays of Cu and SiO_2 lines, we can show that the impurity concentration significantly increases when narrower lines are produced as is clearly observed in Figure 10. A significant increase in the contamination is already observed for the 1000-nm and 500-nm lines and a further increase is seen for the 300 nm Cu lines. These impurity concentrations can have an influence both on the grain size and resistivity of the lines that are produced

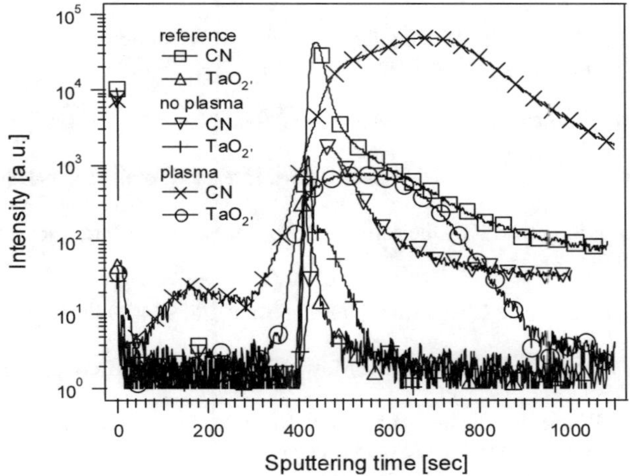

Figure 9. TOFSIMS profile of a Cu/Ta/low-k stack with different treatment of the low-k.

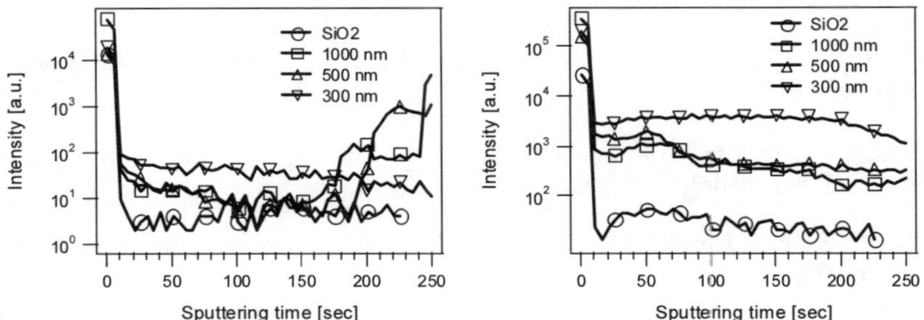

Figure 10. C_2 and Cl TOFSIMS profiles from an array of Cu and SiO_2 lines between 1000 and 300 nm in width.

Conclusion

This short review of the application of TOFSIMS in the field of materials characterisation in microelectronics has shown the versatility of this technique both in terms of the material that can be studied (organic/inorganic, conducting/insulating, *etc.*) or in term of the experimental conditions that can be used (ion types, static *vs.* dynamic mode, imaging, *etc.*). In some areas not covered here, it is also possible to reconstruct full three-dimensional profiles. The main limitation of the technique however remains its difficulty to easily reach quantified information at surfaces or interfaces but also in bulk samples if no reliable standards are available.

References

1. International Technology Roadmap for Semiconductors (ITRS), Semiconductor International Association (2003), http://public.itrs.net/

2. H De Witte., S Degendt., M Douglas., K Kenis., P.W Mertens., W Vandervorst. *et al.,* J. Electrochem. Soc., 147, pp 1915-19, (2000)

3. I. Mowat, P. Lindley, L. McCaig, Appl. Surf. Sci., 203/204, 495-499 (2003)

4. P. Lazzeri, A. Lui, L. Moro, L. Vanzetti, Surf. Interface Anal., 29, 793-803 (2000)

5. F. Zanderico, S. Ferrari, G. Queirolo, C. Pello, M. Borgini, Mater. Sci. Eng. B, 73, 173-177 (2000)

6. M.A. Douglas and P. J. Chen, Surf. Interface Anal. 26, 984-994 (1998)

7. H. Li, D.J. Hymes, J. De Larios, I.A. Mowat, P.M. Lindley. Micro 17, 35-54 (1999)

8. S. Jeon, C. Wong, S. Ohsiek, H.S. Kim, B. Ogle, Diffusion and defect data, part B (Solid State Phenomena) 92, 125-128 (2003)

9. A. Karen, N. Man, T. Shibamori, K. Takahashi, Appl. Surf. Sci. 203-204, 541-546, (2003)

10. A. Roche, C. Wyon, S. Marthon, J.F. Ple, M. Olivier, N. Rochat *et al.,* AIP Conference Proceedings 550, 297-301, 2001

11. T. Conard, W. Vandervorst, J. Petry, C. Zhao, W. Besling, H. Nohira, *et al.,* Appl. Surf. Sci. 203-204, pp400-403 (2003)

12. H. De Witte, T. Conard, W. Vandervorst, R. Gijbels, Appl. Surf. Sci. 203-204, pp523-526 (2003)

13. B. Crivelli, M. Alessandri, S. Alberici, D. Brazzelli, A.C. Elbaz, S. Frabboni *et al.,* CMOS Front End Materials and Process Technology Symposium, Mater. Res. Soc. Symposium processing vol 765, pp65-70, (2003)

14. G. Scarel, S. Ferrari, S. Spiga, C. Wiemer, G. Tallarida, M. Franciulli, J. Vac. Sci. Technol. A, 21, 1359-1365 (2003)

15. S. Balakumar, Chi Fo Tsang, N., Matsuki, Advanced Metallization Conference 2003, 613-620 (2004)

16. Y.Q. Zhu, E.T. Kang, K.G. Neoh, L. Chan, D.M.Y. Lai, A.C.H. Huan, Appl. Surf. Sci, 225, 144-155 (2004)

17. E.T. Ryan, M. Freeman, L. Svedberg, J.J. Lee, T. Guenther, J. Connor *et al.*, Mater Res. Soc. 766, 89-94 (2003)

18. I. Shahvandi, B. Brennan, C. Prindle, D. Denning, T. Lii, J. Iacoponi, Advanced Metallization Conference 2001, 559-566 (2001)

Electronic Properties of the Interface Formed by Pr_2O_3 Growth on Si(001), Si(111) and SiC(0001) Surfaces

D. Schmeißer, P. Hoffmann, G. Beuckert

Brandenburgische Technische Universität Cottbus, Cottbus / Germany

Introduction

The control of interface parameters is the most crucible effort in device technology. With the demand of down scaling to a 65-nm technology node the use of high-k materials and the partial or complete replacement of SiO_2 is emphasized [1,2]. A critical measure of the achieved improvement is the equivalent oxide thickness (EOT) in which the physical thickness of the dielectric layer is scaled by the ratio of DK (SiO_2) / DK (dielectric). In device applications the EOT should be smaller than 1 nm. To do so it is important not only to increase the DK of the dielectric but also to minimize the formation of an interfacial SiO_2 layer. It is evident that this demand is an ultimate challenge for materials science, surface and interface characterization, and technical engineering.

We employ state-of-the-art interface characterization techniques involving synchrotron radiation (SR) excited photoelectron spectroscopy (PES) for the characterization of the hetero-oxide-Si interfaces. We report nondestructive depth profiling by varying the escape depth, give an analysis of intermediate oxidation states and chemical state of interface species to follow the silicate formation Si2p, and report on the use of X-ray absorption spectroscopy (XAS) in materials science. Our contribution to that field is the control of film growth, interface reactivity, and phase homogeneity in ultra-thin oxide films by applying surface science techniques and employing SR based electron spectroscopies for their analysis. We explain our experimental setup, report on the Pr_2O_3 / Si(001) interface, which is one of the few candidates to realize a EOT <1 nm dielectric barrier on Si-surfaces. We compare these data to the corresponding results obtained at Si(111) and at SiC(0001) surfaces. For the latter, the high-k materials would enable considerable progress in high frequencies and high-power applications, while the former are requested by future Si technology nodes.

Experimental

We use the analytical spectroscopy and microscopy (ASAM) end station at the U49/2-PGM2 beam line at BESSY which is schematically depicted in Figure 1, the beam line technical details and specific properties have been described earlier [9]. The end-station is dedicated for spectromicroscopy to be used in materials science.

It consists of two UHV systems are operated in the low 10^{-10} mbar range. The analysis UHV system is equipped with a hemispherical electron energy analyzer (EA125) for photoelectron spectroscopy (PES) and a photoemission electron microscope (PEEM). The preparation system enables the use of deposition techniques [2–4] as used in molecular beam epitaxy as well as high pressure experiments [11]. The sample transfer system is designed to enable a fast access to either the PES or the PEEM system, it also enables the preparation of electrochemical prepared samples [12] and of polymeric samples [13].

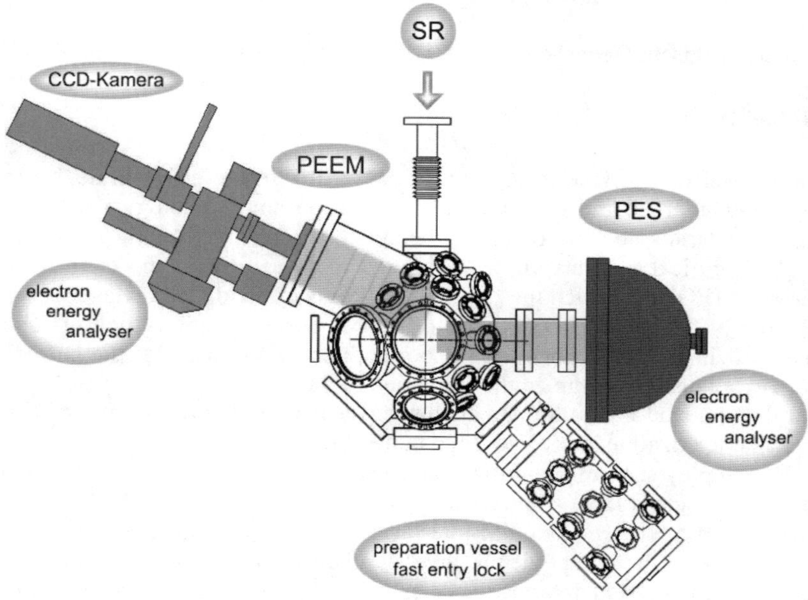

Figure 1. A schematic sketch of the ASAM end station at the BTU-owned U49/2-PGM2 undulator beam line at BESSY, Berlin, dedicated for interface analysis and materials research.

Two types of typical experiments can be performed. One experiment starts with the clean substrate and increase the exposure study the changes of the substrate signals when the film thickness is increased in steps of 0.5 nm. In another experiment we analyze the depth profile or hidden layers through a film prepared in our preparation system by varying the mean free path of the photoelectrons. Both are made possible by the properties of the SR, which gives high sensitivity, high energy resolution, and allows to vary the photon energy. The beam line allows to set photon energies in the range (80–1450 eV) with an energy resolution of $E/\Delta E = 10^4$. The total (combined monochromator and analyzer) energy resolution is set to about 100 meV over the complete spectral range. This allows one to vary the electron mean free path λ for, *e.g.*, SiO_2 between 0.8 nm (hv=150 eV) and 3.5 nm (hv=1400 eV).

Chemical State Analysis: Pr-Silicate Formation

The chemical state analyses of oxidation intermediates and interface species is one of the major advantages of PES in materials science — besides the quantitative elemental assignment – is the determination of chemical state of individual atoms at interfaces, in compounds, and of oxidation intermediates. We start with the O1s level, the binding energy (BE) of which is very close related to the chemical state of the oxygen atoms. In SiO_2 the high covalent contributions cause a large negative charge available for screening the core hole while in the Pr_2O_3 the more ionic nature keeps the electrons at the nuclei. We use the O1s signal for a chemical state analysis at the interface as it appears in SiO_2 at 533.7 eV and in Pr_2O_3 at 529 eV.

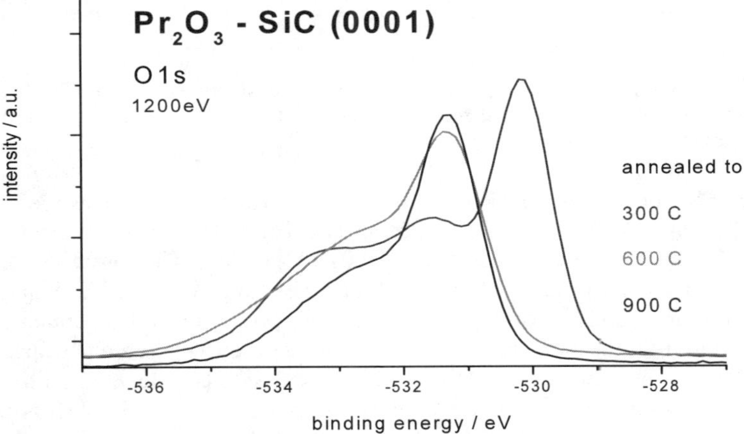

Figure 2. The oxygen 1 s emission as recorded with the bulk sensitive photon energy of 1200 eV of Pr_2O_3 film grown on SiC(001) and annealed to 300°C, 600°C, and 900°C.

As that energy separation is large, the Pr-silicate emission is intermediate as shown in Figure 2. The amount of oxygen and its distribution can be easily detected with increasing distance from the interface by employing photon energies to obtain bulk and surface sensitive data to determine coverage-dependent oxygen concentration and gradients. Consequently, the analysis of the O1s emission gives an easy access to study the phase stability of the Pr-silicate / Pr_2O_3 system [1], the relative intensities of these three components can be used to determine the relative composition of the as grown Pr-oxide layers [4].

In our next example we compare the initial film growth of Pr-oxide on Si(001), Si(111), and SiC(0001) surfaces. We focus on the chemical state analysis of the intermediate oxidation states of the Si2p core levels. At the Si – SiO_2 interface the existence of suboxides is well established [14, 15] with a shift of around 0.9 eV per oxidation state, and the shift of a fully oxidized Si surface amounts to 3.6 eV. With the benefit of high energy resolution the analysis of the surface components in the Si2p core level emission is historical and accepted, their relative intensity and chemical shift in energy depends on the respective reconstruction and number of ad-atoms [15, 16]. Using SR additional benefits involve the high energy resolution, the use of higher sensitivity, and the possibility to vary the probing depth.

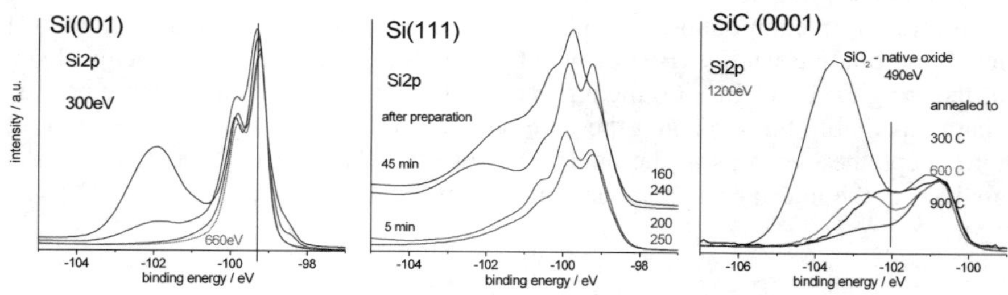

Figure 3. The Si2p emission for Pr_2O_3 at Si(001) and at Si(111) and at SiC(0001) at a coverage below 1 nm. We identify the emission from the Pr – O – Si bonds and of shifted components next to the bulk Si2p emission.

The Si2p core levels for Si(001), Si(111), and SiC(0001) surfaces are shown in Figure 3 with a Pr_2O_3 film thickness of around 1 nm. The emission from the substrate is found at −98.3eV for Si and at −101 eV for SiC. The formation of Si–O–Pr bonds can be followed in all the spectra as the interaction with the Pr_2O_3 causes a new emission band peaking around −102.5 eV which is well separated and distinguishable from the SiO_2 derived emission at 103.6 eV (see reference spectrum in the SiC panel). It compares well to the shift in the Si2p core level upon the formation of other silicate bonds, which is in the range of 1.5 eV to 3 eV. It is reported for a variety of silicates such as La, Zr, Gd, and Dy [14–17]. In all these systems it reflects the formation of more ionic bonds as well as the differences in the electronegativity of the elements associated [17]. It indicates that the Si–O–Si bond is distorted which has a high covalent character right at the Si interface and which causes well established suboxide core level shifts at the SiO_2 / Si interface [18]. That characteristic shifts in between the bare Si value (-99.3 eV) and that of SiO_2 (−104 eV) is rather insensitive on its particular chemical surrounding. This is a consequence of the increased ionicity and the increasing influence of the Madelung sum of the next and second-neighbor atoms in the contribution of the Si2p BE.

At initial Pr_2O_3 coverage there are significant differences in the Si2p data in Figure 3. The following analysis is restricted to the Si surfaces as at the SiC surface the high BE of the Si2p emission and its inherent width limit a further analysis. The spectra of Si(111) show the most significant changes as we take advantage of recording the data with extreme surface sensitivity. This is due to an additional emission which appears around 1 eV above the bulk Si2p emission and which appears on top of the $2p_{3/2}$ spin orbit partner of the substrate emission. This feature indicates a direct bond of the Si atoms to form a Si–O–PrO surface unit. Such a bond is similar to the Si^{+1} oxidation state which is observed upon oxidizing Si(111) and Ge(111) surfaces [14]. With this knowledge we can understand the previously discussed silicate species at -102.5 eV. For its formation the silicon interface must be oxidized to form Si–O–Si – Pr bonds in the Si back bond with a Si–O–Pr bond on top. The oxidation state of that O–Si–O backbone Si atom is fourfold and we expect an emission around -103.6 eV accordingly, which is contained in the tail of the silicate emission. On the Si(001) surface the single bonded Si–O–Pr is not found because of the two dangling bonds. Instead, the Si–O–Pr can occupy two O–Pr groups per Si surface atom. This causes a chemical shift of about 3 eV and makes it undiscernibly from the simple Si–O–Pr bond.

Depending on the surface orientation this breaking of back bonds will give rise to Si atoms which are threefold bonded to the former second layer. On the Si(111) surface the backbone oxidation removes the low coverage single bonded Si–O–Pr group completely as seen by the change in intensities in the middle panel of Figure 3. The backbone oxidation, however, causes a situation where steric arguments need to become considered as some of the Si atoms are not able to find a partner. These atoms then cause the shoulder at the low BE side of the Si substrate emission. On Si(001) it appears as a weak shoulder which is most prominent with surface-sensitive excitation, while on Si(111) the rise of the $Si2p_{1/2}$ emission is very broad because of such unoccupied Si atoms at the interface.

Nondestructive Depth Profiling by Varying the Escape Depth

In silicon oxynitride films we identify the different chemical states of nitrogen incorporated at the Si(001) interface, their relative abundance, and their vertical distribution. This information is again from photoelectron spectroscopy by varying the photoelectron inelastic mean free path which is in the order of the thickness of such ultrathin layers. Oxynitrides are high-k dielectrics with their DK varying between 5 and 7. They can easily be implemented into existing technologies [20]. Oxynitride films are grown by various methods (thermal oxidation and oxinitridation of Si by NO, N_2O, and NH_3 / chemical deposition, atomic layer deposition, physical deposition like N-implantation or magnetron sputtering) [20,21]. The very small thickness (<3 nm) of such layers makes it difficult to analyze them in common techniques like ion scattering techniques Medium Ion Energy Scattering (MEIS) [22]. In Secondary Ion Mass Spectroscopy (SIMS) the depth resolution is limited because of ion mixing effects. But both methods deliver only information about the elemental distribution of, $e.g.$ N, O, or Si within the layer, but no information about the chemical bonding state of these elements.

In the N1s spectra PES of oxynitride films we distinguish two compounds at BE of -398.8 eV and -398.0 eV which are assigned to oxynitrides SiO_xN_y and to silicon nitride Si_3N_4 [26], respectively. The relative intensity of these compounds depends on the selected photon energy. In the surface sensitive spectra the oxynitrides SiO_xN_y emission is stronger while in the bulk sensitive data the nitride emission dominates. In a model that includes the exponential attenuation of the photoelectrons when passing through the top layers we can reconstruct these relative intensities by assuming a layered structure. To do so it is important to know the values for λ. We determined the λ-values very accurately for the system Si/SiO$_2$ [25] and found them in good agreement with previous empirical measurements [23] and calculated values [24]. Knowing the respective electron mean free path λ the thickness s of the intermediate Si$_3$N$_4$ layer can be calculated from the relative intensities and the total thickness d of the film:

$$s = \lambda \cdot \ln\left[\exp\left(\frac{d}{\lambda}\right) + \frac{I^{SiO_xN_y}}{I^{Si_3N_4}} \right] - \lambda \cdot \ln\left[1 + \frac{I^{SiO_xN_y}}{I^{Si_3N_4}} \right] + d$$

Using this formula we determined an oxynitride layer thickness $s = 0.32$ nm and a nitride thickness $d{-}s = 0.30$ nm for a film with a total layer thickness of $d = 0.62$ nm, the latter was calculated from the Si2p spectra [27]. This shows that even for layers with a thickness below 1 nm our method can be applied to determine the hidden interfacial layer properties. The capability of photoelectron spectroscopy to distinguish between different chemical compounds gives here the opportunity to deduce a depth profile of the sample for the bonding state of a specific element.

X-Ray Absorption Spectroscopy

X-ray absorption spectroscopy (XAS) is a powerful technique and used to characterize semiconductors, oxides, and adsorbates [19]. We use it as an implemented tool within the PEEM microscope as we record the camera intensity which is proportional to the total electron yield upon varying the incident photon energy.

First, we like to show the advantages of spectromicroscopy in the thin film analysis. The example shown in Figure 4 is taken after deposition of 1 nm Pr$_2$O$_3$ on SiC(0001). In the PEEM image we realize an inhomogeneous film growth [6], probably because of the existence of graphitic surface clusters we find patches in which Pr$_2$O$_3$ exists (right) beside patches, which are completely free of Pr-oxide (left). This is evidenced by the intensity of the Pr4d absorption edge in the left part of the image (bold lines) which shows the Si2p edge at around 102 eV but vanishing intensities. In contrast, in the right part we find the well-developed Pr4d absorption spectra with the forbidden transitions around 110 eV and the giant resonance at around 120 eV. The three spectra shown in Figure 4 (left) are from different spots within the right part of the image and indicate that the film thickness is not homogeneous in the Pr$_2$O$_3$ covered area.

In the next example we focus on the O1s XAS data and in Figure 4 (right) we demonstrate that there are significant differences for Pr$_2$O$_3$ in the bulk phase and in the Pr-silicate phase. In a detailed analysis of the disappearance of the fine structure and the sharp resonant features we have analyzed the resonant photoelectron spectra around the O1s absorption edge as shown in Figure 5. First we notice a strong enhancement of the Auger intensity at photon energies around the O1s threshold, which is more than three times of the common Auger intensity observed at excitations way above the threshold. It also causes the intensity in the XAS data and is attributed to a resonant Auger decay, which originates from resonant excitation into conduction band (CB) states and involves significant contributions from scattering by (O2p-4f) charge transfer (CT) excitations. They represent electron hole excitations between the valence states (predominately the O2p (ligand) states) and the empty 4f states (529 eV) as well as into the states of the upper Hubbard band (531 eV). In the Pr_2O_3 data, the first two features are due to scattering processes involving CT complexes. The third contribution at 537eV is assigned to scattering at electron-hole pairs created by band-to-band excitations. Consequently, the features in the XAS spectra are not due to transitions into empty CB (π^*) states of the Pr_2O_3 as evident from the Auger yield recorded at a constant kinetic energy of 510 eV [4].

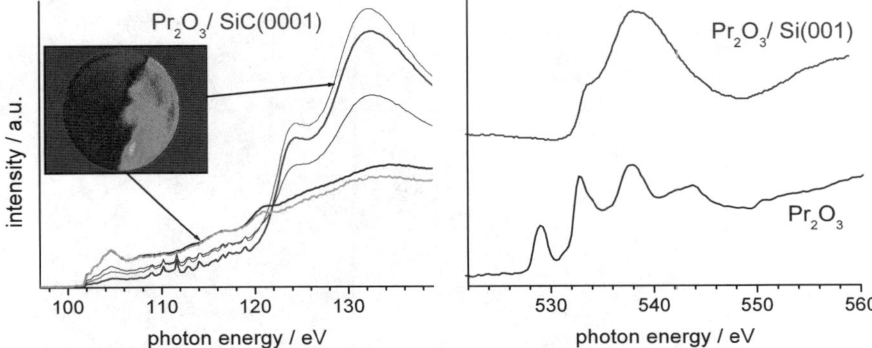

Figure 4. PEEM image (diameter 150 μm) and XAS spectra around the Si2p and Pr4d ionization edges (left panel); The XAS spectra at the O1s absorption edge as obtained for Pr_2O_3 in the bulk phase (bottom) and in the low coverage range (top) on Si(001) (right panel).

Focusing now on the changes that occur in the Pr-silicate we notice that the third contribution is still pronounced in the Pr-silicate spectra (left in Figure 4), while the former two resonant features are tremendously quenched when compared to the bulk Pr_2O_3 phase (right in Figure 4). We explain that reduced contribution by an enhancement of the covalent bonding, which occurs between the O2p states and the Pr4f states with additional contributions from the Si valence states. Awaiting theoretical confirmation we propose that the formation of Pr–O–Si bonds enables the formation of CT excitations by an additional CT from a Si3p electron into the Pr4f or Pr5d empty CB states. At the interface the CT complexes from the Si3p states cause a broadening of the CT states, which are located within the gap. Therefore they are no longer localized but act as gap states which, because of their large concentration, form a band of defects states. The band of defect states causes an almost continuous density of states and consequently, also the unoccupied states must have a similar distribution. The steep rise at 531 eV indicates transitions into these band-like empty states. Consequently, we explain that reduced contribution by an enhancement of the covalent bonding which occurs between the O2p states and the Pr4f states, theoretical studies [6-8] and in agreement with the valence band spectra [2,4,5]. As a consequence, our data indicate that the Pr-silicate has a metallic-like density of states. Indeed, such a metallic-like state is not favorable for the MOS architecture. Even if we consider that the defect states till will have a low mobility because of the Pr4f contribution and the thereby associated effective mass. At present, we still debate the influence of such states, they might contribute to the electrical balance at the interface and cause a metallic-like intermediate layer. However, such behavior depends very sensitively on the amount of covalent admixture and might be influenced by some dopants to remain a localized excitation.

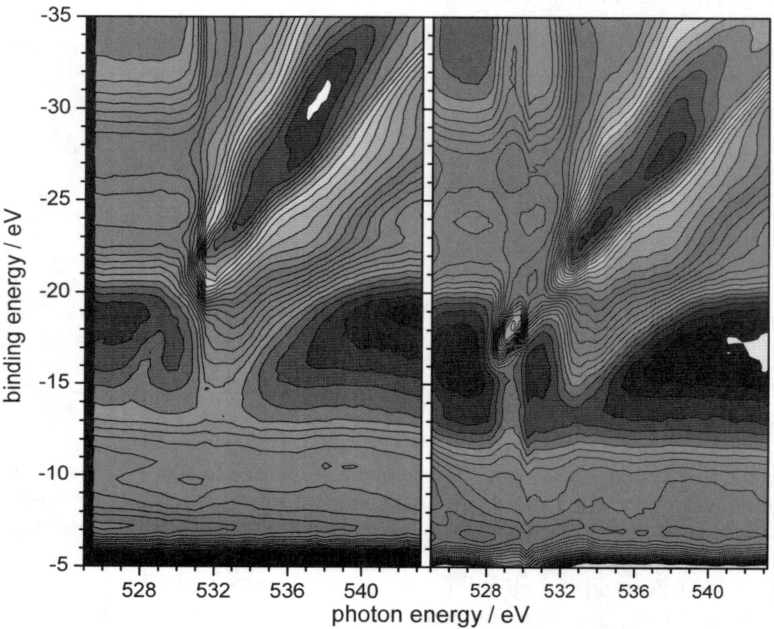

Figure 5. Resonant photoemission data around the O1s edge for 1nm Pr_2O_3 on Si(111) (left) and on Si(001) (right).

Conclusion and Outlook

The development of alternative gate dielectrics includes a series of topics which can be explicitly addressed by our SR-based interface analyses. We have focused here on some results that deliver very detailed information not accessible by other techniques. At this point it should be mentioned that for the Pr_2O_3 / Si(001) system we have recently reported the valence band discontinuities and also determined the oxygen content in a coverage-dependent study [2]. We also like to point out that our techniques are extremely useful for studies at the gate metal / dielectric interface. Here the chemical reactivity, diffusion, and filling of cracks are issues to achieve stable electrical contacts even on further thermal treatments [10]. Besides, the direct analytical information of our data provides the basis for a deeper understanding of complementary studies, when performed on identically prepared samples, in particular for techniques like SIMS and energy-filtered TEM.

Finally, it is important to state that our experimental access gives very detailed and very sophisticated information which is a prerequisite for theoretical models and certainly will help to decide whether Pr_2O_3 is a good candidate to replace SiO_2. The quantitative analysis of the oxygen concentration at the interface enables us to conclude on the phase stability of the Pr_2O_3 and Pr-silicate, which enables the formation of homogeneous interfaces. Phase separation tendency is observed on SiC as long as carbon clusters can not be avoided. The phase segregation is also of importance to avoid the formation of an intermediate SiO_2 layer. This is a very important issue in systems when EOT <1 nm are required. Pr_2O_3 and its Pr-silicate show a very good phase stability and no SiO_2 intermediate is found. However, the role of the 4f electronic state needs to be considered, in particular at the Si interface where a direct covalent contribution by the Si–O–Pr bond is to cause electronic interface states to be formed within the gap. If it is possible to suppress such states (by co-doping with other elements, for instance) a combination of oxynitrides and Pr-silicate might be also a realistic approach to form a dielectric layer with reasonable DK values between 10 and 16.

In summary, the possibilities of synchrotron-based interface spectromicroscopy have been elucidated for the high-k dielectric Pr_2O_3 as grown on the common surfaces of Si(001), Si(111), and SiC(0001). We report on nondestructive depth profiling, on the use of X-ray absorption spectroscopy, and of resonant photoelectron spectroscopy to study the electronic properties at the interface.

Acknowledgements

We acknowledge the excellent support of the BESSY staff. The work on Pr_2O_3 has benefited from our collaboration with H.-J. Müssig, ihp, Frankfurt/Oder (Germany and E. Zschech, AMD Saxony, Dresden (Germany). This work is supported by DFG (Schm749/9-1).

References

1. G.D. Wilk. R.M. Wallace, J.M. Anthony, J. Appl. Phys. 89 (2001) 5243; R.M.C. Almeida, I.J.R. Baimvol, Surf. Sci. Rep. 49 (2003) 1.

2. D. Schmeißer, Mater. Sci. Semicond. Process. 6 (2003)59.

3. D. Schmeißer, H.-J. Müssig, J. Physics: Condensed Matter 16 (2004) S153.

4. D. Schmeißer, H.-J. Müssig, MRS Proceedings Volume 811 (2004) D7.10.

5. D. Schmeißer, G. Lupina, H.-J. Müssig, Mater. Sci. and Eng. B118 (2005) 19-22.

6. A. Goryachko, I. Paloumpa, G. Beuckert, Y. Burkov, D. Schmeißer, Physica Status Solidi C 1 (2004) 265.

7.   A. Goryachko, Y. Yeromenko, K. Henkel, J. Wollweber, D. Schmeißer, Physica Status Solidi A 201 (2004) 245.

8.   R. Sohal, K. Henkel, D Schmeißer, H.-J. Müssig, BESSY annual report 2004, pp. 393-394.

9.   D. R. Batchelor, R. Follath, D. Schmeißer, Nucl. Inst. Meth. Phys. Res. A467(2001)470.

10. M. Torche, D. Schmeißer, BESSY annual report 2004, pp. 395-396.

11. P. Hoffmann, A. Goryachko, D. Schmeißer, Mat. Sci. Eng. B 118 (2005) 270-274.

12. Sidorenko, I. Tokarev, S. Minko, M. Stamm, J. Americ. Chem. Soc. 125 (2003) 12211.

13. R.P. Mikalo, D. Schmeißer, Synthetic Metals 127 (2002) 273.

14. D. Schmeißer, F.J. Himpsel, G. Hollinger, Phys. Rev. B27(1983)7813.

15. F.J. Himpsel, F.R. McFeely, A. Taleb-Ibrahimi, J.A. Yarmoff, G. Hollinger, Phys. Rev. B 38 (1988) 6084.

16. F.J. Himpsel, B.S. Meyerson, F.R. McFeely, J.F. Morar, A. Taleb-Ibrahimi, J.A. Yarmoff, Proc. EnricoFermiSchool, M.Campagna, R.Rosei, Ed., North Holland, Amsterdam (1990) 203.

17. J.J. Chambers, A.L.P .Rotondara, A. Shanware, L. Colombo, Appl. Phys. Lett. 80 (2002) 3183.

18. G. Lucovsky, B. Rayner, Y. Zhang, G. Appel, J. Whitten, Appl. Surf. Sci. 216 (2003) 21; Y. Tu, J. Tersoff, Phys .Rev. Lett. 89 (2002) 86102.

19. J. Stöhr: NEXAFS Spectroscopy, Springer Verlag, Heidelberg (1996).

20. M.L. Green, E.P. Gusev, R. Degraeve, E.L. Garfunkel: J. Appl. Phys. 90 (2001) 2057.

21. E.P. Gusev, H.C. Lu, EL. Garfunkel, T. Gustafsson, M.L. Green: IBM J. Res.& Developm.43 (1999) 265-286.

22. H.C. Lu, E.P. Gusev, T. Gustafsson, E. Garfunkel, M.L. Green, D. Brasen, L.C. Feldman, Appl. Phys. Lett. 69 (1996) 2713-2715.

23. M.P. Seah, W.A. Dench: Surface and Interface Analysis 1 (1979) 2-11.

24. S. Tanuma, C. J. Powell, D.R. Penn: Surface and Interface Analysis 17 (1991) 911.

25. H. Yamamoto, Y. Baba, T.A. Sasaki: Surface Science 349 (1996) L133-L137.

26. P. Hoffmann, R.P. Mikalo, D. Schmeißer: J .Non-Crystal.Solids 303 (2002) 6-11.

27. P. Hoffmann, D. Schmeißer, R.B. Beck, M. Cuch, M. Giedz, A. Jakubowski: J. AlloysComp., in press.

28. D. Schmeißer, H.-J. Müssig, J. Dabrowski: Appl. Phys. Lett. 85/1 (2004) 88-90.

Materials Characterization by Ellipsometry

V. G. Polovinkin[a], M. R. Baklanov[b]

[a]Institute of Semiconductor Physics, Novosibirsk / Russia

[b]IMEC, Leuven / Belgium

Introduction

Ellipsometry is an optical technique that uses polarised light to probe the dielectric properties of a sample (optical system). The common application of ellipsometry is the analysis of thin films. Through the analysis of the state of polarisation of the light that is reflected from the sample, ellipsometry yields information on the layers that are thinner than the wavelength of the light itself, down to a single atomic layer or less. Depending on what is already known about the sample, the technique can probe a range of properties including layer thickness, morphology, and chemical composition. These features make ellipsometry extremely popular and important for microelectronics and other technologies that work with thin functional layers [1–4].

The traditional application of ellipsometry in microelectronics is measurement of thickness and refractive indices of dielectric layers used for gate and interconnect applications. However, traditional dielectric layers that have been playing a key role in microelectronic technology during the past 40 years, no longer meet the strict requirements of modern microelectronics. As transistor geometries scale to the point where the traditional SiO_2 gate dielectric becomes a few atomic layers thick, tunnelling of carriers through the dielectric give rise to leakage current and as such an increase in power dissipation. After a few years of intensive effort, new materials known as high-k dielectrics have been proposed as the gate dielectric. Similar problems exist in interconnect technology. The device physics is not the only limiting factor for continued performance improvement. The challenge will be to carry electric power and to distribute the clock signals, which control the timing and synchronize the operation. The propagation velocity of electromagnetic waves becomes increasingly important due to their unyielding constraints on interconnect delay. The introduction of Cu and low-k dielectrics has improved the situation as compared to the conventional Al/SiO_2 technology by reduction of both the resistivity and the capacitance between wires.

These tendencies have changed materials properties of both gate and interconnect dielectrics. On the one hand, several new materials such as HfO_2, ZrO_2, Al_2O_3 *etc.* are investigated for introduction as high-k gate dielectrics. On the other hand, the need for dielectrics with a reduced dielectric constant requires implementation of hydrophobic porous materials. In both cases, the basic material science of the materials is quite different from that of traditional SiO_2. Metrology of new types of thin dielectric layers requires the development of new instrumentations or appropriate up-dates of existing methods. Although ellipsometry has been recognized as an efficient characterization method for evaluation of thin films, the new types of dielectric layers bring new challenges for critical evaluation including:

- Anisotropic dielectric properties (both high-k and low-k),
- Porous structure (ultra-low-k dielectric layers),
- Difference in the chemical activity of the top layer and bulk materials,
- Sensitive to UV-light (low-k dielectrics and photoresists).

Methodologies based on ellipsometry have already been applied for evaluation of these properties. The main purpose of the paper is to give an overview of new applications of ellipsometry.

Fundamentals of Ellipsometry

The term ellipsometry was introduced by Rothen [5] as a definition of an optical technique allowing characterization of thin films by analysis of polarization state (ellipse of the polarization) of the reflected light. Ellipsometry analyses change of polarization of electromagnetic waves (EW) interacting with an optical system (OS). Theoretical models and experimental data are used to determine the optical system characteristics. Ellipsometry has a number of unique advantages that define its application in various areas of science and technology. This method is non-invasive and non-destructive making it suitable for in-line metrology of technological processes in microelectronics. Ellipsometry is fast and has extremely high sensitivity making it possible to analyze physicochemical processes at sub-monolayer level. This method does not need special conditions, and is applicable over a wide range of pressure, temperature and in various environments. The possibility to measure in various environments allowed the development of new methodologies for the evaluation of porosity and pore size distribution in porous low-k dielectrics, plasma cleaning, atomic layer deposition *etc.*

Basic Definitions.
Monochromatic flat EW is described by two orthogonal components of the electric field vector. In ellipsometry, E_p and E_s are the electric field vectors parallel (p) and perpendicular (s) to the plane of the light incidence onto the OS surface (Figure 1). Interaction of these waves with OS is described within the framework of a formalism of Jones matrices (see E 1) – complex matrices of the size 2×2, describing linear ratio between amplitudes of components on input in and output from OS [1]:

$$\begin{pmatrix} E_{op} \\ E_{os} \end{pmatrix} = \begin{pmatrix} R_{pp} & R_{ps} \\ R_{sp} & R_{ss} \end{pmatrix} \begin{pmatrix} E_{ip} \\ E_{is} \end{pmatrix}, \quad \mathbf{E_o} = \mathbf{R} \cdot \mathbf{E_i}.$$

$$(1)$$

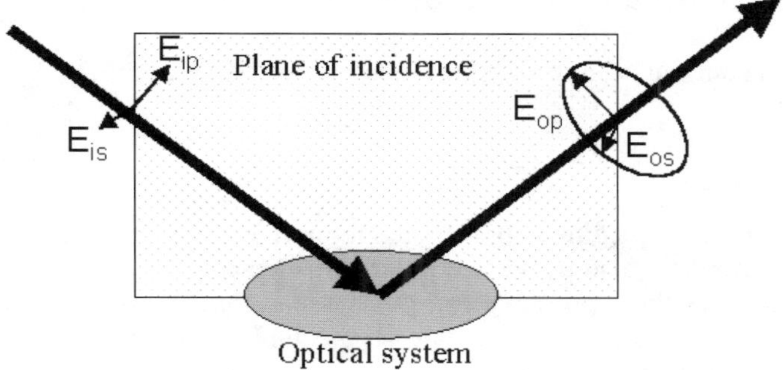

Figure 1. The orientation of incident (input) and reflected (output) beams, plane of incidence and components of polarization.

In the case of the light reflection from a flat isotropic, OS non-diagonal elements of matrix **R** are equal to zero and change of polarization are described by two complex factors of reflection $R_p=R_{pp}$ and $R_s=R_s$ (E 2):

$$E_{op} = R_p E_{ip}, \quad E_{os} = R_s E_{is} \tag{2}$$

The situation is more complex if input and/or output EW is not strictly monochromatic (quasi monochromatic). In this case, EW is described with the help of four real parameters – Stokes vector (E 3):

$$S = \begin{pmatrix} S_0 & S_1 & S_2 & S_3 \end{pmatrix} \tag{3}$$

Components of the Stokes vector can be expressed through the intensities of EW of different polarization (E 4):

$$S_0 = I_o = I_0 + I_{\pi/2} = I_{+\pi/4} + I_{-\pi/4} = I_{lc} + I_{rc},$$
$$S_1 = I_0 + I_{\pi/2}, \quad S_2 = I_{+\pi/4} + I_{-\pi/4}, \quad S_3 = I_{lc} + I_{rc} \tag{4}$$

where I_o is the total intensity of EW, I_{lc}, I_{rc} - left and right circularly polarized and others are intensities of EW polarised at the corresponding angles with respect to the plane of incidence. For totally non-polarized EW, $S_0 = I_o$, $S_1 = S_2 = S_3 = 0$. For the completely polarized EW, $S_0 = S_1 + S_2 + S_3$. Interaction of quasi monochromatic EW with OS is described by formalism of Mueller's matrices (see E 5) - the real matrix of dimension 4×4 describing a ratio between vectors of Stokes of input and output EW.

$$\mathbf{S_o} = \mathbf{M} \cdot \mathbf{S_i} \tag{5}$$

Methods of Measurement and Applications

The general configuration of an ellipsometer is very simple (E 6):

$$L \rightarrow [P] \rightarrow [S] \rightarrow [A] \rightarrow D \tag{6}$$

The ellipsometer begins with a light source (L) and ends with a light detector (D). An ellipsometer also needs two polarising elements. The first is polarizer (P), which establishes the input polarization state and the second one is analyzer (A) that measures output polarization state. The sample (optical system) to be measured (S) is located between P and A.

Zero methods of ellipsometry (*zero-ellipsometry*) are based on finding parameters of a polarizer and the analyzer at which the intensity of the light registered by the photodetector D is equal to zero or is minimal (condition of extinction). Zero-ellipsometry is used, as a rule, to study OS described by a Johnson's matrix (non-depolarizing) systems. High accuracy is an advantage of these methods. However, the presence of moving elements complicates its automation. In the classical zero-ellipsometer, the polarizer represents a combination of a linear polarizing element (a simple polarizer) and a phase-shifting device (compensator), and the analyzer is a simple polarizing element (Figure 2).

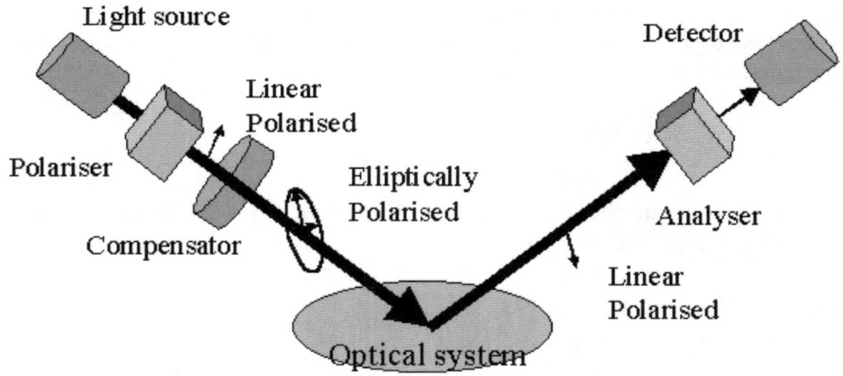

Figure 2. The classical scheme of zero-ellipsometry (PCSA).

The compensator converts linearly polarized light to elliptically polarized light. An ideal compensator is an optical Retarder that has a retardation of exactly 90° (1/4 wave). A rotatable compensator combined with a rotatable polarizer can convert unpolarized light into any elliptical polarisation. Appropriate choice of the rotation angles of P and C creates such polarization that provides linear polarization of the light reflected from the sample (optical system). This light can be extinguished by orientation of the analyzer. The condition of extinction is described by Equation 7:

$$R_p \cos(A) - R_s \sin(A) \exp\left(j \cdot 2(P + \pi/4) \right) = 0 \tag{7}$$

here A and P are the angles of rotation of the analyzer and a polarizer, respectively. This equation can be written as:

$$\rho_S \equiv R_p / R_S = \tan(\Psi) \exp(j\Delta), \tag{8}$$

where Ψ and Δ are so-called ellipsometric angles that describe the module and a phase of relative reflection coefficient ρ_s and simply connected with angles A and P. Equation 8 is known as the fundamental equation of ellipsometry and correlates the measured quantities Ψ and Δ with the sample characteristics, such as film thickness and optical characteristics. Examples of such correlations are presented in Figure 3 and will be discussed later.

Photometric ellipsometry is based on measurement of the light intensity at the detector versus P and/or A parameters. This dependency contains information related to OS characteristics. Different realizations of this method allow determination of all elements of the Jones matrix. In this case, the method is called *generalized ellipsometry* [6] in contrast to *conventional ellipsometry,* where only the ratio of diagonal elements is determined. Photometric ellipsometry can also be applied to depolarizing OS. With a sufficient number of variable parameters of the polarizer and/or the analyzer, it is possible to determine all 16 elements of Mueller's matrix of OS. This method is termed *Mueller's matrix ellipsometry* or *Mueller's ellipsometry* [7]. Photometric methods have had limited applications because of their strict technical requirements for the detectors and the measurement system. However, progress in electronics and optoelectronics is increasing application of these systems.

It is necessary to mention that ellipsometry does not directly measure the thickness and refractive index of films. To obtain these values, it is necessary to solve equations that express measured quantities by model parameters, *i.e.* to solve an inverse ellipsometric problem. This problem may be complicated and is not always unambiguous [8].

Ellipsometry in the VIS, UV and IR Spectral Range

Ellipsometry in the visible spectral range (VIS) is used mainly for the most traditional applications (thickness and refractive index measurements of common films, such as SiO_2, Si_3N_4 *etc.*). Spectroscopic ellipsometry is also efficient for the characterization of new materials. *In situ* ellipsometry allows the study of the kinetics of surface reaction, for instance, during film deposition, etching, curing *etc.*

Figure 3 shows typical Delta-Psi trajectories for films on top of a Si crystal. The values of optical characteristics are typical for materials used in microelectronics. One Delta/Psi point gives two numerical values. With this, one can calculate two unknowns for the film such as thickness and refractive index for a transparent film. It is also obvious that there are regions where Delta/Psi provides much better information than others. Near the period point (or the film free point: $\Delta \approx 179$ and $\Psi \approx 10°$ for the silicon substrate) the curves are not particularly well separated. The Delta-Psi values in this region are not so efficient in determining the index of refraction of the top film. It is also important to note that additional uncertainty can be related to the ellipsometric period equal to (Equation 9):

$$d_0 = \lambda \big/ 2\sqrt{n^2 - \sin^2 \varphi} \tag{9}$$

where λ and φ are wavelength and incident angle, respectively. Therefore, films with thicknesses $d_m = d + m*d_0$ (m = 0, ±1, ±2, ...) have the same values of Delta and Psi. This uncertainty is excluded using variable wavelength and variable angle ellipsometry.

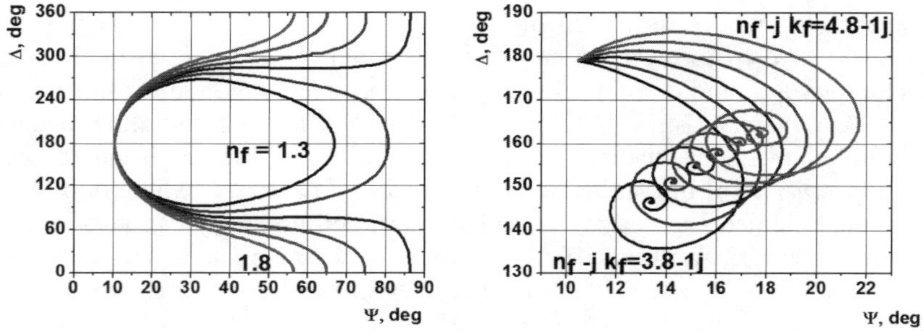

Figure 3. Calculated Delta-Psi trajectories for transparent films having indices of refraction n_f of 1.3-1.8 on top of single crystal silicon (a) and for a non-transparent film on the same substrate with extinction coefficient k_f equal to 1.0 and various indices of refraction ranging from 3.8 to 4.8.

Delta/Psi trajectory for non-transparent films is not periodic. Initially (at low thickness), the curves are well separated. As the thickness increases, the points come closer together and the trajectory begins to spiral into the target value. The Delta-Psi value at the target point correspond to the optical characteristics of the bulk materials having the same composition as the film. One can see that determination of three unknown parameters (*d*, *n* and *k*) is not straightforward because certain points may correspond to films with different characteristics. In this case, the use of variable wavelength and variable incident angle can be useful but it is better to decrease the number of unknown parameters by, for instance, independent measurement of the extinction coefficient.

Ellipsometry in vacuum UV (<190 nm) allows the analysis of materials for the advanced lithography (photoresists, antireflective coatings) at the latest exposure wavelengths (157 nm and 193 nm). The short wavelength also increases the sensitivity of ellipsometric measurements of ultra thin films (<10 nm). Many materials are strongly absorbing in the UV region, which facilitates surface-sensitive measurements such as surface roughness, native oxide coverage, material composition (including surface contaminations) and structural properties.

The near-infrared spectral range (NIR) is used mainly to measure the thickness of single films and layer stacks. Many materials are transparent in this region making this range convenient for measurement. So the depth of EW penetration into the structure (and hence the sensitivity of the ellipsometry) can be extended up to 15–20 µm [9].

These applications of ellipsometry are quite traditional and detailed descriptions of the measurement and calculation procedure are well described in many monographs, handbooks, and papers [1-4].

The numerous bands in the infrared spectrum constitute a highly individual fingerprint of a material, and therefore a sample with a thin surface layer will produce very different contrasts within the infrared spectrum [10]. Moreover, IR ellipsometry can probe the vibrational properties of thin layers and surfaces [11]. Therefore, IR ellipsometry can simultaneously probe film morphology and microstructure based on the vibrational absorption characteristics. A very interesting feature is the sensitivity to free carrier concentration in a material [12].

Progress in the measurement sensitivity in the IR range has been obtained from the systematic combination of spectroscopic ellipsometry and Fourier transform spectroscopy [13]. As a consequence, infrared ellipsometry can be reliably employed for detailed characterization of different layers such as coatings, implantations or surface films in the low nanometer range, *i.e.*, thinner than 1/1000th of the wavelength. In particular, and in contrast to many surface analytical methods, the geometric thickness can be derived. Generally, IR ellipsometry is sensitive to chemical composition and to local structural order of a thin film material [14]. This may be important for characterisation of low-k materials damaged during the plasma etching.

Ellipsometry of Anisotropic Layers

Optical anisotropy of a material is the dependence of EW velocity on the propagation direction. Most crystalline solids are optically anisotropic. A typical result of anisotropy is the phenomenon of birefringence – the existence of two refracted beams. Only some selected directions (axis) can be received by one beam. Crystals with one and two such axes are known as uniaxial and biaxial crystals.

Theoretically, optical anisotropy is described by a dielectric function tensor ε_{ij}, $i, j = x, y, z$. The classification of crystals depends on the values of diagonal components of this tensor in the principal axis presentation $\varepsilon_{ij}=\varepsilon_i\delta_{ij}$:

$\varepsilon_1=\varepsilon_2=\varepsilon_3=\varepsilon$ — isotropic crystal,

$\varepsilon_1=\varepsilon_2=\varepsilon_o$, $\varepsilon_3=\varepsilon_e$ — uniaxial crystal,

$\varepsilon_1\neq\varepsilon_2\neq\varepsilon_3$ — biaxial crystal.

The presence of anisotropic layers in OS means that, in general, non-diagonal elements of the Jones matrix are not equal to zero. Hence, they must use general ellipsometry to characterize OS. The theoretical description of EW propagation in the OS with anisotropic layers uses formalism 4×4 matrixes. This formalism has been proposed by Berreman [15] and has, for example, been applied in [16].

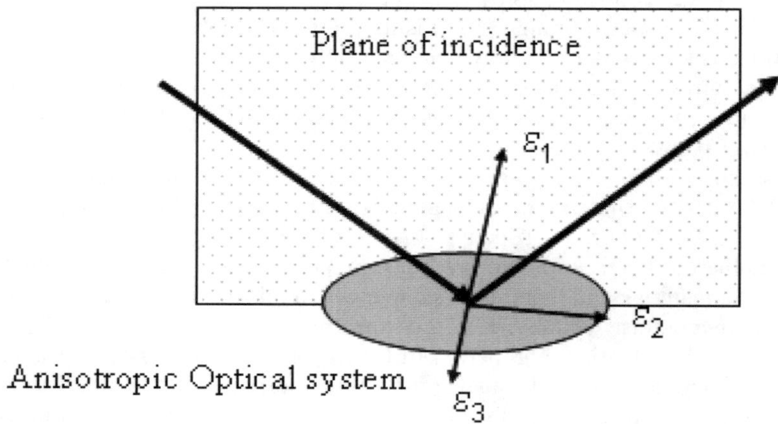

Figure 4. Special (non-permuting) orientation of anisotropic optical system. One principal axis (ε_3) is perpendicular to plane of incidence.

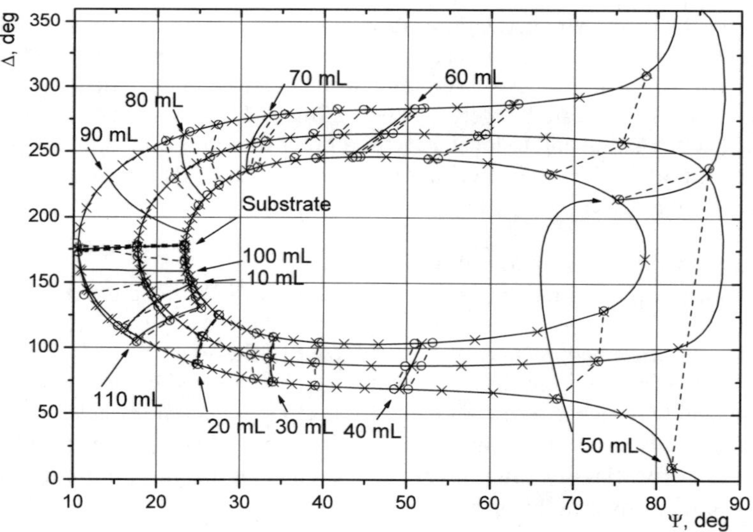

Figure 5. Experimental and calculated ellipsometric angles of the LB films — ○ — experimental data, −×− calculated data. The points corresponding to the number of layers divisible by 2 are marked with the symbol ×. Dotted lines - × - are drawn through the values corresponding to the number of layers divisible by 10. Figures near these lines designate the number of LB layers of the film.

Thus, measurements and calculations of anisotropic systems with arbitrary orientation are quite difficult. However, nondiagonal elements vanish for the same particular orientation of anisotropic layers and it is possible to use conventional ellipsometry. A sufficient condition for that occurs when the orientation of one of the principal axes of crystals is perpendicular to the plane of incidence, as shown in Figure 4. It is possible to do model calculations using a formalism of matrixes 2×2 in this case, but formulas for the latter should take anisotropy into account.

Typical examples of anisotropic optical systems are Langmuir–Blodgett (LB) films prepared by the transfer of monolayers of amphiphilic molecules from a water surface onto a solid substrate [17]. LB films have well-oriented structures formed by nonsymmetric molecules showing optical characteristics of an anisotropic crystal. Taking the structure of molecules into account, we can assume that the LB film is a biaxial crystal and its basic directions are oriented along the transfer direction (x axis) and perpendicular to the interface (z axis). The y axis is located in the plane of the interface; it is orthogonal to the transfer direction. If the plane of light incidence is oriented along the symmetric direction (i.e., in the x, z or y, z planes), the non-diagonal elements of the matrix of reflection, coefficients R_{sp} and R_{ps}, must be equal to zero and it is possible to use the conventional ellipsometry. Figure 5 shows results obtained for films with different thickness of Pb salt of acetylenic acid [18]. Calculations were made with film parameters: $n_x = 1.564$, $n_y = 1.536$, $n_z = 1.586$, $d_{ml} = 2.548$ nm.

For transparent isotropic films it is know that ellipsometric angles are a periodical function of film thickness. In the case of an anisotropic layer periods are different for s and p polarizations and ellipsomeric angles, as we can see in figure 5, and are not periodical functions.

Ellipsometric Porosimetry

Integration of porous low-k dielectrics into ULSI circuits poses a number of challenges, as the materials must meet strict requirements in terms of properties. Pore size and its distribution (PSD) are the most crucial properties of low-k materials. The maximum pore size must be sufficiently smaller than the minimum feature size of device components. All critical properties such as mechanical strength, electrical and thermal stability, and thermal conductivity have a close relationship with the porous structure of the film. Therefore, characterization of pore size and PSD is extremely important because it provides an understanding of the backbone structure. Such information is essential in efforts to improve the properties of porous low-k thin films [19].

Porosimetry is a new application of ellipsometry developed for evaluation of porosity and pore size distribution in thin films. The porosity of the films is calculated from the volume polarizabilities of materials with empty pores and pores filled by toluene. The volume polarizability (B) of two-component systems depends on the refractive indices of each component (indices 1 and 2) and their fractions (Equation 10):

$$B_2 = \frac{n_{re}^2 - 1}{n_{re}^2 + 2} = V \frac{n_1^2 - 1}{n_1^2 + 2} + (1-V)\frac{n_2^2 - 1}{n_2^2 + 2} \tag{10}$$

where n_{re} is the measured refractive index, n_1 and n_2 are refractive indices of the material inside the pores and of the film skeleton, respectively. V is the pore volume.

When the pores are filled by a liquid adsorptive (Equation 11):

$$B_3 = \frac{n_{rl}^2 - 1}{n_{rl}^2 + 2} = V \frac{n_{ads}^2 - 1}{n_{ads}^2 + 2} + (1-V)\frac{n_2^2 - 1}{n_2^2 + 2} \tag{11}$$

where n_{ads} is refractive index of the liquid adsorptive. If B2 is subtracted from B3, (4)

$$V = \left(\frac{n_{rl}^2 - 1}{n_{rl}^2 + 2} - \frac{n_{re}^2 - 1}{n_{re}^2 + 2} \right) \Big/ \left(\frac{n_{ads}^2 - 1}{n_{ads}^2 + 2} \right) \tag{12}$$

This equation shows that the porosity calculation needs the refractive indices of the film with empty pores n_{re}, the pores filled by adsorptive, n_{rl}, and the refractive index of the liquid adsorptive, n_{ads}. The refractive index of the film skeleton is not needed for the porosity calculation. The porosity calculated by equation (12) is a "total" porosity. This porosity includes both the artificial porosity created by porogen and the constitutive porosity (free volume) of the film matrix.

Calculation of the pore size distribution uses the phenomenon of progressive emptying of a porous system initially filled at $P=P_0$. The calculations in mesoporous films are based on analysis of hysteresis loop that appear during the adsorption and desorption [18]. The hysteresis loops appear because the effective radius of curvature of condensed liquid meniscus is different during the adsorption and desorption. The adsorptive vapor condenses in pores at the vapor pressure (P) less than the equilibrium pressure of a flat liquid surface (P_0). Dependence of the relative pressure (P/P_o) on the meniscus curvature is described by the Kelvin equation (Equation 13):

$$\ln\left(\frac{P}{P_0}\right) = -\frac{f \cdot \gamma \cdot V_L \cos\theta}{r_K \cdot RT} \tag{13}$$

where γ and V_L are surface tension and molar volume of the liquid adsorptive, respectively. θ is the wetting angle of the adsorptive, $f=1$ for slit-shaped pores and $f=2$ for cylindrical pores. If the radius of a cylindrical pore is r_p, then $r_p = r_k + t$, where t is the thickness of the layer adsorbed on the pore walls. Values of t are obtained from the adsorption of the same adsorptive on a nonporous sample having a chemically similar surface and are defined by the BET equation [20, 21].

The initial experimental data for the porosity and PSD calculation are the ellipsometric characteristics Δ and Ψ. Particular software allows calculation of the change of n and d during the adsorption and desorption, PSD and specific surface area. The change in the adsorbate volume is calculated using equation (12). The dependence of the adsorbate volume on the relative pressure P/Po is used to calculate the PSD. The specific surface area of each small group of pores δA_i are calculated from the corresponding pore volume and pore radius as $\delta A_i = \delta V_i / r_I$. By summing the values of δA_i over the whole pore system a value of the cumulative surface area is obtained [21].

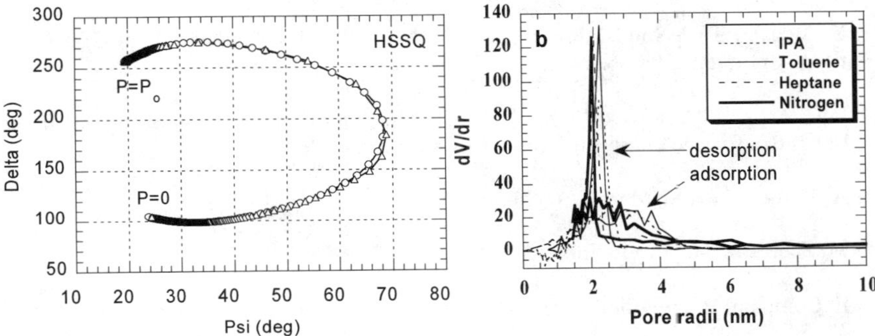

Figure 6. Typical change in ellipsometric characteristics during toluene adsorption in a mesoporous low-k film (a) and pore radius distribution (b) calculated from ellipsometric data. Pore radius distribution demonstrates good agreement between the data obtained with different adsorptives and nitrogen porosimetry.

In micropores with widths of the order of a few molecular diameters, the Kelvin equation no longer remains valid. Not only would the values of the surface tension and the molar volume deviate from those of the bulk liquid adsorptive, but also the concept of a meniscus would eventually become meaningless. To analyze a microporous film, a method based on a theory developed by Dubinin and Radushkevitch is used [21]. This theory takes into account the change in adsorption potential when the distance between the pore walls is comparable with the size of the adsorptive molecules. This analysis can provide the essential i parameters on the micropore structure such as the micropore volume, the average pore width and the isosteric heat of adsorption. Figure 6 shows typical experimental data obtained for a mesoporous film. More detailed information related to evaluation of different types of porous films can be found elsewhere [19–23].

In addition to porosity and PSD evaluation, ellipsometric porosimetry has demonstrated its efficiency for some additional applications. Results of measurements can be used for evaluation of mechanical properties of thin films [22] and the integrity of diffusion barriers deposited on top of low-k dielectrics [23]. Recently, the application of *in situ* ellipsometry for monitoring Cu surface cleaning by hydrogen plasma and alcohol vapors have also been demonstrated [24, 25]. Evaluation of the porosity of low-k films and monitoring of the Cu surface cleaning are typical examples demonstrating importance of ellipsometry for advanced microelectronic technology.

References

1. R.M.A. Azzam, N.M. Bashara, Ellipsometry and Polarized Light, North-Holland, New York, 1977.

2. A. Röseler, Thin Solid Films, Vol. 234 (1993) Proceedings of the 1st International Conference on Spectroscopic Ellipsometry.

3. K. Vedam, Thin Solid Films, Vol. 313-314 (1998), Proceedings of the 2nd International Conference on Spectroscopic Ellipsometry.

4. D.E. Aspnes, Thin Solid Films, Vol. 455-456 (2004), Proceedings of the 3rd International Conference on Spectroscopic Ellipsometry.

5. A. Rothen, Rev. Sci. Instr., 16, 26 (1945).

6. M. Schubert, Thin Solid Films, 313-314, 33–39, (1998).

7. E. Compain, B. Drevillon, J. Hue, J. Y. Parey, J. E. Bourse, Thin Solid Films, 313-314, 47–52, (1998).

8. G.E. Jellison Jr., Thin Solid Films, 313-314, 33–39, (1998).

9. B. Drevillon, Thin Solid Films, 313-314, 625-630, (1998).

10. A. Röseler, Infrared Spectroscopic Ellipsometry, Akademie-Verlag, Berlin, 1990.

11. A. Kasic, M. Schubert, S. Einfeldt, D. Hommel, Vibrational Spectroscopy, 29, 121-124, (2002).

12. T.E. Tiwald, D.W. Thompson, J. A. Woollam, W. Paulson, R. Hance, Thin Solid Films, 313-314, 661-666, (1998).

13. E.H. Korte, A. Röseler, Analyst, 123, 647–651, (1998).

14. E.H. Korte, K. Hinrichs, A. Röseler, Spectrochimica Acta Part B 57, 1625–1634, (2002).

15. D. W. Berreman, J. Opt. Soc. Am., 62, 502, (1972).

16. D. J. De Smet, J. Opt. Soc. Am. 64, 631-638, (1974).

17. G.G. Roberts, Langmuir-Blodgett Films, (Plenum, New York, 1990).

18. I.A. Badmaeva, L.A. Nenasheva, V.G. Polovinkin, S.M. Repinsky, L.L. Sveshnikova, Thin Solid Films 455-456 557-562, (2004).

19. K. Maex, M.R. Baklanov, D. Shamiryan, F. Iacopi, S.H. Brongersma and Z.S. Yanovitskaya. J. Appl. Phys. 93, 8793 (2003).

20. M.R. Baklanov, K.P. Mogilnikov, V.G. Polovinkin and F.N. Dultsev, J. Vac. Sci. Technol. B18, 1385 (2000).

21. M.R. Baklanov and K.P. Mogilnikov. Microelectronic Eng., 64, 1-4, 335 (2002).

22. K.P. Mogilnikov and M.R. Baklanov. Electrochem. Solid State Lett., 5, F29 (2002).

23. D. Shamiryan, M.R. Baklanov, K. Maex. J.Vac. Sci. Technol. B21(1), 220 (2003).

24. M.R. Baklanov, D.G. Shamiryan, Zs. Tokei, G.P. Beyer, T. Conard, S. Vanhaelemeersch et al., J. Vac. Sci. Technol. B, 19(4), 1201-1211, (2001).

25. A. Satta, D. Shamiryan, M.R. Baklanov, C.M. Whelan, Q.T. Le, G.P. Beyer et al., J. Electrochem. Soc. 150 (5) G300-G306 (2003).

Thermal Desorption Spectrometry as a Method of Analysis for Advanced Interconnect Materials

L. Carbonell, A. M. Hoyas, C. M. Whelan, G. Vereecke

IMEC, Leuven / Belgium

Introduction

In semiconductor manufacturing, the increasing demand for smaller, faster, and more powerful devices requires the introduction of novel materials for the fabrication of interconnects. In particular, the thermal stability of such materials and their chemical interaction with various processing conditions must be well understood to achieve a successful integration scheme. Thermal desorption spectrometry (TDS) is a powerful technique that provides quantitative and qualitative information on composition, structure, stability with regard to processing, storage and post-deposition treatments such as annealing and plasma etching [1–3].

Carbon-based aromatic polymers are promising as low–k interlayer dielectrics (ILD) [4]. However, such materials have weak mechanical properties, which can give rise to delamination during the chemical mechanical polishing (CMP) step. Also, due to the porous structure of the material, chemical species such as cleaning agents and metal diffusion barrier precursors may penetrate into the material during processing [5]. One possible solution is the creation of a dense surface sealing layer to improve the adhesion properties of the film and prevent the penetration of contaminants and moisture into the film [5].

Self-assembled monolayers (SAMs) are nanometer-scale organic films that have potentially various applications in very advanced BEoL integration schemes. Thiolate SAMs can be used as effective corrosion inhibitors of copper surfaces for applications such as wafer level packaging and wire bonding [6]. Silane SAMs could be used either to promote adhesion of diffusion barriers onto ILD materials [7] or used as diffusion barriers themselves [8].

In this short review the use of TDS for the characterization of new semiconductor materials for advanced BEoL technologies will be illustrated. Important integration issues such as the hydrophobic/hydrophilic character and surface sealing of low-k ILD materials modified by plasma treatments, the thermal stability of thiolate SAMs on Cu surfaces used as corrosion inhibitors, and the chemical labeling of silane SAMs for process monitoring will be discussed in this paper.

Experimental Set-up

Thermal desorption spectrometry (TDS) measurements were carried out at atmospheric pressure on full 200-mm wafers in a single-wafer rapid thermal processing

(RTP, AST SHS 2800) system, connected to an atmospheric pressure ionization mass spectrometer (APIMS, VG Trace+) (see Figure 1) [9]. The RTP system comprises a single-wafer quartz tube in which the wafers are directly loaded from the cleanroom. The chamber was purged typically for 4–6 min in order to reduce the background levels of impurities, and the TDS heating program was started [10]. The wafer was heated with lamps and the temperature was measured and controlled by a thermocouple molded in one of the pins supporting the wafer. The temperature was ramped up from room temperature to the temperature required with a constant heating rate. Semiconductor-grade nitrogen gas, purified with an active-type purifier (SAES MonoTorr), was used as the carrier gas. The ambient gas inside the chamber was sampled, diluted, and introduced into the APIMS source for analysis in the positive-ion mode, as described elsewhere [9]. The APIMS typically scans over a mass range extending from m/e = 11 to 100, with an integration time of 0.1 s per scanned mass. APIMS is the most sensitive gas analysis technique for bulk gases used in the semiconductor industry (trace gaseous impurities detection on the ppt scale) [9].

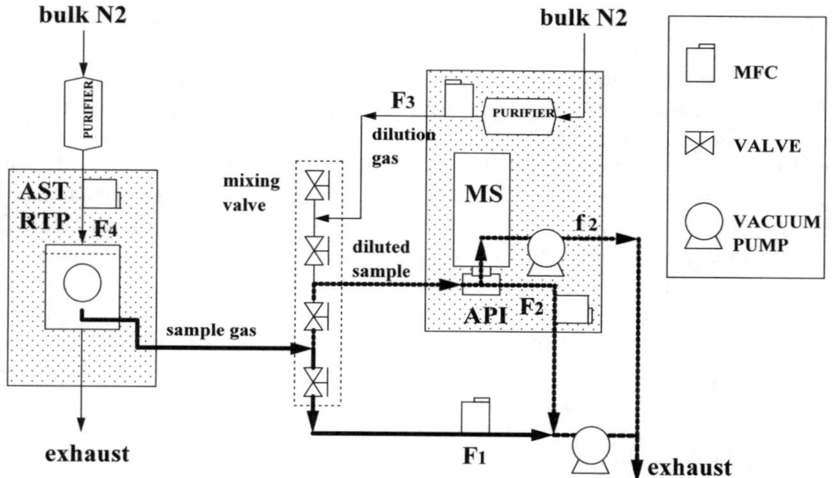

Figure 1. Scheme of the TDS-APIMS set-up.

Impurities are ionized predominantly by charge transfer from the primary ions of the carrier gas molecules created at atmospheric pressure in a corona discharge. Whereas ionization mechanisms only produce mono-charged ions, cluster ions are readily formed. The APIMS response was calibrated using a non-linear method for H_2O, O_2, CO, and CH_4, allowing for a quantitative analysis of these species in the 0.1–1000 ppb range [11]. Before carrying out TDS measurements, the background of residual impurities in the RTP chamber needs to be measured using bare Si_3N_4/Si wafers. This is especially important after the tool has been inactive for a few hours (time of the purge of the quartz line going to the APIMS before starting a run of experiments). The most common impurities include O_2 giving a peak at m/e = 32 (O_2^+), H_2O giving main peaks at m/e = 18 (H_2O^+), 19 (H_3O^+) and cluster ion peaks at 36 ($H_2O.H_2O^+$) and 37 ($H_3O.H_2O^+$), NH_3 with a main peak at m/e = 17 (NH_3^+), and other peaks at m/e = 16 (NH_2^+) and m/e = 35 ($NH_3.NH_4^+$), and CO_2, giving a peak at m/e = 44 (CO_2^+).

Modification of the Hydrophobic /Hydrophilic Character of Plasma Treated low-*k* Dielectric Layers

Due to the low dielectric constant (k value) afforded by their intrinsic porosity, low-*k* constants materials are currently under evaluation for use as interlayer dielectrics in microelectronic devices [12]. The high surface areas of these porous materials make them susceptible to absorption of gaseous and/or liquid species during processing, which increases the dielectric constant. Polymers are inherently hydrophobic due to their large concentration of polar groups. The exposure of the surface/bulk of such a material to certain processing conditions, in particular ashing and cleaning, potentially leads to the creation of a hydrophilic surface/layer susceptible to water absorption and an increased dielectric constant [13]. As a result, the desorption spectra change allowing for a qualitative appreciation of the hydrophilic character of the surface deduced from the amount of desorbed water molecules [14]. The number of H_2O molecules adsorbed or condensed in a layer can be calculated from the area under the main desorption peak for water (m/e=18) after subtraction of the tool background.

TDS proved to be a suitable technique in determining the hydrophilic/hydrophobic character of aromatic-based low-*k* materials exposed to various plasma treatments (N_2/O_2, NH_3, O_2, and Ar) [15]. The H_2O spectra (m/e=18) of the untreated and plasma-treated polymer films are presented in Figure 2. In the low-temperature desorption range (T<300°C) the area of the peaks are much larger compared to that of the background spectrum. As the latter corresponds to true surface adsorption on top of a dense Si_3N_4 layer, the desorption peaks observed for the plasma-treated low-*k* layers actually correspond to water physically adsorbed/condensed in the modified surface layer. The intensity of the desorption peak can thus relate both to the density and thickness of the modified surface layer. A high-temperature desorption range, for T>300°C and below the degradation temperature of the material, corresponds to water released from the bulk of the porous polymer [15].

In the case of the Ar-treated sample, a low-temperature desorption peak comparable to the nontreated film is observed. Some additional water desorption was observed in the high temperature range between 300 and 400°C. As from the reference sample no moisture uptake is expected from the bulk of the material, the desorption of water observed in the temperature range is likely originated from a thick modified layer. The hydrophilic character of this Ar-treated layer determined by contact angle measurements [15] could be due to water molecules trapped in the material under a very dense modified surface layer, *i.e.* very good sealing of the surface.

The desorption spectra of the N_2/O_2, O_2, and NH_3-treated surfaces show only a peak in the low-temperature desorption region (~130°C) and then the modified layer is expected to be comparatively thinner.

The trend deduced from these TDS measurements, *i.e.* hydrophilic character of the plasma-treated samples, is in agreement with lower water contact angle values of 26, 23, 27, and 33° observed for the surfaces created by N_2/O_2, O_2, Ar, and NH_3 plasma treatments, respectively, as compared with 86° measured for an untreated reference sample [15].

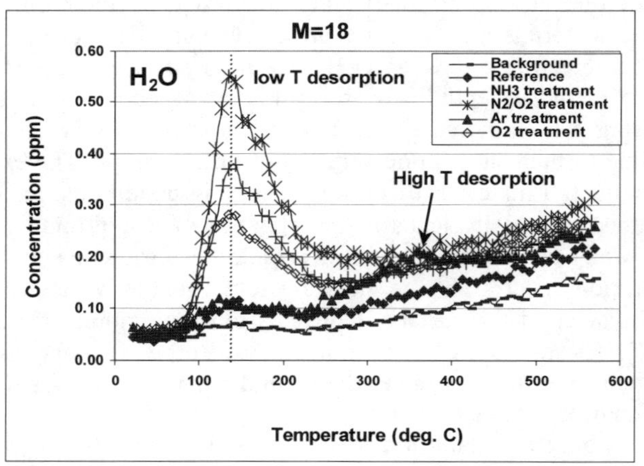

Figure 2. Correlation of the TDS spectra of water at m/e=18 for untreated and plasma-treated aromatic polymer layers.

Surface Sealing of Low-*k* Dielectrics by Plasma Treatments

TDS was used to characterize the surface sealing of an aromatic polymer layer subjected to various plasma treatments, as described in the previous section. Initially the porous matrix of the aromatic hydrocarbon polymer layer allows the penetration of species during processing. The exposure of the surface to a plasma treatment leads to the formation of a densified sealing layer that can be characterized by the changes of RI with respect to the untreated layer. The RI changes when the density and the polarizability of the chemical bonds change. Thus, an impermeable layer, *e.g.* a sealed layer, is characterized by a high density and, assuming that the contribution of the bond polarizability is low, its RI is a measurement of the sealing ability of a layer. Additional information on the sealing character of the modified surface layer is provided by TDS measurements through the desorption of benzene groups at m/e=78 starting at around 400°C, which is the fingerprint of the thermal decomposition of the polymer (Figure 3) [15].

A large increase in RI represents a sealed sample [15]. No degradation at 400°C for the Ar-treated sample combined with a high RI value, suggests strong structural modifications over the total thickness of the layer. If the RI increase is small, as in the case of O_2 plasma treatment, the sample surface is not sealed. Exposure to the O_2 plasma creates a thin densified surface layer, which partially seals the surface. The low-temperature desorption behavior of such a sample indicates that the structure of the surface is quite close to that of the untreated reference layer. No bulk modification was detected as observed in the TDS data and only a small change in RI was measured upon plasma treatment. After exposure to N_2/O_2 and NH_3 plasmas, a densified surface sealing layer is formed. Upon annealing, the desorption of benzene groups and other hydrocarbon fragments, below the degradation temperature of the untreated material, relate to an unstable modified surface layer. Such behavior could be further linked to the unsealing of the surface upon static heating, as observed by other techniques such as RI, and ellipsometric porosimetry [15].

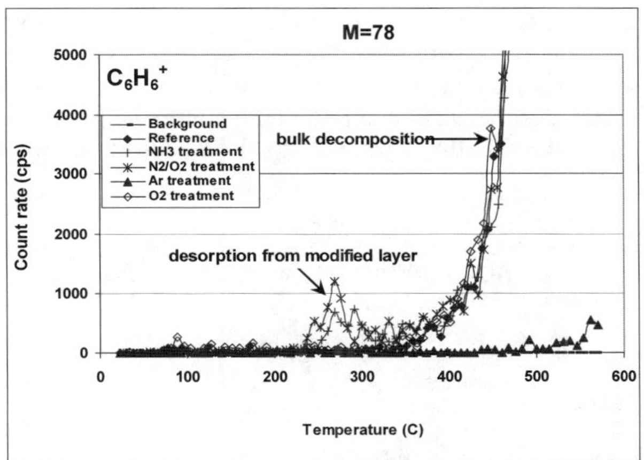

Figure 3. Correlation of the desorption of benzene (m/e=78) measured on the untreated and plasma treated layers [15]. The desorption of benzene groups observed at 250–300°C is characteristic of the unsealing of the modified surface layers formed with N_2/O_2 and NH_3 treatments. Almost no desorption of benzene is observed for the Ar-treated sample.

Thermal Stability of Thiolate Films Formed on Copper Surfaces

Understanding the thermal stability of thin films as a function of the underlying substrate and deposition conditions is essential in developing a successful integration scheme. Thiolate SAMs are of particular interest as they have been found to be effective corrosion inhibitors of copper surfaces [16]. Nevertheless, SAM stability during subsequent processing steps may limit their potential application. The thermal stability of the SAM layer is governed by numerous parameters such as Cu–S bond strength, the packing density of the SAM molecules on the copper surface, the Van der Waals interactions between the chain groups, and the electron density distribution within the SAM molecules. In this work, the thermal stability of 1-

decanethiol (C10) and benzenethiol (BT) SAMs on copper surfaces was investigated by TDS. The influence of the oxidation state of the underlying substrate was investigated by comparing SAMs formed on metallic and oxidized copper surfaces.

C10 and BT SAMs adsorbed on metallic and oxidized copper surfaces decompose at temperatures between 100 and 150°C. The thermal decomposition mechanism involved is dependent on the type of SAM considered and the oxidation state of the underlying substrate. For the alkanethiol (C10), direct interaction between the alkyl group of the thiolate and the metallic copper surface is the dominant pathway for the C–S bond scission. The head group desorbs as oxidized sulfur followed by desorption of the alkyl fragments of the chain adsorbed on the clean copper surface. In contrast, for BT, simultaneous desorption of the head group as oxidized sulfur, and the phenyl group as benzene is observed. Moreover, two distinct SO_2 desorption peaks originating from different SO_2 containing species are observed at ~75 and 150°C. The low temperature SO_2 peak coincides with the desorption temperature of atomic sulfur from the unmodified copper layer (see Figure 4) [17]. This observation may be attributed to chemisorbed phenyl thiolate species that undergo limited desulfurization at room temperature and low surface coverage, with the build-up of the resulting chemisorbed atomic sulfur on the surface [17].

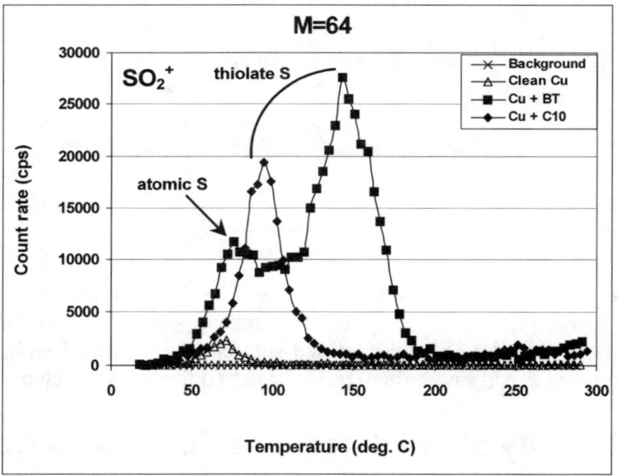

Figure 4. Thermal desorption spectra of SO_2^+ (m/e=34) species for clean Cu surfaces before (△△△) and after surface modification with 1-decanethiol (C10) (◆◆◆) and benzenethiol (BT) (■■■). The background levels of the impurities in the annealing system are given for reference (✕✕✕) [17].

SAM formation on oxidized copper surfaces involves complete removal and/or reduction of the CuO surface [17]. The resulting Cu_2O/Cu surface has enhanced surface roughness (compared with clean Cu) and thus higher SAM surface coverages are achieved. The thiol decomposition mechanisms are modified on oxidized Cu surfaces. The increased surface roughness and the presence of more Cu_2O areas on the surface as compared to clean Cu surfaces reduce the alkyl-copper surface interactions involved in the mechanism of C–S bond cleavage. The resulting SAM layer is thus more stable. Simultaneous desorption of SO_2 molecules related to the thiolate headgroups, and fragments of the alkyl chains observed for C10 modified CuO surfaces indicates that the thermal decomposition mechanism of the SAM layer is dependent of the underlying copper substrate. For BT-modified CuO surfaces, the behavior is also different compared with clean Cu. The limited desulfurization mechanism discussed above is deactivated on oxidized copper surfaces. Although the intensity of the SO_2 desorption peak is dependent on the nature of the underlying substrate, the stability of the SO_2 species related to thiol formation is independent of the substrate, with a desorption maximum around 150°C.

Chemical Labeling of SAMs Formed on SiO₂ Surfaces

As mentioned earlier, TDS can provide useful quantitative and qualitative information on the chemical composition of thin films through thermal degradation mechanisms [7]. This is nicely illustrated through the thermal analysis of organosilane-derived SAMs, with different terminal functionalities, formed on SiO_2 surfaces. Three types of SAMs were considered for this work: undecyltrichlorosilane (UTS, CH_3–$(CH_2)_{10}Cl_3Si$), cyanoundecyltrichlorosilane (CN-UTS, CN–$(CH_2)_{11}Cl_3Si$) and bromoundecyltrichlorosilane (Br-UTS, Br–$(CH_2)_{11}Cl_3Si$). These SAMs have a similar chain length but different terminal functional groups (CH_3, Br and CN). SAMs were formed on SiO_2 following the procedure described elsewhere [7]. The thermal stability and decomposition pathway for these SAMs were investigated using TDS. The decomposition process started at 450°C as evidenced by the desorption of high-molecular-weight hydrocarbon fragments created by alkyl chain C–C bond cleavage, detected for example at m/e=43 ($C_3H_7^+$ fragment). The Br–UTS and CN–UTS films have a comparable thermal stability, with a maximum in desorption at around 550°C. Both types of layers are less stable than the UTS SAM layer (desorption maximum around 600°C). The replacement of methyl by Br and CN terminal groups reduces the Van der Waals interactions between neighboring alkyl chains and even introduces repulsive interactions between the electronegative head groups. This modification in the bonding interactions within the SAM molecule results in lower thermal stability for the Br- and CN-terminated compared with the methyl-terminated SAM (Figure 5).

Interestingly, the SAM terminal functional groups exhibit a specific thermal desorption fingerprint. This is especially true in the case of the CN–UTS SAM layer that decomposes with the formation of HCN^+ (m/e=27) and $C_3H_3N^+$ (m/e=53) (Figure 6). A very low desorption of $C_4H_5^+$ fragments at m/e=53 can also be observed for the various SAM layers.

Thus, the specific chain termination can act as a convenient label to monitor SAM composition before and after eventual surface modifications during processing.

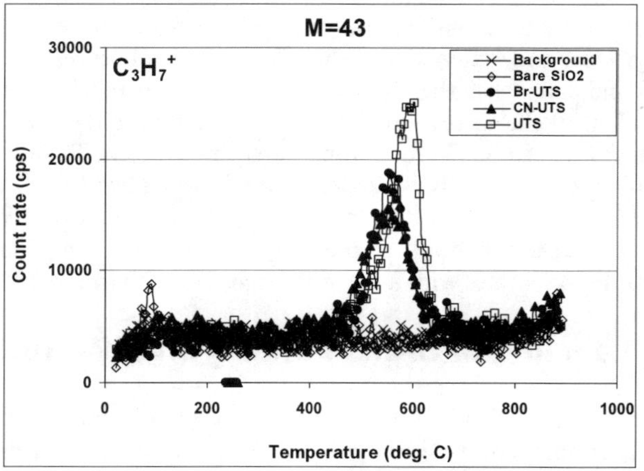

Figure 5. TDS spectra related to C$_3$H$_7$$^+$ (m/e=43) fragments.

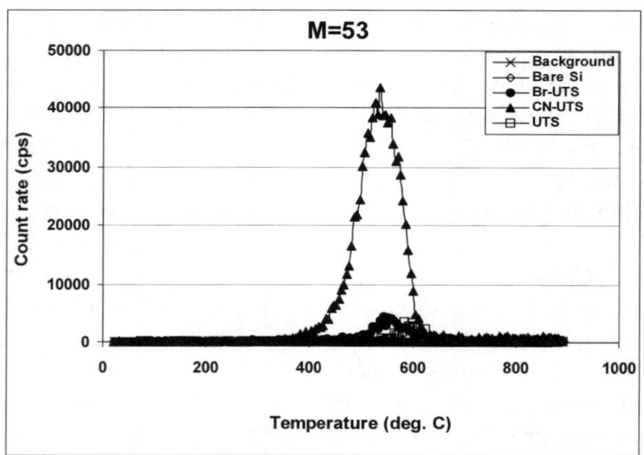

Figure 6. TDS spectra related to C$_3$H$_3$N$^+$ (m/e=53) fragments.

Conclusions

TDS analysis of a plasma-treated aromatic polymer ILD and SAM layers provided essential information with respect to optimization of processing conditions and film properties. The analysis of plasma-treated polymer layers provided valuable information on the surface/bulk modification of the material and the sealing of the surface [15]. Thiolate SAMs as well as silane SAMs were well characterized in terms of thermal stability and chemical composition [7, 17]. Through these experiments, TDS is demonstrated here as a powerful technique in providing chemical and structural information on the surface and bulk of thin films with thicknesses down to the nanometer scale.

References

1. N. Hirashita, S. Tokitoh, and H. Uchida, Jpn. J. Appl. Phys, 32, 1787 (1993)

2. P. W. Lee, S. Mizuno, A. Verma, H. Tran, and B. Nguyen, J. Electrochem. Soc. 143, 2015 (1996)

3. J. Proost, E. Kondoh, G. Vereecke, M. Heyns, and K. Maex, J. Vac. Sci. Technol. B, 16, 2091 (1998)

4. L. Peeters, Semiconductor International, 23, 108 (2000).

5. A. Martin Hoyas, J. Schuhmacher, D. Shamiryan, J. Waeterloos, W. Besling, J. P. Celis *et al.*, J. Appl. Phys. 95, 381 (2004).

6. C.M. Whelan, M. Kinsella, L. Carbonell, H.M. Ho, and K. Maex, Microelec. Eng. 70, 551 (2003).

7. C. M. Whelan, A-C. Demas, J. Schuhmacher, L. Carbonell, and K. Maex, Mat. Res. Soc. Symp. Proc. 812, F2.2.1 (2004).

8. N. Mikami, N. Hata, T. Kikkawa, and H. Machida, Appl. Phys. Lett. 83, 5181 (2003).

9. G. Vereecke, E. Kondoh, P. Richardson, K. Maex and M.M. Heyns, IEEE Trans. Semicond. Manuf., 13, 315 (2000).

10. E. Kondoh, G. Vereecke, M.M. Heyns, and K. Maex, J. Vac. Sci. Technol. A, 17, 650 (1999).

11. H. Li, G. Vereecke, K. Maex and L. Froyen, J. Electrochem. Soc., 148, G344 (2001).

12. K. Maex, M.R. Baklanov, D. Shamiryan, F. Lacopi, S.H. Brongersma, and Z.S. Yanovitskaya, J. Appl. Phys. 93, 8793 (2003).

13. CRC Handbook of Chemistry and Physics, 65th edition, edited by R.C. Weast CRC Press, Inc., Boca Raton, Fl. (1985).

14. L. Carbonell, G. Vereecke, S. Van Elshocht, M. Caymax, M. Van Hove, K. Maex *et al.*, Electrochemical Society Proceedings 2003-03, 150 (2003).

15. A. Martin Hoyas, J. Schuhmacher, C.M. Whelan, M. Baklanov, L. Carbonell, J.P. Celis *et al.*, submitted for publication in J. Vac. Sci. Tech. A.

16. C.M. Whelan, M. Kinsella, H.M. Ho, and K. Maex, J. Electronic Mat., in press

17. L. Carbonell, C.M. Whelan, M. Kinsella, and K. Maex, accepted for publication in Superlattices and Microstructures.

Electron Backscatter Diffraction: Application to Cu Interconnects in Top-View and Cross Section

M. A. Meyer, I. Zienert, E. Zschech

AMD Saxony LLC & Co. KG, Materials Analysis Department, Dresden / Germany

Introduction

With every new generation of microprocessors the number of transistors on a chip increases significantly. Today, there are more than 100 million field effect transistors integrated on a single die. Consequently, the interconnect complexity of such highly integrated circuits increases as well. Currently, at the 90-nm CMOS technology node, advanced designs incorporate up to nine levels of copper interconnects with critical dimensions below 150 nm for the lower layers. Both product performance and reliability are increasingly determined by interconnect design and materials. These challenging requirements can only be met using an advanced process technology and new materials for interconnects and interlayer dielectrics, which result in lower RC values [1]. Aluminium interconnects have been replaced by copper for high-performance microprocessors and, currently, the transition from traditional silicon oxide/nitride to low-ε inter layer dielectrics (ILD) takes place. With shrinking dimensions, the PVD-based barrier deposition will be replaced by new techniques such as atomic layer deposition (ALD), possibly in conjunction with the introduction of new barrier materials. All the process and materials changes mentioned above require increased efforts not only for development and manufacturing but even more so for materials characterization.

A detailed understanding of the effects of geometry, materials and process parameters on the performance and reliability of the copper interconnect system, and ultimately of the microprocessor, are essential in the race for ever faster parts. One important aspect is the microstructure of the interconnects, both globally and locally, in critical regions of the design. The evolution of texture and grain size distribution with processing parameters and interconnect geometry are of particular interest. A correlation of these microstructure data to reliability parameters such as mean time to failure (MTTF) and activation energy (E_a) derived from measurements on suitable test structures and to degradation phenomena observed in *in-situ* experiments [2] will improve the understanding of reliability-limiting mechanisms.

In recent years, electron backscatter diffraction (EBSD) has been proven to be a valuable technique to obtain orientation data from small volumes of crystals such as grains in copper interconnects. EBSD is a scanning-electron-microscope-based (SEM-based) technique which provides orientation information with high spatial resolution. Pole figures, color-coded orientation maps, grain size distributions and misorientation angle distributions among many other representations can be derived from the data.

This paper summarizes the application of EBSD as a tool to evaluate copper interconnects. Experimental consideration such as sample preparation for top-view and cross section investigation will be covered. The impact of ILDs and cover layers will be discussed and the potential for the investigation of passivated metal lines will be highlighted.

The Electron Backscatter Diffraction Technique

The energy of the primary electron beam within the SEM is typically set to 20 keV for an EBSD analysis. Inside the sample, the primary electrons are backscattered and subsequently diffracted at lattice planes of the crystallites of the sample according to Bragg's law [3]. The specimen is typically tilted to an angle of 70°. The steep specimen angle allows a maximum number of backscattered electrons to exit the sample [4]. Kossel cones from diffracting lattice planes are displayed by projection onto a phosphorescent screen. The obtained Kikuchi pattern is recorded by a CCD camera and compared with simulations based on crystallographic data in real time for each measurement spot. The beam is scanned across the sample with a certain step size, collecting Kikuchi patterns for each scan point. Step sizes down to 1 nm can be set on modern systems, and a lateral resolution below 10 nm has been reported [5]. Up-to-date EBSD systems can index up to 100 points per second. However, the speed of the data acquisition depends strongly on the material under investigation. For accurate indexing of a Kikuchi pattern, the image quality (IQ) needs to be sufficiently high. Hence, it may become necessary to increase the exposure time per measurement spot. In addition, the performance of the data acquisition computer and the filter settings for the Hough transformation of the CCD image have an impact on the maximum data collection rate. A Hough transformation algorithm is applied to the Kikuchi pattern in order to detect the individual bands before it can be indexed by comparison with simulated data. Figure 1 shows a principle drawing of the experimental set-up inside the SEM chamber and a typical Kikuchi pattern for copper, overlaid with the simulated data. After a scan is completed, a data file is generated that contains x,y-coordinates, orientation data in form of Euler angles as well as a number of quality parameters describing the sharpness of the diffraction pattern and the confidence in the orientation calculation.

Following the data collection, a second software package is usually used for data treatment and presentation. Grayscale or color-coded maps for different parameters are the most common form of data presentation, *i.e.* unique grain color maps assign a random color to each grain, while inverse pole figure (IPF) maps utilize an orientation unit triangle, which is related to the sample axis as basis (Figure 2). There are differences between the solutions from different vendors of EBSD analysis systems and the possibilities of data treatment and representation have a huge variety. Only a few of them will be described in this paper.

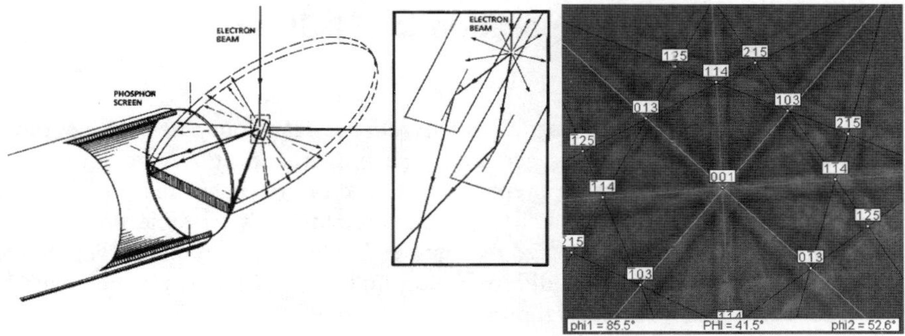

Figure 1. Schematic of EBSD geometry (left) and indexed Kikuchi pattern of copper (right) [6].

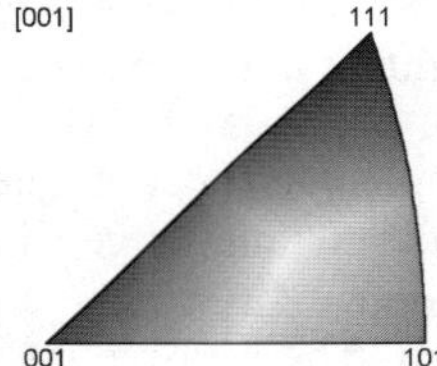

Figure 2. Orientation unit triangle

General Sample Preparation Considerations

EBSD is a surface analysis technique. Consequently, the surface finish of the analysis area has a large impact on the quality of the obtained Kikuchi patterns and the accuracy of the subsequent indexing. It has been shown that only backscattered electrons that lose less than 3% of the energy of the primary beam are able to fulfill Bragg's law, thereby contributing to the formation of the bands in the Kikuchi pattern. Electrons suffering a higher energy loss will only contribute to the noise background of the Kikuchi pattern [4].

This effect implies that only electrons backscattered from a shallow region below the sample surface can contribute to the Kikuchi bands. In addition to the energy loss inside the material itself, there are a number of additional contributions to the overall energy loss possible. Firstly, any cover layer on the copper surface will contribute to the energy loss of the backscattered electrons, and consequently it will reduce the total number of backscattered electrons from the sample. Secondly, shadowing effects on samples with high surface roughness, *e.g.* PVD-deposited Cu seed films, will reduce the energy and the number of electrons reaching the screen due to additional scattering. Furthermore, charging effects on insulating regions of the scan area, such as ILD between interconnect lines, will cause drift, an consequently they will reduce the Kikuchi pattern quality.

Advanced sample preparation for EBSD on Cu interconnects

Specific know-how has to be used to obtain reliable data from narrow copper interconnect lines or even from cross sections of such interconnects. However, there are a number of tricks for sample preparation that allow to detect reasonable EBSD data. Even scanning of passivated copper interconnects is possible under certain conditions. This enables the design of complex studies of interconnect degradation behavior under accelerated conditions, incorporating *in-situ* investigations and before/after comparison of material properties on certain test structures.

Charging Effects

When scanning copper interconnects rather than copper films a main concern is the reduction of drift that arises from a slow charging of the sample along the scan direction. This charging will result in distorted lateral dimensions of the interconnects and in a reduced pattern quality. The effect is more pronounced on narrow interconnect lines compared to wider ones. Although this problem can be very severe, there is a relatively simple way to eliminate drift due to charging by simply coating the sample with a very thin conductive film. Two parameters of the deposited layer are important to control: thickness and roughness. The desired film should be very thin and smooth. The best results have been achieved with an ion beam evaporated chromium coating of less than 5 nm.

Surface Roughness

Reducing the roughness of the sample surface is relatively complicated since the films are usually only a few hundred nanometers thin. Two methods have been successfully applied, low-angle argon ion milling and horizontal FIB polishing. For low-angle argon ion milling, the sample is rotated on an angle of $1°$ to $6°$ to the sample surface with an ion energy of 0.5 keV to 6 keV. Using this technique, relatively large areas up to 1 mm^2 can be polished. Figure 3 illustrates the result of low-angle argon ion milling on a very rough copper film (electro-plated and annealed). The milling was performed in a Fischione ASAP model 1040 tool [7]. This process can be applied to cleaved cross sections, too.

Horizontal FIB polishing is done by aligning the beam parallel to the sample surface and then horizontal cutting down the film thickness across a cleaved edge. This procedure has been described elsewhere [8].

Cover Layers

Depending on the nature of the cover layer, several techniques may have to be applied to remove it. Carbon contamination from previous SEM imaging can be removed using a low energy oxygen plasma. ILD passivation layers can be removed wet-chemically or by reactive ion etching (RIE). RIE can also be used to thin the passivation layer to a thickness that Kikuchi pattern can be obtained, while the interconnect is still passivated. Only a thin layer could be deposited in the first place, but some more complex experiments may require the completion of the entire manufacturing process. A detailed study of the dependence of the image quality (IQ) on the passivation thickness showed that it is possible to obtain reasonably good Kikuchi patterns for thicknesses up to 27 nm. In addition, there is a dependence of the EBSD pattern quality on the line width of the interconnects. Figure 4 shows inverse pole figure maps (IPF) for two different passivation thicknesses.

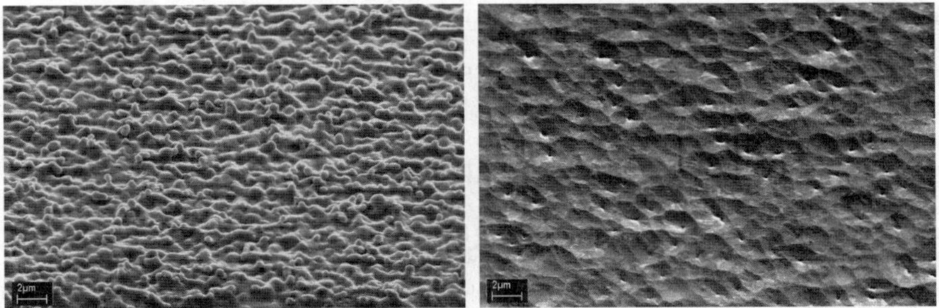

Figure 3. Cu film as deposited (left) and argon ion beam polished (right).

Figure 4. IPF maps of passivated narrow (0.18 μm) and wide (1.8 μm) copper interconnect lines (left: d_{pass} = 27 nm, right: d_{pass} = 34 nm).

For a passivation layer with 27-nm thickness, good indexing was achieved for narrow (0.18 μm) and wide (1.8 μm) lines. However, at 34-nm passivation thickness only poor indexing was achieved for wide lines and none for narrow lines. This observation may be explained by the much smaller grains in the narrow lines. If the passivation becomes too thick, the signal-to-noise ratio is reduced and the image quality (IQ) is dramatically reduced. Consequently, the indexing uncertainty is increased at grain boundaries. Since the grain boundary density is higher for narrow lines, the effect on their indexing is more pronounced.

EBSD on Copper Interconnect Lines

Top-View Studies

Ideally, EBSD analysis' on copper interconnects should be performed after chemical–mechanical polishing, because surface roughness and covering layers will be no issues for such samples. Possible charging problems can be easily resolved by coating the samples as described above. Two main questions regarding the microstructure that can be investigated with EBSD, are usually interesting for the process engineers: the evolution of the grain size distribution and texture with the line geometry, on one hand, and with process parameters and new materials, on the other hand. The following figures show the data of a typical analysis. Orientation and grain size data are presented for several line widths. Figure 5 shows unique grain color maps and inverse pole figure (IPF) maps for an unpatterned film, 1 μm lines and 0.35 μm lines. The step size for these scans was set to 75, 50 and 40 nm, respectively. The different values were chosen considering the resulting scan time, which increases significantly with decreasing step size. The maps show clearly a reduction in grain size with shrinking line width (unique grain color maps). These observations can be more pronounced visualized in grain size distribution plots (as relative cumulative frequency) as shown in Figure 6.

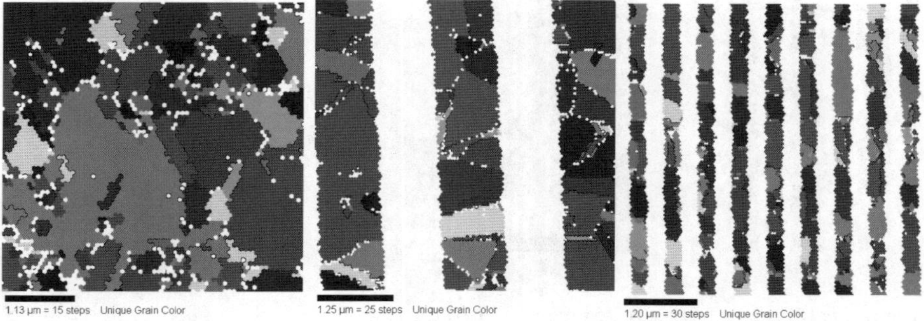

Figure 5. Unique grain color maps: unpatterned film, 1-μm lines and 0.35-μm lines. (Only a part of a larger scan shown for clarity.)

Figure 7 shows pole figures for the different scans. The pole figures confirm the observation from IPF maps (not shown) that the strength of Cu{111} fiber texture becomes significantly reduced as the line width decreases. Additionally, for the 0.35-μm lines, two vertical bands appear in the pole figure together with two distinct maxima at the left and right side of the pole figure. This effect can be explained with sidewall-oriented grains in narrow lines. A large portion of grains along the line are {111} oriented relative to the sidewall of the trench instead to the bottom of the trench [1].

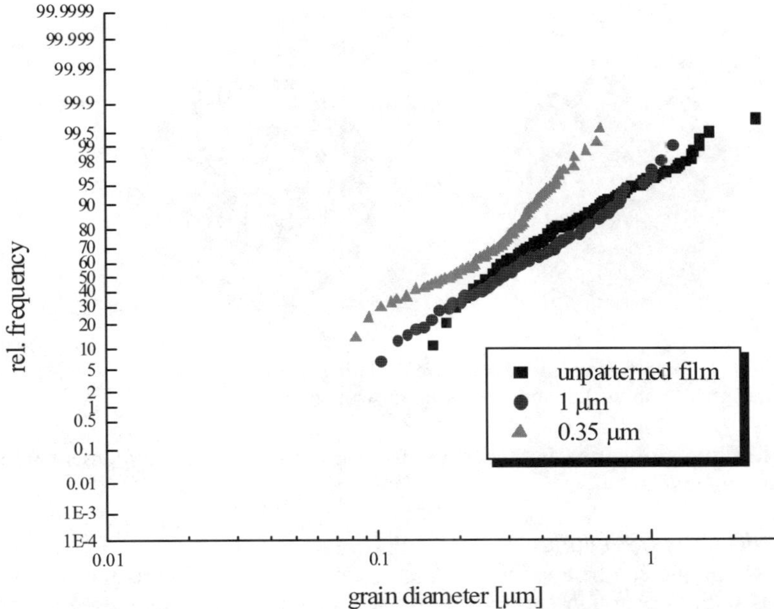

Figure 6. Grain size distributions: unpatterned film, 1-µm lines and 0.35-µm lines (twins as separate grains). Filter criteria: 2° minimum boundary angle, 3 pts per grain.

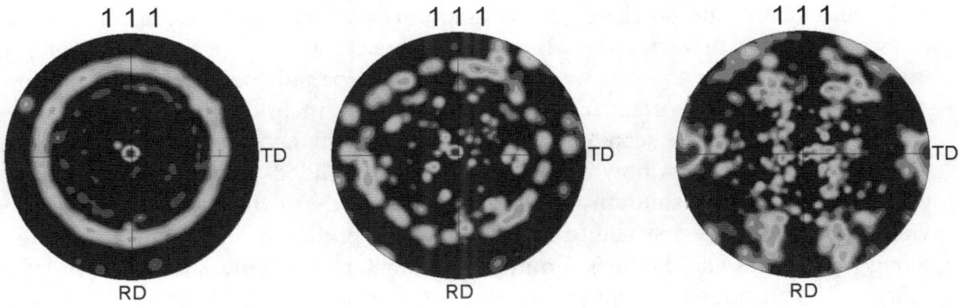

Figure 7. Cu{111} pole figures: unpatterned film, 1-µm lines and 0.35-µm lines.

Such a type of grain growth is considered as a risk for yield and reliability of copper interconnect structures, especially in geometries with high aspect ratios (dual-inlaid technology). More desirable would be a pronounced {111} texture in the trenches. This texture can be achieved by adjusting the electroplating. The grain orientation can be measured with EBSD (Figure 8).

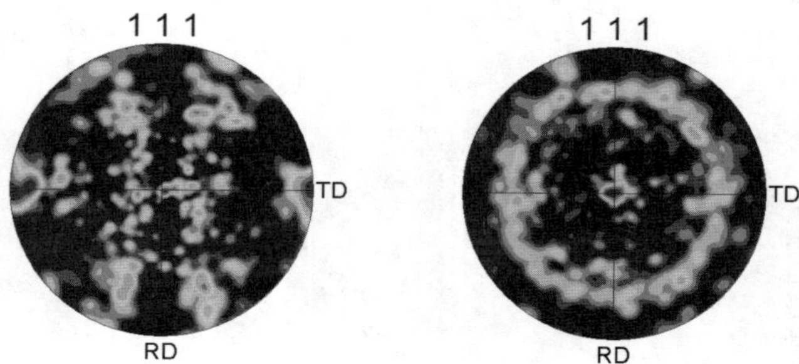

Figure 8. Cu{111} pole figures for two different plating processes (0.35 μm lines). Sidewall dominated growth on the left and bottom-up growth on the right.

The pole figures obtained with EBSD are in good qualitative agreement with XRD measurements [9].

EBSD on FIB Cross Sections
The EBSD technique can be applied to cross sections of copper interconnects, too. However, as with very rough surfaces, cross sections of interconnect stacks need special treatment before the EBSD analysis is performed. Due to the ductility of copper, cross-sectioned copper interconnects will not have a plane surface when cleaved at room temperature. Therefore, further steps are necessary to planarize the scan area. As mentioned earlier, FIB polishing or low-angle argon ion milling are suitable techniques. The method of choice will most probably be FIB polishing, since this is the only technique that allows positioning of the cross section plane with nanometer-scale precision. This is the only way to investigate specific structures of the interconnect system like vias. The advantage of low angle argon ion milling is that larger areas of the cross section can be polished at once, if the specific site is not so important. Figure 9 shows an IPF map together with the corresponding SE image of a scan on a cross section with six layers of copper.

The example shows how EBSD can provide detailed orientation information with a high spatial resolution. A more in-depth study of the microstructure on a via/line dual inlaid test structure revealed that the preferred orientation can be different for the vias and the lines. From complementary XRD measurements on large arrays of metal lines it is known that there is a strong Cu{111} texture. Additionally, there could exist a significant component of Cu{111} sidewall-oriented grains, which can be influenced by the electroplating process parameters and chemistry. However, the study of the via/line test structure showed that the orientation of the grains inside the vias is predominantly Cu{101} (Figure 10). The crystal direction map shows only {111} and {101} orientations with a tolerance angle of 15°. Calculations showed that about 30% of all grains are {101}-oriented and only 12.5% are {111}-oriented. The images also reveal that a larger portion of the {101}-oriented grains are located inside the vias or in the lines just above the vias.

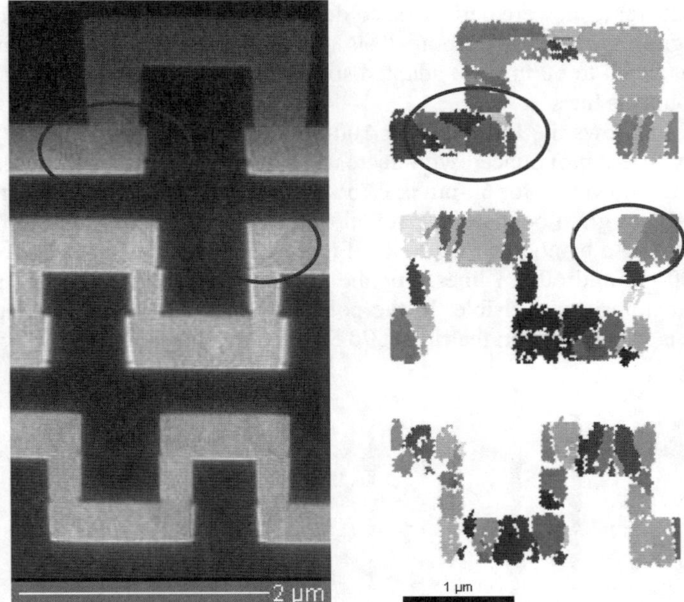

Figure 9. SE image and IPF map for a six-level copper interconnect cross section. (Note that the orientation contrast in the SE image corresponds to the orientation data in the IPF map (marked regions)).

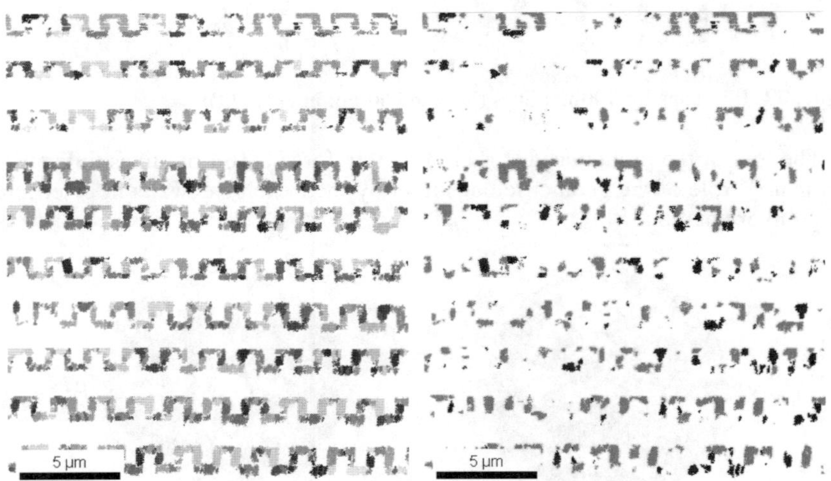

Figure 10. IPF map (left) and crystal direction map for {101} (grey) and {111} (black) oriented grains (right). Ten individual scans have been imported into one file for better statistical evaluation.

Nanointerconnects

EBSD measurements were performed to demonstrate that this technique can be applied for grain size and texture analysis of inlaid copper nanointerconnects with line widths down to 60 nm. An adapted spacer technique was used to manufacture the interconnects lines.

Figure 11 shows the IPF maps for 100-nm and 60-nm lines. The Cu{111} texture becomes less pronounced with decreasing line widths. It is still observable for 100-nm lines, however, for 60-nm lines no preferential texture is apparent. Interestingly, nearly no grain boundaries parallel to the trenches were found in the 60-nm lines, indicating a bamboo structure [8]. Figure 12 shows the Cu{111} pole figures for the 100-nm and 60-nm lines. For the 100 nm lines, the 70.5° ring indicating Cu{111} texture is still visible. In the pole figure for the 60-nm lines, the center maximum is very weak and the ring at 70.5° has almost disappeared.

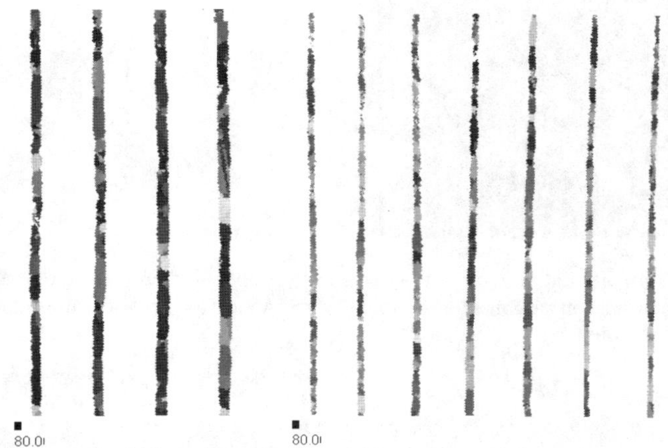

Figure 11. IPF maps for 100-nm lines (left) and 60-nm lines (right).

The loss of the pronounced Cu{111} texture for sub-100-nm lines indicates the dominating role of the geometric constraint by the trench sidewalls in determining the grain orientation [8].

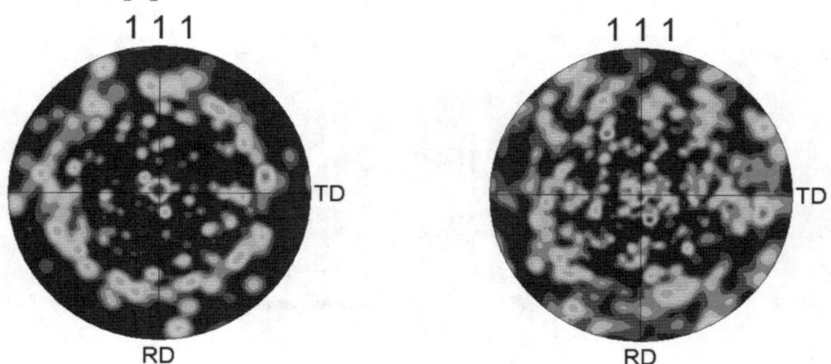

Figure 12. Cu{111} pole figures for 100-nm lines (left) and 60-nm lines (right).

Conclusions

EBSD has been applied to study the microstructure of copper interconnects in detail. Both, top-view and cross-section investigations were performed. FIB polishing was successfully applied to obtain cross-sections of specific structures like vias. Moreover, it has been shown that passivated test structures can be investigated. With the high spatial resolution of EBSD it also becomes possible to study nanointerconnects with critical dimension of 60 nm and below. Complex experiments can be conducted, which will improve the understanding of the microstructure evolution and the impact on performance and reliability of advanced integrated circuits.

Acknowledgements

The authors would like to thank Uwe Kramer (Technical University Dresden, now Infineon Technologies, Dresden), Maelle Duquoc (Ecole National Supérieure de Chimie de Lille) and Felix Geller (Technical University Dresden) for their assistance in performing the EBSD experiments. Discussion with Peter Hübler and Axel Preusse (AMD Saxony, Dresden) and Günter Schindler (Infineon Technologies, Munich) are gratefully acknowledged.

References

1. E. Zschech, W. Blum, I. Zienert, P. R. Besser, Z. Metallkd. 92, 803-809 (2001).

2. M.A. Meyer, M. Herrmann, E. Langer, E. Zschech, Microelectronics Engineering 64, 1, 375-382 (2002).

3. http://www.hkltechnology.com.

4. U. Kramer, Diplomarbeit, Technical University, Dresden (2002).

5. R. de Kloe, EBSD Symposium, BAM Berlin (2003).

6. http://www.edax.com/TSL.

7. http://www.fischione.com/products/model_1030.asp

8. G. Schindler, M.A. Meyer, G. Steinlesberger, M. Engelhardt, E. Zschech, Proc. Advanced Metallization Conference 2003, 205-211 (2003)

9. H. Geisler, I. Zienert, H. Prinz, M.A. Meyer, E. Zschech, AIP Conf. Proc. 683, 469-479, (2003)

X-ray Reflectivity Characterisation of Thin-Film and Multilayer Structures

P. Zaumseil

Innovations for High Performance Microelectronics, Frankfurt (Oder) / Germany

Introduction

Recent progress in growth techniques has made it possible to fabricate different kinds of thin films and multilayer structures. Optimisation of the fabrication process and a physical understanding of the samples require non-destructive methods to study the samples produced. X-ray elastic scattering methods are complementary to direct local probing techniques (*e.g.*, AFM, SEM, TEM, and different spectroscopic techniques). They probe locally the reciprocal space and thus provide information about the statistical properties of the structural parameters averaged over a large sample volume. This is an advantage; however, it is also a disadvantage due to its relatively low lateral resolution. Nowadays, the development of X-ray scattering methods has been enabled by advanced technical equipment becoming widely available, for example, multiple crystal diffractometers. New high-intensity synchrotron radiation sources, which complement to conventional and rotating anode laboratory sources, are advantageously involved in studies of thin-film structures. X-ray reflectivity (XRR) is conveniently applied for the structural studies of both crystalline and amorphous multilayer samples. It is sensitive to the distribution of the refractive index in the sample. XRR is nowadays used for the characterisation of very different materials including surface layers on liquids, organic adsorbate layers, coating layers for optics and surface hardening, and different layers used in micro- and optoelectronics.

Basics of X-ray Reflectivity

X-ray reflectivity experiments are usually performed with a low angle of incidence Θ_i relative to the sample surface (see Figure 1). The reflected beam is analysed at the exit angle Θ_r. We talk about specular reflection, when $\Theta_r = \Theta_i$ is fulfilled. In reciprocal space coordinates

$$Q_x = K\left(\cos\Theta_r - \cos\Theta_i\right), \quad Q_z = K\left(\sin\Theta_r + \sin\Theta_i\right) \quad with \quad K = \frac{2\pi}{\lambda},$$

this includes information about the sample structure perpendicular to the surface along the Q_z axis. The angle between incident and reflected beam under specular conditions is $2\Theta_i$. The typical method to measure reflectivity curves with an X-ray diffractometer is a Θ–2Θ scan (often also called ω–2Θ scan). If the reciprocal space is covered in general, it is a non-specular investigation.

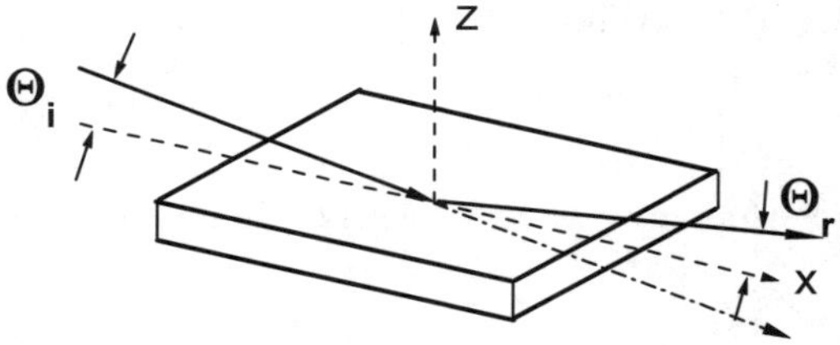

Figure 1. Definition of the angle of incidence and reflection in an XRR experiment.

All formulas describing reflectivity phenomena known from visible light optics, such as Snell's law or the Fresnel equations, are valid for X-rays in the same way. The only, but very important, difference is related to the fact that the index of refraction for X-rays is slightly less than one. The index of refraction n is usually written as:

$$n = 1 - \frac{N_A}{2\pi} r_0 \lambda^2 \frac{Z}{A} \rho = 1 - \delta - i\beta$$

with $N_A = 6.022 \cdot 10^{23} \text{mol}^{-1}$, r_0, the classical electron radius; λ, the wavelength; Z, the atomic number; A, the atomic mass, and ρ, the mass density. δ and β can be expressed by the atomic scattering factors $f = f_0 + f' + if''$ of the present elements:

$$\delta = \frac{N_A}{2\pi} r_0 \lambda^2 \sum_k \frac{\rho_k}{A_k}(f_{0_k} + f'_k), \qquad \beta = \frac{N_A}{2\pi} r_0 \lambda^2 \sum_k \frac{\rho_k}{A_k} f''_k$$

Since n is less than one, there exists a critical angle of incidence Θ_{crit} below which a total external reflection occurs in analogy to visible light optics. This critical angle can easily be derived by a Taylor expansion of Snell's law:

$$\Theta_{crit} = \sqrt{2\delta} \propto \sqrt{\rho}$$

This means that δ is proportional to the electron density of a material. For a typical X-ray wavelength, the absolute values for δ and β are of the order of 10^{-4} to 10^{-5} and 10^{-6} to 10^{-8}, respectively. Figure 2 shows an example of the wavelength dependence of δ and β for silicon.

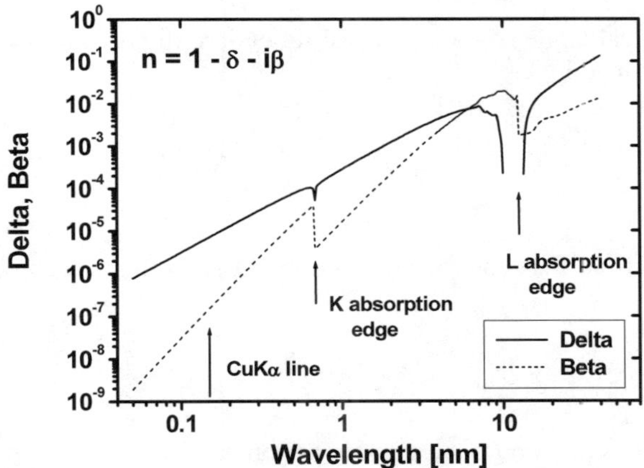

Figure 2. Dependence of the index of refraction for silicon on the wavelength.

Table 1 gives the optical constants δ and β, the density and the critical angle of total external reflection for some materials:

Table 1. Optical constants, density and critical angle of external reflection for several materials

Material	Delta	Beta	Density [g/cm^3]	Θ_{crit} [degree]
Si	7.58E-6	1.73E-7	2.33	0.223
SiO$_2$	7.14E-6	9.23E-8	2.20	0.217
Si$_3$N$_4$	1.11E-5	1.65E-7	3.20	0.270
Al$_2$O$_3$	1.26E-5	1.46E-7	3.96	0.288
Ta	3.99E-5	3.24E-6	16.65	0.512
W	4.65E-5	3.89E-6	19.30	0.553

At incidence angles below the critical angle, an evanescent wave is formed with a penetration depth of only few nanometres. This effect is used for the analytical technique of TXRF (total reflection X-ray fluorescence), which is thus very sensitive to surface contaminations since any signal from the bulk is strongly suppressed due to the low penetration depth of X-rays. Above the critical angle the penetration depth of radiation is determined by the absorption only and can be two or three orders of magnitude larger [1].

Specular Reflectivity at a Surface

The angular dependence of reflectivity and transmission can easily be obtained by solving the Fresnel equations under the given conditions for the index of refraction.

Above the critical angle of total external reflection, the reflectivity of an ideal flat surface is described by

$$R_F \propto \frac{\delta^2}{4\Theta_i^4},$$

and for a rough surface the decrease of intensity with increasing angle is even stronger,

$$R \propto R_F \exp(-(2K\Theta_i)^2 \sigma^2),$$

where σ describes the root-mean-square roughness.

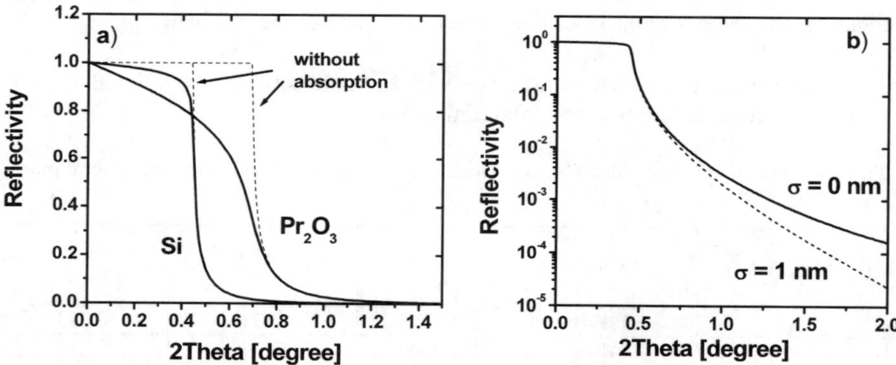

Figure 3. Calculated reflectivity curves for silicon and Pr$_2$O$_3$ (a) (The theoretical curves without consideration of absorption show the critical angle of total external reflection more pronounced) and for silicon with ideal flat (σ = 0 nm) and rough (σ = 1 nm) surface (b).

Figure 3a shows the reflectivity curves of silicon and Pr$_2$O$_3$, representing a light and a heavy material, respectively. The comparison of the calculated curves with and without consideration of absorption demonstrates that the sharpness of the edge of total reflection strongly depends on the absorption. Nevertheless, since the position of this edge depends on the material density, it offers the chance for a very precise measurement of material densities. Furthermore, the influence of the surface roughness on the reflectivity curve (Figure 3b) indicates that this technique is also wellsuited for roughness investigations.

X-ray Reflection by Single Layers

The reflection of a single layer on a substrate can be described as the interference of the reflection at two interfaces. It requires the solution of the Fresnel equations for reflection and transmission and a phase-adjusted superposition including multiple scattering. For a single layer deposited on a semi-infinite substrate the reflectivity is given by

$$R = \left| \frac{r_1 + r_2 \exp(i2K\Theta_i d)}{1 + r_1 r_2 \exp(i2K\Theta_i d)} \right|^2 ,$$

where $r_{1,2}$ are the Fresnel reflectivity coefficients of the free surface and the substrate interface, respectively, and d is the layer thickness.

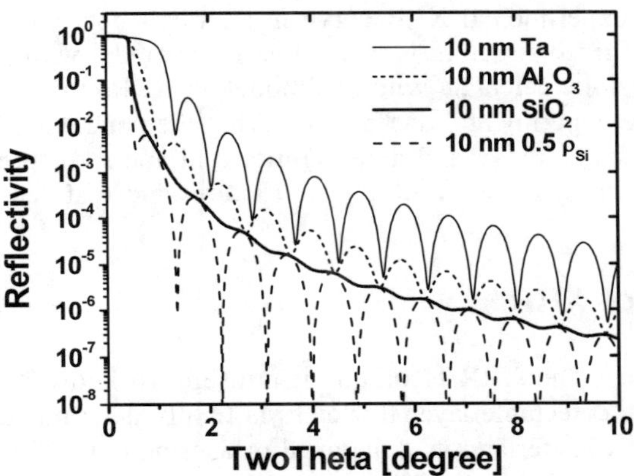

Figure 4. Calculated reflectivity curves for silicon substrate with a 10-nm-thick layer of four different materials on top.

Figure 4 shows calculated reflectivity curves for a silicon substrate with a 10-nm-thick layer on top. The layer materials are Ta, Al_2O_3, SiO_2, and a hypothetical material with 50% of the density of silicon. Above the total reflection range, there are typical interference oscillations visible. The contrast of these oscillations depends on the difference in δ between substrate and layer material. This makes it difficult to analyse, $e.g.$ SiO_2 on Si. On the other hand, the oscillation width for all materials is practically the same. It is important to note that the thickness determination of a single layer does not require any information about the material. The thickness can be estimated using the following approximate form:

$$\Theta_{i,max}^2 - \Theta_{crit}^2 = m^2 \left(\frac{\lambda}{2d} \right)^2 ,$$

where $\Theta_{i,\max}$ is the angular position of the mth oscillation maximum. Sufficiently far away from the critical angle the thickness can be estimated from the oscillation distance $\Delta\Theta_i$ by:

$$d \cong \frac{\lambda}{2\Delta\Theta_i}.$$

X-ray Reflection by Multilayer Structures

The reflectivity features of a multilayer system can be described in the same way as for a single layer. Now it is necessary to solve the Fresnel equations for reflection and transmission at all interfaces. Different algorithms were developed to calculate reflectivity curves [2,3]. An analytical solution of this problem is in general not possible, and thus it is also not possible to calculate a depth profile of electron density from an experimental XRR curve in a straight-forward way. The general way to analyse a given structure is to create a layer model with as much given information as possible, calculate with this model an XRR curve, and fit this calculated curve to the experimental one by modifying the parameters of the individual layers (thickness, roughness, and density) in a trial and error process. The examples given below are generally multilayer structures and will demonstrate this procedure.

Limitations of XRR

For practical use of the XRR technique it is useful to know its limitations. The lower limit in the detectable layer thickness is finally determined by the intensity of the used diffractometer and its dynamical measuring range. This defines the angular range, where at least one intensity oscillation can be measured. The upper limit of layer thickness depends on the angular divergence of the incident beam. If its half-width is on the order of the intensity oscillation distance, they will completely vanish in the convolution of the intrinsic XRR curve and divergence distribution. A rough estimate is $d_{\max} \cong \lambda / 2\ \Delta\Theta_i$, where $\Delta\Theta_i$ is the full-width half maximum of the divergence. In both cases the difference in δ between layer and substrate (oscillation contrast) plays an additional role. Typically the lower and upper limits are of the order of some nanometres and above 100 nm, respectively. The edge of the total reflection range is most sensitive to the density of a layer. A precise measurement requires a perfect sample adjustment and an exact determination of the zero position of the diffractometer. Roughness of the surface or of interfaces causes an attenuation of intensity oscillations and/or a faster decreasing of intensity (see Figure 3b). It is important to know that specular reflectivity is unable to distinguish between a real roughness of an otherwise sharp interface and a smooth variation of the electron density, for example due to a concentration gradient of one component in an alloy. Generally the roughness should not be greater than 2 — 3 nm. A general problem is that curve fitting works only within the limitations of the given layer model. If the model is not appropriate to the real situation, a perfect fitting cannot be expected. To improve a model by adding new layers or the generation of a density gradient within a layer will always remain the task of the user.

Examples

Three examples will demonstrate the abilities of XRR to characterise thin multilayer structures in the field of microelectronics. Figure 5 shows the XRR curve of an about 40-nm-thick $Si_{(1-x)}Ge_x$ layer epitaxially grown on silicon substrate.

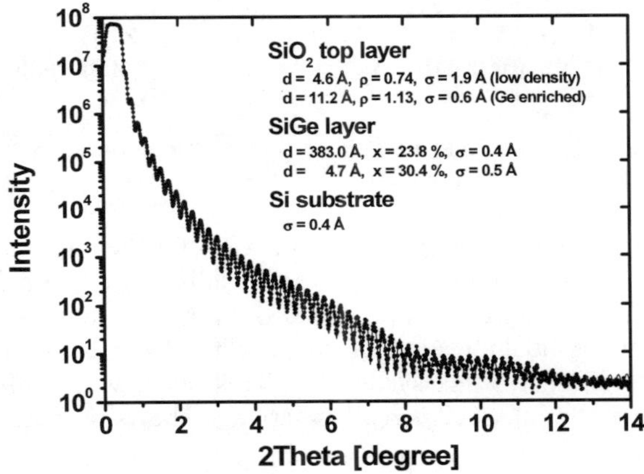

Figure 5. Experimental (dots) and best fitted (line) XRR curve of a $Si_{(1-x)}Ge_x$ layer on Si substrate. Four layers were used to obtain an optimal fit.

A first view indicates that a single layer model is unable to fit the experimental curve, since this would cause a monotonic decrease of the thickness oscillations. A four-layer model gave the best fitting results. The $Si_{(1-x)}Ge_x$ is 383-Å thick with a Ge content of 23.8 % and a roughness of 0.4 Å. At the interface to the Si substrate there is a 5-Å-thick layer of increased Ge content (about 30%). Above the SiGe is a silicon dioxide layer, which splits into a part with increased relative density ρ (compared to the bulk value of SiO_2), obviously due to Ge incorporation, and a low density part on the top (surface adsorption layer). Due to the high number of oscillations, the accuracy of the SiGe layer thickness determination is better than 5 Å.

Figure 6a shows an example of a metallisation layer with nickel deposited on a Si substrate. XRR determines not only the thickness of the Ni layer (100 ± 4 Å), but it is also able to detect a graded interface and a near-surface layer. Since XRR is not an analytical technique as, *e.g.* AES or SIMS, it nevertheless supports the idea of a continuous transition from Si to Ni by a Ni_2Si interface layer. There is a thin adsorption layer at the surface, here described by NiO of low density.

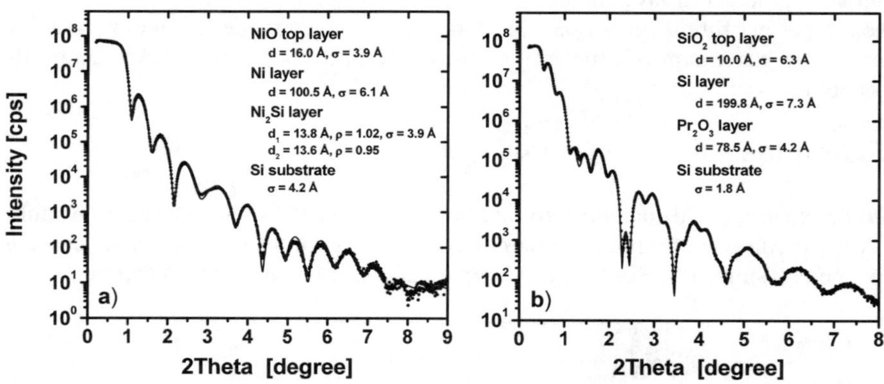

Figure 6. Experimental (dots) and best fitted (line) XRR curve of a Ni layer (a) and a Si/Pr_2O_3 layer stack (b) on Si substrate.

Figure 6b demonstrates the application of XRR for the characterisation of new high-k materials. Here, a Pr_2O_3 layer was epitaxially grown on a Si 100 substrate by molecular beam epitaxy and capped by a Si layer. The important information of XRR is that there is no indication of an interface layer between Si and Pr_2O_3, which is very common for this system [4]. This result was confirmed by TEM, which showed that the interface is really sharp and shows no oxide or silicate-like interface layer.

Non-specular XRR Analysis

Roughness of a surface or interface can generally be described by a correlation function that characterises the correlation of properties between two points. It gives rise to a diffuse scattering with a non-zero lateral scattering vector. Non-specular XRR is the most commonly used method to measure the diffuse X-ray scattering produced by rough surfaces and interfaces. The correlation function of roughness can be obtained from a measured reciprocal-space distribution of diffuse scattering [1]. Its main vertical parameter σ (root mean square roughness) can be obtained from a specular XRR analysis. The in-plane correlation length ξ can be extracted from a Θ scan at fixed detector position 2Θ. Figure 7 gives an example for a 21-nm-thick Pr_2O_3 layer on Si substrate. The Θ scan shows besides the strong specular peak at $Q_x = 0$ and the so-called Yoneda peaks [5], where Θ_i and Θ_r equals the critical angle, additional fringes of diffuse scattering. From their distance, a roughness in-plane correlation length of about 40 nm can be estimated. The σ values of surface and interface is 1.2 nm and 0.4 nm, respectively.

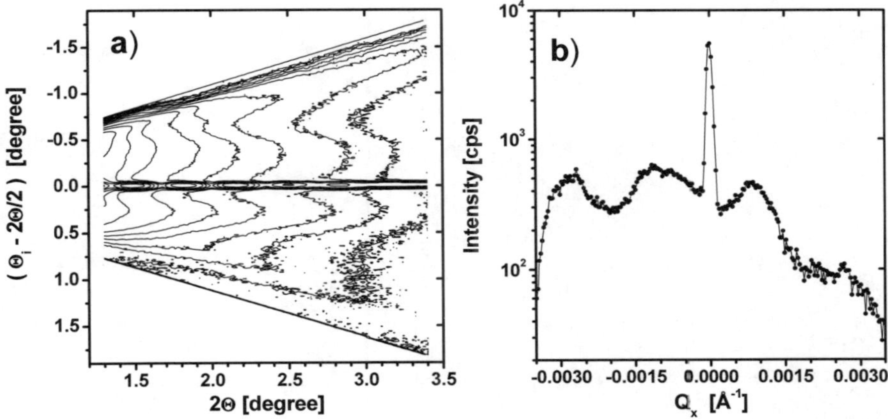

Figure 7. $\Theta - 2\Theta$ mesh scan (a) and one Θ scan at $2\Theta = 2.4°$ converted into reciprocal space coordinates (b) to analyse the diffuse scattering of a 21.4-nm-thick Pr_2O_3 layer on Si substrate.

Conclusion

X-ray reflectivity is a very sensitive method to investigate thin-film and multilayer structures. The main parameters obtained are thickness, roughness, and layer density. Concerning the thickness range of application, it is well suited for many materials used in modern information technologies. It is a non-destructive technique, but due to the very low incidence angles of radiation it requires relatively large sample areas. Nevertheless, it is indispensable for the evaluation and monitoring of future thin film deposition techniques.

References

1.   V. Holý, U. Pietsch, T. Baumbach, High-Resolution X-ray Scattering from Thin Films and Multilayers, Springer-Verlag, Berlin Heidelberg New York (1999)

2.   L.G. Parratt, Phys. Rev. B95, 359 (1954)

3.   L. Nevot, P. Croce, Rev. Phys. Appl. 15, 761 (1980)

4.   P. Zaumseil, E. Bugiel, J.P. Liu, H.J. Osten, Solid State Phenom. 82-84, 789 (2002)

5.   Y. Yoneda, Phys. Rev. 131, 2010 (1963)

Subject Index